Occupational Health Practice

Occupational Health Practice

Third edition

H A Waldron PhD, MD, MRCP, FFOM
Occupational Health Department, St Mary's Hospital, London W2

Butterworths
London   Boston   Singapore   Sydney   Toronto   Wellington

First published 1973
Reprinted 1975
Reprinted 1977
Second edition 1981
Reprinted 1982
Reprinted 1986
Third edition 1989

©Butterworths & Co (Publishers) Ltd. 1989

British Library Cataloguing in Publication Data

Occupational health practice.–3rd ed.
   1. Industrial health
   I. Waldron, A. II. Schilling, R.S.F.
   (Richard Selwyn Francis)
   363.1'1

   ISBN 0–407–33702–4

Library of Congress Cataloging–in–Publication Data

Occupational health practice.

   Includes index.
   1. Industrial hygiene. I. Waldron, H.A.
   RC967.O27   1989   613.6'2   88–7620
   ISBN 0–407–33702–4

Typeset by Scribe Design, Gillingham, Kent
Printed and bound in Great Britain by Anchor Press Ltd,
Tiptree, Essex

Preface to the Third Edition

Under the editorship of Richard Schilling, *Occupational Health Practice* quickly assumed a leading place amongst text books in the field and it was with mixed feelings that I accepted the invitation to take over the responsibility for the third edition. I felt very honoured and flattered to have been asked to do so, but I also had some trepidation about the task, knowing how much the first two editions had become identified with Richard and how much their success owed to him.

These 15 years have seen apparently better control of toxic hazards and a new emphasis on reproductive and behavioural toxicology, stress and the promotion of health. Many of these changes in emphasis have been prompted by research activities in Europe, particularly in some of the Nordic countries where standards of occupational health practice are especially high. Occupational health is also assuming more importance in the developing countries and nowhere is this upsurgence more evident than in China where great strides are being taken.

To the outsider, occupational health appears to be becoming more proactive and less reactive. These changes and improvements, however, may have less substance than seems obvious at first sight. In the United Kingdom the inspectorates and EMAs have been cut back to a point where their effectiveness must be seriously called into question; far too many men and women lose their lives or sustain serious injuries in accidents at work; large companies have been engaged in reducing their occupational health departments as part of the general economic 'rationalization' and small companies have, at best, inadequate provision for their employees.

The state of academic occupational health in the United Kingdom is also giving some cause for concern with serious cutbacks in staff and inadequate funding for research. Some departments are under threat of closure and were this to happen, the harm which would befall academic occupational health would be incalculable and the speciality itself would be harmed since without research of a high quality, the discipline will not be held in the regard which it should by the rest of the medical profession.

The authorship of the present edition has been drawn to only a limited extent from amongst those who were involved in previous editions. The involvement of some new contributors reflects the changes which have occurred and the different emphasis of the new edition. Its aim is to be of direct use to those actively engaged in occupational health practice wherever they may be, and as far as possible, to the problems which may arise from the complex interaction between work and health. Suggestions as to how the occupational health professional should deal with perturbations in the health of the worker and workplace are also included.

I am conscious there are some gaps in the material presented here but I hope that there will be a chance to rectify these in the future.

I hope that occupational health practitioners will find something of value in this new edition. I hope also they will find something of the spirit of the earlier editions which always placed an emphasis on the duty of those in the profession to safeguard the health and well-being of those for whom they have responsibility and see to it that working lives may be spent without coming to harm.

H A Waldron

Preface to the First Edition

The need for a book describing what the physician, hygienist and nurse actually do to protect and improve the health of people at work has become increasingly obvious to the staff of this Institute. Although many books have been written on occupational health, there are none in English which deal comprehensively with its practice. We teach the principles of occupational health practice to postgraduate students in occupational medicine, nursing and hygiene, and the lack of a standard work of reference has made the task of both teaching and learning more difficult.

Our academic staff and visiting lecturers have attempted to fill this gap, which is repeatedly brought to our notice by students. While our primary aim is to meet a need in formal course programmes it is hoped that the book may also be useful to the many whose interests encompass occupational health but who cannot attend a course, and that it will be of some value to medical and non-medical specialists in related fields.

Our students come from all over the world, many from countries undergoing rapid industrialization. We have therefore tried as far as possible to offer a comprehensive, up-to-date, account of occupational health practice, with some emphasis on the special needs of work people in developing countries. Eastern European countries attach great importance to occupational health and provide comprehensive occupational health services and training programmes. We refer to their methods of practice and training as well as to those of the western world because we believe both East and West have much to learn from each other, and the developing countries from both. Terms such as occupational health, medicine and hygiene often have different meanings, particularly in the eastern and western hemispheres. Occupational health in the context of this book comprises two main disciplines: occupational medicine, which is concerned primarily with man and the influence of work on his health; and occupational hygiene, which is concerned primarily with the measurement, assessment and control of man's working environment. These two disciplines are complementary and physicians, hygienists, nurses and safety officers all have a part to play in recognizing, assessing and controlling hazards to health. The terms industrial health, medicine and hygiene have a restricted meaning, are obsolescent, and are not used by us.

The three opening chapters are introductory; the first gives an account of national developments, contrasts the different forms of services provided by private enterprise and the State; and discusses factors which influence a nation or an industrial organization to pay attention to the health of people at work. The second is about a man's work and his health. Everyone responsible for patients needs to realize how work may give rise to disease, and how a patient's ill health may affect his ability to work efficiently and safely. It is as important for the general practitioner or hospital consultant as it is for the occupational physician to be aware of the relationship between work and health. The third chapter outlines the functions of an occupational health service. The chapters which follow describe in more detail the main functions, such as the provision of treatment services, routine and special medical examinations, including 'well-person' screening, psychosocial factors in the working environment and the mental health of people at work. There are chapters on occupational safety and the prevention of accidents and occupational disease which are often the most important tasks facing an occupational health service. Methods used in the study of groups of workers are outlined in sections on epidemiology, field surveys and the collection and handling of sickness absence data; these chapters are of special importance, as it is essential that those practising occupational health think in terms of 'groups' and not just of the individual worker. Epidemiological expertise enables this extra dimension to be added to the investigation and control of accidents and illness at work.

One chapter is devoted to ergonomics while five on occupational hygiene deal with the physical and thermal environments, airborne contaminants, industrial ventilation and protective equipment and clothing. There are concluding chapters on ethics and education in occupational health. Undergraduates in medicine and other sciences frequently lack adequate teaching on this subject and we hope that this book may be useful to them and their teachers.

Although it is not possible to cover fully the practice of occupational health in 450 pages, we hope to convey the broad outlines of the subject to a wide variety of people.

I owe many thanks to many people for help in producing this book, especially to the contributors and to those who assisted them in preparing their manuscripts, and to the publishers for their patience and understanding. For the illustrations I am particularly grateful to Mr C.J. Webb, Miss Anne Caisley and Miss Juliet Stanwell Smith of the Visual Aids Department at the London School of Hygiene and Tropical Medicine, and also to the Wellcome Institute of the History of Medicine, to the Editors of many journals, and to Professors Kundiev and Sanoyski of the USSR.

Manuscripts were read by members of the Institute staff and others who made valuable suggestions; the latter include Professor Gordon Atherley, Professor R.C. Browne, Dr J. Gallagher, Dr J.C. Graham, Dr Wister Meigs, Mr Wright Miller, Mr Andrew Papworth, Miss Brenda Slaney, Professor F. Valic, my wife and my daughter Mrs Erica Hunningher—I am indebted to them all; I am also grateful to Dr Gerald Keatinge and Dr Dilys Thomas for reading proofs, and to my secretary, Miss Catherine Burling for her help and enthusiasm throughout the long period of preparation.

Richard Schilling

Contributors

J R Allan
Director of Research, RAF Institute of Aviation Medicine, Farnborough, Hants

O Axelson
Department of Occupational Medicine, University Hospital, S 581 85, Linköping, Sweden

P J Baxter
Fenners, Gresham Road, Cambridge

A Bernard
Unit of Industrial Toxicology and Occupational Medicine, University of Louvain, Clos Chapelle-aux-Champs 30.54, 1200 Brussels, Belgium

R J Brown
Department of Occupational Health, London School of Hygiene and Tropical Medicine, Keppel Street, London WC1

N M Cherry
School of Occupational Health, McGill University, Montreal, Canada

E N Corlett
Institute for Occupational Ergonomics, Department of Production Engineering and Production Management, University of Nottingham, Nottingham

R B Douglas
Department of Occupational Health, London School of Hygiene and Tropical Medicine, Keppel Street, London WC1

M Floyd
Rehabilitation Resource Centre, City University, Northampton Square, London EC1V 0HB

A Ward Gardner
37 School Road, Eling, Totton, Southampton

R C Hack
Rutherford Appleton Laboratory, Chilton, Didcot, Oxon

R Jenkins
Department of Health and Social Security, Alexander Fleming House, Elephant and Castle, London SE1

J Jeyaratnam
Department of Social Medicine and Public Health, National University of Singapore, Lower Kent Ridge Road, Singapore 0511, Republic of Singapore

J King
126 Ferndene Road, London SE24

R Lauwerys
Unit of Industrial Toxicology and Occupational Medicine, University of Louvain, Clos Chapelle-aux-Champs 30.54, 1200 Brussels, Belgium

J R Lisle
Consultant in Preventive and Occupational Medicine, Joint Research and Health Advisers, 29 Great Pulteney Street, London W1

J A Lunn
Occupational Health Department, Northwick Park Hospital, Harrow, Middlesex

B Malerbi
The Folly, South Stoke Road, Woodcote, Reading

J C McDonald
Emeritus Professor, School of Occupational Health, McGill University, Montreal, Canada

A Newsome
Fellow of the University of Keele and Consultant on Counselling, 11 Danes Court, St Edmund's Terrace, London NW8

C J Purnell
Department of Occupational Health, London School of Hygiene and Tropical Medicine, Keppel Street, London WC1

A G Salmon
RCHAS, California Department of Health Services, 2151 Berkeley Way, Berkeley, CA 94704, USA

R S F Schilling
Emeritus Professor of Occupational Health, University of London

S J Searle
Area Medical Adviser, Post Office Occupational Health Service, 86 Lionel Street, Birmingham

P J Taylor
Late of Unilever PLC, PO Box 68, Unilever House, Blackfriars, London EC4

K M Venables
Department of Occupational Medicine, National Heart and Lung Institute, Brompton Hospital, Fulham Road, London SW3

H A Waldron
Occupational Health Department, St Mary's Hospital, London W2

Contents

Chapter 1

Developments in occupational health

R.S.F. Schilling

Introduction

Indifference to work health and safety has been a feature of both ancient and modern societies until relatively recent times. Rapid and extensive developments in occupational health began in the early 1940s when the Second World War made an impact on manpower. In both developed and developing countries there has been a growing awareness of its importance.

A brief historical review helps to identify those factors that retard and accelerate developments [1]. There are four that have a positive effect on occupational health: the economic need to conserve the efficiency of the work force; changing attitudes of workers and their trade unions towards health and safety; compassion which induces a sense of caring for others: increasing competence of health and safety professionals. Not only governments and industries but also individual workplaces have been influenced by these factors to take more effective action in hazard control and health promotion.

Age of antiquity, Middle Ages and renaissance

Mining is one of the oldest industries and has always been a hazardous occupation. Conditions in the gold, silver and lead mines of ancient Greece and Egypt reveal an almost complete disregard for miners' health and safety [2]. Since the miner of antiquity was a slave, prisoner or criminal, there was no reason to improve working conditions because one of the objectives was punishment and there were ample reserves of manpower to replace those who were killed or maimed.

Agricola and Paracelsus

The first observations on miners and their diseases were made by Agricola (1494–1555) and Paracelsus (1493–1541) in the sixteenth century. During the Middle Ages the status of the miner had changed. From being a feudal enterprise, manned by serf labour, mining in Central Europe had become a skilled occupation, which led to the emancipation of the miner. The growth of trade had created a demand for currency and capital which was filled by increasing the supply of gold and silver from these mines. Mines were made deeper and conditions worsened.

1

After his appointment as town physician to Joachimstal, a flourishing metal mining centre in Bohemia, Agricola described the diseases that prevailed in the mining community [3]. At that time mortality from pulmonary diseases was not recorded, nor were the causes known, but they would have included deaths from silicosis and tuberculosis and from lung cancer due to the radioactive ore in siliceous rock. Mortality must have been high, judging by the evidence of Agricola's statement that 'in the mines of the Carpathian Mountains, women are found who have married seven husbands, all of whom this terrible consumption has carried off to a premature death'. Apart from improvements in ventilation, miners remained without any significant means of protection. However, they organized themselves into societies which provided sickness benefit and funeral expenses, giving them some security and preventing the extremes of social misery [2]. Such improvements as there were followed the changed social status of the miner and the recognition by outstanding physicians like Agricola and Paracelsus of the extent and severity of occupational disease.

Paracelsus [4] based his observations on occupational diseases of mine and smelter workers on his experience as town physician to Villach, Austria, and later as a metallurgist in the metal mines in that area. He relates 'We must have gold and silver, also other metals, iron, tin, copper, lead and mercury. If we wish to have these we must risk both life and body in a struggle with many enemies that oppose us.' Paracelsus realized that the increasing risk of occupational disease was a necessary and concomitant result of industrial development.

Bernardino Ramazzini

During the sixteenth and seventeenth centuries mining, metal work and other trades flourished in Italy following the Renaissance, which had encouraged the transition from feudalism to capitalism [5]. In 1700 Bernardino Ramazzini (1633–1714 (*Figure 1.1*), physician and professor of medicine in Modena and Padua, published the first systematic study of trade diseases [6]. He put together observations of his predecessors and his own, based on visits to workshops in Modena. Rightly acclaimed the father of occupational medicine, he showed an unusual sympathy for the less fortunate members of society and recommended that physicians should enquire about a patient's occupation.

Ramazzini's interest in occupational medicine was inspired by the opportunities it offered to make new observations, and by his sympathy for the common people. As a physician of that time he was probably unique.

> I hesitate and wonder whether I shall bring bile to the noses of doctors—they are so particular about being elegant and immaculate—if I invite them to leave the apothecary's shop which is usually redolent of cinnamon and where they linger as in their own domain, and to come to the latrines and observe the diseases of those who clean out the privies.

Neither his medical colleagues nor other people of standing had any strong humanitarian sense to inspire them to heed his words, nor at that time was there any economic necessity to protect the life and health of workmen. This is in direct contrast to the improved conditions of miners which had taken place in Central Europe a century earlier as a result of the combined effects of an awareness of their hazards and a change in their social status.

Figure 1.1 Bernardino Ramazzini (1633–1714)

The Industrial Revolution in Great Britain

Towards the end of the sixteenth century the manufacture of cotton textiles came to England with the religious refugees from Antwerp. Spinning and weaving thrived as a cottage industry until the latter half of the eighteenth century, when mechanization transferred the making of textiles from people's homes to the new factories. Later the factory system spread to other industries in Europe and North America. This change in method of manufacture so unsettled the traditional routine of family and community life that it became known as the Industrial Revolution. Several forces led to these fundamental changes in the methods of manufacture. Science and technology had enabled the use of steam to be developed for motive power. There were large increases in the population of England and Wales. The breakdown of the strong central government of the Tudors and Stuarts, which had attempted to keep society geographically and socially static, allowed people to move from the country to towns to man the new factories. From her commercial and banking enterprises overseas, Britain had accumulated the financial resources to build new factories as well as towns to house the people.

Thus the eighteenth century brought great technological inventions and laid the foundations, in Europe and North America, of modern society with its factory system. It exposed workers of all grades to the pressures of increasing production and associated physical and psychosocial hazards of work.

Forces which are not dissimilar from those preceding the Industrial Revolution have enabled rapid industrialization to take place in developing countries, where work people have been exposed to the same pressures and hazards. Such forces are the development of hydroelectric and other forms of power, increases in population, the end of colonialism, and the financial and technical assistance made available to them by developed countries.

Effects of industrialization on community health

The more serious effects of health which followed the Industrial Revolution were not directly occupational in origin. Family life was disrupted when men moved into new industrial areas leaving their families behind, a situation that encouraged alcoholism and prostitution. Epidemics followed as a result of overcrowding in insanitary conditions. The change from peasant to town life led to malnutrition, made worse by the poverty and unemployment caused by fluctuations in the economy.

Work people moved from the countryside of rural England to the squalor and ugliness of the new industrial towns, which were described as 'bare and desolate places without colour, air or laughter, where man, woman and child ate, worked and slept' [7]. There were a few sympathetic employers, such as Robert Owen (1771–1858) and Michael Sadler (1780–1833), who provided good working and housing conditions for their employees. But poor housing, overcrowding and lack of sanitation caused by the concentration of an expanding population around the new factories led to the development of the public health services which were designed to control disease and improve the health of these communities.

The health problems arising from industrial progress in developing countries today are, in many aspects, similar to those during industrialization in the nineteenth century; these countries also have to face major threats from endemic disease and generalized poverty. Weakly organized labour with a large work force available places little pressure on employers to provide anything more than wages and basic services.

Effects of industrialization on workers' health

Inside the factories and mines of the nineteenth century the workers were exposed to hazards of occupational disease and injury and the adverse effects of excessively long hours of work.

As manufacturing techniques improved, machines became speedier and more dangerous. Little attention was paid to safety devices and workers were often simple people untrained to handle the new machinery. Toxic hazards increased due to prolonged exposure to a wider range of new chemicals which were introduced without considering their possible effect on workers. In the milieu of his cottage industry, the handloom weaver or the spinner had worked by the rule of his strength and convenience. He could take a break to cultivate his plot of land. In the factory these rules no longer applied, and he became exposed to the pressure of continuous work at a speed imposed by the needs of production—a pressure which dominates society today and against which man so often rebels.

Humanists and public opinion

Man's indifference to his less fortunate fellow men was perhaps assuaged during the eighteenth century by the liberal ideas of men like Rousseau, Voltaire, Kant and Thomas Jefferson. Society was also influenced by the action of humanists like John Howard (1726–90), who led the reform of British prisons; William Tuke (1732–1822) who set an example for the humane treatment of the mentally sick[8] and William Wilberforce (1759–1833), who started the campaign for the abolition of the slave trade. Later, the seventh Earl of Shaftesbury (1801–85) (*Figure 1.2*), evangelist and aristocrat, spent most of his life trying to relieve the conditions of the destitute and deprived in Victorian England. As a Member of Parliament he helped to promote legislation which reduced the hours and improved the conditions of work of women and young persons employed in mines, factories and other workplaces. His reforms were bitterly opposed by employers, but his influence as an acknowledged man of integrity and a leading member of the aristocracy did much to relieve the oppressive conditions created by the Industrial Revolution in Great Britain.

In the nineteenth century, manufacturers generally believed it was economically important to keep their new machines running continuously with cheap labour. Even though the philosophy of the government at that time was to let the people be free and society would take care of itself, it was obliged to interfere because of public reaction to the adverse working conditions of women and children who were subjected to long and arduous hours and were unable to look after themselves. These conditions were compared with those of the negro slaves whose lives had been made more tolerable by the abolition of slavery in 1833.

Figure 1.2 Anthony Ashley Cooper, seventh Earl of Shaftesbury (1801–85)

Insanitary conditions in factories and the dormitories attached to them introduced a risk of infectious disease. Influential people in the vicinity of the factories feared that the mills might be a source of contagion to themselves. Thomas Percival (1740–1804), a Manchester physician called in by the people of Ratcliffe in Lancashire to investigate an epidemic of typhus, went beyond his remit and produced a report on hours of work and conditions of young persons. This report influenced Sir Robert Peel (Sr), a millowner, to introduce into the British House of Commons the first Factory Bill, which became the famous Health and Morals of Apprentices Act of 1802. It limited hours of work to 12 a day, provided for workers' religious and secular education, and demanded the ventilation and limewashing of workrooms. The Act was meant to be enforced by visitors appointed by Justices of the Peace, but it was ineffective. Nevertheless, the principle of government interference was established and this early legislation culminated in the Ten Hour Act of 1847 which restricted the hours of work of women and young persons in factories to 58 in the week. This marked the beginning of the Welfare State—the principle of looking after those who were unable to look after themselves, such as the young, the old, the indigent and the sick.

The factors which encouraged more state control and made a generation of hardbitten employers give way to men who were socially more responsible were the influence of enlightened employers and humanitarians, of medical men and later of trade unions.

Enlightened employers

A small group of perceptive employers, like Sir Robert Peel, Robert Owen and Michael Sadler, influenced Parliament to introduce new legislation to control hours of work of women and young persons.

Robert Owen became a mill manager in Manchester at the age of 20. Later he moved to Scotland to manage the New Lanark Mills where he became famous [9] for his good management and humane treatment of work people. He refused to employ young persons of under 10 years of age. He shortened hours of work, provided for adult and child education, improved the environment and still was financially successful. He persuaded Sir Robert Peel, the architect of the 1802 Act, to introduce legislation to protect young persons in all types of textile mills, to prohibit employment for those under 10 years of age and to limit their hours to 10 a day. The Bill of 1819 was passed in the House of Commons but was emasculated in the House of Lords. Robert Owen's influence was limited by his professed atheism and socialism, which were wholly unacceptable to his manufacturing colleagues.

Medical influence

During the eighteenth and nineteenth centuries facts about the ill-effects of work on health emerged from the observations of a few physicians who followed the example of Ramazzini and took an active interest in the diseases of occupations. In 1775 Percivall Pott (1713–88) had drawn attention to soot as a cause of scrotal cancer in chimney sweeps. The reports of Dr Thomas Percival on conditions in the mills of Ratcliffe influenced Sir Robert Peel to get the Act of 1802 passed through Parliament. Later, Charles Turner Thackrah (1795–1833) (*Figure 1.3*), a Leeds

Figure 1.3 Charles Turner Thackrah (1795–1833)

physician, published the first British work on occupational diseases [10]. Thackrah died of pulmonary tuberculosis at the age of 38, but not before he had made his mark, which earned him recognition as one of the great pioneers in occupational medicine. In 1832 Michael Sadler introduced in the House of Commons a new Factory Bill, later to become the Act of 1833, which created the Factory Inspectorate. During his speech he said: 'I hold in my hand a treatise by a medical gentleman of great intelligence, Mr Thackrah of Leeds.' He then quoted extensively from the text [10].

Measurements of occupational mortality were first introduced in England and Wales in the middle of the nineteenth century by Dr William Farr of the General Register Office. He used census population figures and recorded deaths in certain occupations to calculate mortality rates. This drew attention to the gross risks of injury and disease in factory workers and miners at that time.

Edward Headlam Greenhow (1814–88), one of the outstanding epidemiologists of the nineteenth century, used the unpublished records of the General Register Office to examine occupational mortality in more detail. He compared crude death rates from pulmonary disease in the lead-mining towns of Alston and Reeth in the North of England with those of nearby Haltwhistle, which had no lead mines (*Table 1.1*). He did not consider the possibility of differences in age distributions affecting the rates, but his conclusion that the near four-fold mortality excess in Alston and Reeth was associated with heavy exposure to dust in the lead mines, was almost certainly correct. Greenhow reached a general conclusion that much of the very

Table 1.1 Average annual death rates per 1000 from pulmonary disease from 1848 to 1854 in men and women aged over 20 years, in the mining towns of Alston and Reeth and in the non-mining town of Haltwhistle

	Men	*Women*
Alston	14.4	7.8
Reeth	13.0	7.2
Haltwhistle	3.7	5.8

high mortality from pulmonary disease in the different districts of England and Wales was due to the inhalation of dust and fumes arising at work. Under the influence of his reports, factory inspectors were given powers in the Factory Acts of 1864 and 1867 to enforce occupiers to control dust by fans or other mechanical means [11].

William Farr's successors at the General Register Office have made occupational mortality data more valuable by using purer occupational groups, standardizing for age, comparing the mortality of workers with that of their wives, in order to distinguish between occupational and socioeconomic risks, and most recently, by standardizing for social class.

Thus, medical intelligence stemming from national and local mortality data and from the testimony of individual physicians, had an early influence on the development of health and safety measures.

Medical pioneers have on occasions inspired the hostility of both their colleagues and manufacturers. John Thomas Arlidge (1822–99) an outstanding physician in the pottery district of North Staffordshire, devoted himself to the study of potters' diseases. A medical colleague wrote of Arlidge [12]:

He made an unfortunate beginning of his career by compiling statistics of the people working in the potteries which gravely reflected on the humanity of the manufacturers. He was instrumental in the appointment of factory surgeons for earthenware and china manufacturers, upon whom this entailed much expense. His medical friends were against him, and up to his death this feeling never died out.

Early influence of trade unions

The French Revolution, at the end of the eighteenth century, had a profound effect on Britain [9]. The government, confronted by war abroad and the threat of Jacobite revolt at home, adopted a policy of repression. The Combination Acts of 1799 and 1800, which made trade unions illegal, were not repealed until 1824. Their restraining effect is evident from the large number of unions which then came to life and enabled organized labour to exert its influence to obtain improvements in working conditions. At that time the unions were concerned with reducing hours of work and raising wage levels. Their interest in occupational health and safety came much later.

Development of industrial medical services

Government service

The Factory Act of 1833 introduced two fundamental innovations: the appointment of factory inspectors; and the necessity of certification by a medical man that a child seemed by its strength and appearance to be at least 9 years old, the age below which employment was prohibited in textile mills. Later the Act of 1844 gave inspectors powers to appoint certifying surgeons in each district to introduce more uniformity into certification and prevent parents taking their children from one doctor to another until they got a certificate. With the advent of birth registration in 1837, age certification by the surgeons became redundant. The Factory Act of 1855 gave them new duties: to certify that young persons were not incapacitated for work by disease or bodily infirmity; and to investigate industrial

Figure 1.4 Sir Thomas Morison Legge (1863–1932)

accidents. Thus a rudimentary industrial medical service, the first of its kind, was introduced by law in Great Britain. Towards the end of the nineteenth century, workers in certain dangerous trades were required by regulations to be examined periodically by the certifying surgeons. These regulations applied to those making lead paints, lucifer matches and explosives with dinitrobenzene; and to those vulcanizing rubber with carbon disulphide and enamelling iron plates [13].

To obtain knowledge of important industrial diseases like lead, phosphorus or arsenic poisoning, and anthrax, the principle of notification was introduced in 1895.

The investigation of notified cases of occupational disease was added to the duties of these surgeons, who also had powers to suspend sufferers from work. The high prevalence of lead poisoning in the potteries and in white lead works and the incidence of 'phossy jaw' among match-makers, received a great deal of publicity. These events and the need to deal with notifications and reports from certifying surgeons led to the appointment in 1898 of Thomas Morison Legge (1863–1932) (*Figure 1.4*) as the first Medical Inspector of Factories.

By his own researches and through his axioms for preventing occupational disease, which he evolved during his 30 years of unique experience as a factory inspector, Legge made an outstanding contribution to occupational medicine. He ended his distinguished career in the civil service by resigning when the government refused to ratify an international convention prohibiting the use of white lead for the inside painting of buildings. For a few years, until his death, he was Medical Adviser to the Trades Union Congress and wrote his classic work *Industrial Maladies* [14].

Employers' services

Even before the Industrial Revolution there were isolated examples of occupational health services. In the eighteenth century the Crawley Iron Works in Sussex retained the services of a doctor, clergyman and schoolmaster for the benefit of employers and employees. In the nineteenth century a factory near Stirling in Scotland employed 'a medical gentleman to inspect work people and prevent disease'. The report [15] of Michael Sadler's Select Committee on the employment of children is one of the main sources of knowledge of factory conditions at that time. It describes unusually enlightened actions of John Wood, a Bradford millowner, who employed a doctor and sent his work children to Buxton, or other health resorts, when they were 'overdone'. He had baths on the premises and his works had high standards of ventilation and cleanliness. The motives for setting up the small number of medical services in industry were at that time almost entirely humanitarian.

The first real impetus to the voluntary appointment of doctors by employers came after the passing of the first Workmen's Compensation Act in 1897. The larger firms appointed physicians as a means of protecting themselves against claims for compensation, rather than as a measure to protect employees. Unfortunately the industrial medical officer was often regarded by workmen as the employer's man—a suspicion which, however unfounded it may have been, has died hard [16].

Twentieth century to the outbreak of the Second World War

Great Britain

In Britain the state gradually built up a statutory medical service for factory workers, provided by about 1800 part-time certifying factory surgeons (later called appointed factory doctors). Most of them were general practitioners who were supervised in their work by the Medical Inspector of Factories. They had three main tasks: to examine young persons under the age of 18 for fitness for work when starting employment and at annual intervals thereafter; to undertake periodic medical examinations of persons employed in certain dangerous trades; and to investigate and report on patients suffering from any of the notifiable industrial diseases or injured by exposure to noxious substances. The limitations of this type of statutory service were obvious to the more enlightened employers who made provisions for the medical care of their employees. Later the government recognized the shortcomings of this Appointed Factory Doctor Service. It was abolished by Act of Parliament in 1972 and replaced by the Employment Medical Advisory Service.

The First World War (1914–18) introduced important changes in outlook towards the health of people at work in Britain. Serious shortage of munitions in 1915 was not made good by long hours of work. This led to the appointment of the Health of Munition Workers' Committee which sponsored scientific investigations into the effects of work on health and efficiency. It studied new toxic hazards from handling explosives such as trinitrotoluene and solvents used in making aircraft. There followed a rapid growth in first aid and in industrial medical and nursing services. National survival was the motive for this new interest in occupational health.

The economic slump which followed the war slowed down developments. Nevertheless, the more enlightened and wealthy industries provided their own health services because they realized that legislation laying down minimum standards of health and safety and the statutory medical examinations of the certifying factory surgeons were inadequate. These services broadened in their scope and, generally speaking, their aim was to improve and maintain the employees' health and not merely to protect the employers from compensation claims. As there was no systematic training of doctors and nurses in occupational medicine and the practice of occupational hygiene was almost non-existent, the achievements of these services were limited.

Developments in other countries

Developments in Great Britain illustrate those factors which stimulated changes in attitudes towards the health of people at work. Changes in other countries followed similar patterns for much the same reasons. Developments in the United States of America are contrasted with those in the Union of Soviet Socialist Republics.

United States

The vastness of the United States and the wide range in the origin and culture of its settlers, produced a federation of states in which there was considerable freedom for each state to pursue its own policies for dealing with the problems of rapid industrialization. The State of Massachusetts passed the first Child Labour Law in 1835, and by 1867 had appointed a special police officer to enforce the law prohibiting the employment in factories of children under 10 years of age. Massachusetts was the first state to establish a Bureau of Labor Statistics. Other states followed suit and these Bureaux eventually became State Departments of Labor with responsibilities for enacting and enforcing a growing range of codes to protect workers from long hours, hazardous processes and adverse environmental conditions [17]. The federal government dealt only with the control of working conditions for persons employed by, or on behalf of, the United States government. By its constitution, the main responsibility had to be left to individual states, which varied considerably in the standards of health and safety they demanded for people at work. The federal government created a Bureau of Labor in 1884, a Bureau of Mines in 1910 and the Office of Industrial Hygiene, as part of the United States Health Service, in 1914. These did much to encourage the promotion of occupational health by undertaking research, by their education programmes and by giving advice to individual states on specific problems. The federal government also had an important influence on the development of occupational health through the funds it made available to the various states for setting up occupational hygiene programmes. As a result, in the three years before the Second World War 30 units were established to provide medical and hygiene services for the control of occupational disease. This federal activity created a body of occupational hygienists and enabled the United States to lead the field in environmental measurement and control in the workplace. Railroad, steel and mining companies were among the first industries to set up industrial medical services. Their in-plant health services provided by employers followed much the same pattern as those in the United Kingdom, with many of the large firms employing full-time medical officers.

ALICE HAMILTON

Among the great pioneers in occupational health is Alice Hamilton (1869–1970) (*Figure 1.5*). She spent 40 years of her life searching for occupational hazards which had been overlooked by industry and plant physicians [18]. In 1910 she began her

Figure 1.5 Alice Hamilton (1869–1970)

crusade with a survey of poisoning in the lead industries. She had to face opposition both from employers and from members of her own profession, one of whom described her report on lead poisoning as false, malicious and slanderous. Nevertheless, her investigations led to improvements in working conditions and high standards of medical surveillance. Working for federal and state governments, and finally in the University of Harvard, she continued her investigations so that workmen might be protected against serious risks such as silicosis in the Arizona copper mines, carbon disulphide poisoning in the viscose rayon industry, and mercurialism in the quicksilver mines in California. In 1919 Harvard paid Dr Hamilton the compliment of appointing her Assistant Professor of Industrial Medicine. She was the first woman to be a member of the academic staff and one of the first ever to hold a university post in occupational health. She travelled widely and was able to compare the provisions made for the health of work people in many countries. During her visits to Europe in the 1920s she was surprised by the elaborate provisions for the study and treatment of occupational diseases in the USSR, which she rated as better than in any country she had visited [19].

Union of Soviet Socialist Republics (USSR)

The first important phase in the development of occupational health in Eastern Europe began in the USSR after the October Revolution of 1917; the second took place in other countries, such as Bulgaria, Czechoslovakia, Poland, Romania and Yugoslavia after the Second World War. Before the Revolution Russia, like other European countries at that time, had no organized occupational health services and, generally, there was little or no interest in this subject among members of the medical profession [20]. An exception was F.F. Erisman (1842–1915) (*Figure 1.6*), one of the founders of the science of hygiene in Russia. Erisman pressed hard for improvements in environmental conditions in factories, but his views were not

Figure 1.6 F.F. Erisman (1842–1915)

acceptable. Before 1917 the Bolshevik party had formulated a health policy with two cardinal principles: health services were to be free; and the concentration was to be particularly on prevention. Alexander Semashko, who became the first Commissar of Health in the Russian Soviet Federative Socialist Republic, was one of the architects of this policy. One of his first actions as Commissar was to separate the medical schools from the universities, with the result that the content of teaching programmes was decided at a political level and not by physicians. The first Medical Institute outside the universities for training and undergraduates was set up in Moscow shortly after the Revolution. In 1922 this Institute, later named after Semashko, established a Chair of Hygiene of Labour. A year later a Research Institute of Occupational Health and Safety was set up in Moscow. Health services in workplaces were organized as an integral part of all medical care in the USSR.

From the Second World War onwards

Developments in the provision of health care at work were accelerated by war and economic expansion and by changing patterns of work which can have adverse effects on mental health and well-being. These latter developments emphasized the importance of the health of the organization itself as well as that of its workers. The belief that occupational health services are economically worthwhile, and the demands of workers for better conditions, have stimulated their growth.

Standards of service have been raised by the increasing number of trained health workers, by major advances in the techniques for health and environmental monitoring, and by improved methods of collecting and distributing knowledge about work hazards and their control through national and international agencies.

War and economic expansion

While the First World War had a positive, but somewhat ephemeral, influence on developments in occupational health in Europe and North America, the Second World War and the economic expansion which immediately followed provided a strong impetus for developments in countries all over the world. In the early 1940s there was a rapid expansion of occupational health services. In Britain, for example, the numbers of occupational physicians increased seven-fold. As well as growth in numbers the scope of occupational health expanded to make the fullest possible use of available manpower. There was a new emphasis on assessing ability rather than unfitness for work. The armed forces made special contributions by developing techniques for selecting personnel. They adapted military equipment to suit the soldier, sailor and airman in order to increase his fighting efficiency. This gave a boost to ergonomics. Similarly, the need to get highly trained personnel, such as air or tank crews, back to active service as soon as possible led to substantial improvements in methods of rehabilitating the injured and sick.

The period of sustained high employment which followed the war encouraged further expansion and the development of new concepts of health care.

Reappraisal in recession

Once established, occupational health activity does not appear to have suffered unduly from recession and rising unemployment, but there has been a more critical appraisal of the cost-effectiveness of services. Management has been forced to recognize the need to conserve the health and efficiency of an increasingly skilled work force in both productive and service industries; more health services are being provided in shops, offices, hospitals and universities. It is difficult to put precise monetary values on the benefits of such services and to define links between services and results. Nevertheless, provision of health care demands a declaration of priorities. A choice has to be made in allocating available resources. This means assessing the cost of each health activity in relation to its impact on the community. Thus costs have to be kept, not only to discourage overspending but also in a form that can be related to the effect of specific activities on people's health.

Growing influence of trade unions and workpeople

Trade unions

During the twentieth century the trade unions in many countries began to exert an influence on occupational health by pressing for improvements in legislation and for the extension of compensation laws to cover occupational injuries and diseases. They have since become more directly involved in health and safety.

In Britain the Trades Union Congress in 1968 contributed £125 000 towards the formation of the TUC Centenary Institute of Occupational Health in the University of London. Their object was to expand an existing university department, concerned primarily with teaching and research, to include an information and advisory service for general use, which the unions help to maintain by an annual grant.

Trade unions now participate in formulating national policy and health and safety programmes at workplace level [21]. In several countries contracts between employers and unions carry provision for health and safety [22]. In Britain the Health and Safety at Work, etc. Act 1974 provided for the establishment of a Health and Safety Commission, which consists of representatives of employers, trade unions and local authorities. It is the body which determines policy on health, safety and welfare at work. It proposes regulations and codes of practice for the consent, through Parliament, of the Secretary of State for Employment.

In Israel the General Federation of Labour (Histradut) set up its own Health Insurance Programme (Kupat Holim) in 1911 at a time when Palestine was a backward province of the Ottoman Empire and had totally inadequate health services. In 1945 it opened its first Department of Industrial Medicine in Tel Aviv. It has since developed a comprehensive occupational health service for its members. Today Kupat Holim provides comprehensive medical care for its members who comprise 90 per cent of the working population.

In the USSR the trade unions have extensive responsibilities for health and safety through their technical factory inspectors who are broadly equivalent in their function and powers to inspectors in Labour Ministries in western countries [23].

Co-partnership in responsibility

In developed countries it is recognized that to make work as safe as possible requires the full participation of the workforce. Like the nineteenth-century social reformers, Sir Thomas Legge's axioms were based on his belief that the worker often could not, or would not, take responsibility for his own health and safety. This paternalistic philosophy has become less acceptable in developed countries and is being replaced by the concept of co-partnership—the shared responsibility of management and workers in the prevention of disease and injury. Today's axiom is:

> It is seldom possible to prevent work-related injury and disease without the cooperation of those who are at risk. They must be made fully aware of the hazards of their work and how these may be contained. Workpeople must be given some responsibility and incentive to influence their own work environment towards higher standards of health, safety and job satisfaction.

In many countries it is obligatory for workers to be represented on safety committees and for them to appoint their own safety representatives. In Britain representatives appointed by unions have the legal right to investigate potential hazards, dangerous occurrences, accidents and complaints. They also have the right to carry out periodic plant inspections.

The health of the organization

Management, trade unions and health professionals are becoming increasingly aware of the importance of the relationship between the individual and the organization and the manner in which this may influence health and well-being. Sources of organizational stress, such as role ambiguity, overload and underload, which may adversely affect the mental health of the individual, are difficult to control or even alleviate. The traditional approach is to hope that problems will either go away or resolve themselves. A more constructive approach is to recognize that an organization has its own personality, behavioural problems and state of health. This may be studied by looking at the way in which different departments interact and achieve their goals. For example it is not uncommon for two departments with overlapping responsibilities and areas of mutual interest to be barely on speaking terms, when they should be exchanging information and ideas.

Studies of organizational health aim to elicit from the people concerned the causes of stress and to help them find their own solutions. This may be outside the scope of an occupational health service, but physicians and nurses need to be versed in the various problems which face management and how they may be solved [24]. Rather than attempting to treat the individual with mental illness caused by organizational stress, the doctor or nurse can take the initiative in advising management to treat the cause of the problem by improving human relations or management skills, often by seeking the help of independent experts.

Wider concepts and increasing professionalism

Occupational health services used to be regarded as a luxury rather than a basic need in health care. Their functions were limited to providing first aid and to the control of industrial injury and work-related disease in the larger organizations such as mines, railways and heavy industries with major hazards to the health and safety of manual workers.

The wider aims of occupational health practice in developed countries, stimulated by the need to conserve manpower, were first promulgated by a Committee of International Experts in 1950 as 'the promotion and maintenance of the highest degree of physical, mental and social well-being in all occupations'. Although difficult to achieve, this ideal led to more widespread and effective health

care in workplaces. The setting up of training programmes fo
occupational physicians, nurses and hygienists, but more r
include primary health care workers.

In 1973, WHO listed the three major tasks in occupation
and controlling known or suspected work factors that co
educating management and workers to fulfil their responsi
safety; and promoting health programmes not primarily
related injury and disease. The aim of the occupational health team is
prevent the adverse effect of physical and chemical agents, but also to ensure that
work is suitably adapted to both physiological and psychological needs of the
worker and that, conversely, the worker is fit to do the job [25].

A clearer understanding of the multiple aetiology of disease from the
epidemiological study of risk factors has widened the concept of work-related
disease and injury [26]. They may be grouped into four categories:

1. work the necessary cause, for example, lead poisoning;
2. work a contributory cause, for example, coronary heart disease;
3. work provoking a latent or aggravating an established condition, for example,
   carpal tunnel syndrome, peptic ulcer, eczema;
4. work offering accessibility to potentially dangerous materials, for example,
   keeping a pub or hotel (alcoholism) [27], medical laboratory work (suicide)
   [28], degreasing (solvent abuse).

Occupational health services are undertaking a wide range of preventive
activities not directly concerned with work-related conditions. This new role has
been accepted by governments [29] and by many industries [30]. The workplace
offers opportunities for health education to promote healthier lifestyles by stopping
smoking, modifying diet and increasing physical activity. The cost of unnecessary
ill-health to the organization has stimulated management to pay for these activities.

Professionally trained staff are essential to put these broad concepts of
occupational health into practice. Developed countries have set up university
departments and institutes to provide training programmes [31]. Standards of
service, however, vary because generally the training of physicians and other health
professionals in the occupational health field is not compulsory.

Types of occupational health service

Role of the state

The state's role in providing occupational health and safety varies enormously in
different countries. It ranges from a complete state service to minimal provision,
where the government does little more than set standards through statutory laws.
In the nineteenth century and early years of the twentieth, minimal provisions
prevailed and were almost completely inadequate because of difficulties in
enforcing the law.

Extensive state involvement has become necessary because of the increasing
costs of providing health care for workpeople and the demands for higher
standards.

Britain's occupational health programme was drastically reorganized as a result
of the Health and Safety at Work, etc. Act 1974. Health and safety legislation
became stricter and more comprehensive, covering almost all workers including

...elf-employed. The Employment Medical Advisory Service (EMAS) advises ... government, management, union and staff of occupational health services. ...imilarly, the federal government of the United States, through its Occupational Safety and Health Act 1970, set up a new administration to enforce the law and an Institute (NIOSH) to recommend hygiene standards, sponsor and undertake research and to finance training programmes for health workers.

There is no single government system applicable to all countries, though the aims of occupational health may be similar. Methods of achieving these aims vary according to the form of government and the type of health service provided outside the workplace. Two distinct systems exist. In one the state provides the service, as in Eastern Europe. In the other the state plays an advisory and supervisory role, encouraging or making it a statutory obligation for employers to provide their own services. This second method has been adopted by most countries in the European Economic Community (EEC).

Comprehensive services in Eastern Europe

In Eastern European countries where governments are responsible for providing comprehensive health care, services at the workplace are planned as an integral part. This helps to avoid unnecessary overlap and uses resources where they are most needed. It encourages the exchange of information about the health of the individual and the community. There may be difficulties in maintaining standards, of encouraging flexibility and adapting the system to meet local needs. Within Eastern Europe there are at least two methods for providing occupational health services.

In the USSR, Czechoslovakia, Romania and Bulgaria health services are organized into separate branches of therapeutic and prophylactic medicine. Therapeutic services are provided by the hospitals, polyclinics and medical departments of large plants and preventive services by the sanitary and epidemiological stations (*sanepids*), which are located in towns, rural areas and in large workplaces. The physicians in the hospitals, polyclinics and large plants are responsible for all forms of medical treatment, including the diagnosis and treatment of occupational diseases; whereas the physicians in the *sanepids* assess and control the working and general environment and are in charge of the prevention of communicable and non-communicable disease. This system encourages integration of preventive medical services and, presumably, avoids duplication of treatment. The separation of treatment and preventive services has a possible disadvantage in that treatment may show where prevention is needed.

In Yugoslavia the organization differs from the above in two respects. The aim is to decentralize responsibility, giving cities more freedom to develop occupational health services according to local needs. Industrial plants are free to set up their own type of health service provided they stick to certain basic standards. At all levels there is closer integration of therapeutic and preventive services.

Services in the European Economic Community

Services in EEC countries differ from those in the socialist countries of Eastern Europe. The state does not provide an occupational health service as an integral part of health care; instead the aim is that they should be state-controlled but not state-run, embodying a compulsory do-it-yourself system. The Council of the EEC has made increasing demands on its member countries. In 1978 it adopted an Action Programme on Safety and Health [25]. It listed priorities which included: promotion of research into the aetiology of accidents and diseases; protective measures against dangerous substances, such as carcinogens; preventing the harmful effects of machines, particularly noise and vibration; development of monitoring and inspection methods to promote better protection of workers' health, especially expectant mothers and adolescents.

The situation in Europe and in countries all over the world will change as a result of the International Labour Organization's Convention 161 and Recommendation 171 of 1985 [32]. They recognize that occupational health includes medical, nursing and hygiene services, as well as related disciplines such as ergonomics. The Convention which Member States of the ILO are asked to ratify, requires that 'services should be developed for all workers . . . in all branches of economic activity and all undertakings'. The Recommendation which supplements the Convention, stipulates that occupational health services have sufficient staff who are trained in their specialty and keep themselves up to date in scientific and technical knowledge.

These documents provide a detailed blueprint for the future of occupational health services. They should stimulate governments and undertakings to make radical improvements in the health, safety and welfare of working people.

Conclusions

After the period of rapid expansion which followed the Second World War, occupational health practice has continued to develop and consolidate. Group occupational health services have been set up to meet the needs of small plants. Most of the heavy industries and public organizations have established services. Shops, offices, universities and hospitals have followed suit. Services which had previously provided only primary care, for example those for hospital staff and students, now include an occupational health element. The scope of occupational health services has widened to meet health care needs which are not directly related to the effects of work on health. This is for the convenience of workpeople for whom community health services are absent or inadequate. In this way the organization may derive economic benefits from reductions in unnecessary sickness and disability.

In developing countries, through a combination of failure to recognize needs and inadequate resources to meet them, services have often been either insufficient or non-existent. Recently primary health care services have taken on the provision of occupational health care.

Even in highly industrialized countries examples of good and bad practice may be seen side by side. Standards are dictated as much by attitudes of management and workers as by statutes. An important factor is the extent to which employers and workers want to have safer and healthier workplaces.

Where there is a health service much depends on the training and experience of its staff. In the past, when so little was known about the ill-effects of work on health, a few pioneers encouraged or cajoled governments and employers to make improvements by revealing facts about loss of life and limb and of sickness, caused by disregarding working conditions. Standards of health and safety now depend on a more positive approach and more widely available expertise in occupational health.

References

1. Sigerist, H.E. Introduction. *In the History of Miners' Diseases*. (ed. G. Rosen) Schuman, New York (1943)
2. Rosen, G. *The History of Miners' Diseases: a Medical and Social Interpretation*. Schuman, New York (1943)
3. Agricola, G. *De Re Metallica* (1556), trans. H.C. Hoover and L.H. Hoover; *Mining Magazine*, London (1912)
4. Paracelsus. On the miners' sickness and other miners' diseases. In *Four Treatises of Paracelsus* (ed. H.E. Sigerist), Johns Hopkins Press, Baltimore (1941)
5. Bernal, J.D. *Science in History: 2 The Scientific and Industrial Revolutions*. Penguin, Harmondsworth (1969)
6. Ramazzini, B. *De Morbis Artificum*, 2nd edn (1713), trans. W.C. Wright, Chicago University Press, Chicago (1940)
7. Hammond, J.L. and Hammond, B. *The Town Labourer*. Longmans Green, London (1917)
8. Glendenning, L. *Source Book of Medical History*. Dover, New York (1960)
9. Fay, C.R. *Life and Labour in the Nineteenth Century*. Cambridge University Press, London (1945)
10. Thackrah, C.T. *The Effects of Arts, Trades and Professions and of Civic States and Habits of Living on Health and Longevity*. 2nd edn. Longmans, London (1832). Reprinted in A. Meiklejohn, *The Life, Work and Times of C.T. Thackrah*. E. and S. Livingstone, London and Edinburgh (1957)
11. Munitions, Ministry of, Health of Munition Workers Committee, *Final Report*; Cmnd 9065. HMSO, London (1918)
12. Posner, E. John Thomas Arlidge (1822–99) and the Potteries. *British Journal of Industrial Medicine*, **30**, 266–270 (1973)
13. Schilling, R.S.F. Developments in occupational health during the last thirty years. *Journal of the Royal Society of Arts*, **3**, 933–984 (1963)
14. Legge, T.M. *Industrial Maladies*. (ed. S.A. Henry), Oxford University Press, London (1934)
15. Sadler Committee: Select Committee of House of Commons, on Bill relating to labour of children in mills and factories (1832)
16. Meiklejohn, A. Sixty years of industrial medicine in Great Britain. *British Journal of Industrial Medicine*, **13**, 155–162 (1956)
17. McKiever, J. *Trends in Employee Health Services*. US Department of Health, Education and Welfare, Washington (1965)
18. Hamilton, A. *Exploring the Dangerous Trades*. Little Brown, Boston (1943)
19. Grant, M.P. *Alice Hamilton, Pioneer Doctor in Industrial Medicine*. Abelard-Schuman, New York and London (1967)
20. Union of Soviet Socialist Republics: Ministry of Health. *The Systems of Public Health Services in the USSR*. Ministry of Health, Moscow (1967)
21. Jaques, P. Trade unions and the working environment. *Journal of the Royal Society of Arts*, **125**, 672–683 (1977)
22. World Health Organization. *Health Aspects of Well-being in Working Places*, Euro Reports and Studies 31. WHO, Copenhagen (1980)
23. World Health Organization. *Occupational Health in Four Countries, Yugoslavia, the USSR, Finland and Sweden*. WHO, Geneva (1963)
24. Arthur Young Books Ltd. *The Manager's Handbook* (ed. Erica Hunninger). Sphere, London and Sydney (1986)
25. Murray, R. Health of the worker in the 20th century. In *Hunter's Diseases of Occupations*; (eds P.A.B. Raffle, W.R. Lee, R.I. McCallum and R. Murray) pp. 156–203 Hodder and Stoughton, London and Sydney (1987)
26. World Health Organization. *Identification and Control of Work-related Diseases*: Report of a WHO Expert Committee. Technical Report Series 74. WHO, Geneva (1985)
27. Registrar General. *Occupational Mortality Supplement. OPCS*. HMSO, London (1978)
28. Harrington, J.M. and Oakes, D. Mortality study of British pathologists. *British Journal of Industrial Medicine*, **41**, 188–191 (1984)

29. Health and Safety Executive, Medical Division, Employment Medical Advisory Service, *Health at Work*. Report. HMSO, London (1983–85)
30. Webb, T. and Schilling, R. with B. Jacobson and P. Babb, *Health at Work? A Report on Health Promotion in the Workplace*, Health Authority, London (1988)
31. World Health Organization. *Training and Education in Occupational Health, Safety and Ergonomics*. Technical Report series 663. WHO, Geneva (1988)
32. International Labour Organization. *Convention 161 and Recommendation 171 on Occupational Health Services*. ILO, Geneva (1985)

Chapter 2

Pre-placement screening

H.A. Waldron

Introduction

Pre-placement screening has a long history in occupational health practice and formerly took up a great deal of the occupational physician's time. In recent years it has become increasingly clear that the wholesale clinical examination of predominantly healthy men and women has little to commend it although this is a view which may not always be shared by employers.

A discussion of pre-employment screening has to recognize that for the three groups of people involved—the occupational health practitioner, the employer and the prospective employee—it has widely differing objectives. The doctor or nurse wishes to ensure that individuals are both physically and mentally suited for the job for which they have applied, so far as is reasonably practicable. This by no means requires them to be in perfect health. The employer, however, wishes the pre-employment screen to act as a guarantee that all workers newly engaged do have perfect health and will continue to do so in order to maximize efficiency and minimize time lost for reasons of illness. Finally, prospective employees tend to view pre-employment screening as a hurdle to be overcome on their way to a job and if they are aware of anything in their medical history which may be thought to be a hindrance in this task, they may well be economical with the truth when it comes to filling in a questionnaire or answering direct questions from a nurse or doctor. In some cases, another doctor may collude with an applicant for a post by minimizing the effects of an existing medical or psychiatric condition, seeing his first duty as securing a job for his patient. It is obvious that these different objectives have little in common.

Reasonable objectives

The most reasonable objective of pre-employment screening from the point of view of occupational health practice is to ensure that individuals and their jobs are as well suited, in all respects, as can be expected. It should always be remembered that the occupational physician or the occupational nurse is in no position to deny or to promise an individual a job except when it is within their own department and they are actually making the appointment. The responsibility of the occupational health department is to give advice in general terms to management on the medical suitability of a candidate and it is the prerogative of management to accept that advice or not as they think fit, taking all circumstances into consideration.

The approach

There are few conditions which are an absolute disbarment to any kind of work, and what is required of the occupational health practitioner is firstly to determine as accurately as possible the physical and mental state of the applicant and then see how these measure up against the requirements of the job. In order to do this it should be obvious that those requirements must be well known to those making the assessment; this calls for an intimate knowledge of the working practices within the organization which is not likely to be achieved by a doctor or nurse who does not regularly visit the various places of work. Any legal requirements, such as those which relate to diving or to exposure to lead or asbestos in the United Kingdom, for example, will necessarily also have to be taken into account. In some cases the doctors making pre-employment assessments must be appointed by the Health and Safety Executive. This applies under the terms of the Work in Compressed Air Special Regulations 1958, the Diving Operations at Work Regulations 1981, the Merchant Shipping (Diving Operators) Regulations 1975, the Control of Lead at Work Regulation 1980 and the Asbestos (Licensing) Regulations 1983.

For each type of job within an organization any physical or chemical hazards must be known about and quantified. It is hardly likely, however, that any occupational health practitioner would overlook these so long as he was as familiar with the working practices within his company as he ought to be.

There may be other requirements which are not so obvious, however, and it is useful to have not only the job description which applies to an applicant, but also the manager's assessment of those attributes which he considers are essential for someone in the post in question. For example, is it a job which requires a great deal of manual dexterity; of physical strength; is it a task which requires considerable mental agility? Drawing up what one might call a managerial protocol requires a good deal of collaboration between the occupational health department and management but it is an opportunity to clarify thought, to develop strong ties between the occupational health department and heads of other departments and each side has an opportunity to educate the other about their own approaches to the task. Out of such deliberations may come some written company policies in relation to persons with particular conditions. At present there is much heart-searching about the suitability of employing those who might have AIDS or be HIV-positive. Some companies consider that the risks of employing such individuals are too great and will not employ them, whereas others take a more liberal view and have a policy which does not discriminate against them. Whatever policy is arrived at, it is important that the occupational health department has an input to the decision-making.

With all the relevant information available, the occupational health department must choose how to implement its pre-employment screening procedures. There are, broadly speaking, three choices of assessment: by questionnaire alone; by questionnaire and health interview with a nurse; by questionnaire and clinical examination. Until comparatively recently, the third option was widely used with the result that occupational physicians spent much time examining well people to no great advantage to the applicant, the company or themselves. Scarcely anyone would advocate such an approach nowadays.

The questionnaire

The simplest and most cost-effective method of pre-employment screening is by the use of a simple questionnaire. An almost infinite variety of these must have been developed over the years but none can be completely generalized since companies vary in the specific requirements of those they choose to take on their payroll. An example of a screening questionnaire is shown in *Figure 2.1*; it is not meant to be a definitive model but is one which has worked well in practice. Prospective employees should be sent a questionnaire to complete only when they are being seriously considered for a post. This point is worth emphasizing. If questionnaires are sent to *any* individual who applies for a job the occupational health department will find itself carrying out a large number of unnecessary assessments. The questionnaire can usually be sent out with the invitation to attend for an interview.

The forms may be sent to applicants by the personnel department or by departmental heads; it does not matter who, so long as they are sent out at the appropriate time. They must always be returned—preferably in a reply paid envelope—to the occupational health department, however. It is absolutely essential that applicants are assured that the information in the questionnaire is entirely confidential to the occupational health department and that management is advised about health matters in general terms only. Some forms require the applicant to sign a declaration that the information given is true and that they understand that falsehoods may lead to dismissal. This quasilegal declaration appears to be included on the basis that it will produce more honest answers, but an applicant who wishes to conceal information will almost certainly not be dissuaded from doing so because he has to put his name to the document and, since such a declaration has no standing in law, it is much better to leave it out altogether. What the applicant should be asked to sign, however, is a consent form to allow further information to be obtained from his own medical advisers if this is considered to be necessary to provide a fully informed opinion to management. Most applicants will readily agree to this.

On their return the forms can be scanned by an occupational nurse who should have the authority to advise that an applicant is *suitable* for employment. At times there may be some urgency in advising a manager about the suitability of a candidate and it is reasonable to give a verbal opinion; this must always be confirmed in writing, however.

Where there is any doubt about a candidate's fitness for work, the form should be referred to the occupational physician. At this stage it is possible to advise management against employment in some well-defined cases. For example, it would be unwise to consider anyone with neurological or renal disorders as suitable for exposure to heavy metals; for those with neurological or hepatic disorders to be exposed to solvents; for those with asthma to be allowed to come into contact with known allergens; or for those with a history of contact dermatitis to work with skin sensitizers. In those cases, although some further information may be required from the applicant's own doctor to confirm a diagnosis, it may not be necessary for the occupational physician to see the individual concerned. When the matter is not entirely clear cut, however, then the applicant must be seen before advice can be given one way or the other.

The majority of cases can be dealt with satisfactorily in this way. Where there are special risks, or where the health and safety of others may be affected, an

IN STRICT CONFIDENCE

Surname .... First names................................

Address ....

Date of birth .... Marital status ....

Proposed employment ....

Proposed employer ....

Please complete this form as fully as possible and return it as soon as you can in the envelope provided. All the information is strictly confidential and will not be disclosed to anyone without your written permission.

Have you *ever* had any of the following conditions? Please give further details where appropriate.

	Yes or No	*Date*	*Details*
1. Persistent, productive cough?			
2. Asthma or hay fever or any other condition?			
3. Any skin disorders?			
4. Unusual shortness of breath on exertion?			
5. Persistent chest pain?			
6. Palpitations?			
7. Any other heart disease?			
8. Fits or faints?			
9. Any nervous or mental illness?			
10. Jaundice?			
11. Kidney or bladder infections?			
12. Dysentery, food poisoning or gastroenteritis?			
13. Stomach or duodenal ulcers?			
14. Persistent pain in the joints?			
15. Severe back pain?			
16. Diabetes?			
17. Do you have any problem with your hearing?			
18. Do you have good vision?			
19. Do you wear glasses?			

20. Have you ever had any illness which required admission to hospital?
YES/NO

If 'YES', please give further details: ....

....

21. Have you ever had any major operations? YES/NO

If 'YES', please give further details: ....

....

22. Have you ever had an accident which required admission to hospital?
YES/NO

If 'YES', please give further details: ....

....

23. Are you *at present* having any treatment from your doctor?
                                    YES/NO

If 'YES', please give further details: ....

....

24. Are you on the Disablement Register?   YES/NO

If 'YES', what is your disability? ....

....

25. When did you last have a chest X-ray? ....

26. Do you consider that you are in good health at present? YES/NO

27. Have you had a medical examination in the last five years for an insurance policy or for any other purpose?   YES/NO

If 'YES', what was the outcome? ....

28. Do you smoke?   YES/NO

If 'YES', how many cigarettes, or how much tobacco do you smoke a day? ....

....

29. Do you drink alcohol?   YES/NO

If 'YES', how much do you drink per week? ....

Figure 2.1 Pre-placement health questionnaire

interview with the occupational nurse as a follow-up to the questionnaire may be advisable. The health interview would be used to obtain specific information. For example, what are the standards of personal hygiene of those who may be handling food? Is the vision of prospective company drivers adequate? As before, the nurse should have the authority to recommend acceptance but must refer doubtful cases to the occupational physician. It is preferable to conduct the health interview on the day of the applicant's job interview, but in busy departments this may not always be possible.

From what has been said so far, it will be clear that pre-employment medical examinations should be the exception rather than the rule. Some are obligatory in order to obtain a licence to undertake the job in question—airline pilots and heavy goods vehicle drivers, for example—while the demands of some other occupations may make a medical examination desirable. Candidates for the fire and police services might come within this category. Some employers may require it for some or all new employees; this is particularly the case for senior appointments and the occupational physician will have to comply in those instances. He may choose to try to influence against such a policy if he feels that nothing useful is served by it, but in the end the employer must be free to exercise his preference.

Pre-employment testing

Pre-employment tests may be required as part of the assessment of fitness for a particular job to ensure that applicants meet certain prescribed standards, or they may form the baseline for a programme of continuous biological monitoring. Examples of tests in the first category would include tests of vision in crane drivers or VDU operators where emphasis on depth of vision and middle vision respectively might be important; chest X-rays in those who are going to work in close contact with patients in a children's hospital; hepatitis B antigen screening in those on renal dialysis units, or liver function tests in those who might be exposed to solvents. As a general rule, pre-employment tests should be carried out only when there is a good reason to do so—blunderbuss screening has little to commend it.

Tests which anticipate participation in a biological monitoring programme might include a chest X-ray where exposure to fibrogenic dust occurs in the job; lung function tests where there may be exposure to allergens, or audiometry where the work is noisy. The aim at this stage is partly to exclude anyone with an abnormal result on the grounds that exposure may exacerbate the condition, but also to provide a baseline against which future test results can be judged.

Conclusions

Pre-employment screening, like all forms of screening, should have a clearly defined end-point which is to ensure the best fit between employees and their jobs. Occupational health practitioners must be familiar with the demands of each job, they must be aware of any special hazards associated with it and any managerial requirements. It is this special knowledge about the nature of the work carried out in their organization which makes occupational health practitioners better able to carry out this task than their colleagues in other specialties.

Routine medical examinations are unnecessary in the majority of cases which can be dealt with adequately by a questionnaire supplemented, in some instances, by a health interview with an occupational nurse. The questionnaire can be assessed in the first instance by a nurse who can advise acceptance; advice against employing an individual on health grounds must come only from the occupational physician.

Pre-employment testing should be carried out only if there is a special reason for doing so or if the assessment will form the baseline in a programme of continuous biological monitoring.

Confidentiality must be assured at every stage and the prospective employee must be made to feel that the occupational health department has his or her interests foremost in their deliberations while not abrogating responsibilities to other employees and to the company.

Further reading

Health and Safety Executive. *Pre-employment Health Screening (Guidance Note MS20)*. HMSO, London (1982)

Zenz, C. (ed.) *Occupational Medicine. Principles and Practical Application*. 2nd edn, Chapter 8. Year Book Medical Publishers, Chicago (1988)

Chapter 3

Principles of toxicology

A.G. Salmon and H.A. Waldron

Introduction

Toxicology has developed from the early stage of being a science of pure observation and reporting, to one where various underlying principles and concepts may be discerned. Some of these concepts are unique to toxicology and developed by analysis and comparison of data on the toxicity of various compounds to various species. Many were imported from related areas of science, particularly those of biochemistry and pharmacology. These subjects are of fairly recent development, but preceded the modern expansion of knowledge and interest in toxicology. In fact the application of biochemical techniques and concepts has been a major factor in the recent development of scientific toxicology, and this process continues, for instance with the use of genetic engineering techniques to investigate carcinogenesis. Toxicology has also relied extensively on developments in other areas including histology and electron microscopy. Modern high-sensitivity analytical chemistry is not only essential to the forensic and environmental branches of toxicology, but is a prerequisite for any serious study of toxicokinetics and foreign compound metabolism.

Basic definitions

Dose and dose–time integrals

Like most specialized areas, there are some commonly used words or concepts which must be accurately defined. One of the most basic, and least consistently described of these is the *dose*. The intuitive definition of this word, as the amount of material entering the body of an exposed person or experimental animal, is adequate and unambiguous for single short-term exposures or treatments and immediate effects. However, for chronic or repeated exposures, or when considering the dose received by a particular tissue allowing for pharmacokinetic factors, both the amount or concentration present, and the time for which it is present must be considered. This is expressed as *dose–time integral*, and it is this quantity which is the fundamental toxicological measure. Some authors use the word 'dose' to refer to this integral of time and concentration, but although this has the advantage of brevity and logic it conflicts with non-specialist uses of the word: in this chapter it will not be used in this way.

A particular dose–time integral is the result of a particular pattern of exposure, defined as the amount of material in contact with the body, and capable of being absorbed, by a person or experimental animal. The time factor is again of fundamental importance, and in describing real exposures not only is the concentration–time integral important, but the actual variation with time may be relevant, for example, several successive discrete exposures or a single but continued exposure. In occupational contexts, and in toxicological experiments designed to model these, the exposure will usually be an amount applied to the skin, or a concentration in air or diet, over a defined period.

Measuring lethality

Toxicology uses many measures of effect, depending on the nature of the process described, the techniques available to detect it, and the purpose of the investigation. One of the simplest and oldest approaches to quantitative toxicology is to measure lethality. This is conventionally expressed as the LD_{50}, the dose (for discrete single exposures) or concentration–time integral (for continued or repeated exposures) causing on average 50 per cent of exposed organisms to die within a stated timescale. The similar term LC_{50} is used for exposure to an inhaled substance. A fundamental assumption in this definition is that the lethality response in a population of exposed organisms follows a statistical distribution (usually assumed to be 'normal'). A global population value may thus be inferred by statistical analysis of the results of a particular experiment.

Carcinogenesis

Another important end-point in toxicology is the observation of carcinogenesis. A carcinogen is a substance which causes tumours, or unrestrained growth of some cells in the body. Normal cells of most tissues have the capacity to grow and divide so as to produce an increase in the mass of that tissue; this is important both for overall body growth and for repair and adaptation of specific organs. However, this growth potential is tightly constrained, so that the amount and location of a tissue is always determined by the requirements of normal body structure and function. In tumours, this regulation is broken down so that growth is continuous and unrestrained: the cells may also lose their limitation to an appropriate site and spread by invasion of surrounding tissue and by metastasis (spread of tumour cells to remote sites). Specialists distinguish different degrees, properties, or origins of tumours, but for the purposes of this definition any abnormally growing (*neoplastic*) cells would be counted. Carcinogenesis is envisaged as a multistep process, the first step of which is usually the alteration of a regulatory gene in a somatic cell. This first step is called *initiation*. The proposal that a somatic mutation is the origin of cancerous change has been a preferred hypothesis for many years, but recently has been strongly supported by the discovery of similar alterations in certain genes (*oncogenes*) in cells from both viral and chemically induced tumours which perform specific functions in the growth regulation of normal cells. The initiation event is followed by a series of further changes, described as *promotion*, whereby other chemical, physical or biological agents may trigger the change, and *progression*, where changes in the character and number of tumour cells occur spontaneously. Some of these changes appear to involve further mutational events, and carcinogens such as the polycyclic hydrocarbons which are effective initiators are often effective at the promotion stage also. On the other hand many promoters are not mutagens, at least in the usual assays, and some promotional steps must be non-mutagenic in character, even if they result in genetic alterations. At the end of this multistage process a clone of tumour cells results which is significantly altered from normal cells not only in its growth regulation (or lack thereof), but frequently also by the loss of normal differentiated functions, and sometimes by expression of inappropriate functions also.

Specific effects on body systems

Specific effects on major functional systems of the body constitute another important class of toxic actions; these are obviously numerous, and various specific examples are discussed in the description of effects of specific compounds. However, there are two groups of effects which are sufficiently widespread and distinctive to merit definition in this general discussion.

The immune system

The immune system is an important target of toxic actions, which may either increase or decrease its level of activity. Changes in overall responsiveness to immunological stimuli have been reported after various chemical exposures. Many chemicals will also evoke specific responses such as sensitization. This term is used to describe the situation where, after one or more exposures to a chemical, humoral or cell-mediated immune reactions occur so that subsequent exposures even to very low levels of the activating chemical produce a dramatic response, such as dermatitis or asthma.

Reproduction

Toxic effects on reproduction are also characteristic of certain chemicals. The best-known of these result from exposure to drugs, such as thalidomide, but certain industrial chemicals and pesticides also display this type of toxicity. Effects may be classified as mutation, a phenomenon already considered as a factor in carcinogenesis, but affecting germ cells rather than somatic cells, fertility effects on either male or female parent, embryolethality and teratogenesis (induction of structural or functional abnormalities in the offspring as a result of exposure *in utero*). Several cases in which male sterility or reduced fertility is produced by industrial chemicals have recently been identified. Dibromochloropropane (a pesticide) was found to have caused reduced fertility or sterility among a number of workers handling it. The glycol ether 2-methoxyethanol has been subjected to considerably more stringent regulation since it was observed to cause testicular atrophy and infertility in exposed male rats. Possible occupational teratogens such as halocarbons have also received some attention, in addition to the well-known harmful effects of lead exposure during pregnancy.

Routes of entry

Toxic materials may enter the body in various ways, and the applicable route affects the way in which the material becomes distributed. In the case of drugs, the oral route is one of the most important, along with artificial routes such as injections into various sites. These routes are also widely used for experimental studies of toxic compounds. They are much less common for occupational exposures to toxic chemicals, which more often involve inhalation of vapours or aerosols, and absorption of materials through the skin. It should be noted, however, that many exposures to inhaled dusts produce oral exposures since much of a material deposited in the nose, throat and lungs is trapped in mucus which is eventually swallowed.

As a result of having evolved to deal with diets (especially those of vegetable origin) which contain many exotic and potentially toxic chemicals, the mammalian body is well equipped to deal with foreign compounds, especially those entering orally. The detoxifying enzyme systems in the intestinal mucosa and in the liver are highly active, and these tissues are well protected biochemically from reactive intermediates and capable of rapid recovery from damage. Substances entering the body by the oral route, and absorbed in the intestine, are carried directly to these sites of active metabolism by the hepatic portal system, and many ingested compounds are entirely converted to harmless and excretable products before they ever reach the general circulation. By contrast, inhaled substances and those absorbed through the skin are likely to enter the general circulation directly and thus may affect sensitive tissues in any part of the body. The skin and the lungs also have some detoxifying enzymes, but these are less likely to be able entirely to remove an incoming foreign compound.

Because of these marked differences in the internal distribution of toxic materials produced by different routes of entry, it is important when considering experimental models of possible occupational toxicity to choose experimental routes of exposure which resemble those applicable to the conditions of human exposure. Observed effective toxic concentrations may vary very widely for some compounds according to the route of entry, and in a few cases even the nature of effects produced is different. This may be true either as a result of the protective effect of hepatic metabolism, or because a particular route of entry preferentially exposes a sensitive tissue. The polycyclic hydrocarbons such as benzo[a]pyrene illustrate this well, being capable of causing skin tumours after dermal exposures, lung tumours after inhalation or intratracheal instillation, liver tumours (at very high doses only) after oral ingestion and even, after ingestion by strains of mice in which the enzymes responsible for hepatic metabolism of the compounds are not inducible, aplastic anaemia.

Interspecies variations

The major concerns when considering the toxicity of industrial chemicals are the human toxicity, and also toxicity to some wild or domesticated animal species which may be exposed if the material is liable to be released into the general environment. However, except in the case of well-known materials which already have a long history of both use and misuse, there will be no data on toxicity to man or other actual target species. Any judgement must therefore be made on the basis of experiments on laboratory animals. All these animal species respond differently to toxic chemicals, but it has generally been found that the differences are not so great that interspecies comparison is impossible. This is especially so if more than one suitable animal species has been studied, and this is the standard procedure for carcinogenicity, where interpretation can be particularly difficult. The responses of all these species may vary as a result of different sensitivities of the target tissues to the toxic effect of a compound. More usually the reason is that there is variation in the metabolism and distribution of the compound; there are marked differences between mammalian species in some cases. Selection of a test species appropriate for modelling the human response requires knowledge of the metabolic processes involved, and their characteristics in different species. It has been suggested that more effective models of toxicity to man would be obtained by studying animals, such as primates, which are more closely related to man than the usual laboratory test species, but this has not been found to be true in practice.

Dose–response relationships

Toxic damage to a living organism by a chemical is the result of a specific biochemical or physiological change which impairs the function of certain cells in a particular tissue and organ. There may be only one single target organ and sensitive site, or the chemical nature of a toxin may be such as to damage several different sites. In the latter case the toxic effects seen will essentially be the sum of several distinct processes; for the purposes of discussion the simpler case where there is a single target will be considered. Since the fundamental process is some kind of chemical reaction, such as irreversible reaction with a certain protein, or reversible binding to a receptor, toxic processes at the basic level obey the law of mass action. The extent of damage caused by a chemical depends on the concentration of this material reaching the target tissue. Since the reaction process and its damaging consequences occur at a finite rate, the time for which the chemical is present in the target tissue is also significant. This is the origin of the relationship between degree of effect and the dose–time integral noted previously. Where both the dose received by the target tissue and the damage caused can be adequately quantitated, this relationship can be readily demonstrated.

Unfortunately, there are not many cases where this level of knowledge is available. More usually there are several complicating factors to be considered. Usually the actual dose–time integral applicable to the target tissue is not known, and has to be inferred from the exposure data. In this case the bioavailability (that is, the fraction of that present as an exposure which actually enters the body tissues) by the relevant route or routes of absorption into the body has to be considered. Also there are important processes responsible for distribution of material around the body both to target tissues and to other tissues, which do not suffer any damage but may for example act as storage reservoirs for toxic materials. In many cases the damaging material is not the substance present in the environment but an intermediate produced from this by metabolism, and in this case the rate of metabolism producing this intermediate, and the rate of its removal by decay or further metabolism, are important. Finally the rate at which toxicants or their metabolites are removed by the various excretory processes, especially via urine, bile and exhaled air, will play a part in determining the dose–time integral experienced by the target tissue.

Evidently, dose–response relationship in reality may be very complicated, due to the interaction of all these processes. There is often found to be a linear relationship between the logarithm of the externally applied dose–time integral and effect, at least at intermediate levels of effect. This may apply as a consequence of the law of mass action which applies to chemical equilibria such as the reversible binding of a molecule to a receptor. This log–dose relationship also approximates the central part of the cumulative normal distribution describing individual discrete responses among a randomly varying population. However, the internal processes and responses of the exposed organism impose some limits on this linearity. At high dose levels there may be saturation either because metabolic processes involved in toxicity are running at their maximum sustainable rate, or because all the susceptible sites in the target tissue have already been affected. Similarly a threshold is often seen below which no adverse effects are seen, either because metabolism is clearing all the toxic material before it can reach the target tissue, or because repair (or adaptation) are able to sustain the normal function of the target tissue in spite of its receiving some damage or effect.

For simple acute effects all these processes can be measured, and a dose–response curve constructed which describes all the possible consequences of an exposure. A problem arises when important parts of the dose–response curve are not accessible to experiment either in man or in animals. This is the case for carcinogens, where large numbers and high doses are often required to demonstrate their effects in animal experiments due to the statistical properties of tumour induction. The exposures of interest in the workplace and in the general environment tend on the other hand to be very low. This low-exposure range cannot be observed directly, although in animal experiments the dose–response curve usually continues uniformly downwards to the limit of measurement. The dose dependence of carcinogenic effect could in theory take a number of different forms other than linearity with log–dose seen for many acute effects. The carcinogenic process is assumed to be based on the combination of a highly reactive chemical intermediate, present at low concentrations but reacting as soon as it is formed, with any one of a very large number of possible sites on cellular DNA. In the absence of other factors, this 'single-hit' model suggests a linear dependence of effect on absolute value (not logarithm) of the effector concentration. This results from the fact that this is a kinetic rather than an equilibrium model; the process is limited not by the law of mass action, but by the kinetics of generation of the proximate, reactive carcinogen. However, another theoretically possible model envisages several discrete DNA damaging events being required for neoplastic transformation of a cell to occur ('multi-hit' model), suggesting dependence of effect on some power of absolute dose. Further possible effects might produce thresholds at very low doses, or alternatively increase the effectiveness of very low doses above that predicted by the linear model; purely statistical models have also been used to try to predict dose-dependence of carcinogenesis.

While these theoretical models have many interesting features, the crucial fact is that the data available from most animal experiments and epidemiological studies are unable to distinguish the various possibilities. It is therefore conventional to assume the linear model is applicable: this produces an estimate of risk at low doses which is more or less a worst case among the reasonable alternative hypotheses. If this linear trend is assumed to continue downwards indefinitely, no exposure to a carcinogen is safe, and if large numbers of people receive a low exposure to such a compound the aggregate risk is large. If on the other hand there is a threshold there may be no such risk.

Metabolism

As applied to foreign compounds (*xenobiotics*) such as industrial chemicals, the term *metabolism* describes the conversion by the body's enzyme systems of the initial material to other products, which are usually more readily excreted in the urine or faeces. Many xenobiotics are rather unreactive, lipophilic compounds which are difficult for the body's excretory systems to handle, since these systems are primarily oriented to handling compounds in solution in water. These compounds are also difficult to involve in the sort of chemical addition and transfer reactions which are the stock-in-trade of intermediary metabolism. The system which has evolved to deal with this consists of two steps, an activation process ('phase 1') which allows a second step ('phase 2') to proceed by the addition of some endogenous group to the foreign compound to render it more water-soluble and thus more readily excreted. In the majority of cases this strategy is successful in both detoxifying and facilitating excretion of foreign chemicals. However, a minority of substances on activation to a more chemically reactive form are either so reactive as to be indiscriminately damaging, or are resistant to the phase 2 reactions which are designed to eliminate them before they cause harm. These cases include many of the toxic xenobiotics; it is often the reactive intermediate rather than the initial compound which causes toxic effects.

Phase 1

The enzyme system responsible for the metabolism of xenobiotics are of several types, but perhaps the most important phase 1 system is that referred to generically as the *cytochrome P_{450} system*. This is an extremely widespread, and therefore evolutionarily ancient, enzyme group, being represented in virtually all eukaryotic organisms (animals, plants, protozoa and fungi) and quite a few prokaryotes (such as bacteria). Its functions include the catalysis of several important reactions of endogenous substrates and the activation of many xenobiotics. It is a membrane-bound enzyme, located on the cell membrane of prokaryotes and the endoplasmic reticulum (that is, the 'microsomes') of eukaryotes. There is a catalytically active haem protein which binds both oxygen and the main substrate, and an electron transport system which provides reducing equivalents from NADPH, the general redox carrier for biosynthetic reactions. The name P_{450} refers to the fact that the reduced haem complex with carbon monoxide has an absorption peak at 450 nm, an unusually long wavelength compared with other haem proteins (for which around 420 nm is typical). This spectral peculiarity is the result of the unusual coordination environment of the haem iron atom, which is hypothesized to have the sulphur atom of a protein cysteine residue as the fifth (axial) ligand.

The typical reaction catalysed is the 'mixed function oxidase' type, in which one atom of an oxygen molecule is attached to the substrate and the other is reduced to water. The oxygen attached to the substrate is inserted, either into the pi electron orbitals of an olefinic double bond or aromatic system to produce an epoxide, or into the sigma electron orbital or a carbon–hydrogen bond (or sometimes a nitrogen–hydrogen bond) producing an hydroxyl group. Further reactions and rearrangements of the products formed may generate phenols or dealkylated products. Substrates include aromatic hydrocarbons such as the polycyclic hydrocarbons widespread as pollutants and contaminants. Also many drugs and natural products, aromatic and aliphatic halogenated hydrocarbons, amines and other materials of industrial and toxicological interest are metabolized by cytochrome P_{450}. The system consists of a number of similar enzymes with different substrate specificities; some of these are present constitutively whereas others are normally absent or at low levels but may be induced by compounds which are, or resemble, substrates for that particular enzyme. Most tissues normally contain, or can produce on induction, a certain level of some variants of this enzyme system. The highest levels are found in the liver but the kidney and intestinal mucosa are also very significant sites. The enzyme system is also active in the skin and in lung tissue, although the variants normally present and the response to inducers are different from the liver.

Phase 2

Phase 2 reactions are of several types, each resulting in a characteristic type of excretory product. The epoxide intermediates produced by cytochrome P_{450} are often converted to diols by hydration, a reaction catalysed by epoxide hydrase. These and other reactive phase 1 products are also subject to detoxification by reaction with glutathione, a tripeptide (gamma-glutamyl-cysteinyl-glycine). This reaction is catalysed by a group of (mainly soluble) enzymes collectively described as glutathione transferases. The initial product is the xenobiotic residue linked to the sulphur atom of the cysteinyl residue by a straightforward addition reaction. Depending on structure, the product may further rearrange, for instance by dehydration and rearomatization where the epoxide is derived from an aromatic ring system. These conjugates may be excreted in the bile as such, or else the glycine and glutamate may be removed and the cysteine N-acetylated. These latter operations, occurring in the kidney, produce the mercapturic acid type of metabolite which is a characteristic urinary product when reactive intermediates are being generated. Excretion of metabolites in urine is characteristic of small to medium-sized materials, but if the molecular weight is in excess of about 300 (in the rat) they may also appear in the bile and will do so exclusively if much larger.

The other important types of phase 2 reaction also result in materials readily handled by the kidney. Many phenols and aliphatic hydroxylated compounds (either absorbed as such or produced by phase 1 reactions) are conjugated with glucuronic acid. Sulphation is also common, and amines are often acetylated. The latter reaction does not necessarily result in improved solubility (amines generally being at least as soluble as amides), but it seems to be important as a route of detoxification of physiologically active amines of exogenous origin, including some drugs. All these processes are catalysed by specific enzymes, which are generally induced by the presence of appropriate substrates in the same way as the cytochrome P_{450} and other phase 1 enzymes: indeed the different enzyme groups may be genetically linked so that the corresponding phase 1 and phase 2 enzymes necessarily appear together in appropriate ratios.

The full range of possibilities for metabolism of xenobiotics is very wide, and many other examples can be found by considering specific examples dealt with elsewhere. However, these limited descriptions exemplify the general nature of the routes involved. The formation of reactive intermediates by phase 1 reaction, which are often but not invariably oxidative, is not only a key step in the systems resulting ultimately in detoxification and excretion, but also the source of the proximate toxic agent in both acute toxicity and in chronic effects such as carcinogenesis.

Bioaccumulation and excretion

The removal of a xenobiotic from the body is obviously an important process in determining the severity and duration of any toxic effect. The mechanism by which such compounds are converted to materials capable of being excreted by the body has been described. The role of the kidney and liver in handling the main pathways of excretion of non-volatile toxicants and their metabolites is critical. In some cases toxic actions on these organs can compromise their excretory function and thus prolong the persistence of chemicals in the body. The extent of this persistence can also be considerably prolonged where the material is highly fat-soluble, and thus accumulates preferentially in fatty tissues of the body. This is observed for a number of very inert materials where metabolism is slow, such as DDT and its first metabolite DDE, and for some of the isomers present in polychlorinated biphenyl mixtures. The highly publicized environmental contaminants known as 'dioxins' also belong in this category. These compounds are not necessarily highly toxic, although dioxins appear to be. DDT and its metabolites have low mammalian toxicity, although their adverse effects on birds, resulting in disturbance of calcium metabolism, thin eggshells and reproductive failure are well known. However, the fact that they are highly persistent means that even a very low level exposure, if continued, can build up a relatively high body burden of the material. This accumulation occurs at all stages of environmental food chains, so that organisms near the top of the chain receive a much larger dose than those at the bottom. The most susceptible species are thus the top-level predators such as the raptorial birds, and man in some ecosystems. The effect tends to be more extreme in marine and other aquatic environments where food chains are often longer than on land.

Occupational exposures to persistent compounds are obviously a potential source of toxicity problems due to this tendency of the body burden to build up over an extended period of time. In the case of organics, it has been hypothesized that if the body fat were to become highly contaminated by, for instance, a chlorinated pesticide, a sudden episode of toxicity could be precipitated by 'slimming', illness or some other abrupt change in metabolic balance which resulted in a significant proportion of the body fat reserves being mobilized. Cumulative toxicity is also observed when persistent toxic metals are present in the working environment. The typical chronic toxicities of mercury, lead and cadmium which are at least historical features of the working environment all show the effect of slow excretion allowing progressive accumulation of metal and resultant toxic damage. Bioaccumulation is also an important determinant of the effect of radioactive elements, and it is generally reckoned that strontium and iodine which accumulate selectively in specific tissues thereby present a much greater hazard than those with short biological half-lives.

Uptake and excretion of volatile materials

Whereas non-volatile materials will be absorbed primarily by the oral and skin routes, and excreted mainly in the urine and bile, volatile materials will enter the body via the lungs, as well as by other routes such as skin absorption. They will also leave by this route, and this process is in competition with conversion in the liver or other sites of active metabolism to non-volatile metabolites which are then excreted in the usual ways via kidney and liver. The process which predominates depends on the relative rates of excretion from the lungs and of metabolism. The chemical identity of the material is an important determinant of the latter rate: thus inhaled trichloroethylene is excreted to a significant extent in the urine as trichloroacetate and trichloroethanol glucuronide, whereas 1,1,1-trichloroethane which is physically rather similar but chemically less susceptible to oxidation is only metabolized to a small extent (a few per cent under real occupational conditions). These distinctions are an important constraint on the usefulness of urinary metabolites as biological indices of exposure. Such biological indices, and where necessary the analysis of exhaled air by mass spectrometry, are important tools in determining the actual dose received as a result of occupational exposures, especially where the possibility of skin absorption means that simply monitoring the workplace atmosphere is an unreliable guide to the extent of uptake of a material.

The rate of uptake of a volatile material such as a solvent from the atmosphere via the lungs, and the rate of excretion of the material from the lungs, depend on a number of factors. These include the nature of the solvent itself, and particularly its solubility in blood. The rate of pulmonary ventilation and blood flow, and hence the extent to which the subject is exercising, may also be important. The effect of these variables on uptake depends on whether the solvent has a high solubility in blood (for example, xylene, styrene, alcohols and ketones) or a low solubility (for example, toluene, chlorinated solvents). For high solubility solvents the uptake increases markedly with exercise, and correlates with the increase in pulmonary ventilation. This is because the capacity of the blood and tissues for these materials is large. Because of this a substantial mass of material has to enter and be distributed around the body before equilibrium is approached, and the rate-limiting steps in this equilibration are the pulmonary ventilation and perfusion. Since these factors are increased by exercise, the uptake of these solvents is increased when this occurs. The extent of mass transfer required to reach equilibrium for solvents of low solubility in blood is small, so for these the approach equilibrium is more rapid at least among the watery tissues which constitute most of the body. It is not limited by ventilation or perfusion rates, so there is little effect of exercise on uptake of these materials. Only the fatty tissues have a high capacity for uptake of these solvents, and therefore equilibration of these tissues is slow. There is a small increase in perfusion of such tissues with exercise, which results in a slight increase in uptake, but this effect is much less marked than the increase in lung perfusion. Exhalation rates will eventually match uptake rates in any situation where equilibrium is reached, but will similarly be limited by ventilation and perfusion rates where substantial mass flows are involved. This is often seen for insoluble vapours which dissolve in adipose tissue and are not metabolized rapidly, such as halocarbon vapours. Removal of such materials from the blood and from the main kinetic compartment is rapid; however, there may be a long 'tail' in the exhalation profile after an exposure which saturates the body. This is probably associated with slow removal of the material from the relatively poorly perfused adipose tissue. Uptake and exhalation rates will also show the influence of metabolism as a competing elimination process for those solvents where the rate of metabolism is significant compared with the uptake and elimination rates.

Repair and adaptation

The damage caused by a toxic agent may be repaired by the body's natural defences at a finite rate, either continuously, or at least once the toxic influence is removed.

Damage may be repaired partially or completely over a widely variable time scale. On the other hand some toxic effects are reversible like the responses seen to drugs and natural pharmacologically active substances, so recovery may be virtually instantaneous once the toxic material is removed. There is an important conceptual difference between repair of actual damage, and an adaptation to some perturbing influence which does not exceed the normal range of cellular homeostatic mechanisms. It is not always easy to make this distinction in practice, and some of the disputes about what constitutes an acceptable occupational exposure limit derive from this difficulty. This is especially true of behavioural effects where the underlying system is extremely flexible and difficult to evaluate quantitatively. One would tend to suppose that any damage would have to be quite severe before it was reflected in a qualitative change in behaviour, but this is often the only reliable parameter, especially in animal experiments.

Repair and adaptation may take many forms: alterations in the internal milieu and actual damage may be remedied at the biochemical level by synthesis of more intermediates, or by increases in the amounts of active proteins such as enzymes. Both these changes are seen when the introduction of a potentially toxic xenobiotic provokes the synthesis of a cytochrome P_{450} enzyme which is capable of metabolizing the compound. In most cases this will succeed in removing the offending material, and no damage results. The enzyme induction is therefore an adaptive process rather than a case of toxicity, but it does warn of the possibility of damage at higher doses or in other test systems. Where the dose and nature of the compound are such as to be damaging, toxic effects may result not only from the direct action of the compound and its metabolites, but from the overstretching of adaptive responses which are unable to cope. This may be the origin of the porphyriogenic action of dioxins and similar chlorinated aromatic hydrocarbons, which at lower doses are powerful inducers of cytochrome P_{450} but are also resistant to metabolism and hard to excrete.

Repair of damage to DNA is an important special case which may have large, but poorly understood, effects on the dose–response and kinetics of carcinogenesis. There are several repair processes active in living cells to preserve the integrity of the DNA against attack by chemical agents, and ionizing radiation such as cosmic rays or ultraviolet light. The impressive stability of the huge amount of information in a typical genome is testimony to their success. The most common mechanism is the 'excision–repair' process, where a single-strand area of damage is identified and removed by endonucleases. The resulting single-strand gap is repaired by synthesis using the remaining complementary strand as a template, and joining of the 'nicks' to restore the complete double strand. This process is error free, and so restores the damaged gene without mutation. Where damaged sites are very closely spaced, however, or where replication of damaged DNA is attempted before excision repair has occurred, other error-prone mechanisms may be called into action, which can fix a mutation in the affected gene. This appears to be one element of the process of mutation which is important in reproductive toxicity and carcinogenesis, although there are several other less well-understood systems controlling the arrangement and interaction of genes on the chromosome, which probably contribute to the overall phenomenon. DNA damage and repair are interesting as a clear and relatively well-studied example of the drastic effects of individual genetic variation on susceptibility to health effects. The various repair mechanisms are as one would expect under genetic control, and a mutant or aberrant gene will confer weakness or absence of one of the mechanisms. This is known to cause hypersensitivity to chemicals, ultraviolet light or other mutagens, depending on the deficient repair system, in bacterial cells and mammalian cells in culture. There are also several human genetic diseases (xeroderma pigmentosum, for example) which result from DNA repair deficiency. It may be that these represent the extreme and highly visible end of a broader but less drastic spectrum of variation in the human population.

Conclusions

The science of toxicology has progressed from its early observational beginnings to a stage where some underlying principles have emerged. These are useful, and indeed necessary, not only to interpret observations and problems in the toxicological sphere, but also to guide and inform decisions in many areas of occupational health. Although practical and political considerations are bound to play their part, no effort to regulate workplace exposures so as to avoid adverse health effects, or to monitor the state of health of chemically exposed workers, can succeed without a basis of toxicology. This brief survey of some of the underlying principles is intended to serve as an introduction to this subject, along with some of the specific cases described elsewhere.

Further reading

Homburger, F. and Hayes, J.A. *A Guide to General Toxicoloty*. Karger, Basel (1983)
Timbrell, J.A. *Principles of Biochemical Toxicology*. Taylor and Francis, London (1982)

Chapter 4

The introduction and monitoring of occupational diseases

P.J. Baxter and H.A. Waldron

Introduction

The general principles which have led to the gradual discovery and control of occupational diseases have been slow to emerge, many of the causes of ill-health in the workplace having been discovered more by serendipity than through planned scientific investigations. Despite the enormous progress that has been made in developed countries, continuing vigilance is required not only to maintain the effective control of established hazards but also to rapidly detect the emergence of new or hitherto unsuspected disorders which might appear as a consequence of changing technologies or work practices. The approach of the occupational physician differs in many important respects from that employed by physicians in other fields, for example knowledge of epidemiology and occupational hygiene principles, industrial relations and legislation are all required. Very often judgements have to be made in the face of inadequate scientific information being available on risk given the vast array of often poorly understood hazards to be found in modern workplaces. In this chapter we outline some general principles for the clinical monitoring and investigation of the more well-understood disorders as examples of occupational medicine practice. We have avoided the detailed description of clinical presentation and diagnosis which may be found in specialized texts.

Cardiovascular disease

Much is known about the environmental determinants of coronary heart disease (CHD) and the possibility that causal factors might exist in the workplace has not escaped attention. Yet the few positive finding that have emerged so far have done so in a sporadic or chance way and not as a result of a systematic search in possible high-risk industries. It is therefore likely that our knowledge of occupational factors in the development and outcome of CHD and other cardiovascular disorders is incomplete. The general risk factors for cerebrovascular diseases, such as cerebral infarction, are likely to differ in some important ways from myocardial infarction; and because so little is known about cerebrovascular diseases in relation to occupation this account will confine itself to coronary heart disease.

Recognized occupational factors

Physical activity

Epidemiological studies have demonstrated a protective effect of jobs requiring high levels of physical activity, but technological developments have resulted in the steady decline in the number of manual jobs involving heavy exertion. Accordingly, leisure-time exercise is becoming more important for maintaining physical fitness in industrialized countries.

Carbon disulphide

A raised mortality from CHD in viscose rayon workers was first shown in a British company in 1968, since when further investigations have confirmed a causal link between carbon disulphide exposure in the working atmosphere and both fatal and non-fatal CHD. Carbon disulphide is also used as a solvent in other industries, for example, chemical and rubber. The metabolism of carbon disulphide is complex and it has been postulated that exposure could promote atheroma formation or blood-clotting abnormalities, but recent work suggests that it may precipitate clinical events in workers who already have advanced CHD from other causes. Exposure in the viscose rayon industry has declined over the years though it has yet to be shown that the risk has been reduced by adoption of the current control limit (10 ppm, 8 h time-weighted average, TWA).

Dynamite manufacture

Dynamite is an explosive consisting of nitroglycerine or ammonium nitrate mixed with kieselguhr, sawdust or wood pulp. Nitroglycerine is a viscose liquid and exposure may arise through inhalation of its vapour or by direct absorption through the skin. Reports of sudden death or angina in dynamite workers after a period of 36–72 h away from work have led to the risk becoming recognized. Throbbing headaches have long been known to be a response to heavy exposure to nitroglycerine and characteristically these improve during the working week only to return with renewed intensity on resumption of work after a weekend away. Nitrate esters act as vasodilators, and are used for this reason in the treatment of angina. It has been postulated that the effect on the heart in dynamite workers could be due to a rebound constriction of the coronary arteries at the weekends when away from exposure, or a high risk on Monday mornings may also result from re-exposure when sudden hypotension could be caused by vasodilation.

Solvents and aerosol propellants

Heavy exposure to chlorinated hydrocarbon solvents in poorly ventilated work-places, or as a result of solvent abuse, can result in a sudden loss of consciousness and death. Solvents such as trichloroethylene or 1,1,1-trichloroethane may sensitize the myocardium to the action of endogenous catecholamines and result in ventricular fibrillation. Many of the commonly used solvents can be monitored biologically as well as by air monitoring. Chlorofluorocarbons (CFC) are commonly used as propellants in aerosol cans and as refrigerants; one study reported cardiac arrhythmias in workers in a hospital pathology department where they used a CFC spray to prepare frozen sections. Deaths have also occurred from abuse or sniffing of aerosols. However, there does not seem to be an increased risk of sudden death among workers engaged in the manufacture of aerosol cans or refrigerants.

At least one death from myocardial infarction has been reported as a result of working with methylene chloride (dichloromethane), a readily volatile and widely used solvent which is a common constituent of paint remover. Methylene chloride has the unusual property of being metabolized to form carboxyhaemoglobin (CoHb) in the body, and levels as dangerously high as 50 per cent have been reported under working conditions, far in excess of that found in heavy cigarette smokers (5–10 per cent). In workers with pre-existing CHD, these high levels are likely to precipitate angina or even infarction. CoHb can be detected in the blood at the current control limit (100 ppm 8 h time-weighted average value) and its measurement is useful in monitoring exposures, at least in non-smokers.

Stress and type A behaviour

Stress is the most often quoted causal factor for CHD in the workplace yet it may come as a surprise to many that the evidence is far from conclusive. Despite numerous articles and studies over recent decades, the role of occupation or indeed general psychosocial stress remains elusive and is unlikely to be as straightforward as is frequently supposed. There is no evidence that monotonous and repetitive work, for example, on a production line, on the one hand, or the varied and onerous tasks performed by senior executives on the other *per se* place the employee at any special risk. However, the interaction of personality and other life stress factors (for example, social isolation, family and financial problems) with occupation could be important, particularly for those who have already sustained a myocardial infarction and are therefore at a higher risk of cardiac death in the future. People with aggressive, competitive personalities, who are overcommitted to work and have a strong sense of time urgency (that is type A personalities) have been thought to be at a higher risk of CHD than their type B opposites; yet after three decades and numerous studies this attractive theory still awaits wide scientific acceptance.

Hypertension and cardiomyopathy

Experimental exposures to lead, mercury, cadmium, cobalt and arsenic have all been shown to have toxic effects on the cardiovascular system of test animals. Epidemiological studies in man have failed to demonstrate a consistent effect of occupational exposure to mercury, cadmium or lead in the causation of hypertension, cerebrovascular disease or cardiomyopathy. Cobalt and arsenic have been implicated as causing cardiomyopathy in two separate epidemics involving the consumption of beer contaminated by each of these substances, but cardiomyopathy has not been demonstrated to occur with any frequency in the workplace. Shift workers and workers exposed to excessive noise have also been cited as being at higher risk of hypertension and these findings require confirmation.

Outcome of heart attacks

The overall survival rate from heart attacks is about 60 per cent, being higher in first attacks and lower in subsequent ones. The majority of deaths take place before the patient reaches hospital, and some of these deaths could be prevented if cardiac resuscitation was begun early and maintained until defibrillation was performed at the hospital or emergency facility. In some countries, notably the United States, special programmes have been instituted in certain cities widely to educate the public in cardiac resuscitation techniques so that the patient can be supported until skilled help arrives. Some workplaces have installed facilities for performing cardiac defibrillation, but as the number of treatable cases which would be seen over the years is likely to be quite small the benefit of such facilities remains unproven.

Most young survivors are able to return to work and to the same job, but a major cause of invalidism has been inadequate medical instruction about fitness and when to return to work. Most people without complications should be able to return to work within weeks and at least before three months. Confidence can be instilled by providing appropriate advice over the person's capabilities and through cardiac rehabilitation involving progressive, supervised programmes of physical activity. The outcome for patients who have coronary artery bypass grafting for angina is also very good; most can return to even physically demanding jobs after 3 months following the operation.

Health monitoring

Surprisingly, there is no legally based medical guidance on health surveillance of viscose rayon or dynamite workers in the United Kingdom, despite the evidence of the increased risk of their jobs. As a minimum, a pre-employment screening examination and a regular health surveillance programme should be instituted among these workers so that at least people with known CHD are excluded from exposure. Workers with known CHD exposed to solvents should be warned about the risks of excessive exposure and in any event the recommended exposure limits should be adhered to. Where appropriate, warnings should be given about exposure through the skin. As a general precaution, not just to protect against CHD, exposure to heavy metals should be regularly checked by biological monitoring as well as by environmental measurements.

A European multicentre trial has shown that intervention programmes in the workplace do alter CHD risk factors and myocardial infarction rates where such programmes have been accepted. There is now considerable public demand for general health screening, and occupational health departments have had seriously to consider how these demands can be met without jeopardizing the more traditional aims of their service; indeed some managers may perceive health screening to be the main role of occupational health, particularly in non-manufacturing industries. Health promotion is more than health education, as it seeks to create conditions in which healthy ways of living can develop; and the workplace is increasingly seen as an important area for its application. Thus a coronary screening programme could be combined with company initiatives on antismoking, healthy eating and stress reduction. Apart from improving well-being, stress reduction may have an important role for those who have already suffered a myocardial infarction or have angina. The screening programme should aim at providing cholesterol-lowering dietary advice, cessation of smoking, weight reduction for those overweight, encouraging exercise, and the control of hypertension. There is probably no justification in performing resting or exercise electrocardiograms to detect CHD in asymptomatic individuals.

Investigation of a hazard

Despite the limited number of occupations where a real risk of coronary heart disease has been demonstrated, many occupational groups at one time or another believe that they are at special risk and will point to a cluster of heart attack cases or deaths among their members, or to the premature coronary deaths of retirees as evidence. Occupational stress is frequently cited to be responsible. These claims are usually unfounded as the evidence provided often merely reflects the high prevalence of CHD in the general population, or, as sometimes occurs, that people with pre-existing CHD have been selected or selected themselves into less demanding jobs. Nevertheless, not all concerns should be lightly dismissed and clusters of CHD cases should be investigated. A case definition should first be established (the criteria should include characteristic history of chest pain, typical ECG and/or serum enzyme changes and, in fatal cases, the finding of recent infarction or coronary occlusion at necropsy).

Diagnostic criteria for each attack in the cluster should be obtained and the validity of the diagnoses then assessed, as other conditions may be wrongly labelled as heart attacks. The age, sex, and whether it is the first or subsequent attack should be noted. It is essential to survey medical records or other sources to identify any further cases which have not come to light. For each attack, the precise relations with occupational exposure should be noted, such as the time of day and the day of week of the onset, type of job, etc. to see if the attacks form a recognizable pattern. Reports of palpitations or angina should receive similar attention.

A simple descriptive study of the attacks along these lines may be useful in excluding an acute effect of occupation, as has been shown with dynamite and solvent workers, but will be of little value for exposures suspected of causing long-term changes or having more subtle effects. Epidemiological studies of CHD incidence may be essential to exclude or confirm the existence of a problem. Because of their frequency and available criteria for case definition, heart attacks are the best manifestations of CHD to study and in the United Kingdom heart attack rates for the general population are available in some areas for comparison. More elaborate studies using questionnaires, electrocardiograms and measuring coronary risk factors are important for understanding underlying mechanisms once a CHD risk has been established but have limited value initially in determining the presence or absence of a risk compared with a study of heart attacks. Ambulatory electrocardiography has not been shown to have a place in occupational studies in general though its use should be considered to elucidate complaints of arrhythmias arising at work.

Reproductive system

Early interest in reproductive hazards of occupation focused on the effects of work itself and the role of chemical and radiation exposures on the health of the developing fetus. Studies exploring the relation between paid employment during pregnancy and the outcome of pregnancy have had conflicting results. Over the last two decades animal and human investigations have led to a wide range of adverse reproductive outcomes being considered, and numerous chemicals have fallen under suspicion because of their mutagenic, carcinogenic and teratogenic properties in animal or *in vitro* systems.

Some chemical exposures may cause endocrine and neurological abnormalities, or have direct toxic effects on the reproductive organs of both sexes. Of the range of possible adverse effects, infertility, spontaneous abortions, congenital defects, low birthweight, chromosome abnormalities and childhood cancer have been most amenable to epidemiological study. The results of animal studies in this field are not readily applicable to man, and to date the number of chemicals which have been convincingly shown to have an effect on human reproduction is small.

Known occupational or related hazards to reproduction

Lead

Occupational lead poisoning reached epidemic proportions in the nineteenth century at a time when women were widely employed in lead industries. Infertility, spontaneous abortion and stillbirth, amongst other problems, were attributed to lead exposure and lead compounds were in fact widely used as illicit abortifacients. Traditionally women have been employed in large numbers in repetitive assembly or manufacturing processes involving lead, such as battery manufacture and enamelling. In the United Kingdom all workplaces where potentially hazardous exposure to lead may occur should be under the Lead Regulations which prohibit the employment of pregnant women in lead work and require the regular biological monitoring for lead among all workers; a lower suspension level for blood lead has been set for women of reproductive age compared with men. There is some evidence that excessive lead exposure in male workers may result in lowered sperm counts, and other sperm abnormalities.

Organic mercury

Evidence for the teratogenic effect of organic mercury has come from two epidemics of mercury poisoning caused by the consumption of contaminated food and not from the study of occupational exposure. At Minimata Bay in Japan in 1953–65, where fish had become contaminated with mercury released into the sea from a polyvinyl chloride (PVC) factory, 22 infants were born with congenital damage to the central nervous system; further cases came to light in the Iraq epidemic in 1972 due to the mass consumption of home-made bread prepared from wheat treated with methyl mercury fungicide. The congenital syndrome includes cerebral palsy, deafness, blindness and microcephaly. Organic mercury is still used as a seed dressing in many countries, usually as the less toxic phenyl mercury. In practice, exposure to mercury compounds has not been on anything like the scale of lead and, partly in consequence, there are no regulations in the United Kingdom on the employment of women in work involving either inorganic or organic mercury. Very low exposure to inorganic mercury, for example, among female dentists or dental assistants in clinics where mixing of dental amalgam is well controlled, is not believed to be hazardous to the fetus.

Polychlorinated biphenyls

Another food-borne epidemic provided the first evidence for the effects of polychlorinated biphenyls (PCBs) in pregnancy. Contamination of rice oil with PCBs occurred in Japan in 1968; PCBs were used in the heat exchange system and a pinhole leak in the piping resulted in these chemicals becoming inadvertently mixed with the cooking oil. About 130 000 people consumed the oil before the accident was discovered and babies born to women pregnant at the time were found to be pigmented and have low birth weight; stillbirth was also reported. Manufacture of PCBs has ceased in most countries as a result of growing concerns about their toxicity and persistence in the environment.

Chlordecone

In 1974 workers at a US pesticide plant manufacturing chlordecone (Kepone) showed evidence of a general acute neurological illness ('Kepone shakes') resulting from heavy exposure to the pesticide. As part of the general investigation of these workers, oligospermia with abnormal and non-motile forms predominating was found. Kepone manufacture ceased with closure of this plant in 1975.

Dibromochloropropane

Shortly after the Kepone episode a group of men working in a California plant manufacturing agrochemicals recognized that they shared an inability to father children, a finding which led to a detailed investigation. Workers who formulated pesticides containing 1,2 dibromo-3-chloropropane (DBCP)—a soil and grain fumigant with nematocidal activity—were shown to be aspermic or have low sperm counts, a finding which in some has subsequently been shown to be irreversible. DBCP was already known to cause testicular damage in rats, guinea-pigs and rabbits, and was shown to be carcinogenic in both rats and mice; its commercial use has now been abandoned.

Ionizing radiation

Exposure to ionizing radiation in man can cause teratogenic effects, fetal death and an increase in malignancy in children. Under the UK Ionizing Radiation Regulations (1985) exposure of pregnant women, such as hospital radiographers, is strictly controlled.

Other possible occupational hazards to reproductive outcome

Anaesthetic gases

Anaesthetic gases have been suspected of causing an increased incidence of spontaneous abortion and congenital abnormalities among operating room staff since a report on the health of anaesthetists in the USSR was published in 1967. Since then at least 15 epidemiological studies have been undertaken in different countries; these mostly retrospective studies have been shown to have methodological problems which, taken with the findings of a recent prospective UK study, suggest that there is no clear evidence of a hazard after all.

Cytotoxic drugs

Certain chemotherapeutic substances used mainly in the treatment of malignancy in man are known to be mutagenic *in vitro* and carcinogenic in animals, and there is some epidemiological evidence for an increased spontaneous abortion rate in nurses handling these drugs. Occupational exposure to these drugs should be avoided.

Ethylene oxide

Ethylene oxide is used as a sterilizing agent in some hospitals. It is a mutagen and animal carcinogen, and there is some evidence that it is associated with leukaemia and spontaneous abortion in exposed workers.

Dioxin

Dioxin is the name given to tetrachlorodibenzodioxin (TCDD), one of several isomers of varying degrees of toxicity which occurs as a contaminant of herbicides manufactured from 2,4,5-trichlorophenol. One such herbicide (Agent Orange) containing 2,4,5-acetic acid (245-T) was used as a defoliant agent by US forces in the Vietnam war. Reports of congenital abnormalities in exposed communities in Vietnam and among US veterans who were involved in its use have not been confirmed. In the Seveso incident in Italy in 1976, when several hundred grams of dioxin were released in an accident in a factory manufacturing 245-T, over 30 000 people were living in the contaminated area at the time, but offspring born to exposed mothers were not found to be at special risk.

Vinyl chloride and other chemicals

Vinyl chloride monomer (VCM) is mentioned here because it is typical of modern chemicals which have been suspected of causing reproductive problems, though evidence in man has been found wanting. An increased incidence of congenital defects was reported in a study of births round a PVC plant in the United States and in another study offspring of VCM exposed workers were reported to have an increased incidence of central nervous system abnormalities; neither finding has been sustained. In animal studies angiosarcoma of the liver was induced not only in adult animals but also in the offspring of females exposed during pregnancy, and this effect of transplacental carcinogenesis has been shown for many other animal carcinogens as well. Fat soluble chemicals such as chlorinated hydrocarbon pesticides (for example, DDT, hexachlorobenzene) and solvents can be excreted in breast milk and pose a hazard to breast-fed babies. Results of some epidemiological studies suggesting a teratogenic role for organic solvents in exposed workers await further confirmation. Recent work suggests a risk of stillbirth in leather workers, but the exposure which may be responsible is unclear.

Visual display units (VDUs)

Early alarm about a risk of miscarriages among VDU operators has not been sustained. Measurements of electromagnetic radiation emanating from these very widely used machines have been found to be very low. A relationship between posture, stress and miscarriage has been proposed but current opinion is that there is no risk. Nevertheless, many companies have adopted a policy of finding alternative jobs for pregnant workers if requested on the grounds that excessive worry about the hazard could in itself be harmful.

Investigation of a hazard

As with other clinical occupational health problems complaints of infertility, miscarriages or other adverse reproductive outcomes need to be considered against the epidemiological background of incidence in working populations not exposed to the putative factor. For example, about a third of cases of infertility in the general population are caused by low counts of motile sperm (less than 20 million) and thus this finding in a single worker may have no occupational significance.

Clinical and epidemiological studies of male workers where a risk is suspected should include: a physical examination which concentrates on the reproductive and endocrine systems; and examination of semen specimens for ejaculate volume, sperm count and sperm motility and morphology. Serum testosterone, follicle stimulating hormone and luteinizing hormone should also be measured, though these tests should be done in collaboration with endocrinologists as their interpretation may not be straightforward. In spouses of the male workers (and for any exposed women workers) details of pregnancy rate, course of pregnancy and pregnancy outcome should be obtained.

Epidemiological questionnaire studies with a retrospective design (that is, asking about past reproductive histories) are an obvious approach in the workplace, but they usually suffer from serious methodological drawbacks when used without some means of verifying the responses. Low response rates, recall bias, non-ideal controls and difficulties in standardizing for confounding variables such as parity, smoking, previous miscarriages, are only some of the problems. Spontaneous abortions are common end-points for study as they are frequent events and may be manifestations of several different adverse effects on the reproductive system, but their study is open to all the difficulties mentioned. Prospective occupational studies are far less open to criticism but are expensive to mount and are not often undertaken. Alternative approaches have included routine monitoring studies of births and birth defects in hospitals and obtaining occupational as well as other data (for example, drug exposures during pregnancy) for all births and then performing case control studies for the outcome of interest. In some Scandinavian countries use of record linkage data between hospital, birth registers and occupational histories of workers has been made possible by every citizen having a unique identification number.

Studies of chromosomal damage in individuals are of undoubted use in evaluating accidental exposures to ionizing radiation but their value in toxic exposures is less clear. The frequency of chromosomal aberrations and/or sister chromatid exchange in peripheral blood specimens is sometimes studied in groups of exposed workers and compared with unexposed controls matched for age, sex and social class, but the implications of any positive findings for the health or reproductive capabilities of the individuals concerned are unknown. The results may also be influenced by smoking habits, drug usage and recent viral infections. Chromosomal studies cannot be recommended as a routine method of surveillance. Testing for mutagenicity of urine in chemical workers or nurses handling cytotoxic drugs has not been shown to be a valid method of monitoring exposure.

When presented with a cluster of spontaneous abortions the characteristics of each case should be studied, for example, whether pregnancy was actually verified and the approximate week of pregnancy in which the abortion occurred. The population at risk needs to be ascertained and the abortion rate compared with rates available for the general population specific for maternal age, parity and week of gestation; it is normally only possible to consider the spontaneous abortion rate in first pregnancies by this method. Other factors should always be evaluated, including social class, medication, maternal illness, smoking, and alcohol consumption.

Health monitoring

There is a dearth of information on the hazards to human reproduction for most substances and processes and the results of animal studies cannot be readily extrapolated to man. There are no routine epidemiological methods of monitoring for such hazards in the workplace and, in consequence, where a risk is suspected the advice to most women workers intending to become pregnant will have to be guarded and by necessity based upon general considerations. With the known risk of lead the course of action in the United Kingdom is laid down by law and the pregnant worker is automatically suspended from the job if it falls under the Control of Lead at Work Regulations (1985).

The occupational health practitioner will need to discuss the risks on an individual basis with the woman concerned and, if necessary, establish whether she is intending to become pregnant or not, rather than proceed on a blanket categorization of her being of reproductive age.

Environmental, and where possible, biological monitoring, should be done to establish the safety of working practice involving chemical exposures. If working with substances strongly suspected of having harmful effects on pregnancy the woman should be advised not to do the work during pregnancy, as an important consideration even when exposure is normally minimal is the possibility of accidental overexposure. Where the risk is documented, for example, organic mercury or vaccinia virus, the woman should not contemplate continuing with the work if she is trying to become pregnant let alone when she becomes pregnant. Where the risk seems slight (for example, anaesthetic gases in operating rooms) alternative work should be sought for the occasional woman who, for whatever reason, may be anxious about the pregnancy outcome. Difficulties may arise, however, if another job is not readily available. In many cases the workplace hazard may theoretically apply to the reproductive system of male as well as female workers and in practice it is difficult to proffer any firm advice other than limit exposure as much as is practically possible and certainly to within the recommended exposure limits, rather than make special guidelines for women. Some companies may attempt to avoid these issues altogether by employing women who are beyond the reproductive age or have undergone sterilization, expedients unlikely to be as acceptable as the adequate control of exposure.

Occupational lung diseases

The occupational health department involved in the routine task of detecting new cases of lung disease and monitoring its prevalence must have as basic equipment a standard questionnaire based on the Medical Research Council or American Thoracic Society respiratory symptom questionnaire, a dry wedge spirometer (for measuring vital capacity (VC), forced expiratory volume in 1 s (FEV_1) and forced vital capacity, FVC) and access to chest X-ray and other hospital diagnostic facilities. The diagnosis of occupational lung disorders is of such importance for the employee as well as the organization that it should only be made in collaboration with a chest physician with the necessary specialized diagnostic facilities. The questionnaire and spirometry should be used in conjunction, as by itself spirometry is of limited value. For the detection of occupational asthma it is the questionnaire which will elicit a history of asthmatic symptoms related to work whereas a one-off test of lung function may show no abnormality unless the worker is having an attack at the time. Chest X-rays have little to offer in the detection of occupational asthma though in a diagnostic work-up they are important to exclude other pathology. On the other hand, the chest X-ray is a central tool in fibrogenic lung disease, both in its early detection and as a means of determining its severity. In pre-placement screening and for diagnosis knowledge of the previous and current jobs held, and hobbies, is essential, including whether any involved exposure to dust, fumes or chemicals.

Where a respiratory hazard is suspected the type of dust or fume needs to be established and its concentration in the workplace atmosphere measured. Personal as well as static samples should be taken over a variety of working conditions as there may be a poor correlation between the two methods, the most useful being the former as it reflects the worker's exposure best. For dust it is essential to measure total inhalable and respirable fractions (less than 10μm). Observation of working conditions without these measurements can be quite misleading, particularly as a risk from fine dust particles may not be obvious and is easily overlooked by inexperienced investigators. The results of air monitoring should be compared with the currently recommended exposure levels where they exist.

Some mineral dusts containing less than 1 per cent of quartz and free of asbestos fibres have traditionally and unfortunately been labelled as 'nuisance dusts', a label which implies they hold no health hazard. These include cement, marble and, until recently, wood dusts. Recommended exposure levels for nuisance dusts are 10 mg/m^3 for total dust and 5 mg/m^3 for the respiratory fraction, yet there may be little incentive for these levels to be enforced. There is usually no good scientific basis for assuming that these dusts are completely safe and gross exposures to them should never be condoned.

Fibrogenic dusts

Exposure to certain mineral dusts can cause fibrogenic reactions of the lungs; common examples are coal miners' pneumoconiosis, silicosis and asbestosis. Even today these disorders are not uncommonly found in hospital chest clinics in industrialized areas in the United Kingdom, though most cases will have arisen as a result of exposure to the dustier conditions extant at least one or more decades ago. Greater knowledge of the effects of dust exposure and technological changes in production have, among other reasons, resulted in gradual improvements in conditions in the heavy industries, but it would be incorrect to suppose that these diseases can be relegated to the past, as any lapse of the stringent controls necessary will result in cases of disease continuing to occur. This danger exists in times of economic recession, particularly in small workplaces where ignorance and lack of government inspection can be seen even today to result in conditions comparable to some of the worst found in heavy industries in the past.

Coal miners' pneumoconiosis

This is classified according to agreed radiological criteria. Simple pneumoconiosis is of little health significance on its own except that progression to the disabling progressive massive fibrosis (PMF) can occur. In the United Kingdom, coalminers are offered chest X-rays every 4 years and if simple pneumoconiosis, category 1, is found the worker is informed and offered appropriate advice. With category 2, simple pneumoconiosis, the worker is referred to the DHSS pneumoconiosis panel for assessment, and if the disease is confirmed a less dusty or non-dusty job is offered and the man kept under regular medical supervision. PMF may nevertheless arise many years after exposure to coal dust has ceased; elevated concentrations of quartz mixed in with the coal dust may be a factor in its development. There is increasing evidence that coal workers are also at greater risk of developing occupational bronchitis (chronic airflow obstruction).

Silicosis

The traditional hazards of silicosis in industries such as quarrying, foundries, ceramics, stonemasonry, and so forth are well known, but siliceous minerals are so common that the presence of crystalline silica should be suspected in any dusty processes. The commonest form of crystalline silica is quartz (the other forms are tridymite and crystobalite) and its presence in respirable fractions of dust in excess of 1 per cent is indicative of a hazard. A frequent misapprehension arises over the silica content of minerals. SiO_2 (silica) is commonly present in minerals in a combined form—silicates—which should not be confused with the usually much smaller amount of free or crystalline silica. Recommended limits of exposure to mineral dusts containing crystalline silica can be calculated from its percentage weight in the total inhalable and respirable fractions using standard formulae. High concentrations of quartz may arise in finely ground sand known as silica flour and heavy exposure to this dust can give rise to acute silicosis within only weeks or months of exposure. Otherwise silicosis has a slow progression. There is no legal guidance in the United Kingdom on medical monitoring, but workers at potential risk should be offered chest X-rays every 2–4 years.

Asbestos-related diseases

Asbestos comprises six naturally occurring substances: chrysotile (which belongs to the serpentine group of minerals), crocidolite, amosite, anthophyllite, tremolite and actinolite (all belonging to the amphibole group of minerals). Most of the world usage of asbestos has been chrysotile and much smaller amounts of crocidolite and amosite, whereas the other three amphiboles are important as contaminants of other minerals or agricultural soil. In the United Kingdom, new exposure limits were introduced in 1984, but it is predicted that about 2500 asbestos-related deaths will continue to occur annually in the United Kingdom until the end of the century as a result of past conditions.

Asbestosis, mesothelioma and bronchial carcinoma result from exposure to amphiboles, as these survive indefinitely in the lung, whereas chrysotile does not, though it may be contaminated by the other fibres. In addition, asbestos workers can develop hyaline pleural plaques, benign pleural effusions and skin corns. An occupational history is the key in distinguishing asbestosis from other causes of diffuse interstitial pulmonary fibrosis, as the radiological appearances and results of lung function tests may be identical; a history of prolonged heavy exposure makes lung biopsy unnecessary. Asbestosis, like the malignant changes, may become manifest and progress years after exposure has ceased yet there is little value in regular surveillance of ex-workers as the clinical course of these conditions cannot be altered at present even if they are detected at an early stage.

In accordance with UK regulations and recommendations, workers at potential risk should have chest X-rays, spirometry and an abbreviated clinical examination every 2 years. The earliest evidence of the development of asbestosis on routine examination may be the presence of crackles at the lung bases which occur typically at the end of full aspiration. The key question is what advice should be proffered to the asbestos worker who is found to have early lung changes on X-ray as the condition may progress or arrest whether or not the worker is removed from exposure at the current exposure limit. Information on the natural history of the disease at this level of exposure is lacking, and there is no official guidance. Most physicians would advise on general grounds on a change of job, though the worker may not agree if there is a loss of income or employment as a result. An important consideration for those with any type of cardiorespiratory disease is whether or not they are fit enough to wear a respirator, an essential requirement for work involving asbestos stripping.

No adequate screening methods exist for the early diagnosis of lung cancer or mesothelioma. In view of the multiplicative relationship which exists between cigarette smoking and asbestos exposure in the development of lung cancer (but not mesothelioma) the periodic examination should include advice on cessation of smoking.

Man-made fibres

There is good experimental evidence that any straight mineral fibre of a certain size (diameter of under $0.25–3\,\mu m$ and a length of at least $5–10\,\mu m$) may cause malignancies and lung fibrosis. The question arises whether man-made mineral fibres (MMMF), used as a substitute for asbestos and other purposes, may have similar properties. Epidemiological studies of MMMF workers are under way and until further information is available these materials should be treated with respect; in the United Kingdom a special control limit of $5\,\mu g/m^3$ 8 h time-weighted average is recommended for bulk material and a recommended limit of one fibre per ml has been agreed for superfine man-made mineral fibres. No medical surveillance has been officially recommended for these workers.

Occupational asthma

Although accurate prevalence figures do not exist this condition is probably increasing in frequency as the list of organic and inorganic agents which can give rise to it grows. Even a mild form of occupational asthma can affect well-being and reduce exercise tolerance, and with some agents attacks may be severe enough to be life-threatening. In addition, repeated exposure may lead on to irreversible disability; this serious outcome is suspected to occur in workers exposed to toluene diisocyanate (TDI), solder fumes, western red cedar wood dust and crab processing, a list which will undoubtedly expand to include other agents when the appropriate follow-up studies of affected workers have been done. The risk of developing disabling asthma is probably related to the length of exposure after symptoms have developed and it therefore behoves the occupational health team to undertake regular medical surveillance.

Health monitoring and investigation of a hazard

The clinical features of asthmatic reactions may be 'immediate', with a fall in peak flow rate shortly after starting work, or 'late', with a response several hours after exposure has begun, or both ('dual') depending upon the mechanism of sensitization. With 'late' responses' symptoms which become more severe several hours after exposure has ceased frequently obscure the causal link with occupation for both patient and doctor. The symptoms should relate to work exposure, with a history of improvement at weekends or with longer absences from work in some patients. Many dust and fume exposures, including some allergens, have irritant properties so that airways narrowing can be a reaction to irritation rather than a true sensitization; common examples of substances with both properties are phthallic anhydride, colophony fume, and formaldehyde vapour. Nasal and eye symptoms may similarly have an irritative or hypersensitivity basis. It is essential that the clinical investigations are undertaken to make a firm diagnosis between the two responses as their management is quite different.

Affected individuals may present themselves for advice on their respiratory symptoms, but in practice this appears to be an exceptional event rather than the rule. Fears of dismissal may dominate the workers' thoughts, or a failure to link the symptom to the job is common. Others may phlegmatically tolerate the condition for fear of being switched to a job of lower income. Regular monitoring programmes are therefore essential to detect cases.

A detailed clinical history should be obtained using a standard respiratory questionnaire. Lung function testing is important in monitoring the progress of workers with symptoms, though it has limited use in the initial diagnosis. A positive or suspect clinical history may be followed up by requesting the worker to use a mini peak-flowmeter to record his peak respiratory flow rate every 2 h when awake for a period of 3–4 weeks. A pattern of work-related asthma may be found; the clinician should plot the maxima and minima and means of peak flow rates and then eyeball the graph to look for variability over the observation period. In practice, however, the patterns in some patients may not be as clear as could be wished and a further difficulty can be enlisting a worker's cooperation so that the recordings are completed successfully over this long period.

Further investigation requires the facilities of a laboratory which specializes in occupational immunology. The specialist will utilize skin prick tests for a range of common inhalant allergens and an extract of the suspect occupational allergen; a blood sample will be taken for measurement of specific IGE antibody against the allergen (RAST or radioallergosorbent test). A typical clinical history together with a positive skin prick and/or RAST test should be sufficient to make a diagnosis, though it is important to stress that a positive immunological test can frequently be found in exposed workers without it having any clinical significance. As an adjunct, a histamine or methacholine challenge for airways responsiveness is helpful in defining true allergy, though ultimate confirmation may only be possible through a bronchial provocation test or challenge test. In the latter the worker is deliberately exposed to the suspect dust or fume under control conditions in a hospital setting. As well as being important as a research tool for identifying new occupational allergens challenge testing can be important in the precise identification of the allergen, information which may be essential for advising the person about his job.

When a diagnosis of occupational asthma has been made and the causal agent identified, the worker should be advised to avoid or at least severely limit further exposure. This seemingly simple advice may create numerous difficulties in practice. For the semiskilled worker (for example, a solderer in the electronics industry) this may mean losing their job if alternative work is not available in the same factory; for the highly trained and specialized worker (for example, animal toxicologist) the prospect may be a broken career. In many patients the symptoms may be adequately controlled by modification of work practices, improving environmental conditions to reduce antigen levels in the air or, if these fail, by the wearing of respiratory protection, for example, high efficiency lightweight masks or air stream helmets. However, symptoms and lung function should be regularly monitored and if abnormalities persist there may be no alternative other than to advise a change of job to be away from exposure.

At a pre-placement health review, potential employees with asthma or other chronic lung disorders should be excluded. Whether or not to exclude atopic persons, on the other hand, is more controversial. Atopic individuals are identified by skin testing (not on a history of allergy alone) but atopy is not in practice a powerful enough predictor of which workers are most at risk becoming sensitized to an occupational allergen. In consequence it would be wasteful to exclude workers on the basis of atopy alone as about a third of the adult population would be found positive.

Workers exposed to occupational allergens should in general be monitored on an annual basis with a questionnaire and lung function test as a minimum requirement. For workers with the more potent sensitizers such as isocyanates and platinum salts, more rigorous guidelines requiring more frequent monitoring have been issued by the Health and Safety Executive and these should be adhered to. Because of the rapidity with which new occupational allergens are being identified most do not have published recommended exposure limits. With some agents, for example, asthma in laboratory animal workers (rodent urine proteins) or in solderers (colophony fume) the methods for their measurement in air may still be at an early stage of development. What advice on exposure can be offered? Environmental controls to reduce airborne levels to as low as reasonably practicable should be adopted; but it is essential that the efficacy of such measures is checked by a programme of health monitoring. It is important to emphasize that any recommended exposure limits set for allergens are intended to prevent sensitization occurring in the first place; they are of little value for sensitized workers, who may develop asthma on exposure to concentrations which are fractions of those limits.

Byssinosis

Byssinosis is best regarded as a type of asthma which, in its initial stages of development, is experienced at the beginning of the week only and then with continued working gradually extends to include other days of the week. A disease of cotton, flax and soft hemp workers, the precise causal agents in the vegetable dust await confirmation, though bacterial endotoxin seems a likely candidate. In advanced disease chest X-ray appearances are normal and there are no characteristic pathological findings; the diagnosis is therefore principally made on the basis of responses to a respiratory symptom questionnaire. Epidemiological surveys conducted in the United Kingdom in the 1950–60s showed that continuing exposure over some years could lead to long-term disability, but there are now considerable doubts that this risk exists at current low exposure levels. Regular monitoring for the development of symptoms of byssinosis is advisable, though the value of this to the individual worker who develops early symptoms is now open to doubt if long-term disability cannot be shown to result at current exposure levels.

Extrinsic allergic alveolitis (hypersensitivity pneumonitis)

The diagnosis of occupational asthma may be confused with alveolitis in workers exposed to organic dusts, particularly spores, avian and animal proteins; farmer's lung, mushroom picker's lung and bird fancier's disease are familiar examples. By contrast with asthma, there is typically an acute systemic response with malaise and fever as well as breathlessness, occurring about 4–8 h after exposure, a pattern which is repeated when exposure reoccurs. Diagnosis is made on the history, chest radiography (in an attack the X-ray shows shadowing in the lower zones) and by demonstrating an acute fall in lung function with a reduction in transfer factor. Precipitating antibodies to the antigen may be found on serum testing. With repeated exposure there may be progression to chronic interstitial fibrosis. The best remedy is avoidance of exposure, but failing that respiratory protection backed up by regular monitoring of lung function, as for asthma, can be tried.

Disorders associated with heating, ventilation and air conditioning systems

Reports of vague symptoms often associated with complaints of deficient ventilation have become common in modern offices and the term 'building sickness' has been coined for a syndrome comprising tiredness and lassitude, eye and throat irritation, itchy skin, malaise, headache, sinusitis, often in association with mild chest symptoms of cough and subjective sensations of difficulty in breathing. The symptoms improve rapidly away from work, though feelings of fatigue may persist. There are no objective signs on clinical examination and the chest X-ray and lung function tests are normal. Characteristically the majority of the workforce in an area will complain of one or more of the above symptoms. Careful examination of the ventilation system usually shows no major defects, though occasionally the control of mixing of recirculating air with outside air may be inadequate, or the system is poorly balanced or designed so that some office areas may have reduced ventilatory flow rates despite the overall exchange rate being adequate.

In systems which involve humidification it is essential to exclude humidifier fever and extrinsic allergic alveolitis, both of which can be caused by a growth of microorganisms in the recirculating water used to humidify the air. Symptoms of humidifier fever are worse at the beginning of the working week and may be misdiagnosed as a Monday morning feeling, as they recur after a weekend away from work. A common mistake in the investigation of these problems is not to obtain samples of reservoir water for examination before attempts are made to clean or sterilize the system. Extracts may be prepared from the microorganisms for antibody testing in blood samples from the workers in order to confirm the cause of the problem. Thermophilic actinomycete organisms have been associated with outbreaks of extrinsic allergic alveolitis from humidifiers in the United States; in the United Kingdom protozoa (mainly amoebae) and in Sweden bacterial endotoxin have all been implicated as causal agents in humidifier fever. When steps have been taken to remove the organic growth and prevent its recurrence, the symptoms subside. Building sickness on the other hand tends to be intractable despite attempts at improving ventilation, and reassurance is only possible after a range of investigations is complete.

Cancer of the respiratory tract

Occupational cancers of the respiratory system include carcinoma of the nasal sinuses, larynx and bronchus, and mesothelioma of the pleura. Chemicals or processes associated with bronchial carcinoma include arsenic, hexavalent chromium compounds, asbestos, bischloromethyl ether, and radon; and for nasal cancer, wood dust (furniture-makers), nickel-refining, and isopropyl alcohol manufacture. The prognosis for these cancers is poor and is not improved by screening to detect cases at an early stage; there are therefore no published recommendations on medical surveillance, even for those workers in industries at high risk. Stringent control of exposure to known carcinogens is therefore the key preventive measure. Nasal cancer is an uncommon tumour and one case in a workforce should attract immediate investigation. Cases of lung cancer will arise in a workforce inevitably as a result of cigarette smoking, but clustering of cases with an uncommon cell type (such as oat cell carcinoma) or arising in young workers (in their 30s and 40s) should alert to the presence of a possible occupational carcinogen.

Occupational bronchitis

There is increasing evidence (for example, in coal miners) that prolonged exposure to high dust concentrations may lead to chronic bronchitis, with cough and sputum, in some workers. The symptoms are most common in smokers but non-smokers may also be affected.

Liver disease

The classic chemicals capable of causing diffuse or massive hepatic necrosis were first identified as a result of very heavy exposures to workers during the First and Second World Wars and are now of mainly historic interest only. These chemicals were the chlorinated hydrocarbon tetrachloroethane (used as dope in airplane manufacture) the chlorinated naphthalenes (electrical insulation) and the nitro-aromatic compounds which were used mainly in explosives manufacture, dinitro-benzene and trinitrotoluene (TNT). Zonal damage, the other main pattern of injury, probably arises from a different pathogenic mechanism and carbon tetrachloride and chloroform are important industrial and experimental examples, both causing centrilobular necrosis. This effect of carbon tetrachloride has been shown experimentally to be enhanced by the administration of phenobarbitone or alcohol; both induce liver enzymes. The more hepatotoxic solvents have been largely replaced by safer ones or are handled with due regard to their hazard, and occupational liver disease is now uncommon in modern industry.

Polychlorinated dibenzofurans and dibenzodioxins

These substances are highly toxic and are found as trace contaminants in PCBs (furans), pentachlorophenol and the herbicide 245-T (dibenzodioxins). Most attention has focused on the isomer tetrachlorodibenzodioxin (TCDD, known as dioxin). In the Yusho incident in Japan people who ingested food contaminated with PCBs developed chloracne and had transiently elevated serum triglycerides. Workers accidentally exposed to TCDD develop chloracne and evidence of liver damage such as elevation of blood lipid levels has been reported.

Evidence for the development of toxic porphyria in TCDD-exposed workers is conflicting. While PCBs have been phased out, 245-T continues to be manufactured in several countries, and formulation of herbicides containing 245-T is carried out in many more. Workers should be carefully monitored for chloracne, the hallmark of exposure to these compounds, on a regular basis. TCDD has a half-life in blood of several years and is persistent in body fat. Blood tests and fat biopsies have been used to evaluate exposure, but they cannot be recommended for routine monitoring.

Hexachlorobenzene and methylene diamine (syn 44-diamino diphenyl methane or DADPM)

The first conclusive human evidence that an identical clinical picture to porphyria cutanea tarda could be produced by an environmental agent emerged in the toxic epidemic in Turkey in 1956 when people ate bread prepared from wheat which had been treated with the fungicide hexachlorobenzene. The clinical picture of toxic porphyria includes skin photosensitivity, blistering and scarring with pigmentation and hypertrichosis; there is also evidence of liver dysfunction.

In Epping in 1965 an outbreak of cholestatic jaundice occurred in 84 people who consumed bread made from flour which had accidentally become contaminated by methylene diamine during its transport. The number of workers exposed to methylene diamine is not large in the United Kingdom and current usage does not appear to be associated with toxic hepatitis.

Vinyl chloride

Although numerous industrial chemicals are suspected of being hepatic carcinogens (for example, DDT, chloroform, carbon tetrachloride, TCDD) the only one which has so far been conclusively implicated in man is vinyl chloride monomer (VCM), used to manufacture the widely used plastic polyvinyl chloride or PVC. Since 1974 when the first human cases of hepatic angiosarcoma (HAS) were reported in VCM workers, over 130 cases of this otherwise rare tumour have now been recorded in the PVC industry worldwide, virtually all of whom have received heavy exposure in the polymerization process. The clinical presentation and prognosis of HAS resembles primary liver carcinoma, with upper abdominal pain as the most common symptom. Clinical evidence of liver failure is usually apparent at initial presentation when life expectancy may be merely a matter of a few weeks or months. Arsenic is also a recognized cause of HAS as a result of its use in medications in the past (for example, Fowler's solution), a practice which continued in the United Kingdom until the 1960s. The only workers exposed to arsenic so far reported to have developed HAS were vintners who used a fungicidal spray containing arsenic, but they imbibed the contaminated wine as well. Hepatic angiosarcoma is so rare that the finding of a single case should alert the clinician to search for an aetiological agent.

Other effects of VCM exposure include acro-osteolysis of the terminal phalanges, and occasionally of other bones, and a scleroderma-like affliction of the skin. The resorption of bone is reversible in some cases. Acro-osteolysis has also been observed to occur in the epidemic of hexachlorobenzene poisoning in Turkey and the toxic oil syndrome in Spain in 1981. Heavy exposure to VCM may also cause non-cirrhotic fibrosis of the liver and result in a presinusoidal portal hypertension with oesophageal varices, a condition closely resembling Banti's syndrome and which has been found associated with chronic arsenic ingestion. Despite the liver damage, remaining liver cell function is usually good and these patients do well after shunt surgery.

Health monitoring and hazard investigation

The value of the currently available battery of liver function tests as pre-employment screening tools or as a means of monitoring toxic occupational exposure, or screening for alcohol abuse, is very limited. Sensitive markers of liver function such as serum gamma-glutamyl transferase (γ-GT) have drawbacks because of their lack of specificity. In a routine screening procedure abnormalities are more likely to be due to non-occupational factors, and very sensitive tests such as serum bile acids are only of value in clinical research. On the other hand standard liver function tests may be normal in the presence of even advanced liver damage. Hepatic imaging methods have been found wanting in the detection of early liver disease for similar reasons; ultrasonic scanning did not live up to its early promise as a screening tool in VCM workers for the detection of early hepatic fibrosis. Liver function testing for occupational disease should only be done in conjunction with a clinical examination and knowledge of occupational exposures, together with a full medical and social history which includes details of alcohol consumption.

Although the indications for routine liver function testing in industrial workers are few, where special investigations are required (for example, after an accidental overexposure to chlorinated hydrocarbon solvents or other hepatotoxins) testing should include liver function tests, serum lipids and urinary porphyrins and the results should be evaluated on a group basis. Elevated gamma-glutamyl transferase levels in a group of workers compared with matched controls may be an indication of microsomal enzyme induction consequent upon the occupational exposure, but on an individual basis may be of no significance or reflect recent alcohol consumption. The mainstay of hepatologists is the percutaneous liver biopsy, but as this carries an appreciable morbidity and mortality it cannot be advised as a routine follow-up investigation to the finding of abnormal liver function tests in otherwise fit workers.

For workers who have been heavily exposed to VCM in the past there are no tests which can be currently recommended to detect liver damage or early HAS, and there are no therapeutic strategies which have been devised to lower the risk of subsequent malignant complications. Occupational physicians and family practitioners should keep these workers under periodic review so that the worker is aware that any decline in health needs to be investigated with the occupational history in mind.

Neuropsychiatric disorders

Toxic neuropsychiatric disease has become rare in the developed countries, but it is probably true to say that it is attracting almost more attention—amongst research workers at least—than for many years. The reason for this is the concern that some chemicals may have adverse effects at levels of exposure which do not produce clinical signs or symptoms, so-called subclinical effects. There has also been some concern that some exposures, especially to solvents and heavy metals, may be implicated in the production of organic brain disease. The practising occupational physician may, justifiably, find some of these premises somewhat difficult to accept.

Peripheral neuropathy

An occupational neuropathy may result from exposure to neurotoxic chemicals, from accidents or from nerve entrapments. Accidental damage to the peripheral nerves rarely occurs in isolation, rather it is a complication of trauma to a limb which is still too common an occurrence in industry. The distal part of a transected nerve will undergo Wallerian degeneration, resulting in a mixed motor and sensor deficit. The peripheral segment will regenerate but even with skilful surgical repair, the prospects of anything but partial recovery of function are poor and the rehabilitation of employees with peripheral nerve injuries may prove time-consuming and difficult.

Peripheral nerves passing close to the surface of the body or contained within a tight compartment are the most likely to suffer from entrapment. Thus the ulnar nerve at the elbow, the median nerve at the wrist and the lateral popliteal nerve at the knee are the most commonly affected. Carpal tunnel syndrome is the classic condition among this group of neuropathies. It has a number of causes including obesity, pregnancy, myxoedema, acromegaly and rheumatoid arthritis but occupational factors may not be considered. However, it is becoming clear that carpal tunnel syndrome may be a relatively frequent component of repetitive strain injury especially amongst those engaged in tasks which require forceful flexion at the wrist. It may be a difficult condition to treat satisfactorily especially if the patient is unable or unwilling to change his job.

Toxic neuropathy

Neurotoxic chemicals which may be encountered in the workplace include some heavy metals, organic solvents, pesticides and a number of miscellaneous compounds.

Heavy metals

The most important of these are lead, mercury (especially the organic compounds), thallium, and (stretching the definition somewhat) arsenic. Neuropathy caused by any is now rare. Lead causes a predominantly motor neuropathy, mercury a sensory neuropathy and the other a mixed form. Lead has a number of effects on the nervous system; it causes axonal degeneration and demyelination and it blocks postsynaptic transmission by local competition with calcium ions. Prolonged motor conduction velocities and distal latencies have been described in lead workers without clinical signs or symptoms but the relationship with blood lead levels is by no means clear cut. Patients with mercury poisoning may complain of paraesthesiae, and in fatal cases axonal degeneration and demyelination may be found in the dorsal and ventral roots of the spinal cord.

Arsenic and thallium produce a dying back neuropathy caused by interference with glycolysis within the neurone. Arsenic inhibits the conversion of pyruvate to acetyl coenzyme-A and the arsenate ion blocks the synthesis of 1:3 diphosphoglyc-erate from glyceraldehyde-3-phosphate. Thallium disrupts the transport of electrons during intracellular respiration such as also occurs in riboflavin deficiency.

Under normal conditions of work neuropathy due to heavy metal poisoning should never occur and cases are now extremely rare. With regard to lead and mercury, however, one must guard against being too complacent. Lead exposure occurs in many occupations and it may not always be self-evident. The risk to those who demolish buildings or cut down painted iron work ought to be well known by now, but many doctors still appear unaware of it. There is also something of a misconception that the risk to those at work arises predominantly from inhalation. This may be so for a good many—perhaps the majority—of lead workers but for a substantial minority the more serious hazard is from the ingestion of lead on the hands. In these occupations—which include all those in which the metal itself is handled—control by airborne monitoring serves no useful function and blood lead concentrations are the only way to control the hazard.

With regard to mercury, we are frequently amazed by the cavalier fashion in which it is handled in laboratories and elsewhere. Many of those who come into contact with it seem unaware that the metal volatilizes at room temperature or that some mercurial compounds can be absorbed through the skin. When used without strict supervision, mercury can be a very real threat to health.

Organic solvents

Only four solvents are well known to cause peripheral neuropathy: trichloroethy-lene, n-hexane, methyl butyl ketone (MBK) and carbon disulphide (CS_2). Trichloroethylene has been in use as a degreasing agent for dozens of years, but it was only shortly after it was introduced as an anaesthetic in the early 1940s that patients were noted with marked and persistent sensory disturbances of the face, mouth and lips. A number of heavily exposed workmen have also been reported with severe symptoms in the territory of the Vth cranial nerve and motor paralysis of the Vth, VIth, VIIIth, IXth and XIIth cranial nerves. At autopsy, gross degeneration of the Vth nerve and its nuclei was found with less severe damage in other cranial nerve nuclei. The underlying mechanism is obscure although similar changes can be produced in animals exposed to dichloroacetylene, the toxic metabolite of trichloroethylene. Given the numbers of workers exposed to trichloroethylene, it is perhaps surprising that less severe cases are not more common.

The pathological changes seen in the neuropathies consequent upon CS_2, n-hexane and MBK exposure are identical. Swellings are found on the axon around the node of Ranvier. Proximally to the node the myelin sheath is thinned or absent over the swelling. The characteristic feature of this neuropathy is the accumulation of 10 nm neurofilaments which tend to disrupt the normal distribution of other organelles within the axon and impede axonal flow. CS_2 probably exerts its effects by crosslinking axonal proteins through the mediation of its highly active sulphur group; in the case of n-hexane and MBK, a metabolite common to them both, 2,5-hexanedione is the proximate neurotoxin. This compound owes its neurotoxicity to its gamma di-ketone spacing, a feature it shares with other compounds which incorporate a similar structure.

One interesting feature of solvent neurotoxicity is the ability of different solvents to interact with synergistic or protective effects. For example, MBK may potentiate the neurotoxic effects of methyl ethyl ketone which is only weakly neurotoxic, whereas toluene appears to be able to protect against the harmful effects of *n*-hexane. These interactions may be particularly important in solvent-using industries where exposure is generally to a mixture of compounds.

Triorthocresyl phosphate (TOCP)

Triorthocresyl phosphate causes axonal degeneration and secondary demyelination probably as the result of the inhibition of a specific esterase called neuropathy target esterase (NTE). Although TOCP is used in great quantities as a plasticizer, most cases of TOCP neuropathy have occurred as the result of using contaminated edible oil or drinks which have been contaminated. Other organophosphorus pesticides impair cholinesterase and NTE to a lesser or greater degree, and while they may produce symptoms of autonomic dysfunction, they do not seem to produce peripheral neuropathy.

Acrylamide

This is a vinyl monomer which readily undergoes polymerization and copolymerization. It is used in the manufacture of flocculators and it may be absorbed through the skin where it may locally affect peripheral nerves.

Toxic organic psychosis

A toxic organic psychosis may be produced by some metals (including arsenic, lead, manganese and mercury) and by some solvents including most notably carbon disulphide and toluene. When mercurial poisoning was common in the hat trade, erethism, the name given to the psychiatric symptoms, was so well known that the phrase 'mad as a hatter' entered the language. It is rare nowadays as are psychotic symptoms following exposure to other metals. In organic lead poisoning the picture is dominated by psychotic manifestations but this condition is excessively rare in industry although it may arise when leaded petrol is inappropriately used as a solvent or when it is sniffed.

The hallucinogenic properties of several solvents are sufficiently pleasant to encourage some individuals to sniff them. Those most commonly abused are trichloroethylene, 1,1,1-trichloroethane and toluene. Generally, the intoxication wears off with no apparent sequelae but there is increasing evidence that toluene sniffing may produce permanent brain damage and dementia. The evidence that occupational exposure to solvents causes dementia is extremely unconvincing but some permanent effects, manifested by lassitude, poor memory and inability to concentrate do occur in a small number of those with prolonged and heavy exposure. Acute symptoms of intoxication, vertigo, headache and irritability are very common and entirely reversible.

Parkinsonism

Parkinsonism is known to follow heavy exposure to manganese. The structural changes are somewhat different than in idiopathic parkinsonism in that, whereas in the idiopathic condition the substantia nigra is most affected and the striatum and pallidum tend to be spared, the converse is the case in manganese poisoning. The discovery that methylphenyl tetrahydropyridine (MPTP) can induce parkinsonism has excited speculation about the possible role of other as yet undiscovered environmental contaminants in the aetiology of this condition.

Investigation of neurological disease in the workplace

Peripheral neuropathy occurring in someone at work is much more likely to be due to factors other than an occupational neurotoxin; diabetes and alcoholism must be amongst the commonest causes. However, complaints of paraesthesiae and incoordination or muscle weakness must be taken seriously in workers exposed to recognized neurological hazards. If they occur in several workers in the same areas of a factory, then the possibility that they are exposed to an unrecognized neurotoxic material should be considered. A full medical, social and occupational history must be taken; factors which are particularly important are a history of neurological disease in the past, drinking habits, medication and any previous exposure to potentially neurotoxic substances. An enquiry should be made of any hobbies or any other work which may entail exposure.

For epidemiological work, the clinical examination should be standardized and signs graded on an ordinal scale; if more than one investigator is involved in a study, then it is essential that some preliminary training is undertaken to ensure comparability of results. Sensory testing is the most difficult to quantify because it relies heavily on the subjective appreciation of stimuli. Because of the great interest in supposed or actual environmental and occupational hazards, some manufacturers are developing sensory testing kits which can deliver standard sensor stimuli— weighted bristles of different thicknesses, for example. These kits are in the early stages of development and more work is required before they can be generally applied. Machines to test vibration sense are available commercially and are generally reliable. They are quite widely used in the United States and in some of the Scandinavian countries in epidemiological work. There is some suggestion that vibration sense may be impaired earlier than other sensory modalities.

Electrophysiological studies are most important in the assessment of suspected causes of occupational neuropathy whether on an individual or a population basis. Nerve conduction studies are the simplest and most widely used measures and may show deficits before there is any other evidence of neurological involvement. In any study, motor and sensory conduction velocities should be measured and amplitudes and distal latencies determined. The procedure is straightforward and involves little discomfort to the subject. It is very important to pay careful attention to technique. More than one nerve should be tested and it is preferable to test at least one in the arm and one in the leg. In the arm the ulnar and median nerves are readily accessible; if there is any suspicion of carpal tunnel syndrome both must be tested in order to help confirm the diagnosis.

Careful attention must be given to placing the electrodes in order to maximize stimulation and to pick up the signal, and the test must be carried out in a warm room. If the skin temperature on the limb to be tested is below about 32 °C, the nerve conduction velocity will appear to be slow because the conductivity of the skin decreases as its temperature falls. Thus the skin temperature must always be recorded and the limb warmed if the temperature is too low. It is also important to obtain age-matched reference values since nerve conduction velocity declines somewhat with age. In a study where subclinical effects are being examined and the difference between exposed and non-exposed is likely to be small, the errors which may be brought about by poor technique may well be greater than any difference due to exposure to the putative neurotoxin.

The sensory system, both central and peripheral, can be tested using evoked responses. The most commonly used is the visual evoked response (VER) in which the speed at which a visual stimulus reaches the occipital cortex is recorded; it is often used clinically in the assessment of patients with multiple sclerosis. More complicated apparatus is required to carry out these tests but they are relatively simple and have the advantage of providing objective sensory data. There is some evidence that evoked responses are slowed before other signs of neurological impairment become apparent.

At a more complex level yet, EEGs may be recorded. This has been done in some groups of solvent workers and abnormalities have been noted in workers exposed to styrene, for example, changes which correlate with levels of mandelic acid in the urine.

For most investigations at work, nerve conduction studies, perhaps supplemented by VER would suffice both on grounds of usefulness of the data and of ensuring subject participation.

The investigation of psychological abnormalities can be undertaken by using one of a number of test batteries which are available. Most of these derive from clinical psychology and have been validated against patients with proven lesions; all test different areas of higher cerebral functioning. The twin trends are towards developing a standard test battery which it is hoped may be applied anywhere in the world and towards computerization of the test procedures. Since some of the tests included in psychological test batteries show considerable crosscultural variation, it is not likely that one battery will ever be universally applicable although the one now being developed by the World Health Organization is being strongly promoted and has a number of different language versions.

When testing subjects it is important to include a measure of premorbid (or pre-exposure) ability since many psychological tests are highly correlated with intelligence. Controls must be age-matched since—as most people will be aware—performance worsens with age, especially in tasks which measure or rely on memory.

Many investigators make the error of interpreting a decrement in performance in psychological tests as being necessarily indicative of brain damage or of psychiatric morbidity. This has been particularly the case in discussion about the possible chronic effects of solvent exposure in which painters and others have been diagnosed as having dementia on the basis of poor performance in psychological tests. To determine psychiatric morbidity requires an input from a psychiatrist—something most occupational epidemiologists forget. There are a number of screening tests for determining psychiatric morbidity in a population; the best known is the General Health Questionnaire developed by Goldberg. A more comprehensive questionnaire which has been developed for psychiatric epidemiology is the present state examination (PSE) which requires the investigators to be trained prior to its use in the field.

How far any of these methods can be used for monitoring workers exposed to neurotoxic substances depends to some extent on the credence given to reports of chronic effects. The most satisfactory way to avoid the appearance of neurological disease in the workforce is to control exposure rigidly to known neurotoxic agents and ensure that biological standards are not exceeded (*see* Chapter 5). If this is done then frank neurological disease will almost certainly not appear. Our view is that psychological tests are not sufficiently sensitive or reproducible yet to provide a basis for monitoring for supposed subclinical effects and that they serve no useful routine function at present.

The kidney

There are many potential renal toxins both inorganic and organic, but in practice, improvements in hygiene within factories have decreased the prevalence of occupational renal disease to a low level.

Heavy metals

Many heavy metals, including lead, mercury, cadmium, gold, thallium and uranium damage the kidney, but only the first three are of anything but passing interest to the occupational physician. Prolonged and heavy exposure to lead may produce an interstitial nephropathy with the production of secondary hypertension. Early effects may be directed against the proximal tubule and in children with lead poisoning, a Fanconi-like syndrome with glycosuria, aminoaciduria and hyperphosphaturia is not uncommon. The most usual effect of mercury on the kidney is to produce tubular damage and necrosis but the glomerulus may also be damaged and some workers with long exposure have albuminuria. Rarely the nephrotic syndrome may appear in those exposed to inorganic mercury, probably as the result of an idiosyncratic response.

Cadmium predominantly affects the renal tubule causing tubular proteinuria. The appearance of β_2-microglobulins and retinol binding protein and the lysosomal enzyme *N*-acetyl-*B*-*D*-glucosaminidase (NAG) in the urine is an early indication of cadmium-induced renal damage. Cadmium does not appear to damage the kidney until the concentration in the renal cortex exceeds about 200 parts per million (ppm) and at levels of this order about 10 per cent of the exposed workforce is likely to develop renal dysfunction. This concept of a critical concentration applies to the total cadmium content of the kidney but it is only that fraction which is not bound to metallothionein which is toxicologically active and this is about 1 per cent of the total. It is likely that factors such as the inability to produce metallothionein and ageing decrease the critical concentration which is associated with renal damage.

Arsine

Arsine is formed whenever nascent hydrogen comes into contact with an inorganic arsenic compound in solution. The gas may be generated by wetting hot scrap metal or metal dross, during the smelting of many metals and by the action of crude sulphuric acid on metals. Arsine is intensely haemolytic and produces profound intravascular haemolysis. The sludging of cell debris in the renal tubules may induce acute renal failure.

Organic compounds

Any hepatotoxic organic compound is likely also to damage the kidney but the renal effects tend to be of secondary importance compared with hepatic ones. In massive accidental exposure, acute tubular necrosis may occur; this is especially liable to happen with chlorinated hydrocarbon solvents.

Ethylene glycol is metabolized with the production of oxalic acid and if taken either deliberately or accidentally in large doses may cause tubular obstruction through the deposition of calcium oxalate crystals.

Malignancies of the genitourinary tract

The bladder is the most susceptible part of the genitourinary tract to the action of occupational carcinogens and of these, the best known is β-naphthylamine formerly found as a contaminant of antioxidants used in the rubber industry. This compound, also present in the fumes from coke ovens, was probably also responsible for cases of bladder cancer in coke and coal gas workers. Benzidine and 4-aminophenyl are also bladder carcinogens. The use of substances containing these classical bladder carcinogens is now restricted or banned. Epidemiological studies have shown that rubber workers no longer have an enhanced risk of the disease. There is some rather inconclusive evidence that hairdressers may also be at risk from the use of hair dyes. In countries where schistosomiasis is rife, those who are liable to contract the disease as the result of their work have a greater than average risk of bladder cancer.

Although lead salts can induce renal tumours in laboratory animals there is no evidence that this is the case in man.

At one time cadmium was thought to be implicated in the development of prostatic cancer but recent studies do not support this view. Those exposed to acrylonitrile, however, have been found to have a significant excess of prostatic tumours as have printing workers.

Investigation and health monitoring of genitourinary disorders

Renal disease presenting in someone at work is more likely to have a non-occupational than an occupational cause and the first line of investigation must be to concentrate on making a non-work-related diagnosis.

The pointer to an occupational cause here, as with all other systems, would be the occurrence of a cluster of cases in one part of a single factory or amongst workers engaged in the same type of work in different areas of a factory or factories.

Where there is exposure to cadmium, a biological monitoring programme must be instituted. Urine and blood cadmium levels are of limited value since they do not accurately reflect the renal concentration, so monitoring is also based on the measurement of β_2-microglobulins in the urine (see Chapter 10). Some recent work has suggested that equally good control may be achieved by measuring retinol binding protein or NAG levels but it is still too early to advocate this approach unreservedly. The measurement of renal or hepatic concentrations by non-invasive methods based upon neutron activation probably has no place outside research projects. Any biological or medical supervision of workers exposed to mercury should include regular simple tests for proteinuria in addition to those which are used exclusively to measure exposure.

The British rubber industry has for many years undertaken exfoliative cytology testing of the urine on its employees with the aim of detecting bladder tumours at an early stage and thus improving the prospect of successful early treatment. The benefit of such monitoring remains disputed and the use of cytological testing is unlikely to spread to other recognized bladder carcinogens (MOCA, for example) until firmer evidence of its worth emerges.

The skin

Skin disease is the most common condition with which occupational health practitioners have to deal, and the number of agents which can affect the skin and the number of jobs in which skin disease is found are so many that great oversimplification is needed when discussing them. Broadly speaking, occupational skin disease can be caused by physical agents, by sensitizers and by irritants. There are also some compounds which may alter the pigmentation of the skin, produce folliculitis and acne, or skin cancer.

Physical agents

Minor cuts and abrasion are amongst the most common of all skin conditions in industry; almost anyone whose work involves moving machines can expect at least a few cuts during the course of their working life. Frequently there will be no sequelae but there is an ever-present risk of secondary infection or of the development of dermatitis. Most secondary infections will be staphylococcal in origin but butchers and other meat handlers may develop orf if they injure themselves at work.

Extremes in the ambient temperature may induce prickly heat or cold urticaria and burns will result from contact with hot objects. In cold, dry conditions the skin may become chapped and fissured whereas in hot, humid conditions, intertrigo may result.

Irritant dermatitis

Irritant dermatitis, which is now sometimes referred to as non-immunological contact dermatitis (NICD) is characterized by redness, scaling and fissuring of the affected parts. In acute cases the skin is red, swollen and itchy and there may be blisters present. The condition tends to resolve spontaneously but the healing time increases when there is further contact with the irritant and it may eventually become permanent.

Any of the physical conditions noted previously can cause NICD; other common causes are exposure to mineral oils, degreasing agents, detergents and cement. Building workers are in a particularly unfortunate position for not only are they liable to injure themselves on bricks and other building materials, but they are prone to an irritant dermatitis from their contact with cement dust and, if the dust contains chromates, then they may subsequently develop an immunological contact dermatitis.

Immunological contact dermatitis

Immunological contact dermatitis (ICD) is a type IV delayed hypersensitivity reaction brought about through the mediation of T-lymphocytes. In general, repeated exposure to a sensitizer has to take place before symptoms appear, but with some very potent sensitizers a short exposure time is all that may be required.

There are many sensitizers in the workplace, the majority of which are low molecular weight compounds which form a hapten with a protein after they have penetrated the skin. The most important occupational skin sensitizers include nickel and chrome salts, epoxy resins, accelerators and oxidants used in rubber mixes (paraphenylenediamine derivatives are especially potent), antibiotics, some skin disinfectants, paraphenylenediamine dyes used by hairdressers and a number of compounds produced by plants, particularly by the Primulaceae. It is probably true to say that there is almost no occupation in which potential exposure to a skin sensitizer will not occur at some time or another.

Changes in pigmentation

Hyperpigmentation of the skin may result from exposure to pitch and tar, to compounds containing mercury and it is a classic feature of arsenic poisoning. It may also be related to photosensitivity such as may occur in some agricultural workers following exposure to some plant species.

Depigmentation (vitiligo) is caused by exposure to some substituted phenols and catechols; it is indistinguishable from the naturally occurring type.

Oil folliculitis and acne

Those exposed to cutting oils may develop follicular papules, pustules and comedones on the areas in contact with the oil such as the forearms, thighs and abdomen. Chloracne is caused by exposure to certain chlorinated hydrocarbons, especially those containing chlorinated dibenzodioxins or dibenzofurans as contaminants, PCBs, for example. Comedones also occur in this condition as do cysts and inflammatory lesions. The face and neck are most commonly affected and the lesions may itch by contrast with those of oil folliculitis or teenage acne.

Malignant disease

Skin cancer is a considerable hazard to fair-skinned people who work long hours in the sun; thus farmers and fishermen have a higher than average risk. The same is also true for those exposed to pitch and tar, to soot and to mineral oils. As is well known, cancer of the scrotum in chimney sweeps was the first occupational cancer to be described; despite the fact that this has been known about for over 200 years, there is still an enhanced risk of this tumour in sweeps. Scrotal cancer was once common in the cotton industry but the introduction of non-carcinogenic oils has reduced the risk. Exposure to ultraviolet light also enhances the risk of melanoma, especially in those with freckles or who tan badly.

Investigation and monitoring of skin disease

The occurrence of dermatitis in someone shortly after they have started a new job or changed their work practices so that they may have become exposed to an agent which they have not encountered before should alert the occupational physician to the probability that the disease is work-related. If there are clusters of disease within a factory this should serve to increase the index of suspicion.

It is always helpful to go to see the patient at work so that some idea can be gained of the type of physical or chemical hazards to which he might be exposed and also to study work practices. For example, it is by no means uncommon for painters to clean their hands with white spirit or for rubber workers to clean rubber mix off their hands with toluene or some other solvent.

Where ICD is suspected, then patch testing may be helpful in arriving at a diagnosis. In this test, a variety of known sensitizers in appropriate concentrations is applied to the skin in a suitable vehicle to determine whether or not they elicit an eczematous reaction. Patch testing is not without some risk to the patient and should be carried out only by someone with experience and preferably in a dermatological clinic. The test substances are applied to the back and patients must be warned that the skin on the back may become red and itchy. In some cases the whole of the back may become red and oedematous and patch testing may cause recrudescence of existing eczema; some unfortunately individuals may actually become sensitized as the result of inexpert patch testing.

There are a number of standard test kits available which contain 20 or 30 sensitizers. Some kits are designed to be used specifically in some occupations; there is one available for testing nurses, for example. The patches are read at 48 and 96 hours after application and the reaction graded according to the degree of erythema produced and whether it is associated with infiltration and papules or vesicles. False negative reactions occur if the test compound has been applied in too low a concentration or in the wrong vehicle; false positives occur if the concentration is too high or the material is incorporated into an irritant vehicle.

To prevent dermatitis, workers should be encouraged to minimize contact as far as possible, to be scrupulous in their personal hygiene, to dry their hands carefully after washing and not to use irritant cleansing agents. The use of protective gloves is often helpful although, paradoxically, some individuals may be sensitive to chemicals in the gloves especially if they are made of rubber. Barrier creams are of doubtful value.

Those with a history of contact dermatitis should not be permitted to work in an occupation in which exposure to sensitizers is known to occur.

The prevention of skin cancer depends upon shading the skin as far as is possible from the harmful effects of the sun and, in those who are exposed to cutting oils, in scrupulous hygiene and the use of protective clothing and frequent changes of overalls. Where there is a known risk of scrotal cancer some form of regular medical inspection should always be carried out. The Mule Spinning (Health) Special Regulations 1953 actually prescribe medical examinations but the law has not yet caught up with the engineering industry. By the time it does, changes in the composition of cutting oils and improvements in hygiene will probably have all but eliminated the disease.

In the investigation of vitiligo in fair-skinned workers it is often helpful to use a Wood's lamp, particularly during the winter months when the skin is not tanned. The ultraviolet emissions from the lamp are absorbed by melanin but are reflected back from depigmented areas and thus they appear bright under the lamp.

The musculoskeletal system

Musculoskeletal disorders are extremely common in industry and account for an enormous amount of time lost from work—some 14 million working days are lost because of back pain, for example. There can scarcely be an occupational health practitioner in the country who has not frequently had to deal with one or other of these disorders.

Back pain

Back pain is so prevalent in the population as a whole that it is sometimes difficult to determine exactly how much has an occupational cause. About 2 per cent of the adult population seeks medical advice each year for back pain and the prevalence increases with age. Approximately 10 per cent of the male population at any one time will complain of pains in the neck or lower back and up to 20 per cent of the female population will do so. Radiological evidence of disc degeneration is common and also increases with age; in random cross-sectional surveys, the prevalence of cervical and lumbar disc degeneration in males is 40 per cent and 60 per cent respectively. In females the figures are slightly lower, 37 per cent and 44 per cent respectively. Against these very high prevalence rates it is often difficult to discern an occupational cause but there is little doubt that those who perform heavy manual labour have an increased prevalence of back pain, especially if the work involves a lot of twisting or bending or if sudden movements have to be made. Long distance lorry drivers have an increased frequency of back pain and prolapsed intervertebral disc and the vibrations emitted from the driving seats of slow moving, heavy vehicles may cause problems especially as the vibration is at the natural resonant frequency of the body (*see* Chapter 6). Ergonomic factors are also important and complaints of back pain are common from those who have to work at poorly designed work stations. Back pain amongst nurses is one of the most frequent causes of lost time in the profession.

Bony changes

In most cases of back pain, there are no radiological abnormalities but bony changes do occur in some occupations. Most of these are well known; the stress fractures and tarsal injuries in ballet dancers, avascular necrosis in divers and tunnellers working under conditions of increased atmospheric pressure, carpal cysts in some of those using vibrating tools and acro-osteolysis in workers exposed to high concentrations of vinyl chloride monomer can all be included. Whether osteoarthritis can ever be directly attributable to occupational factors, however, seems less certain, although an increased incidence of osteoarthritis in the hands has been noted in some cotton spinners. What is more likely is that occupational factors may help to determine in which joints the disease is expressed in those who are susceptible to it.

Repetitive strain injuries

The oldest of these conditions is writer's cramp which is characterized by progressive difficulty in making repetitive movements for which there is no neurological or rheumatological cause. There is often a large psychological component to the condition which may be extremely resistant to treatment.

In more recent years it has become apparent that there are many more of these so-called repetitive strain injuries which may affect the joints, the muscles, the tendon sheaths or cause entrapment neuropathies.

Repetitive strain injuries may affect anyone who is called upon to make repeated, rapid and perhaps forceful movements of the hands for the greater part of the working day. Workers on semiautomated tracks are often affected and so are those who use keyboards, especially since the advent of the word processor where very high typing speeds can be attained; musicians have also been found to suffer from the condition. It may present as a true tenosynovitis or as tendinitis crepitans, as muscle or joint pains in the arms or shoulders or as carpal tunnel syndrome.

Investigation and monitoring of musculoskeletal disorders

When people at work complain of back pain it is most important to observe them performing their task wherever possible. Following this it may be evident that what is required is some instruction in proper lifting techniques, or that they are attempting to lift objects which are too heavy for them. If this is so, then they should be advised to do their work in conjunction with a colleague or with lifting apparatus. There is no consensus about what is a safe maximum load to lift although a limit of 25 kg has been suggested for youths and 58 kg for others. When lifting heavy objects the knees should be bent and the back kept straight. The object should be lifted from between the feet and kept close to the body and never lifted above shoulder height. Spinal torsion should always be avoided when lifting heavy objects.

It may also be obvious from simple observation that individuals are working at a high desk with a low chair or vice versa and simple steps can generally be taken to remedy this, usually with much relief. All office chairs should have good lumbar support and it should be possible to adjust the back of the chair both vertically and horizontally. Buying cheap office furniture is always a poor investment since it is a sure way to promote backache, poor efficiency and sickness absence amongst the staff.

There is little prospect at present of being able to screen out individuals who are likely to develop back pain prior to employment unless they have a history of the condition. Radiology is extremely unhelpful and can scarcely be justified on ethical grounds. Some experimental use has been made of ultrasonic measurement of the vertebral canal diameters and it has been suggested that those with a narrow canal are more likely than others to develop back pain. The test is non-invasive and technically not very difficult but its usefulness in individual cases seems doubtful at present and it is likely only ever to come into use in those industries such as coal-mining where the prevalence of back pain is particularly high.

The prevention of repetitive strain injuries can only be done if it is possible to ameliorate the pattern of work which produces it by rotating different tasks or by taking frequent breaks from repetitive movements. Typists, for example, can be encouraged to break their work up so that they are not at the keyboard for long stretches at a time but it is not much help to a violinist to tell him that he should find some other way of fingering or to ask workers on a track to work less hard when their bonus and that of their colleagues may depend upon achieving preset targets.

Repetitive strain injuries are extremely difficult to treat. They improve with rest but will frequently return when the original pattern of use is resumed. Many patients have no course but to change their work which is an unhappy and unsatisfactory outcome for everyone concerned with the problem. Pre-employment screening offers very little prospect of reducing the prevalence of the conditions although—obviously—anyone with a previous history should not be permitted to enter another occupation where it is likely to recur.

The bone lesions in those who work at increased atmospheric pressure can largely be prevented by strict compliance with decompression schedules but regular radiological monitoring is advisable. Whether a person who has avascular necrosis should be allowed to continue to work depends upon the site of the lesion. Lesions on the shaft or the neck of a bone are clinically unimportant but those in a juxta-articular position may progress to involve the joint with the production of pain, stiffness and secondary osteoarthritis. Those who are found to have this kind of lesion are best advised not to allow themselves further exposure.

Diving and working in compressed air are subject to a number of regulations in the United Kingdom and those who undertake them must be under medical supervision; in the cases of divers, a certificate of fitness is required each year. Those who are aged over 40 should only be allowed to take up this kind of work if there are the most pressing reasons for allowing them to do so. There are numerous conditions which would debar an applicant from diving, including obesity, chronic disease of any kind and severe visual defects.

Conclusions

All those who are involved with occupational health practice must maintain a high index of suspicion when considering illness which may arise in people at work. Many of the classical occupational diseases are uncommon and likely to become more so, and they may tend to be overlooked. This is, of course, especially true if the practitioner is inexperienced or seeing people outside a work context. Occupationally related illnesses, by contrast, are relatively common and seem to be increasing in prevalence and it is these which will come to occupy more of the occupational health practitioner's time. This changing pattern will eventually come to alter methods of practice and, sooner rather than later one hopes, the form of postgraduate training.

Further reading

Adams, R.M. *Occupational Skin Disease*. Grune and Stratton, New York (1983)
Barlow, S.M. and Sullivan, F.M. *Reproductive Hazards of Industrial Chemicals*. Academic Press, London (1982)
Cronin, E. *Contact Dermatitis*. Churchill Livingstone, London (1983)
Health and Safety Executive, *Occupational Exposure Limits 1987* (Guidance Note EH 40). HMSO, London (1987)
Health and Safety Executive, *Guidance Note MS Series*. HMSO, London
Johnson, B.L. (ed.) *Prevention of Neurotoxic Illness in Working Populations*. John Wiley and Sons, Chichester (1987)
Mitchell, J.C. and Rook, A. *Botanical Dermatology*. Greengrass, Vancouver (1979)
Parkes, W.R. *Occupational Lung Disorders*, 2nd edn. Butterworths, London (1982)
Raffle, P.A.B., Lee, W.R., McCallum, R.I. and Murray, R. (eds) *Hunter's Diseases of Occupations*. Hodder and Stoughton, London (1987)
Scott, J.T. (ed.) *Copeman's Textbook of the Rheumatic Diseases*, 6th edn. Churchill Livingstone, Edinburgh (1986)

Chapter 5

Mental health of people at work

R. Jenkins

Introduction

Employers, trade unions, occupational doctors and the informed public have, until recently, concentrated their attention on three major work related areas: unemployment; physical and chemical health hazards; and absenteeism. Mental illness in the work force has been of subsidiary interest, and attention has largely focused on the separate issues of whether work is an aetiological factor in mental illness and on the rehabilitation of the mentally ill. Nevertheless, whether or not work is an aetiological factor in mental illness, the prevalence of mental illness in employees makes it an important issue in its own right. During periods of occupational stability, prevalence rates of over 30 per cent have been found in executive officers [1], industrial workers [2], pharmacists and dispensers [3], journalists [4] and air traffic controllers [5]. During periods of occupational instability prevalence rates may be higher, rates of over 37 per cent being reported during a labour dispute [5] and during and immediately after a period of threatened redundancy [4]. These prevalence rates are as high as those found in studies of general practice attenders [6], rather higher than those found in community surveys [7], and much higher than those found in studies deriving figures from general practice certification [8] and diagnosis by non-psychiatric physicians of attenders at occupational clinics [9].

The reasons for the discrepancy between studies using direct psychiatric examination of the workforce and studies relying on general practitioner certification are not hard to find. Epidemiologists are aware that rates of diagnosed or treated illness are underestimates of rates of illness in the entire population, since they are affected by the individual's readiness to recognize illness in himself and to seek medical care for his symptoms, by the availability of medical services, and by the primary care doctor's ability to diagnose illness and treat it. General practitioners' certificates are notoriously unreliable, and it is known that between a third and a half of psychiatric disorders presenting in general practitioners' surgeries remain undetected by the general practitioner [10]. In addition, since some stigma and discrimination may accrue to receipt of a psychiatric diagnosis, the general practitioner may avoid citing such a diagnosis on the certificate of an employed person, and probably, therefore, figures derived from general practitioners' certificates are considerable underestimates of the extent of minor psychiatric morbidity in the workforce.

This chapter aims to discuss the epidemiology and presentation of mental illness in people at work, and to detail the essential skills of eliciting common psychological symptoms. The causes, consequences and treatment of mental illness at work are discussed and the case is made for the establishment of mental health programmes in industry.

Prevalence of mental illness in the community

Mental illness may be broadly classified into the psychoses (or major psychiatric morbidity), the neuroses (so-called minor psychiatric morbidity), the personality disorders, organic brain syndromes (including senile dementia), psychological aspects of physical disease, alcoholism and drug dependence and mental retardation. In the general population, psychoses are relatively rare, occurring in about 2 per cent of the adult population. Mental abnormality occurs in rather less than 1 per cent of the adult general population. Senile dementia occurs in about 4.2 per

Table 5.1 Prevalence of mental illness in the community

Category of disorder	Prevalence	Disorder
Organic disorders	10 per cent of acute medical and surgical inpatients	Acute brain syndrome (delirium)
	10 per cent of population over age 65	Chronic brain syndrome (dementia) Korsakoff's syndrome
Functional psychoses	1–3 per cent	Schizophrenia + paranoid psychoses manic depressive psychosis psychotic depression
Non-psychotic disorders	10–30 per cent	Depressive illness Anxiety disorders anxiety state specific phobias agoraphobia panic disorder
		Maladaptive behaviours obsessive compulsive disorder alcoholism drug dependence anorexia nervosa and bulimia nervosa self-poisoning self-mutilation
		Abnormal illness behaviours hypochondriasis hysteria malingering factitious illness
Personality disorders	unknown	For example, schizoid, cyclothymic sensitive, anxious, depressive, obsessional, histrionic, antisocial

cent of the general population, mostly in those who are aged over 80; it is extremely rare in adults under age 65. No community surveys have been done of personality disorders, and no one knows how common they are. Prevalence figures will obviously depend on the strictness of criteria used to define a disorder. In general practice, they form less than 1 per cent of recorded episode rates. In a similar vein, the prevalence of alcohol abuse depends on the definition employed, but most authorities agree that the prevalence of actual alcoholism in the United Kingdom is at least 1 per cent of the adult population and rising. The prevalence of different degrees of alcohol abuse is clearly much higher. By contrast with the foregoing categories, the neuroses are much, much more common and occur in 10–25 per cent of the adult general population (*Table 5.1*).

Selection factors in occupational settings

How far do these prevalence figures help us in estimating prevalence in occupational settings? Obviously different levels of selection factors operate in different industrial professional settings, but some broad principles can be immediately obtained. In general, pre-employment screening will tend to select out individuals who are chronically mentally disabled, so prevalence rates of chronic schizophrenia, manic depressive psychoses and mental retardation will be lower in the employed population than in the general adult population. However, some employers may operate a deliberate policy of recruiting a proportion of disabled, and this would serve to reverse that trend. Furthermore, many individuals' first episodes of illness will occur when they are already in work. Whether or not they remain in work will depend not only on their degree of recovery, but also on employment policy, and on the degree of support from occupational health services, general practitioners and the local psychiatric services.

The current high level of unemployment has made occupational rehabilitation of chronic schizophrenics and the mentally handicapped increasingly difficult, and in general it seems reasonable to estimate that the prevalence of psychoses, dementia and mental handicap is lower in employed populations than in the general population. However, for neuroses the picture is very different. Neuroses or minor psychiatric morbidity occur in 10–25 per cent of the adult general population at any one time. These are relatively brief conditions, at least half lasting a few months only and, in many cases naturally self-terminating. They occur in most individuals at one time or another during their life. Their prevalence is therefore not reduced by pre-employment screening, and they can be expected to form the bulk of mental disorder in an employed population.

Epidemiology and symptomatology of the major categories of mental illness

This section briefly describes each of the main categories of mental illness, but most attention will be given to the neuroses (minor psychiatric morbidity), particularly depression and anxiety states which are by far the commonest disorders in the community and in the workforce, and will form the bread and butter of the occupational doctor's practice.

Schizophrenia

At some time in their lives 1 per cent of adults are diagnosed as schizophrenic. The peak incidence is between 25 and 30 years of age, with a smaller peak in the over 65s. Rates are higher in those of lower social class and living in inner city areas. These associations occur as a consequence of the illness rather than the cause. The social class distribution of fathers of patients with schizophrenia approximates to that of the general population and the downward social drift that occurs in people with schizophrenia is accounted for by the many handicaps that accompany the illness.

Symptoms

The symptoms of schizophrenia can be divided into the positive and negative symptoms. Positive symptoms include:

1. The reduced contact with reality, whereby the normal mental processes which allow clear differentiations between subjective experiences and the outside world break down, and involve experiences of thought possession, passivity phenomena and thought withdrawal.
2. Hallucinations, usually auditory, voices occurring in the third person, or thought echo.
3. Thought disorder, as evidenced by unusual logic: idiosyncratic use of words, neologisms, knight's move thinking and derailment.
4. Delusions, which are mistaken beliefs, held with conviction, not shared by others of the same cultural or social background and intellect, and which persist despite all evidence to the contrary.
5. Incongruous affect, as evidenced by outbursts of laughter or anger without any appropriate stimulus.

The so-called negative symptoms include poverty of speech, slowness of thought and movement, emotional flatness, loss of volition, underactivity and social withdrawal.

Course

The course of schizophrenia is very variable, 10 per cent show a rapid and permanent deterioration, 35 per cent show mild and persistent symptoms which require continuing care, either in a hostel or a patient's home, 35 per cent appear 'cured' for long periods of time, but have relapses, and 20 per cent are apparently cured and stable. Repeated episodes of psychoses, together with periods of hospitalization lead to a set of secondary handicaps in addition to the negative symptoms of the illness itself. The person's job will be disrupted by repeated absences through illness, the onset of psychotic symptoms at the place of work will further increase the chance that he may lose his job and find further employment difficult to obtain.

Prognosis

Poor prognostic factors include:

1. A family history of schizophrenia.
2. A premorbid schizoid personality.

3. A high level of expressed emotion in the home environment.
4. A poor previous work record.
5. A gradual onset of the illness without a clear precipitating cause.
6. The absence of prominent affective catatonic symptoms.
7. The presence of pronounced negative symptoms.

Mania

Here, mood is elevated out of keeping with the person's circumstances and may vary from carefree joviality (hypomania), to almost uncontrollable excitement. Elation is accompanied by increasing energy which results in overactivity and a decreased need for sleep. Normal social inhibitions are lost, and attention cannot be sustained. Self-esteem is inflated and grandiose ideas, which may be delusional, are freely expressed. The subject may embark on extravagant and impracticable schemes, spend money recklessly, or become aggressive, anxious or facetious in wholly inappropriate circumstances. In some manic episodes the mood is irritable rather than elated and delusions are persecutory rather than grandiose. If the disorder is mild the subject may appear, to those who do not know him, merely to be boorish, garrulous and conceited; and if the person is only seen at the height of his illness, widespread delusions, incomprehensible speech or violent excitement may result in an erroneous diagnosis of schizophrenia. The first attack occurs most commonly between the age of 15 and 30 but may occur at any age from late childhood to the seventh or eighth decade.

Depression

Here mood is lowered, and there is reduction of energy and decrease in activity. Capacity for enjoyment, interest and concentration are impaired, and marked tiredness after even minimum effort is common. Sleep is usually disturbed, and appetite is diminished. Self-esteem and self-confidence are almost always reduced, and even when mild some ideas of guilt or worthlessness are often present. The future seems bleak and suicidal thoughts and acts are common. The lowered mood varies little from day to day, is unresponsive to circumstances and may show a characteristic diurnal variation as the day goes on. In some cases, anxiety, distress and motor agitation may be more prominent at times than the depression, and the mood change may also be marked by added features such as irritability, excessive consumption of alcohol, histrionic behaviour, exacerbation of pre-existing phobic or obsessional symptoms, or by hypochondriacal preoccupations.

If the depression is severe, it may be accompanied by the so-called 'biological symptoms' such as motor retardation, weight loss (often as much as 5 per cent of the body weight in a few weeks), marked diurnal variation, early morning waking and loss of libido, loss of self-esteem and ideas of worthlessness, guilt, sin and eminent disasters may become delusional. There is an increased danger of successful suicide.

Generalized anxiety disorder

The essential feature is anxiety which is generalized and persistent but not restricted to, or even strongly predominating, in any particular environmental circumstances (that is, it is free floating). As with other anxiety disorders the dominant symptoms are very variable but complaints of feeling nervous all the time, trembling, muscular tension, sweating, lightheadedness, palpitations, dizziness and epigastric discomfort are common. Fears that the subject or a relative will shortly become ill or have an accident are often expressed, together with a variety of other worries and forebodings. This disorder is commoner in women, and often related to chronic environmental stress. Its course is variable but tends to be fluctuant and chronic.

Mixed anxiety and depressive disorder

This category consists of people with symptoms of both anxiety and depression to a mild or moderate degree. It is very common in the community and in primary care. If both sets of symptoms are present but severe, then both anxiety and depressive disorders should be regarded as present.

Agoraphobia

This includes fears of open spaces, the presence of crowds, and the difficulty of immediate easy escape back to a safe place (usually home). The term therefore refers to an interrelated and often overlapping cluster of phobias embracing fears of leaving home, or remaining at home alone; entering shops, crowds and public places, or travelling alone in trains, buses or planes. Although the severity of the anxiety and the extent of avoidance behaviour are variable, this is the most incapacitating of the phobic disorders and some subjects become completely housebound. Many are terrified by the thought of collapsing and being left helpless in public. The lack of an immediately available exit is one of the key features of these agoraphobic situations. Women are more affected than men, and the onset is usually early in adult life.

Social phobias

These often start in adolescence, and are centred around a fear of scrutiny by other people in comparatively small groups (as opposed to crowds) leading to avoidance of social occasions. Unlike most other phobias, social phobias are equally common in men and women. They may be discrete, for example, restricted to eating in public or to public speaking or to encounters with the opposite sex, or diffuse, involving almost all social contact outside the family circle. A fear of vomiting in public may be important. Social phobias are usually associated with low self-esteem and fear of criticism. They may be present as a complaint of blushing, hand tremor, nausea or urgency of micturition, the patient being convinced that one of these secondary manifestations of their anxiety is the primary problem; symptoms may progress to panic attacks. Avoidance is often masked, and in extreme cases may result in almost complete social isolation.

Specific (isolated) phobias

These are phobias restricted to highly specific situations such as proximity to particular animals, heights, thunder, darkness, flying, closed spaces, urinating or defaecating in public toilets, eating certain foods, dentistry, the sight of blood or injury, and the fear of exposure to specific diseases. Specific phobias usually arise in childhood or early adult life and, if they remain untreated, can persist for decades. The seriousness of the resulting handicap depends on how easy it is for the subject to avoid the phobic stimulus. Fear of the phobic object tends not to fluctuate, in contrast to agoraphobia. Radiation sickness and venereal infections are common objects of disease phobias and, more recently, AIDS.

Panic disorder (episodic paroxysmal anxiety)

The essential feature is recurrent attacks of severe anxiety which are not restricted to any particular situation or set of circumstances, and are therefore unpredictable. As with other anxiety disorders the dominant symptoms vary from subject to subject but sudden onset of palpitations, chest pains, choking sensation, dizziness and feelings of unreality (depersonalization or derealization) are common. There is almost invariably also a secondary fear of dying, losing control or going mad. Individual attacks usually last for minutes only, though sometimes longer. Their frequency and the course of the disorder are both rather variable though it predominates in women. Patients in a panic attack often experience a crescendo of fear and autonomic symptoms which results in an exit, usually hurried, from where they are. If this occurs in a specific setting, such as on a bus or in a crowd, the patient may subsequently avoid it; similarly, frequent and unpredictable panic attacks produce fear of being alone or going into public places.

Obsessive-compulsive disorder

The essential feature is recurrent obsessional thoughts or compulsive acts. Obsessional thoughts are ideas, images or impulses which enter the subject's mind again and again in a stereotyped form. They are almost invariably distressing (because they are violent or obscene, or simply because they are perceived as senseless) and the subject usually tries, unsuccessfully, to resist them. They are, however, recognized as the subject's own thoughts even though they are involuntary and often repugnant. Compulsive acts or rituals are stereotyped behaviours which are repeated again and again. They are not inherently enjoyable, nor do they result in the completion of inherently useful tasks. The patient often views them as preventing some objectively unlikely event, often involving harm to or caused by the subject. Usually this behaviour is recognized by the subject as pointless or ineffectual and repeated attempts are made to resist it; in very longstanding cases, resistance may be minimal. Anxiety is usually present. There is a close relationship between obsessional symptoms, particularly obsessional thoughts, and depression. Patients with obsessive-compulsive disorder often have depressive symptoms, and patients suffering from depressive disorder may develop obsessional thoughts during their episodes of depression.

Obsessive-compulsive disorder is equally common in men and women and there are often prominent obsessional features in the underlying personality. Onset is usually in childhood or early adult life. The course is variable and more likely to be chronic in the absence of significant depressive symptoms.

Anorexia nervosa

This is an illness characterized by deliberate weight loss, induced and/or sustained by the patient herself. The illness occurs most commonly in adolescent girls and young women, but adolescent boys and young men may be affected more rarely, as may children approaching puberty and older women up to the menopause.

Although the fundamental causes of anorexia nervosa remain elusive, there is growing evidence that interacting sociocultural and biological factors contribute to its causation, as do less specific psychological mechanisms and a vulnerability of personality. The illness is associated with undernutrition of varying severity with resulting secondary endocrine and metabolic changes and disturbances of body function. There remains some doubt whether the characteristic endocrine disorder is entirely due to the undernutrition and the direct effect of various behaviours that have brought it about (for example, restricted dietary choice, excessive exercise and alterations in body composition, induced vomiting and purgation, and the consequent electrolyte disturbances), or whether other uncertain factors are also involved.

Precise criteria can be laid down for the diagnosis of anorexia nervosa, so that the disturbances listed below are all necessary for the diagnosis.

1. There is a significant weight loss.
2. The weight loss is self-induced by:
   (a) avoidance of fattening foods and one or more of the following:
   (b) self-induced vomiting;
   (c) self-induced purging;
   (d) excessive exercise;
   (e) use of appetite suppressants and/or diuretics.
3. A specific psychopathology whereby a dread of fatness and/or flabbiness of body control persists as an intrusive overvalued idea, and the patient imposes a low weight threshold on herself;
4. A widespread endocrine disorder involving the hypothalamic–pituitary–gonadal axis, manifest in the female as amenorrhoea, and in the male as loss of sexual interest and potency. There may also be elevated levels of growth hormones, raised levels of cortisol changes in the peripheral metabolism of the thyroid hormone, and abnormalities of insulin secretion.

Bulimia nervosa

This is a syndrome characterized by repeated bouts of overeating and an excessive preoccupation with the control of body weight, leading the patient to adopt extreme measures so as to mitigate the 'fattening' effects of ingested food. The age and sex distribution is similar to that of anorexia nervosa, but the age of presentation tends to be slightly later. The disorder may be viewed as a sequel to persistent anorexia nervosa. A previously anorexic patient may first appear to improve as a result of a gain in weight and possibly a return of menstruation, but a pernicious pattern of overeating and vomiting then becomes established. Repeated vomiting is likely to give rise to disturbances of body electrolytes and physical complications (tetany, epileptic seizures, cardiac arrhythmias, muscular weakness). Diagnostic criteria include:

1. There is persistent preoccupation with eating, irresistible craving for food, and the patient succumbs to episodes of overeating in which large amounts of food are consumed in short periods of time.
2. The patient attempts to counteract the fattening effects of food by one or more of the following: self-induced vomiting, purgative abuse, alternating periods of

starvation, use of drugs such as appetite suppressants, thyroid preparations or diuretics. When bulimia occurs in diabetic patients they may choose to neglect their insulin treatment.

3. The psychopathology consists of a morbid dread of fatness and the patient sets herself a sharply defined weight threshold, well below her premorbid weight, which she then considers her optimum as 'healthy' weight.

4. There is often, but not always, a history of an earlier episode of anorexia nervosa, the interval ranging from a few months to several years.

Premenstrual tension (PMT) syndrome

This is a cyclical disorder with symptoms of sufficient severity as to interfere with some aspects of living and occurring with a consistent and predictable relationship to menstruation. The condition may not appear with the onset of puberty or in early adult life but more commonly arises in the third and fourth decades, following childbirth or other hormonal disturbances. One or more of the following symptoms characteristically occur:

1. Irritability which, while marked, is not usually accompanied by aggressive or violent behaviour.
2. Depression of mood which is not persistent but commonly alternates with normal mood or even periods of mild euphoria; subjective feelings of bloating and weight gain, occasionally accompanied by objective signs of swelling of hands and/or feet, abdominal distension and/or weight gain.
3. Breast tenderness and/or swelling.
4. Marked tension, sometimes accompanied by anxiety.
5. Headaches, backache, and/or generalized aches and pains.
6. Difficulty concentrating.
7. Physiological changes, including sleep disturbance, appetite loss or increase and food cravings.

A remarkable number and variety of symptoms and behaviours repeatedly occur or are exacerbated during the premenstrual phase, including asthma, acne, epilepsy and alcohol abuse, but their precise status in terms of the definition of the syndrome is disputed. The symptoms appear during the initial phase, increase in severity as the menstrual period approaches and are at their peak of severity 1 or 2 days prior to the onset of the period. The symptoms are relieved with the appearance of the menstrual flow or shortly afterwards.

Abnormal illness behaviours

Illness behaviour refers to the ways in which given symptoms may be differentially perceived, evaluated and acted upon. It includes both the lifelong tendency to have a low or high threshold for consulting doctors and the behaviour at a particular time. There is evidence that traits are acquired in childhood, by learning patterns of illness behaviour characteristic of the culture and the family. Abnormal illness behaviours refer to illness behaviours which are inappropriate and includes hypochondriasis, hysteria, factitious illness, malingering and the denial of disease.

Hypochondriasis

This means a preoccupation with disease, either physical or mental, and may occur as part of a personality disorder, a neurotic disorder or as part of a psychosis. Hypochondriacal personality traits include a lifelong tendency to be overconcerned with health, food fads and physical fitness, as well as fear of illness. Hypochondriacal neurotic symptoms include the presence of overvalued ideas about illness and vague but pervasive ideas that he is suffering from an unidentified disease despite extensive normal investigations. Hypochondriacal delusions can occur in any psychotic syndrome, including schizophrenia, depression and dementia.

Hysteria

This term has been used in different ways and it is particularly important to clarify its meaning. 'Hysterical' is sometimes used by lay people to describe dramatic or histrionic behaviour; however, it is preferable to use these latter terms instead. The term hysteria should be reserved to describe the unconscious simulation of the signs of disease.

Common hysterical symptoms include astasia-abasia (inability to stand and walk), fits, tremor, anaesthesia, paralysis, amnesia, and dysphoria. Conversion is the psychological defence mechanism regarded as responsible for the production of simulated 'physical' symptoms, which are therefore called conversion symptoms.

Dissociation is the defence mechanism thought to result in the simulation of mental symptoms, for example amnesia, which are therefore called dissociative symptoms.

Hysterical symptoms should fulfil the following three criteria:

1. They mimic neurological and psychological disorders but reflect the patient's concept of how such disorders present. For example, a hysterical paralysis of a limb will not have the appropriate neurological signs of an upper or lower motor neurone lesion. Similarly a hysterical anaesthesia will not reflect a dermatome.
2. The lost function is often intact in conditions which the patient does not associate with the symptoms. For example, a person with hysterical dysphasia may phonate normally when coughing or laughing.
3. The simulation is unconscious, that is, the patient is not malingering.

Hysterical symptoms, like hypochondriacal symptoms, have little diagnostic specificity. They can occur in almost any diagnostic setting and the diagnosis depends on the identification of the associated symptoms and signs. They occur especially in affective disorders, personality disorders, schizophrenia, and organic brain disease. It is unusual to have a first onset after the age of 40, and neurological disorder such as multiple sclerosis (MS) should be excluded. Hysteria is of course not incompatible with organic disease.

The syndrome of hysteria is characterized by the presence of hysterical symptoms in the absence of any other demonstrable physical or psychiatric disorder. The symptoms should have some adaptive value for patients, which is usually the relief of anxiety when the development of the symptom removes the patient from the conflict. The relief of anxiety is termed the primary gain of the illness. Because of its value, such patients are often not distressed, and this is termed *'la belle indifference'*. The subsequent use of symptoms to manipulate others is called secondary gain, and can of course be the consequence of any physical or mental illness.

Basic essential skills for eliciting psychological symptoms

From the above brief resumé of the spectrum of mental illness, it becomes clear that of the occupational doctor's clients, less than 1 per cent will be suffering from a psychotic illness, but 10–30 per cent at any one time will have a neurotic disorder, most likely depression or anxiety or a mixture of the two. It is therefore most important to be familiar with the common symptoms of depression and anxiety, and to be at ease with eliciting them, establishing their severity and frequency, and charting their consequences for everyday working and domestic life. The questions suggested below are based on those used in the clinical interview schedule.

Well people may of course experience one or two of the following symptoms at any one time, but when a constellation occurs in one individual to the extent that serious interference happens to the individual's life, then we think in terms of illness.

Poor concentration

This can range from some difficulty in performing the more complex tasks of one's job (for example, the civil servant who starts to put the most difficult files in a drawer to deal with at a later date—he tends to come to his colleagues' notice when the drawer overflows), through to inability to concentrate on simple tasks at work, or to read a novel or newspaper at home. The person may watch the news on television, and then be unable to remember the main items a few minutes later. This is an important symptom, partly because it is often one of the first symptoms to arise in a depressive episode, so it may be helpful if both the individual and those around him are alerted by it to take preventive action. It is also important in that it is often the factor in an illness which can cause most damage to performance and productivity.

Useful questions about concentration include the following:

Do you find it difficult to concentrate?
Do you get muddled or forgetful?
If yes, has it caused any difficulty at home or at work?
Can you concentrate on a newspaper or on a play on television?
Has it stopped you from doing anything?
Which activities are affected?

Fatigue

Fatigue or tiredness may range from simply feeling rather more tired than usual at the end of the working day, through to collapsing in front of the television in the early evening, unable to move for the rest of the evening, right through to feeling overwhelmed with exhaustion through the day. This is a particularly debilitating symptom for individuals who carry large domestic responsibilities as well as their occupational load. Parents of young children are at particular risk.

Questions you may find useful about fatigue are:

Have you noticed that you get tired easily?
Or that you seem lacking in energy?

If yes, what sort of things do you find most tiring?
 Do you feel completely tired out in the evening?
 Has it stopped you from doing anything you've wanted to do?

Irritability

Irritability ranges from the occasional, uncharacteristic snapping at colleagues or family through to extreme irritability or extreme violence, which occurs more commonly in the home than in the workplace. Regular alcohol consumption may exacerbate irritability and the tendency to violence, and it is helpful to advise those suffering with irritability to cut out alcohol completely *pro tem*.

Sometimes, rather than be irritable with others, the person may bottle things up and become oversensitive, taking chance remarks too much to heart. This symptom can be less obvious to colleagues than poor concentration, but it causes more distress to the individual.

Useful questions are:

Do you think that you are easily upset or irritable with those around you?
 If yes, what sort of things upset you?
 Have you had any rows?

Low depressed mood

Feeling low may range from one or two spells of low mood a week lasting an hour or two at a time, through to a persistently lowered mood which is difficult to snap out of. The person may feel like crying, or may in fact be tearful, particularly when shown a little sympathy. Upsetting items on television may elicit tears. (For cultural reasons, men on the whole tend to cry less than women. It follows that crying in a man should be taken particularly seriously, although it can sometimes be an indication of the emotional lability consequent upon a history of heavy drinking rather than an indicator of severe depression.)

Rather than feeling low, the person may simply describe feeling flat, feeling no pleasure in life, family or occupation. Life may be described as empty, bleak or even hopeless. The person may start to feel that he would not mind going to sleep and never waking up, or being run over by a bus, or catching a fatal illness. This is the start of suicidal ideation, which may proceed to a feeling that one would rather be dead, and then to a wish to kill oneself, prevented only by the desire not to upset family and friends. The next step is to start thinking about how one might kill oneself, for example, by pills or by more violent means. This may be followed by active planning, including buying the means to kill oneself. Anyone who is entertaining actual plans of suicide is at serious risk and should be supervised in hospital by staff who are alerted to the danger.

It is crucial for every doctor to be familiar with the assessment of suicidal risk. It is a common but dangerous myth that those who talk about suicide and those who have previously taken an overdose do not kill themselves. The best predictor of suicide is an episode of self-poisoning. Increased risk of suicide is associated with a personal history of depressive illness, alcoholism, schizophrenia, dementia, epilepsy, antisocial personality or chronic painful physical disease. Suicide is also associated with increasing age, particularly after age 45, the single state, social isolation, unemployment and bereavement. Men are more likely to kill themselves than women.

Lowered mood or depression may sometimes be accompanied by feelings of guilt or self-blame. These are usually exaggerated or even completely unjustified, and may concern events years ago, such as a family death. The person frequently sees the future as empty and non-existent, and feels that there is nothing to look forward to.

Useful questions include:

Have you had spells of feeling sad or miserable?
If yes, how bad does it get?
Do you feel like crying?
Can you snap out of it?
Do you sometimes feel hopeless?
Have you felt like making an end to it all?
Have you thought how you might do it?
Have you made actual plans?
Do you ever blame yourself for being like this?
Do you ever find yourself feeling guilty?
Do you sometimes feel inferior to other people?
How do you feel about the future?

Anxiety

This is psychological 'worry' about either trivia or major items, and may or may not be accompanied by physical manifestations caused by fluctuations in autonomic activity. This commonly includes increased arousal characterized by sweating, palpitations, dry mouth, and frequency of micturition. There also may be hyperventilation, raised blood pressure, raised pulse and increased gastrointestinal mobility ('butterflies in the tummy').

Useful questions include:

Do you get anxious or frightened for no good reason?
Do you worry a lot about things?
If yes, what kind of things do you worry about?
How does it feel when you are anxious?
Do you get sweaty palms, palpitations, butterflies in the stomach?

Phobias

A phobia is an excessive persistent and groundless fear of an object or situation. These are common and only cause difficulties or cause the person to alter or modify his or her activities. For example, a phobia of snakes usually causes no problems for the city dweller except perhaps on walking holidays. However, a phobia of heights may be particularly troublesome for a bank clerk walking in a high-rise building, or for a builder who needs to be able to climb ladders (and may be the unadmitted cause of alcohol and benzodiazepine abuse while at work).

Phobias of mice, spiders and insects are common.

Agoraphobia (*see also* p. 78) is a fear of crowded places and may be troublesome in busy department stores. Claustrophobia is a fear of small, confined spaces such as lifts, which can again be troublesome for employees who need to be able to use lifts several times a day. Some people are frightened of the sight of blood. Social phobias (*see* p. 78) can be very troublesome and may lead to a pattern of drinking alcohol in order to allay anxiety before a meeting or party.

Useful questions include:

Are there any special things or situations that you find frightening or upsetting?
What about being alone in the house?
What about going out by yourself?
What about travelling on buses or trains?
Animals? Insects? Heights? The dark?
If yes, how bad is it?
Do you have to go out of your way to avoid such a situation?
Do you alter your usual activities in any way?
Do you take a drink to calm your fears?

Obsessional thoughts and activities

Obsessional thoughts are preoccupations which are accompanied by feelings of subjective compulsion, which the person regards as irrational and unwelcome and which he strives to resist. At its simplest it is an irritating tune going over and over in one's mind which will not go away. Much more distressing are unwelcome thoughts of aggression and violence. Some people have to live with such thoughts for years.

Obsessional activities are repeated actions which are again accompanied by the feeling of subjective compulsion which the person regards as irrational and which he strives not to do. Mild examples are checking that the door and windows are locked two or three times before retiring to bed or leaving the house. For some people, such behaviour is habitual, for others it only arises when they are depressed or anxious. The employee who checks his work five or ten times is considerably slower than his colleague who only needs to check once before moving on. Some people are obsessional at work, but not at home, or vice versa; others are obsessional in both areas. A few people become so handicapped by their obsessional activities that there is little time left in their day for anything else, for example, the person who has to wash their hands 60 or 70 times a day.

Useful questions include:

Do you ever find you have to do things over and over to make sure you've done them right?
Or that you keep having unwelcome thoughts that you can't get rid of?
Do you find it hard to make decisions?
If yes, how many times do you find yourself checking? (for example, your work, the locks, the gas taps)
Do you check it even though you know it is all right really?
Are there any things you find yourself doing a number of times?
For unwelcome thoughts:
Can you describe them?
Difficulty with decisions:
Is it just on important issues or does it affect trivialities as well?
For all the above:
Do you try and struggle against it?
How much does it distress you?
How much of your time does it take up?

Somatic symptoms

These are usually somatic correlations of stress, anxiety or depression such as headache, backache, and abdominal pains where there is a clear relationship with psychological stress. Furthermore, some organic conditions are aggravated by psychological factors. Duodenal ulcers, asthma and ulcerative colitis are common examples.

Useful questions include:

Have you had any headache, backache or indigestion recently?
Is it brought on by stress?
How much does it distress you?

Causes of mental illness

It is helpful to think of aetiological factors in mental illness in terms of predisposing factors, precipitating factors and maintaining factors. Predisposing factors are those which increase an individual's vulnerability to a particular illness at any time in the future. They may be biological, social and psychological. Biological variables include genetic constitution (for example, a family history of schizophrenia), intrauterine damage, birth trauma, and personality disorder. Relevant social predisposing factors include physical or emotional deprivation in childhood due to family discord, bereavement or separation, chronic difficulties at work or in the domestic sphere, and lack of supportive relationships. Relevant psychological predisposing factors include poor parental models, for example of violence or of normal illness behaviour, and low self-esteem which may be culturally induced, particularly in women.

Precipitating factors are those which determine when the illness starts, and again these may be biological, social or psychological. Biological precipitating factors include recent infections, disabling injury, and malignant disease. Social precipitating factors include recent life-events particularly those which involve the threat of a loss, or an actual loss itself. Threat of redundancy, unemployment, major illness in the family, a child leaving home, separation or divorce and the loss of a supportive relationship are common examples of life-events preceding illness. Psychological precipitating factors refer to the subject's maladjustive response to the biological and social factors by developing feelings of helplessness and hopelessness.

In addition, there may be maintaining factors which prolong an illness if it continues for longer than would be expected. Biological maintaining factors to be considered include chronic pain or disability, side-effects of medication or failure to take medication. Social maintaining factors include chronic social stress, no intimate, confiding relationships at home and lack of support from colleagues. The principal psychological maintaining factors are low self-esteem and lack of expectation of recovery. For the psychotic disorders, biological factors are usually just as important as the environmental (social) variables. However for the non-psychotic disorders, biological factors usually play a lesser role and environmental factors are much more important. Environmental factors may be classified into six social domains: occupational, marital, financial, housing, leisure (social life), and family (parents and children). It is thus immediately clear that each domain has to be considered in its own right, and each can be either stressful or supportive or both. Thus work in itself is not necessarily stressful. For many people, it may be the most supportive area of their lives. However, for others, a great deal of stress may be encountered at work which, if not balanced by sufficient support, may result in depression or anxiety or other illness [11].

Occupational stresses

There has now been a considerable amount of psychological research into occupational stresses, and these have been broadly classified into factors intrinsic to the job, role in the organization, career development, relationships at work, and the organizational structure and climate [11].

Factors intrinsic to the job

These include poor physical working conditions, shift work, work overload, work underload, physical danger, person–environment (P–E) fit and job satisfaction. For example, an important stress factor in the Three Mile Island accident was the distraction caused by excessive emergency alarms [12]. Shift work is a common occupational stressor and is known to affect neuropsychological rhythms such as body temperature, metabolic rate, blood sugar levels, mental efficiency, work motivation, and may ultimately result in stress-related disease. Air traffic controllers were found to have four times the prevalence of hypertension as second-class airmen, and also more diabetes and peptic ulcers. The study identified shift work as a major occupational stressor [13]. However, it has been argued that shift work becomes less stressful as individuals habituate to it, although it inevitably decreases their access to supportive relationships at home and in their social life. Job overload can be either quantitative (having too much to do), or qualitative (having too difficult tasks), and has been associated with cigarette smoking, low self-esteem, low work motivation and escapist drinking [14]. Job underload associated with repetitive, routine, boring and understimulating work has also been associated with ill health [15]. There are certain occupations which have been identified as particularly dangerous, for example, miners, soldiers, firemen and the police. However, the stress induced by the uncertainty of physical danger is greatly ameliorated if the individual is adequately trained and equipped to cope with emergency situations [16]. Person–environment (P–E) fit is a helpful concept which can be defined as the interaction between an individual's psychosocial characteristics and his or her environment at work. It is hypothesized that stress can occur and result in anxiety, depression and physical disease if there is a P–E misfit [17].

Role in the organization

This includes such issues as the degree of role ambiguity (lack of clarity about goals), role conflict (conflicting job demands) as well as responsibility for people and conflicts stemming from organizational boundaries. These role stresses have been linked to coronary heart disease, particularly in non-manual occupations, and to job satisfaction, but correlations with measures of mental health are weak [18].

Career development

This refers to the impact of overpromotion, underpromotion, status incongruence, lack of job security and thwarted ambition. The degree to which there is job advancement has been found to be negatively related to the incidence of psychiatric disorders [18].

Relationships at work

Relationships at work which include the quality and degree of social support from colleagues, superiors and subordinates have also been related to job stress. French and Caplan [14] suggest that role ambiguity may cause poor relationships with colleagues which in turn may produce psychological strain. On the other hand, strong social support from colleagues will relieve job strain [11].

Organizational structure and climate

This includes such factors as office politics, lack of effective consultation, lack of participation in the decision-making process, and restrictions on behaviour. Greater participation has been found to lead to higher productivity and performance, lower staff turnover and lower levels of physical and mental illness [14].

Burnout

It is impossible to conclude this section on the causes of mental illness at work without mentioning 'burnout'. Burnout is an increasingly recognized and popularized phenomenon in the last 15 years. The term, coined by Freudenberger and first applied to volunteer workers in a free health clinic, was derived from street slang for excessive drug use [19]. It has since come to describe a variety of physiological and psychological symptoms observed in many different groups including nurses, social workers, health administrators, police officers and counsellers [20]. Cherniss [21] has provided a succinct and workable definition when he described it as a process beginning with high, sustained levels of job stress that produce subsequent feelings of tension, irritability and fatigue ending with a defensive reaction of detachment, apathy, cynicism or rigidity. The relationship between the concept of burnout and depressive illness and anxiety states has not been systematically examined.

Physical consequences of mental illness

Mental illness has both physical and social consequences. There is both a primary and a secondary association of physical illness with psychiatric illness. Eastwood and Trevelyan [22] demonstrated the primary association in a London group practice during a health screening programme on 1470 individuals who received psychiatric and physical examinations in a general practice population. The authors found that individuals with psychiatric disorder had a significant excess of ischaemic heart disease and other physical disorders than controls. Eastwood and Trevelyan [22] suggest that their findings support the notion that individuals have a generalized propensity to disease, whether physical or psychological. Simms and Prior [23] reported an increased mortality from organic disease in patients with severe non-psychotic disorders in a 10 year follow-up study after psychiatric hospital treatment. It is possible that some physical illness arises through the self-neglect and malnutrition that may accompany psychological disturbance.

The secondary association between physical and mental illness occurs for a variety of reasons. The individual may have had a physical complaint for some time, but in a period of emotional stress and perhaps depression the physical symptom may seem to worsen and is therefore presented to a doctor, sometimes instead of the emotional problem which exacerbated it. Depressed people are frequently more introspective than usual, and examine their internal body sensations more closely than normal.

There is still a stigma attached to mental illness, and it is more socially respectable to have a physical illness than a psychiatric one. Friends, relatives and doctors often share this view. Depression can be secondary to a painful or worrying physical illness or symptom, and while it is of course appropriate to the individual to offer the physical problem to the doctor, the onus remains on the doctor to be aware of the likelihood of the secondary depression, to detect it and offer appropriate therapy [24].

Occupational consequences of mental illness

It is of considerable practical importance to industry and the occupational medical services to understand the effect of minor psychiatric morbidity on sickness absence, relations with colleagues, work performance, accidents and labour turnover.

Sickness absence

Three methods have been used to determine the contribution of mental disorders to sickness absence. The first is based on the examination of the diagnoses given by general practitioners on sickness certificates [25]. These figures, although high, are considerable underestimates since, like all figures derived from rates of diagnosed or treated illness, they are affected by the individual's readiness to recognize illness in himself and to seek medical care for his symptoms, by the availability of medical services, and by the primary care doctor's ability to diagnose mental illness and treat it.

The second method, used by Fraser [2], is based on retrospective attribution of spells of absence to neurosis made by research doctors on the basis of lengthy personal interviews with the subjects, and access to their medical records. Using this method, Fraser and his collaborators found that neurotic illness caused between a quarter and a third of all absence from work. Such a method overrides the major disadvantages associated with simply basing estimates on sickness certificates. The method, however, is based on the notion that an episode of sickness absence may indeed be attributed to one particular cause, and it ignores the overwhelming evidence that most absence is voluntary behaviour that has been shown to be affected not only by demographic and environmental factors, but also by the individual's attitude to his work as well as by the presence or absence of a physical or psychological disorder.

The third method, used by the present author, makes no attempt to attribute one particular episode of absence to any one cause, but rather to make comparisons of the annual absence taken between individuals with identified minor psychiatric morbidity and those without [1]. Using this method, it was found that the presence of minor psychiatric morbidity does make an important contribution to both retrospective and prospective sickness absence, and that this contribution is greater for certified absence than for uncertified absence, and is greater for duration than frequency of absence. Similar results have been reported by Ferguson who found that telegraphists and mail sorters who had suffered neurosis during their service with an Australian mail communications organization had a greater frequency of certified absences in the preceding 2½ years, but no greater frequency of uncertified absences [26].

Labour turnover

Labour turnover is costly to the employer in terms of wasted training resources and work experience, and is potentially costly to the individual in terms of disruptive career pattern, attendant social disruptions such as loss of peers, a break in income, insecurity for the family, and the risk of unemployment. Attempts to understand the causes of labour turnover have largely concentrated on the relation between occupational attitudes and labour turnover. Porter and Steers [27] concluded that job satisfaction is consistently and universally related to labour turnover but no such relation was found by Talaachi [28]. Pettman [29] suggests that although job dissatisfaction may be a sufficient condition for high labour turnover, it is not a necessary condition.

The relation between the mental health of an employee and labour turnover has only rarely been studied. Some suggestive evidence that neurotic workers might be more prone to change their jobs than those enjoying normal mental health was found by Cherry who retrospectively examined data from the National Survey of Health and Development and found that young men and women who had several jobs before the age of 18 had more psychiatric problems between ages 18 and 25 than more occupationally stable young workers [30]. A more recent study by the present author [31] showed that psychiatric score was twice as high in men and women who subsequently left the organization within the next 12 months than in those who stayed. It seems that mental health is just as important as occupational attitudes in determining labour turnover.

Other occupational consequences

There are indications that the mental health of an employee may be an important determinant of performance, relations with colleagues and management and accidents. Further studies in these areas are required.

Domestic consequences of mental illness

Mental illness may cause a variety of problems in the social domains of marriage, finance, housing, family life and social life. For example, continued depression or irritability may place a great strain on the understanding and tolerance of a spouse, eventually leading to marital disharmony if prolonged, this may cause more serious disruption of the relationship and may occasionally lead to divorce, with far-reading long-term sequelae. The depressed person may become unable to manage his financial affairs, bills may remain unpaid and letters unanswered. Essential house repairs may not be carried out, leading to further deterioration in the property which could easily have been prevented. Parental illness can lead to conduct disorders and emotional problems in the children. (Children are affected if both parents are depressed, but usually remain unscathed as long as one parent is well and fully functional.) There is a tendency to withdraw from friendships when depressed, thus losing opportunities for social support. All these social consequences form a vicious circle which in turn acts as further stress on the individual, maintaining the illness [24].

Treatment of depression

By far the commonest disorder that the occupational doctor may be called upon to treat is depression. Studies in general practice and occupational settings have shown that while up to 50 per cent of illnesses resolve spontaneously within 12 months, the other 50 per cent remain unwell beyond that time [1,32]. Factors predicting chronicity include the initial severity of the illness, and unchanging adverse social conditions [32].

The first step in treatment of any disorder is to detect it. The general practitioner detects only a half to two-thirds of psychiatric illness presenting in his surgery, and probably such rates of detection may also occur in the occupational health services. Should this level of detection be improved? Is the outcome of conspicuous psychiatric morbidity better than that of hidden morbidity? Goldberg and Blackwell showed that patients with 'hidden' illnesses have as many symptoms as those with conspicuous 'illnesses', and that hidden illnesses do not have a better prognosis [33]. Johnstone and Goldberg [10] showed that if a family doctor were made aware of these hidden illnesses, then the patients were more likely to improve more quickly and would have fewer symptoms when seen at follow-up a year after initial consultation.

Active listening is an important part of the therapeutic process. The patient often feels better after the opportunity to review and express emotions associated with events in his life. Simple supportive psychotherapy consists of the various ingredients of active listening, allowing ventilation of feelings, particularly anger, frustration or grief that has previously been suppressed, reassurance, explanation and advice.

Antidepressants

Antidepressants are effective in the treatment of moderate and severe depressive illnesses, and can usually effect recovery within 6 weeks. There are four major classes of antidepressant drugs: the tricyclics, second generation antidepressants, monoamine oxidase inhibitors (MAOIs), and tryptophan.

Tricyclics

Tricyclic antidepressants have been in clinical use for 30 years and they are still the most important antidepressants. They have characteristic effects on a number of central and peripheral synapses. According to the monoamine theory of affective disorders, the antidepressants relieve depression by correcting a deficiency in central monoaminergic neurotransmission; this is achieved by an increase in the synaptic concentrations noradrenaline and/or by 5-hydroxytryptamine (5-HT) via reuptake blockade.

More recently it has been proposed that a more relevant common mode of action of these drugs is the down-regulation of central adrenoreceptors; the development of this effect has the same time-course as the development of the antidepressant effect, and antidepressants of other classes also share this effect. The blockade of central or adrenoreceptors, muscarinic and histamine receptors is believed to be responsible for the sedative effects of tricyclic antidepressants; the blockade of these receptors in the periphery results in characteristic autonomic side-effects of dry mouth, blurred vision, constipation and urinary retention. It is important to administer a therapeutically effective dose, to warn in advance of the usual autonomic side-effects, and to recommend that sedative drugs be taken at bedtime. It is also recommended that antidepressant medication should be taken for at least 3 months after symptoms remit because of the increased risk of relapse in the months following a depressive episode. A large number of tricyclic antidepressants are available on the market. *Imipramine* is the prototype drug with established effectiveness. *Desipramine* is less sedative than imipramine and has more marked sympathomimetic activity. *Amitriptyline* is a widely used and effective drug with marked sedative and peripheral side-effects. *Clomipramine* does not cause appreciable sedation, and it has been claimed to have beneficial effects in obsessional states.

Atypical

Atypical antidepressants differ structurally from the tricyclic antidepressants, and they also have varied pharmacological properties. Most are relatively free from peripheral anticholinergic and cardiotoxic side-effects, so have been especially recommended for the treatment of elderly patients and for those suffering from heart disease. However, clinical experience with these drugs is still rather limited compared to the tricyclic antidepressants. *Maprotiline* can cause convulsions, *mianserin* and *nomifensine* can produce potentially lethal blood dyscrasias, and *trazodone* can produce sedation, orthostatic hypertension and impairment of male sexual function.

Monoamine oxidase inhibitors

The monoamine oxidase inhibitors are an old-established class of antidepressant; their use is much more limited than the tricyclics because they are both less effective and more toxic. The antidepressant action of these drugs has been directly related to their main pharmacological action—the inhibition of the enzyme monoamine oxidase, which results in an increased concentration of cerebral monoamines. There are three monoamine oxidase inhibitors in clinical use; *tranylcypromine*, *phenelzine*, and *isocarboxazid*. They are used as a second line of treatment for patients whose depression does not respond to treatment with tricyclic antidepressants, for some atypical forms of depression including people with marked hypochondriacal features; and for phobic anxiety states. They are potentially toxic drugs, and side-effects are common and troublesome, including agitation, hypomania, convulsions, orthostatic hypertension, and rarely, hepatocellular jaundice. Drug interactions are important. The potentiation of the effects of sympathomimetic amines can lead to a hypertensive crisis (as in the cheese reaction following the ingestion of tyramine-containing foodstuffs such as cheese, broad beans, pickled herrings, avocado pears, Marmite and Chianti).

Treatment of anxiety

The treatment of anxiety is principally behavioural.

Benzodiazepines

Benzodiazepines may be used in the treatment of short-lived situational anxiety, but their use in the management of anxiety neurosis is more controversial, because of the likelihood of addiction, and because they do not effect a cure. The current recommended practice is to use these drugs for only a limited period (2–4 weeks) during which the patient will become more amenable to other forms of treatment, behavioural or social, for example. *Diazepam*, a long-acting drug may be more suitable for the management of persistent high levels of anxiety, whereas shorter acting drugs such as *lorazepam* or *oxazepam* can be used for the management of episodic or situational anxiety. Several benzodiazepines are suitable hypnotics: *triazolan* and *temazepam* are short-acting drugs suitable to induce sleep and relatively free from hangover effects during the daytime, whereas longer acting drugs such as diazepam, *nitrazepam* and *flurazepam* can both induce sleep and maintain sleep throughout the night. Almost all the side-effects of the benzodiazepines are related to their actions on the central nervous system (CNS). Daytime sedation, drowsiness, impairment of psychomotor performance can occur both in conjunction with antianxiety treatment and the use of long-acting hypnotics ('hangover effect'). Increased hostility and aggressive behaviour can occur as a rare but unpleasant complication. Physical dependence is common, especially in association with large dose long-term medication.

Withdrawal symptoms

Withdrawal symptoms are common; these are more severe following the use of short-acting potent drugs (such as lorazepam) than long-acting drugs (such as diazepam). They include increased anxiety, rebound insomnia, dysphasia, anorexia; rarely more severe symptoms (severe agitation, panic, depression, delusions, hallucinations, convulsions) can occur. These withdrawal symptoms can take several weeks to settle. Benzodiazepines should therefore be withdrawn *extremely slowly* if dependence is a problem. They are relatively safe in overdosage, unless used in combination with other CNS depressants.

Behavioural treatment

Behavioural treatments are environmental manipulations based on principles derived from experimental psychology in order to bring about changes in overt behaviour, thoughts or autonomic functions and are useful in a wide range of disorders including anxiety, depression, obsessional states, sexual problems and abnormal illness behaviour. Specific techniques include relaxation, desensitization, flooding, biofeedback, response prevention, cognitive therapy, and social skills training.

Relaxation

In relaxation therapy, the patient is taught to recognize the signs of tension in his body and how these can be altered by voluntary control. He is then given an audiotape which describes exercises to relax the body and the mind and takes this home to practise. The technique of relaxation must become sufficiently rehearsed so that it can be applied even in situations which have previously caused anxiety.

Desensitization

In desensitization treatment, the patient constructs a hierarchy of anxiety-provoking stimuli. These stimuli are then presented to the patient, starting with the most innocuous. As each stimulus is mastered, the patient progresses to the next one. The stimuli may be presented either in imagination, with the patient imagining anxiety-provoking scenes while in a relaxed state, or *in vivo*, with the patient encountering real life anxiety-provoking stimuli, starting with those which provoke least anxiety. The latter method is more effective.

Flooding

Whereas desensitization is a gradual approach, flooding is more rapid, but it is difficult to persuade patients to undergo it. The patient is straightaway confronted with a situation at the top of his hierarchy and is required to remain in that situation until his anxiety has spontaneously dissipated ('habituation').

Biofeedback

Biofeedback is often used as an adjunct to relaxation training, but can also be used in the treatment of specific symptoms such as tension headache, or hypertension. An auditory or visual 'feedback' signal is used to bring visceral or neuromuscular activities, of which we are normally unaware, into consciousness so that some voluntary control may become possible.

Response prevention

Response prevention is used in the treatment of compulsive rituals. The patient feels compelled to perform some act such as washing hands but is firmly persuaded from doing so. This must be maintained for prolonged periods to be effective. It is best performed while the person is repeatedly faced with a situation in which the compulsion occurs, such as touching dirt. In this way it is similar to flooding—the patient is faced with intense anxiety which gradually wanes as it is not relieved by the habitual means of giving in to the compulsion.

Cognitive therapy

Cognitive therapy is of particular value in depressive states. It can be used in conjunction with antidepressants, and can be especially useful if antidepressants are ineffective. Cognitive therapy aims to correct the gloomy thoughts that depressed patients have about themselves, their circumstances and their future and also to demonstrate to the patient that he is capable of tackling difficulties which, because of his low self-esteem he would not normally face. Thus the emphasis is often on very practical goals.

Social skills training

Social skills training is usually undertaken in groups, and aims to provide explicit training of a full repertoire of appropriate sophisticated social responses. Mock social situations are set up, and the patients practise by role-playing, after the therapist has provided a model of the appropriate behaviour. The patient's performance may be videotaped, so that particular problems can be discussed and corrected in debriefing sessions. Assertion training is done in a similar way.

Conclusions: implications for occupational health services

The high prevalence of so-called minor psychiatric morbidity, largely made up of the depressions and anxiety states, found in working populations raises the issues of how far employers should take responsibility for the mental health of their workforce, and whether occupational medical services should concern themselves with the detection, treatment and prevention of minor psychiatric morbidity. Research studies have shown that at least half of these illnesses last longer than 12 months and are associated with handicapping and costly consequences such as increased rates of sickness absence and labour turnover [1,31]. Furthermore, there are indications that the mental health of an employee may be an important determinant of his performance [34], job satisfaction [35], relations with colleagues and management [36], and accidents [37]. It would seem therefore that there are persuasive economic arguments why employers should allocate resources for occupational medical services to make adequate provision for the detection, treatment and prevention of the high prevalence of minor psychiatric disorders, despite the fact that it is not a very common diagnostic category on sickness absence certificates.

Pre-employment screening to avoid employment of potentially unstable individuals cannot provide a solution because most minor psychiatric illnesses are discrete episodes with no premonitory signs that might be detected perhaps years before the development of symptoms. If pre-employment screening did detect the presence of a depressive illness or an anxiety state, to exclude such a person from the workforce might be to lose a potentially productive and able worker who might not have another episode of psychological disturbance.

Compulsory medical retirements are also not a good solution. They are costly exercises, and include the expense of finding and training a replacement. Prevention, early detection and treatment are the most economic and the most constructive solutions.

Traditionally, employers have argued that mental health problems are the responsibility of the individual and not of the company. Underlying this belief is the assumption that mental health problems arise independently of the job. However, we have seen that the origin of mental ill-health is multifactorial, and work is as important as domestic life in carrying the potential for both stresses and supports, and in bearing the consequences of illness.

The last few decades have seen a slow, albeit patchy, often reluctant but nonetheless inexorable move towards appreciating the importance of mental health in the workplace. While those concerned with health and safety have largely hitherto concentrated on the physical and toxicological environment, there is now an increasing upswell of interest in 'stress', 'burnout', depression and anxiety at work, with corresponding interest in its causes and its consequences for the individual, the employer and society at large.

Concern for mental health at work has been much greater over the last century in the United States than in this country. Since the First World War, American industry has gradually and steadily responded to the challenge of mental health at work and has explored differing innovative programmes which, although by no means universal are sufficiently common to make a substantial contribution to the general climate of opinion surrounding mental health issues at work. On the other hand, in this country, Sir Aubrey Lewis' vision of psychiatrists visiting the large workplaces in their area is completely unfulfilled. Psychiatry as a whole seems to be showing much less interest in the possibilities for primary prevention which industry offers than it did in the years immediately following the Second World War.

There is a pressing need to set up and evaluate the effectiveness of occupational mental health services and provisions. So far, research in this field is scanty, and there is now no adequate reason why this should remain so. Considerable progress has been made towards resolving the difficulties noted by Lewis in undertaking psychiatric research in occupational settings [38]. Reliable, standardized and structured psychiatric instruments have been developed, that are acceptable to individuals in occupational settings who do not usually perceive themselves as ill, and which function adequately in working environments and are short enough to be used in working hours [33,39–41].

References

1. Jenkins, R. Minor psychiatric morbidity in employed young men and women and its contribution to sickness absence. *British Journal of Industrial Medicine*, **42**, 147–154 (1985)
2. Fraser R. *The Incidence of Neurosis among Factory Workers*. HMSO (Industrial Health Research Board report No. 90), London (1947)
3. Jenkins, R. Minor psychiatric morbidity in employed men and women and its contribution to sickness absence. Preliminary communication. *Psychological Medicine*, **10**, 751–757 (1980)
4. Jenkins, R., Macdonald, A., Murray, J. and Strathdee, G. Minor psychiatric morbidity and the threat of redundancy in a professional group. *Psychological Medicine*, **12**, 799–807 (1982)
5. Macbride, A., Lancee, W. and Freeman, S.J.J. The psycho-social impact of a labour dispute. *Journal of Occupational Psychology*, **54**, 125–133 (1981)
6. Goldberg D. and Huxley, P. *Mental Illness in the Community: the Pathway to Psychiatric Care*. Tavistock Publications, London (1980)
7. Henderson, S., Byrne, D.G. and Duncan Jones, P. *Neurosis and the Social Environment*. Academic Press, Sydney (1981)
8. Taylor, P.J. Aspects of sickness absence. In *Current Approaches to Occupational Health* (ed. A. Ward Gardener) John Wright and Sons, London, pp. 322–338 (1979)
9. Rosen, B.M., Locke, B.Z., Goldberg, I.D. and Babigian, H.M. Identifying emotional disturbance in persons seen in industrial dispensaries. *Mental Hygiene*, **54**, 271–279 (1970)
10. Johnstone, A. and Goldberg, D. Psychiatric screening in general practice. *Lancet*, **i**, 605–608 (1976)
11. Payne, R. Organisational stress and social support. In *Current Concerns in Occupational Stress* (eds C.L. Cooper and R. Payne) John Wiley and Sons, Chichester, New York pp. 269–298 (1980)

12. Davidson, M.J. and Cooper, C.L. A model of occupational stress. *Journal of Occupational Medicine*, **23**, 564–574 (1981)
13. Cobb, S. and Rose, R.H. Hypertension, peptic ulcer and diabetes in air traffic controllers. *Journal of Australian Medical Association*, **224**, 489–492 (1973)
14. French, J. and Caplan, R. Organisational stress and individual strain. In *The Failure of Success* (ed. A.J. Harrow) Amacon, New York pp. 31–66 (1972)
15. Cox, I. Repetitive work. In *Current Concerns in Occupational Stress* (eds C.L. Cooper and R. Payne) John Wiley and Sons, Chichester, New York, pp. 23–42 (1980)
16. Davidson, M.J. and Veno, A. Stress and the policeman. In *White Collar and Professional Stress* (eds C.L. Cooper and J. Marshall) John Wiley and Sons, London (1980)
17. McMichael, A.J. Personality, behavioural and situational modifiers of work stresses. In *Stress at Work* (eds C.L. Cooper and R. Payne) John Wiley and Sons, Chichester, New York, pp. 147–147 (1978)
18. Cooper, C.L. and Marshall, J. Occupational sources of stress. A review of the literature relating to coronary heart disease and mental ill health. *Journal of Occupational Psychology*, **49**, 11–28 (1976)
19. Freudenberger, H.J. Staff burnout. *Journal of Social Issues*, **30**, 159–165 (1974)
20. McDermott, D. Professional burnout and its relation to job characteristics, satisfaction and control. *Journal of Human Stress,* **Summer**, 79–85 (1984)
21. Cherniss, C. *Staff Burnout: Job Stress in the Human Services.* Sage Publications, Beverly Hills, California (1980)
22. Eastwood, M.R. and Trevelyan, M.H. Relationships between physical and psychiatric disorders. *Psychological Medicine*, **2**, 363–372 (1972)
23. Simms, A. and Prior, P. The pattern of mortality in severe neuroses. *British Journal of Psychiatry*, **133**, 299–305 (1978)
24. Jenkins, R. and Shepherd, M. Mental illness and general practice. In: *Mental Illness: Changes and Trends* (ed. P. Bean) John Wiley and Sons, Chichester (1983)
25. Wyatt, S. *A Study of Women on War Work in Four Factories.* London: HMSO (MRC Industrial Health Research Board report No. 88) (1945)
26. Ferguson, D. A study of neurosis and occupation. *British Journal of Industrial Medicine*, **30**, 187–198 (1973)
27. Porter, L.W. and Steers, R.M. Organisational, work and personal factors in employee turnover and absenteeism. *Psychological Bulletin*, **80**, 151–176 (1973)
28. Talachi, S. Organisation size, industrial attitudes and behaviour: an empirical study. *Administration Science Quarterly*, **5**, 398–420 (1960)
29. Pettman, B.O. Some factors influencing labour turnover: a review of research literature. *Industrial Relations Journal*, **4**, 43–61 (1973)
30. Cherry, N. Persistent job changing—is it a problem? *Journal of Occupational Psychology*, **49**, 203–221 (1976)
31. Jenkins, R. Minor psychiatric morbidity and labour turnover. *British Journal of Industrial Medicine*, **42**, 534–539 (1985)
32. Mann, A.H., Jenkins, R. and Belsey, E. The twelve month outcome of patients with neurotic illness in general practice. *Psychological Medicine*, **11**, 535–550 (1981)
33. Goldberg, D.P. and Blackwell, B. Psychiatric illness in general practice. A detailed study using a new method of case identification. *British Medical Journal*, **ii**, 439–443 (1970)
34. Markowe, M. Occupational psychiatry: an historical survey and some recent researches. *Journal of Mental Science*, **99**, 92–102 (1953)
35. Kasl, S.V. Mental health and the work environment: an examination of the evidence. *Journal of Occupational Medicine*, **15**, 509–518 (1973)
36. Tredgold, R.F. Mental hygiene in industry. In *Modern Trends in Psychological Medicine* (ed. N.G. Harris) Butterworths, London (1948)
37. Adler, A. The psychology of repeated accidents in industry. *American Journal of Psychiatry*, **98**, 99–101 (1941)
38. Lewis, A. Research in occupational psychiatry. *Folia Psychiatrica, Neurologica et Neurochirurgica Neerlandica*, **56**, 779–786 (1953)

39. Goldberg, D.P. *Manual of the General Health Questionnaire*. Windsor, NFER (1978)
40. Goldberg, D.P., Cooper, B., Eastwood, M., Kedward, H.B. and Shepherd, M. A standardised psychiatric interview for use in community surveys. *British Journal of Preventive and Social Medicine*, **24**, 18–23 (1970)
41. Jenkins, R. Sex differences in minor psychiatric morbidity. *Psychological Medicine Monograph Supplement No. 7*, 1–53 (1985)

Chapter 6

Vibration

B. Malerbi

Introduction: definition

Vibration is an oscillating motion about a central fixed position. However, in the context of occupational health practice, 'vibration' implies the motion of a solid object, and concern is centred on vibrational frequencies and amplitudes which are likely to affect the comfort and well-being of a person exposed to them. Vibrations at frequencies outside the audible range may also be of interest where they give rise to audible harmonics, or where they can be eliminated from the investigation. Only vibrational frequencies up to 80 Hz are considered to be harmful to the whole body, whereas for the hand and arm, frequencies up to 2 kHz may be damaging.

Properties

Figure 6.1 illustrates the simplest form of oscillation when a particle is displaced from its resting position (●). Its displacement (x) increases to a maximum as its velocity (v) falls to zero. As the particle passes through the midpoint of its oscillation, where the displacement is zero, its velocity reaches a maximum. Since the velocity is not constant, the particle must be accelerating. The acceleration (a) varies in proportion to the displacement, but in the opposite direction, since the particle decelerates towards maximum displacement, and accelerates towards minimum displacement. At this point, where maximum velocity occurs, acceleration is zero before becoming negative again.

The three parameters are related by differentiation/integration:

Parameter	*Unit*	*Symbol*
Displacement	(m)	x
Velocity	(m/s)	dx/dt
Acceleration	(m/s^2)	d^2x/dt^2

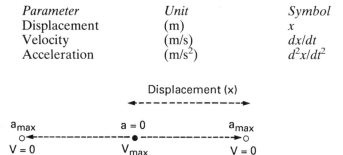

Figure 6.1 Particle oscillating with displacement (x), velocity (v), and acceleration (a)

The relationship between acceleration and displacement is:

$d^2x/dt^2 \propto -x$ which is the equation for simple harmonic motion
$\therefore d^2x/dt^2 = -cx$ where c is a constant equal to $(2\pi f)^2 = 40f^2$

Free vibrations

A body free to move in space is said to have 6 degrees of freedom, 3 degrees in linear directions and 3 degrees in twisting directions. In practice, movement (for example, of a machine) is restricted to prevent damage. Theoretical calculations are based on a simple vibrating system with one degree of freedom, namely a mass attached to a spring. This represents a machine on resilient mountings.

Natural frequency

If the mass is disturbed from its resting position by extending or compressing the spring, and allowed to vibrate freely, in theory it will oscillate indefinitely, at a frequency known as the natural frequency (f_0) of the mass/spring system.

From the equation for simple harmonic motion (above), the natural frequency (f_0) of a body can be calculated, that is:

$$f_0^2 = a/40x$$

Damping

In practice, there is always some dissipation of energy due to the stiffness of the spring, or to friction between contacting surfaces, so that the oscillation decays exponentially, and the mass eventually comes to rest. This effect is called 'damping', and can be deliberately applied as a means of controlling vibrations. Most materials have some inherent damping, therefore when constructing a system which will be subject to vibration, it is desirable to select materials with high internal damping, and design the system to incorporate high structural damping.

Damping capacity

Internal damping is measured by recording the percentage loss of energy per cycle after a freely suspended sample bar of material has been set into vibration by an impact. It is a measure of the material's capability to absorb vibrations, and is a means of comparison when selecting structural materials. For example, the damping capacities [1] for steels, aluminium and brass are <5 per cent, whereas the values for manganese–copper alloys are >40 per cent.

Critical damping

This is just sufficient damping to prevent free oscillations, so that the system comes to rest in the shortest possible time without overshooting. A typical example is the movement of an analogue indicating instrument.

Damping ratio (D)

The degree of damping in a vibrating system, expressed as a fraction of the critical damping, is known as the damping ratio. Its value depends on the construction of the system, for example, on the type of joint. One-piece or welded structures have lower damping ratios (0.01–0.02) than riveted or bolted ones (0.03–0.05) [1].

Forced vibration

In an industrial setting, machines are subject to a regular, applied force which is usually rotational or reciprocal. A machine will vibrate at the forcing frequency (dependent on the rpm of the motor) until the force is removed when the machine is switched off. Vibrations will then revert to the natural frequency of the system, with gradually reducing amplitude, before coming to rest.

Resonance

Should the forcing frequency (f) coincide with the (undamped) natural frequency of the system (f_0), the phenomenon of resonance occurs, whereby the amplitude of vibration becomes very great (as shown in *Figure 6.2* where $f/f_0 = 1$), and may result in damage to the machine. This situation occurs momentarily during start-up and run-down of a machine at the point where the movement changes from being stiffness-controlled at the lower frequencies, to being mass-controlled at the higher frequencies. At that point there is no such control, as the two effects cancel each other out. The only control that can be effective is applied damping.

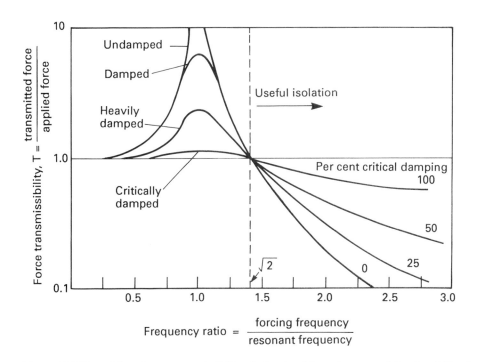

Figure 6.2 Effect of damping on transmissibility of vibrational force (courtesy of Bruel and Kjaer)

Health effects of vibration

Infrasound is defined as sound below 20 Hz, and is said to be inaudible. However, since the hearing threshold rises from 40 dB at 32.5 Hz to 140 dB at 1 Hz, it is thought that if the intensity were high enough, infrasound would be heard, but because it would also be felt, the two effects would be difficult to distinguish.

Where the vibrating object which produces the infrasound is actually in contact with the body, either because a person is standing, sitting or lying on it, or is holding a vibrating tool, there may be unpleasant, or even harmful effects.

The various parts of the body have their own natural frequencies. Suspended organs have higher natural frequencies than structural parts, but if the body is in contact with vibrations at frequencies which coincide with these natural frequencies, those parts of the body will resonate. In *Figure 6.3* the body is shown as a mass/spring system with incorporated damping.

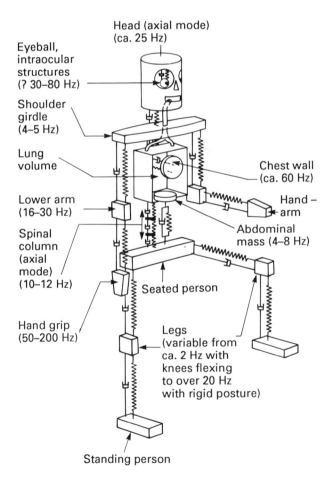

Figure 6.3 Mechanical system representing the human body (courtesy of Bruel and Kjaer)

Prolonged whole body vibration has been reported to cause giddiness, headaches, nausea, spinal disorders, varicose veins, blurred vision, lung damage, rectal bleeding, haematuria, weight loss, and heart failure in extreme cases. Each of these effects is associated with vibration at a particular (resonant) frequency. For example at 60–70 Hz, the eyeballs resonate, affecting vision; 10–20 Hz (depending on activity) is the frequency of the α-wave of the brain; at <1 Hz, motion sickness is induced. Most effects are reversible when exposure ceases, and are frequently experienced by shipboard personnel, lorry-drivers and tractor-drivers.

When vibration is localized in the hands over long periods, it can cause Raynaud's phenomenon, or vibration-induced white finger (VWF) as it is more commonly called when induced in this way.

Hand/arm vibration arises when swaging, spinning, fettling, using a grinding wheel, chainsaw, or pneumatic tool, that is, when either pressing the work-piece to a rotating tool, or pressing the tool to the work-piece. About half those regularly using hand-held tools suffer from vibration-induced white finger (VWF) to some degree. It is inadvisable for heavy smokers to use vibratory tools, since nicotine similarly reduces the blood supply to the extremities. VWF has been added to the list of Prescribed Industrial Diseases [2] and is disease No. A11.

Measurement of vibration

Vibration amplitude can be measured in terms of displacement, velocity or acceleration. For occupational health purposes, acceleration (a) is preferred, because it represents the vibrational force applied, also it is numerically greater than displacement (x) or velocity (v) as shown in the relationships:

$$a = (2\pi f)^2 \cdot x = 2\pi f \cdot v$$

where a, v and x are in comparable units, and f is in Hz.

If any one of these parameters is measured, the others can be calculated from the above relationships; for a displacement of 0.001 m at 10 Hz, the corresponding velocity would be 0.065 m/s and acceleration 4 m/s^2.

Random vibration is assumed to comprise many sinusoidal components, at a range of frequencies, therefore the use of filters allows the principles of simple harmonic motion (on which vibration theory is based) to apply. For this reason, vibration is usually measured via octave, or one-third octave filters.

An accelerometer is a piezoelectric transducer, capable of converting the acceleration of vibration into a small alternating voltage which is proportional to the acceleration over a wide range of frequencies (*Figure 6.4a*). Piezoelectric materials set up a voltage when they are deformed, as is the case when the accelerometer is held against a vibrating surface. The piezoelectric element may be deformed either in compression, or in shear and the many different types of accelerometer are characterized by their mass, sensitivity and dynamic range. By this means, vibrations can be measured over very wide dynamic and frequency ranges (160 dB and 20 kHz respectively), therefore, as with acoustic measurements, logarithmic scales are used. The accelerometer can be connected either to a vibration meter incorporating an integrator, or to a separate integrator replacing the microphone on a sound level meter (*Figure 6.4b*).

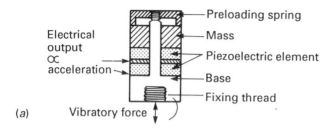

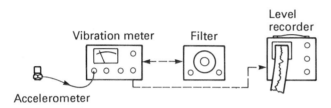

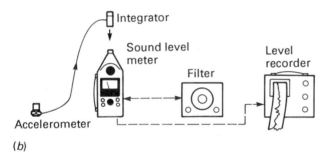

(b)

Figure 6.4 (a) Configuration of compression type accelerometer. (b) Alternative systems for measuring vibration (courtesy of Bruel and Kjaer)

For measurement of continuous vibration, the rms value is relevant, since it is directly related to the vibration energy and therefore to the damage potential, whereas the peak value indicates the maximum acceleration level of single impacts.

From Ohm's Law, $V^2 = W \cdot R$; thus acceleration energy $\propto V^2$
∴ accel. energy/ref. energy = $[(\text{accel. voltage})/\text{ref. voltage})]^2$
accel. level (dB) = $10\log_{10}[(\text{accel. voltage})/(\text{ref. voltage})]^2$
= $20\log_{10}[(\text{accel. measured})/(\text{ref. accel.})]$

When using a sound level meter (SLM) as the indicating device (*Figure 6.5*), the dB readings need not be converted to units of acceleration where comparison of measurements before and after remedial treatment of a vibrating machine is the objective. However, for assessment of human response to vibration, acceleration amplitude in m/s² must be calculated from acceleration levels in dB, using the standard reference acceleration of 10^{-6} m/s² in the above formula.

Figure 6.5 Measuring acceleration using the sound level meter (SLM) (courtesy of Bruel and Kjaer)

Evaluation of vibration

Whole-body vibration (WBV)

Whole-body vibration (WBV) is vibration transmitted to the body usually via the buttocks or feet, occasionally via other points of contact where the body is reclining horizontally. There is a simple method of assessing repeated shocks and intermittent vibration in terms of a vibration dose value (VDV), calculated from the acceleration, frequency-weighted over the range 0.5–80 Hz, and integrated over the exposure period.

For continuous, steady vibration, the estimated vibration dose value $(m/s^{1.75})$ is

$$[(1.4a)^4 \times b]^{14} \text{ where } a = \text{rms acceleration } (m/s^2)$$
$$\text{and } b = \text{duration (s)}$$

This relationship (*Figure 6.6*) holds provided that the crest factor is <6.

Crest factor − peak/rms (weighted acceleration values)

No opinion is given on the precise relationship between vibration dose value and risk of injury, but severe discomfort is usually experienced around $15 \, m/s^{1.75}$, and it is assumed that as the exposure increases, so does the risk of injury.

Like noise, vibration may have a pleasing or unpleasant effect according to the circumstances, but in general no discomfort is felt at accelerations $<0.315 \, m/s^2$, and for 50 per cent of those exposed, the threshold of perception is $0.015 \, m/s^2$. Hand control and manipulation, also vision, start to be affected above $0.5 \, m/s^2$.

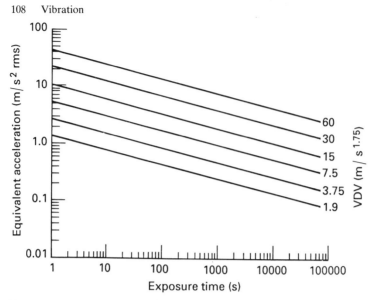

Figure 6.6 Estimation of vibration dose value from rms acceleration and exposure time, where vibration is steady and continuous (from BS 6841:1987, *Guide to Measurement and Evaluation of Human Exposure to Whole Body Mechanical Vibration and Repeated Shock*, reproduced by permission of BSI)

The frequency range considered does not extend beyond 80 Hz, because at higher frequencies the body's response is influenced by so many factors, and as illustrated in *Figure 6.3*, whole body effects are associated with frequencies up to 80 Hz.

Motion sickness, which is induced by vibration at frequencies between 0.1 and 0.5 Hz, is considered separately. The likelihood of its occurrence increases with increasing duration of exposure, up to a point (after a few hours or days) where adaptation occurs. One method of evaluating such exposure is:

$$MSDV_z - (a^2 \cdot t)^{1/2}$$

where $MSDV_z$ = motion sickness dose value (z axis)
and t = period of exposure (s)

Hand-transmitted vibration

Hand-transmitted vibration exposure is measured and evaluated over a frequency range of 8–1000 Hz in three axes. The rms acceleration obtained can be expressed as a frequency-weighted value, or in the form of a frequency analysis. For intermittent exposure, or fluctuating vibration amplitudes, an integrating meter must be used to estimate the cumulative value over an 8 h day.

For normal tool usage, it is unusual for finger blanching to occur where the frequency-weighted rms acceleration remains below 1 m/s², and it is assumed that the daily exposure time required to produce such symptoms is inversely proportional to the square of the frequency-weighted acceleration. This relationship is shown in *Table 6.1*.

If the hand is in direct contact with the vibrating tool handle, then the accelerometer should be attached to the handle when measurements are taken, but if there is a protective resilient layer between the hand and the handle, then a thin, moulded, metal sheet may be inserted in contact with the hand, and the

Table 6.1 Frequency-weighted vibration acceleration magnitudes (m/s² rms) which may be expected to produce finger blanching in 10 per cent of persons exposed (from BS-6842:1987, *Guide to Measurement and Evaluation of Human Exposure to Vibration Transmitted to the Hand,* reproduced by permission of BSI. Complete copies can be obtained from them at Linford Wood, Milton Keynes)

Daily exposure	Lifetime exposure					
	6 months	1 year	2 years	4 years	8 years	16 years
8 h	44.8	22.4	11.2	5.6	2.8	1.4
4 h	64.0	32.0	16.0	8.0	4.0	2.0
2 h	89.6	44.8	22.4	11.2	5.6	2.8
1 h	128.0	64.0	32.0	16.0	8.0	4.0
30 min	179.2	89.6	44.8	22.4	11.2	5.6
15 min	256.0	128.0	64.0	32.0	16.0	8.0

(1) With short duration exposures, the magnitudes are high, and vascular disorders may not be the first adverse symptom to develop.
(2) The numbers in the table are calculated, and the figures behind the decimal points do not imply an accuracy which can be obtained in actual measurements.
(3) Within the 10 per cent of exposed persons who develop finger blanching, there may be a variation in the severity of symptoms.

accelerometer attached to that. A similar insert will be necessary when measuring whole body vibration if the person normally sits on a cushion or stands on a mat, otherwise the accelerometer is attached to the floor or stool close to the point of contact with the body.

Measurements are made in the directions of the *x*, *y* and *z* axes, for whole-body vibration in one-third octave bands, and for hand-transmitted vibration in octave or one-third octave bands. Weighting factors for octave and one-third octave bands are listed in the standards and are not the same as the 'A' weightings used when measuring noise.

For vertical movement, the human body is most sensitive to vibrations between 4 and 8 Hz, and for horizontal movement, sensitivity is at a maximum between 1 and 2 Hz, whereas the hand/arm system is most affected by vibration frequencies between 4 and 16 Hz.

When von Békésy analysed the ability of the human body to attenuate vibration, he found that for vibration at 50 Hz, the attenuation from the feet to the head was 30 dB, and from the hand to the head, about 40 dB.

Vibration control

As with control of noise (*see* Chapter 7), the first priority is to study the source of vibration, and ensure that the machine or tool is well-maintained. Rotating parts may become unbalanced, resulting in worn bearings, and excessive vibration. Flexible shaft couplings, which eliminate vibration and accept misalignment, have been developed.

Reduction of hand/arm vibration can be achieved with antivibration handles on hand-held tools, by substituting turning or milling for grinding, and by holding the work-piece in a clamp. Swedish manufacturers have designed a pneumatic chipping hammer which is recoilless, and thus cuts vibration to a minimum. Suitable gloves, which are flexible and warm, should be worn in cold conditions. Although gloves are a hindrance in manipulating a tool, and their usefulness as vibration isolators is limited because increased grip pressure may be necessary, it is important that the hands are kept warm.

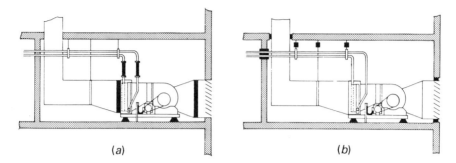

Figure 6.7 Alternative means of machine isolation (courtesy of Sound Research Laboratories)

For control of whole-body vibration in vehicles, efficient suspension of the cab or driver's seat is essential.

Having reduced the vibrations at source where possible, the next objective is to prevent the remaining vibrational energy from being transferred elsewhere. In industry, where vibration is transmitted from a machine via the floor to the feet of the operator standing nearby, the remedy is efficient vibration isolation between the machine and the floor.

To complete the isolation, associated ducting, pipes and conduits must be flexibly connected to the machine, or supported along their length by resilient hangers or clamps (as illustrated in *Figure 6.7*), otherwise vibrations will travel by those routes to the building structure, or to other large surfaces such as radiators or tanks, which will act as sounding-boards, and introduce noise into areas far removed from the vibration source. Where they pass through a wall, an oversized hole should be made, and the surrounding gap plugged with mineral wool and sealed with mastic, if the gap would create a noise problem.

Vibration isolation

The magnitude of the vibrational force transmitted into the structure is directly proportional to the displacement amplitude of the machine on its mounting. In introducing vibration isolation, the objective is to reduce the force on the machine mounting, and prevent transmission to the building structure. This can be achieved by interposing a resilient material.

The theory of vibration isolation is based on the relationships illustrated in *Figure 6.2*.

Vibration isolators

Machines, fans, pumps etc. should always be mounted on vibration isolators (antivibration mounts). When in position, with the machine at rest, isolators deflect by an amount which is governed by their stiffness and the weight that they support. This is known as the static deflection, and the natural frequency of the system depends solely on this parameter, regardless of the load applied, therefore isolators are selected on the basis of the desired static deflection (d), calculated as follows:

$$f_0 = 15.8/\sqrt{d} \text{ Hz} \qquad \text{where } d \text{ is in mm.}$$

Selection of isolators

The first stage is to decide how much transmissibility is acceptable. For this purpose, guide tables derived from practical experience are available [3]. These take into account the type of machine, the sensitivity of the location, and the type of floor. The next stage is to calculate the lowest forcing frequency in the system. The method of mounting must be considered, and the weight of the machine plus additional equipment comprising the total load on the isolators must be estimated. It is also necessary to locate the position of its centre of gravity so that the proportion of the total load on each isolator can be estimated. Suitable isolators which will give the required deflection are selected from the supplier's catalogue.

There are five main classes of isolators as follows:

1. Solid material such as rubber, cork, felt, in the form of a pad or mat, is useful for audiofrequency isolation. These have a high resonant frequency, and compress with a small deflection (up to 5 mm).
2. Neoprene- or rubber-in-shear isolators with deflections up to 15 mm, must be fixed to the machine feet, and usually also to the floor.
3. Preformed glass fibre pads with a neoprene coating are tough, and cover a wide range of duties. The air trapped between the fibres provides viscous damping applicable to shock isolation, in addition to vibration isolation. Natural frequencies down to 8 Hz are controlled by the material thickness rather than the static deflection.
4. Steel springs in the form of a coil are made with static deflections up to 175 mm, and natural frequencies down to 2 Hz. Since unlike the other types they have little internal damping, thus allowing high frequency vibrations to be transmitted along the coils to the floor, a neoprene pad is usually incorporated to prevent metal-to-metal contact.
5. Air springs, or air mounts, are the best means of protecting sensitive equipment such as electron microscopes, lasers etc. from vibrations in the building structure. An air-filled bag acts as a spring, and the air pressure is varied to give the desired static deflection. Natural frequencies down to 1 Hz can be achieved.

Antivibration mounts do not prevent the machine from vibrating; they control the transmission of vibrational energy to the floor beneath, and if they are not selected and fitted correctly, vibrations may be magnified rather than reduced. It is important to ensure that the isolators are not bridged (bypassed). The difference in acceleration readings taken on the floor close to the mounts when they are bridged, and when they are not, is a measure of isolator efficiency.

References

1. *Kempe's Engineers Year-Book*, 91st edn, B4/67. Morgan-Grampian Book Publishing Co. Ltd, London (1986)
2. *DHSS Leaflet* NI.2 April (1987) *Industrial Injuries Disablement Benefit*.
3. Webb, J.D. *Noise Control in Industry*, 2nd edn. Sound Research Laboratories Ltd, Colchester (1978)

Further reading

The following International Standards provide guidance based on practical experience and laboratory experiments, but do not define safe exposure limits:

ISO 2631:1985 *Guide for the Evaluation of Human Exposure to Whole-body Vibration.*
ISO 5349:1986 *Guidelines for the Measurement and the Assessment of Human Exposure to Hand-transmitted Vibration.*

Two similar British Standards have been in the consultative stage for many years under the following titles:

DD 32:1974 *Guide to the Evaluation of Human Exposure to Whole Body Vibration.*
DD 43:1975 *Guide to the Evaluation of Exposure of the Human Hand/Arm System to Vibration.*
NB DD denotes Draft for Development.

Limits in terms of acceleration (m/s^2), with respect to frequency and exposure time, are recommended therein. However, these proposed standards have recently been withdrawn. Replacing them are:

BS 6841:1987 *Guide to Measurement and Evaluation of Human Exposure to Whole Body Mechanical Vibration, and Repeated Shock.*
BS 6842:1987 *Guide to Measurement and Evaluation of Human Exposure to Vibration Transmitted to the Hand.*

Chapter 7

Noise

B. Malerbi

*

Introduction: definitions

The usual definition of noise is 'unwanted sound'. However, this is a subjective definition and must be qualified with respect to the listener, the time and the place. Any sound, regardless of its volume, or characteristics, could be described as noise by someone, sometime or somewhere. Moreover, it is noise only if someone is within earshot. Even for one individual, however, the degree of tolerance will vary according to both time and place, and consequently the definition defies the quantification necessary for the development of comprehensive legislation, and is only relevant when considering whether noise is a nuisance.

Current and proposed legislation is aimed at preventing permanent damage to hearing caused by occupational noise. The legal definition of 'an excessive sound level' refers to the measurable properties of sound, such as its amplitude, frequency and duration but has been the subject of fierce debate, so the term 'excessive' has different values in different countries.

Where the sound level is reduced to meet legal requirements, an added advantage is that communications affecting safety (instructions, warning signals and so on), are more likely to be audible.

Properties of sound

Sound is a form of energy, generated when a surface vibrates and sets the adjacent air molecules into sympathetic vibration, creating pressure fluctuations above and below atmospheric pressure. As each successive molecule collides with its neighbour and rebounds, waves of compression and rarefaction propagate the energy in all directions and through any other elastic medium in its path. The smallest pressure change detectable by the human ear is 2×10^{-5} pascals ($20\,\mu$Pa), used as a baseline for comparative measurements and as the standard hearing threshold at 1000 Hz. [NB 1 Pa $\equiv$ 1 N/m^2]

The oscillating air molecules execute simple harmonic motion, thus the resulting acoustic pressure fluctuations are basically sinusoidal. The graph of a sine wave can trace the change in amplitude of acoustic pressure either with time at a fixed point, as in *Figure 7.1a*, or with distance at a particular instant (*Figure 7.1b*). From this information, frequency and velocity can be derived.

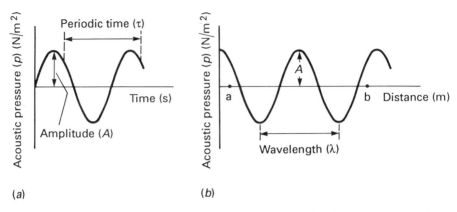

Figure 7.1 (*a*) Change of acoustic pressure with time at a fixed point. (*b*) Change of acoustic pressure with distance at one instant (courtesy of Blackwell Scientific Publications Ltd.)

The simplest type of sound, known as a pure tone, is sound of a single frequency. Most industrial noise comprises many different frequencies from a variety of sources. Some components are fundamental frequencies and associated harmonics, others are of a random nature.

Frequency

The number of complete vibration cycles per second is the frequency of the sound in units of hertz (Hz), and remains constant throughout the propagation path unless the frequency of the sound source changes. The distance between corresponding points on successive waves of compression or rarefaction is the wavelength expressed in metres, and the time taken for each wave to pass a fixed point is the periodic time (*T*) in seconds.

Pitch is the subjective property of a tone which enables the ear to identify its frequency.

Velocity

The velocity of sound depends upon the elasticity and density of the medium through which it is transmitted, and is therefore affected by temperature [1]. Examples of the wide variations at 20 °C are 344 m/s in air, 1410 m/s in water, and 5200 m/s in steel. The basic formula relating to these parameters is:

$$\lambda = cT = c/f \qquad \text{where } T = 1/f,$$

λ = *wavelength (m)*, c = velocity (m/s), f = frequency (Hz)

Wavelength

Since the frequency is constant, and can be measured, or calculated for an identifiable source, and the velocity can be obtained from tables [1], it is possible to calculate the wavelength in any medium and at any temperature.

Audible frequencies

The normal, healthy, human ear can detect and discriminate between a wide range of frequencies—from 20 to 20 000 Hz in a young person—and this is known as the audiofrequency range.

Inaudible frequencies

Vibrations at frequencies below 20 Hz (infrasound) are not perceived as sound in air, but can be felt through any part of the body which is in contact with a solid or liquid vibrating at these frequencies.

Ultrasound is beyond the upper limit of the human audiofrequencies, but some other species can hear ultrasonic frequencies.

Sound power

The sound power (W) of a sound source is the total sound energy emitted per unit time, measured in watts. This property is independent of environmental conditions.

Sound intensity

The sound intensity (I) in watts per square metre (W/m^2) is the quantity of sound energy passing through unit area per unit time, and is relevant only when related to a particular location under stated conditions. It is the sound intensity at the ear which has been shown to govern the degree of hearing loss, therefore measurements are made at a defined distance (r) from the source, usually at the ear of any exposed individual.

The lowest intensity of sound that is detectable by the human ear is $10^{-12} W/m^2$ and is the established 'threshold of hearing'. By contrast, the highest intensity that the ear can bear without feeling pain is between $1 W/m^2$ and $10 W/m^2$, therefore the 'threshold of pain' is in this region.

For a plane source

Where the sound source is encased in a tube of dense material and the sound can travel only in one direction, and no energy is lost through the walls, the sound intensity will be equal to the sound power divided by the cross-sectional area of the tube, that is:

$I = W/\pi R^2$ (where R is the inside radius of the tube)

and will not decrease with distance from the source. In the usual case where the sound is free to travel away from the source in both directions along a tube, the sound power will be divided equally between the two parts of the system, and the intensity at any point will be half the original value ($W/2\pi R^2$).

In practice, sound energy is lost through the walls, and there is a gradual reduction in intensity with distance from the source. This can be calculated from tables and graphs [2]. Where the duct or pipe is lagged to prevent sound breaking out as it passes through sensitive areas, the high sound level inside the duct will be maintained over a long distance, and it is necessary to ensure that outlet grilles are not positioned close to where people are working.

For a point source

Where the sound is free to travel equally in all directions from a stationary source of relatively small dimensions (that is, where there are no reflecting barriers), any location may be considered to lie on the surface of an expanding sphere of radius r, with the sound source at the centre of the sphere. The intensity at distance r metres in terms of sound power of the source is given by:

$$I = W/4\pi r^2 \qquad \text{(surface area of a sphere} = 4\pi r^2)$$

and the decrease in sound intensity as distance r increases obeys the inverse square law—that is, the intensity varies inversely with the square of the distance from the source and therefore falls off rapidly.

For a line source

A linear sound source is analogous to the axis of a cylinder with radius r, and length l. In free space, the sound would be propagated cylindrically, therefore at any distance r from any point on the line source, the sound intensity associated with unit length of the source is given by:

$$I = W/2\pi r \qquad \text{(surface area of a cylinder} = 2\pi rl)$$

Here the decrease in intensity as r increases obeys the inverse law, that is, the intensity varies inversely with the distance from the source—a less rapid fall in intensity than from a point source.

In general

The examples given above are ideal conditions rarely met with in practice, but form a basis for predicting sound levels at any distance from the sound source, in any environment. Modifications of these formulae can be made where the sound source is at ground level, close to a wall, or in a corner.

Since the surfaces in most industrial buildings are efficient reflectors of sound, by visualizing the reduced volume occupied by the same amount of sound power, it is apparent that for a point or line source, the intensity at distance 'r' from the source will be increased concomitantly, and can be calculated by introducing a directivity factor (Q) into the formula for intensity to give:

$$I = QW/4\pi r^2 \qquad \text{where } Q \text{ has the value 1, 2, 4 or 8 according to the fractional volume of a sphere represented.}$$

Figure 7.2 illustrates this for a point source. For a line source, the calculation is based on the fractional volume of a cylinder.

Sound pressure

Sound energy gives rise to acoustic pressure fluctuations which are detected as sound and the smallest detectable pressure change is 2×10^{-5} Pa (where 1 Pa $\equiv$ 1 N/m^2). This corresponds to a sound intensity of 10^{-12} W/m^2 at the normal threshold of hearing.

Similarly at the threshold of pain, the sound pressure is between 20 and 60 Pa. At pressures above this, protective mechanisms are alerted, and the ear responds

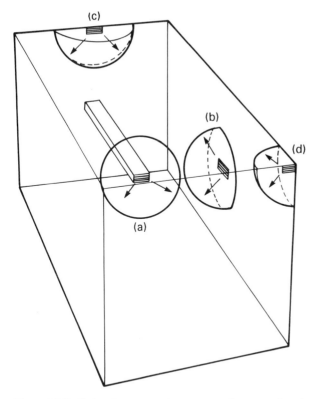

Figure 7.2 Radiation from a sound source at various room locations (courtesy of Woods of Colchester). (*a*) Source in centre of room; (*b*) source in centre of one wall; (*c*) source in junction of two room surfaces; (*d*) source in junction of three room surfaces

in a different manner, therefore the normal audible range lies between these two thresholds.

$$I = p^2/\rho c \qquad \text{where } p = \text{rms acoustic pressure}$$
$$\rho = \text{density of the medium}$$
$$c = \text{sound velocity in the medium}$$
$$(\rho c = 420 \text{ in air})$$

Thus sound intensity is directly proportional to the square of the sound pressure.

Derived parameters

Sound intensity level (sound level)

Because human perception of sound covers such an enormous range of intensities, it is more convenient to use a logarithmic scale when making measurements, particularly since the ear responds logarithmically, rather than linearly, to changes in intensity.

Sound is measured in terms of the logarithmic increase in intensity relative to that at the threshold of hearing, and this value is known as the sound intensity level, or the sound level and is measured in decibels (dB).

Intensity level (dB) = $10\log_{10}I/I_0$
where I = intensity of interest
and I_0 = reference intensity (10^{-12} W/m^2)

Thus at the hearing threshold, the value is 0 dB, and at the pain threshold it is between 120 and 130 dB.

The inverse square law can be expressed in decibels by taking logarithms to base 10 of the formula for intensity as follows:

$$10\log_{10}I_1 = 10\log_{10}W - 10\log_{10}4\pi - 20\log_{10}r_1$$
$$10\log_{10}I_2 = 10\log_{10}W - 10\log_{10}4\pi - 20\log_{10}r_2$$

By subtraction, where distance $r_2 - 2r_1$

$$10\log_{10}I_1/I_2 = 20\log_{10}2 = 6\,\text{dB}$$

Thus in theory, the sound level falls by 6 dB for each doubling of distance from the source. In practice the difference may be nearer 5 dB because of environmental factors.

Sound pressure level (SPL or L_p)

The sound pressure can be similarly related to pressure at the threshold of hearing using the relationship between intensity and pressure given above ($I = p^2 \times$ constant).

$$\text{SPL (dB)} = 10\log_{10}(p/p_0)^2 = 20\log_{10}(p/p_0)$$

where p = pressure of interest
and p_0 = reference pressure (2×10^{-5} Pa)

Table 7.1 Relationships between various measures of sound

Stimulus		Response		
Intensity (W/m^2)	Pressure (Pa)	Sound level meter (SLM) (BEL)	(dB)	Physiological
10^3	600	15	150	Instantaneous damage
10^2	2×10^2	14	140	
10	6×10	13	130	Threshold of pain
1	2×10	12	120	
10^{-1}	6	11	110	
10^{-2}	2	10	100	
10^{-3}	6×10^{-1}	9	90	Permanent hearing loss
10^{-4}	2×10^{-1}	8	80	Discomfort
10^{-5}		7	70	Annoyance
10^{-6}	2×10^{-2}	6	60	
10^{-7}		5	50	Comfortable audibility
10^{-8}	2×10^{-3}	4	40	
10^{-9}		3	30	
10^{-10}	2×10^{-4}	2	20	
10^{-11}		1	10	
10^{-12}	2×10^{-5}	0	0	Threshold of hearing

When measured in air, the dB values are numerically the same as for intensity.

Although for hearing conservation purposes it is the sound intensity (representing sound energy) which is relevant, this is difficult to measure, so the instruments in common use are designed to detect the air pressure changes induced by a sound source, and convert them to decibels. The notation for sound level measurements is therefore SPL (dB).

Table 7.1 clarifies the relationships between the measures of sound described as the stimuli, and their effects in terms of the microphone's and the ear's response.

Sound power level (SWL or L_w)

This is derived as for sound intensity level

$$SWL = 10\log_{10}W/W_0$$

where W = power of interest
and W_0 = reference power $(10^{-12}\,W)$

Properties of decibels

Decibels cannot be added or subtracted arithmetically. To combine sound levels, dBs must be expressed in logarithmic form, so that intensity values can be calculated. After adding or subtracting intensities, the combined value can be converted back to dBs, which should always be corrected to the nearest whole number.

Table 7.2 Rule of thumb for adding or subtracting decibels (derived from *Figure 7.3*)

Addition of two values

For a difference of:	The higher value is increased by:
0 or 1 dB	3 dB
2 or 3 dB	2 dB
4 to 9 dB	1 dB
10 dB or more	0 dB

Subtraction of two values

For a difference of:	The higher value is reduced by:
10 dB or more	0 dB
6 to 9 dB	1 dB
5 or 4 dB	2 dB
3 dB	3 or 4 dB
2 dB	4 or 5 dB
1 dB	5 to 10 dB
0 dB	10 dB or more

When adding more than two levels, group in pairs and add successive totals as in example below:

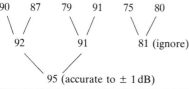

95 (accurate to ± 1 dB)

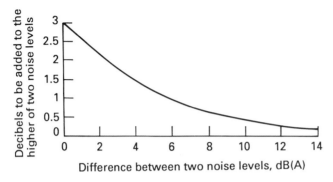

Figure 7.3 Noise level addition chart (courtesy of Bruel and Kjaer)

In practice, there is no need to use this cumbersome method of calculation, since it is easy to remember a simple 'rule of thumb' as outlined in *Table 7.2*, which is derived from *Figure 7.3*. This information is useful for estimating the noise level from one of a number of noise sources where some or all of the others cannot be switched off.

Practical use of parameters and formulae

A relationship between SWL and SPL for a small source can be derived from the formula for sound intensity as follows:

$$\text{SPL} = \text{SWL} - 20\log_{10}r - 10\log_{10}4\pi$$

In the case where $r = 1\,\text{m}$, $\text{SPL} = \text{SWL} - 11\,\text{dB}$, a relationship which can be used as a starting point to predict sound levels anywhere in the environment within which a small machine is to be installed, provided that the SWL is known. Alternatively, when planning to relocate an existing machine with unknown SWL, measure the SPL at 1 m to obtain an estimate of its SWL, and proceed to predict sound levels in the proposed environment, taking into account the directivity patterns, and any other relevant factors.

Note that when working in decibels, Q values of 1, 2, 4 and 8 translate to 0, 3, 6 and 9 dB. For example, a small (point) source is located in the centre of the concrete floor of a large workshop (hemispherical propagation, $Q = 2$), and the SPL measured 1 m away at the operator's ear is 87 dB. It is proposed that the machine be relocated to the corner of the workshop ($Q = 8$), but this would result in an increase of 6 dB, that is, 93 dB at the operator's ear, which would be unacceptable and necessitate some form of noise control.

An alternative proposal is to purchase two more identical machines and install them around a central position to be occupied by the operator, allowing him to control all three. Using the rules for adding decibels (*see Table 7.2*), the SPL at the operator's ear can be predicted $(87 + 87 + 87)\,\text{dB} = 92\,\text{dB}$.

In the unlikely event of the machine being suspended in midair (spherical propagation, $Q = 1$), the SPL at the same distance (1 m) would fall by 3 dB to 84 dB, and the SWL, which is independent of position or environment, would be $(84 + 11)\,\text{dB} = 95\,\text{dB}$.

These predicted values are only approximate, but give adequate guidance for practical purposes. In practice there are influences which are ignored in the theoretical conditions described above, but will be considered later in this chapter.

Effects of noise

General effects of different levels of noise

For hundreds of years, many people have been paying a high price for industrial noise, in permanent loss of hearing, and other disorders. Levels of noise well (*see Table 7.1*) below those which produce hearing loss can cause annoyance, and interfere with concentration and communications in the workplace, whereas in the community, noise from any source (not necessarily industrial), can cause annoyance and loss of sleep.

Noise is unique among health hazards in that it would be undesirable to eliminate it altogether, since man relies on sounds to control his actions. If a machine suddenly changes its normal noise pattern, this is a warning that something is wrong, and evokes a quick response to avert serious damage to machine or man. It is necessary, therefore, to establish levels of noise which will be acceptable to the majority under various conditions but will not produce undesirable effects.

Quantifiable effects

These include:
1. Temporary threshold shift (TTS) which is recoverable in 40 hours;
2. Permanent threshold shift (PTS) which is irreversible;
3. Noise-induced hearing loss (NIHL) = PTS minus presbyacusis loss; and
4. Occupational deafness which is NIHL due to noise exposure at work.

Of the two main types of hearing damage, conductive and perceptive, only the latter is indicative of NIHL, therefore it is important to be able to distinguish between them (*see* Chapter 12).

Non-quantifiable effects

TINNITUS

Tinnitus is a subjective sensation of noise in the ears or head, and can be either a high-pitched ringing, hissing or whistling, or a low-pitched rushing or buzzing. Low-pitched tinnitus is not normally associated with NIHL [3]. Short periods of high-pitched tinnitus are often experienced before PTS is established, and can therefore be taken as a warning sign of impending hearing damage. In advanced cases of NIHL, where tinnitus is present continuously, it may obscure the extent of the hearing damage. Although tinnitus is incurable, some relief can be obtained by playing masking noise into the ear from an adjustable sound generator worn like a hearing aid.

Kemp [4] (*see also* [5]) demonstrated that objectively measurable sound could be produced by the cochlea. When the hair cells are activated by certain types of noise, mechanical energy is reflected back and forth along the cochlea, further stimulating the hair cells, and leading to a self-sustaining oscillation (tinnitus) known as the 'Kemp echo'. Because it was noted that when subjects with normal hearing entered an anechoic chamber they experienced tinnitus, it was suggested that tinnitus may always be present, but is masked by external noise, except in cases of hearing loss, when the perceived external noise is too low to provide masking.

VERTIGO

Vertigo is a sensation associated with disturbances of the organ of balance; it is included here because it is often mistakenly suggested as a symptom of NIHL.

The two organs of hearing and balance comprising the inner ear are so closely connected that a disorder of one may affect the other, particularly where disease or injury is concerned. The fluid in the cochlea is continuous with that in the semicircular canals, and these organs share the same blood supply. The vestibular nerve is also in close proximity to the auditory nerve and disturbances of the vestibular apparatus produce a sensation of giddiness, or vertigo. Although vertigo may be associated with deafness, as in Ménière's disease, the underlying cause of the deafness stems from the organ of balance [6].

LOUDNESS RECRUITMENT

Loudness recruitment occurs where there is a high hearing threshold. As noise intensity is increased beyond this threshold, there is a rapid growth in perceived loudness. This effect is only associated with damage to the organ of Corti.

MASKING

Masking occurs when the level and frequency content of background noise renders speech unintelligible, or other sounds inaudible. This is an advantage where it is necessary to maintain speech privacy in open-plan offices, or in suites of offices with lightly constructed partition walls. The deliberate introduction of masking noise is known as 'sound conditioning'. The sound spectrum of the masking noise can be designed to smooth out a low level intrusive noise in a quiet area, and make the environment more pleasant.

In other circumstances, where it is necessary to hear speech, or a warning signal, above background noise, masking is a disadvantage, and must be reduced. *Figure 7.4* illustrates how a narrow band of noise exerts a masking effect over a wider frequency range as the intensity increases, and shows the limited extent of masking at frequencies below that of the masking noise, even at high intensities, compared with the wider influence at higher frequencies. This effect should be borne in mind when setting the level of warning signals, since in many industries warnings and buzzers are at sound levels well above 110 dB(A), and are numerous enough over the working day to contribute a major portion of the workers' noise exposure (*Table 7.3*).

Infrasound and ultrasound

Although unpleasant physiological and psychological effects, such as headaches, dizziness and nausea, are said to be associated with exposure to high levels of infrasound and ultrasound, there has been no evidence of hearing loss.

Infrasound can be generated in ships' engine rooms, compressor rooms, and some motor vehicles. Ultrasonic cleaning baths, mixers and welders are commonly used in industry. It is generally agreed that for both infrasound and ultrasound, levels of exposure should not exceed 130 dB.

Evaluation of noise

Subjective units of loudness (phons)

Equal loudness contours (*Figure 7.5*) were obtained empirically by asking large groups of young people with normal hearing to adjust the levels of pure tones at

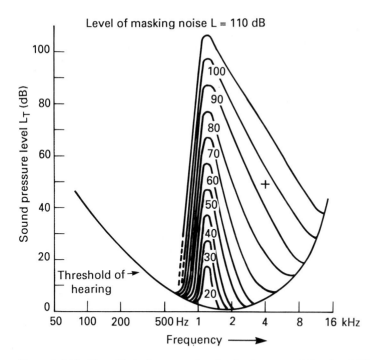

Figure 7.4 Masking effect of a narrow band noise (courtesy of Bruel and Kjaer)

Table 7.3 Relationships between measures of exposure to sound

Noise dose per cent	Sound exposure (Pa².h)	L_eq (8 h) (dBA)	Maximum allowable exposure time (h)*
	0.01	65	
	0.1	75	
	0.32	80	↑
12.5	0.4	81	
	0.5	82	
	0.64	83	↑
25	0.8	84	Unlimited
	1.0	85	24
	1.27	86	
50	1.6	87	16
	2.0	88	
	2.54	89	
100	3.2	90	8
	4.0	91	
	5.0	92	
200	6.4	93	4
	8.0	94	
	10.0	95	
400	12.7	96	2
800	25.4	99	1
1600	50	102	½
3200	100	105	¼

*It must be emphasized that the maximum exposure times are those in force at present, and they are likely to be reduced as a result of future legislation.

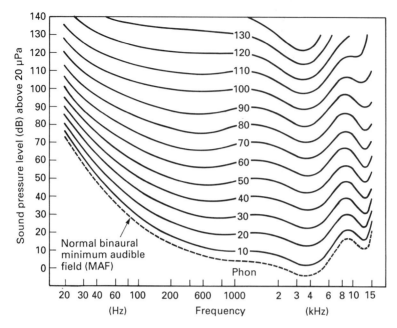

Figure 7.5 Normal equal loudness contours for pure tones (courtesy of Bruel and Kjaer)

a series of frequencies, until they were judged of equal loudness to a 1 kHz reference tone [7]. The exercise, carried out by Robinson and Dadson [8], was repeated at different sound levels, in 10 dB steps, using 1 kHz as the reference tone throughout. Thus the loudness level, in phons, is numerically equal to the sound level at 1 kHz.

This subjective assessment of loudness revealed that neither a doubling of sound energy (3 dB increase), nor of sound pressure (6 dB increase), was judged as a doubling of loudness for which a 10 dB increase in sound level was necessary. Loudness is dependent not only on both the frequency and the amplitude of the sound, but also on its duration.

Dips in the contours at around 4 kHz and 12 kHz are due to resonances in the ear canal. Sound-measuring equipment is designed to simulate the ear's response to sound by means of weighting networks which attenuate the high and low frequencies in an analogous manner.

'A', 'B', 'C', 'D', and 'E' weightings have been internationally standardized. The first three represent the equal loudness contours at 40, 70 and 100 phons respectively, and 'D' is used for measuring aircraft noise (*Figure 7.6*). However, established practice is to use the 'A' weighting for industrial noise at any level, because experience has shown that it gives the closest correlation with subjective reactions and it is important to use the designation dB(linear), dB(A) etc. when recording a sound level.

The sound spectrum

Broad-band noise contains a variety of frequency components at differing amplitudes. Although there are many instances of sound at a single fixed frequency

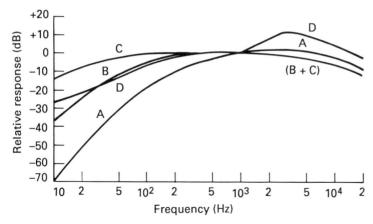

Figure 7.6 Frequency response of the weighting networks (courtesy of Bruel and Kjaer)

in an industrial environment, these 'pure tones', if present, are usually superimposed on a background of apparently random noise.

For dealing with complaints about annoyance, identifying sound sources, and designing or selecting means of noise control, or hearing protection, information about the frequency content of the noise may be essential. High frequency sound behaves differently from low frequency sound, and is more easily controlled.

For a frequency analysis, the sound spectrum is divided into 'octave bands' and the total dB level in each band is measured. Octave bands are described by their 'centre frequency', which is the geometric mean of the band limits, calculated as follows:

$$\text{Centre frequency } (f_0) = (f_1 \times f_2)$$

where f_1 and f_2 are the lower and upper band limits, and $f_2 = 2f_1$.

In international (ISO) and British (BS) standards, the 'preferred' octave bands are the series of octaves between 11 Hz and 22 500 Hz [9] with centre frequencies from 16 Hz to 16 kHz.

For the more detailed one-third octave band analysis, each octave band is divided into three geometrically equal sections, where $f_2 = f_1 \cdot \sqrt[3]{2}$ and the centre frequency is the geometric mean of the band limits. For the one-third octave band of centre frequency 63 Hz, the limits are 57 Hz and 71 Hz.

It follows that the bands become progressively wider with increase in frequency, and they are said to have a 'constant percentage bandwidth', which for the octave band is 70.7 per cent. Although an octave band analysis is used for specifications of noisy equipment, and of hearing protective devices, the one-third octave analyser, with its 23.1 per cent bandwidth, is thought most closely to resemble the response of the hearing mechanism to noise.

Interior and exterior noise criteria

Noise criteria (NC) curves

Original work carried out in the United States by Beranek [10] in the 1950s, provided an indication of the maximum levels allowable in the eight octave bands, 63 Hz to 8 kHz, for minimal interference with telephone conversations between females. A family of curves (*Figure 7.7*) numbered from 15 to 70 (which refers to the octave band sound level at 1 kHz in 5 dB steps), known as the noise criteria (NC) curves, was produced for use in office environments. In general they are intended for noise spectra of a similar shape, and should not be used for pure tones. They follow a similar pattern to the equal loudness contours. Background noise measured in dB (linear) octave bands is plotted on the graph. The NC rating is the number of the curve immediately above all plotted points.

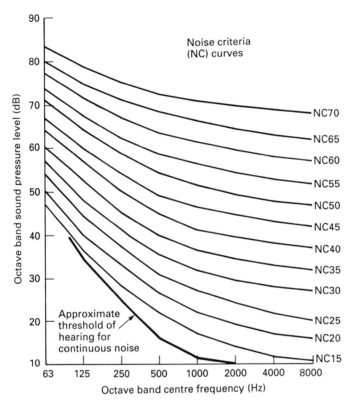

Recommended NC levels (112) for various environments

Factories (heavy engineering)	55–75
Factories (light engineering)	45–65
Kitchens	40–50
Swimming baths and sports areas	35–50
Department stores and shops	35–45
Restaurants, bars, cafeterias and canteens	35–45
Mechanized offices	40–50
General offices	35–45
Private offices, libraries, courtrooms and schoolrooms	30–35
Homes, bedrooms	25–35
Hospital wards and operating theatres	25–35
Cinemas	30–35
Theatres, assembly halls and churches	25–30
Concert and opera halls	20–25
Broadcasting and recording studios	15–20

Figure 7.7 Noise criteria (NC) curves (courtesy of Woods of Colchester)

Noise rating (NR) curves

The noise rating (NR) curves (*Figure 7.8*) are a similar family of curves, produced by Kosten and van Os [11], and adopted by the ISO (International Organization for Standardization). They give a rating similar to NC curves, but have a wider range, from 0 to 135, to cover industrial use and community response.

Both sets of curves are accompanied by a list of recommended levels likely to be acceptable in various environments.

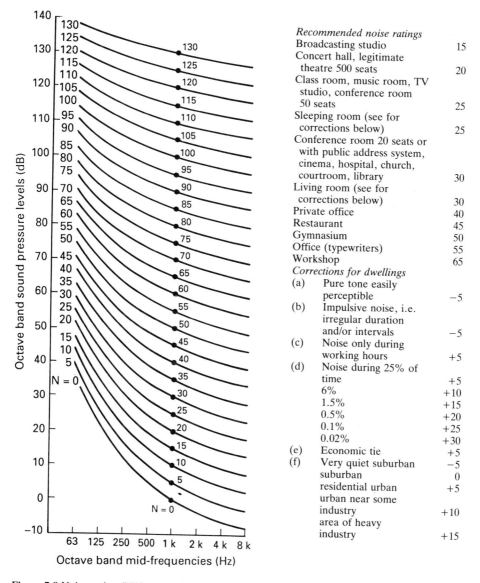

Recommended noise ratings	
Broadcasting studio	15
Concert hall, legitimate theatre 500 seats	20
Class room, music room, TV studio, conference room 50 seats	25
Sleeping room (see for corrections below)	25
Conference room 20 seats or with public address system, cinema, hospital, church, courtroom, library	30
Living room (see for corrections below)	30
Private office	40
Restaurant	45
Gymnasium	50
Office (typewriters)	55
Workshop	65

Corrections for dwellings		
(a)	Pure tone easily perceptible	−5
(b)	Impulsive noise, i.e. irregular duration and/or intervals	−5
(c)	Noise only during working hours	+5
(d)	Noise during 25% of time	+5
	6%	+10
	1.5%	+15
	0.5%	+20
	0.1%	+25
	0.02%	+30
(e)	Economic tie	+5
(f)	Very quiet suburban	−5
	suburban	0
	residential urban	+5
	urban near some industry	+10
	area of heavy industry	+15

Figure 7.8 Noise rating (NR) curves (courtesy of Blackwell Scientific Publications Ltd)

The advantage of using criteria which require a frequency analysis is that by comparing the noise spectrum with the standard curves, the frequencies giving most concern are identified, whereas with the dB(A) value, the frequency content is incorporated, but not revealed. The NC and NR numbers are usually found to be up to 7 dB less than the related overall dB(A) level.

Speech interference levels

A slightly different approach was taken by Webster [12] when developing the preferred speech interference levels (PSIL) from his own work, combined with information obtained from Beranek's survey. The arithmetical average of the ambient sound levels in the 500 Hz, 1 kHz and 2 kHz octave bands, the critical speech frequencies, is compared with average values for different voice levels (*Figure 7.9*), which are related to distance between speaker and listener. The PSIL is the maximum background level which will not interfere with speech communication. An ISO recommendation [13] prefers to include the 4 kHz octave band in the average.

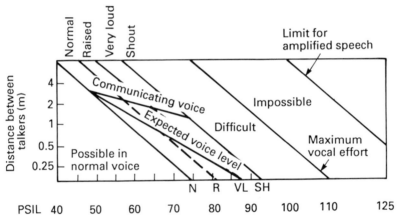

Figure 7.9 Communication limits in the presence of background noise (courtesy of Bruel and Kjaer)

Damage risk criteria (DRC) curves

For assessing risk of hearing damage, the damage risk criteria (DRC) curves (*Figure 7.10*) supplement the dB(A) value, particularly where there are strong pure tones [14]. As with the NC and NR curves, the dB(linear) octave band levels are measured and plotted onto the DRC curves. Each curve, by taking account of the duration of exposure and the frequency, represents the maximum allowable noise exposure over the working day (*see Table 7.3*), so by contrast with the above criteria for nuisance noise, no tolerance is permitted.

Sound measurement

Root–mean–square value

Sound is a fluctuating pressure graphically described as a sine wave. The acoustic energy is a summation of positive and negative quantities to an alternating current and is best estimated from the root–mean–square (rms) value, which is seven-tenths of the maximum (peak) value of the amplitude (*Figure 7.11*).

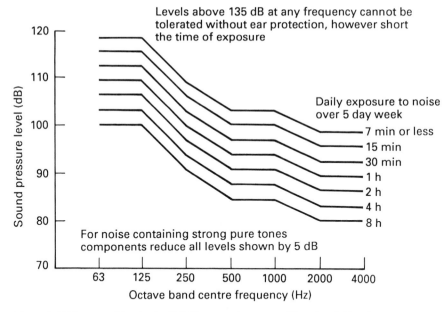

Figure 7.10 Damage risk criteria (DRC) curves (courtesy of Bruel and Kjaer)

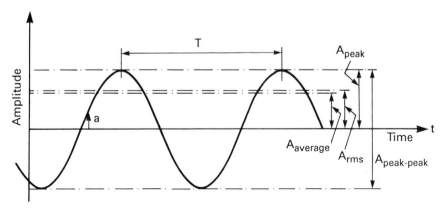

Figure 7.11 Relationship between rms, peak and average values for a sinusoidal signal (courtesy of Bruel and Kjaer)

The concept of L$_{eq}$ and SEL

Maximum permitted noise levels are in terms of dB(A) L$_{eq}$(8 h). The L$_{eq}$ (equivalent continuous sound level) is a single value having the same energy content as a fluctuating, or intermittent sound over the same period.

The sound energy of a single event can be measured in terms of the constant level over a 1 s period, having the same energy content, and is called the sound exposure level (SEL). The symbols used for this parameter are variously L$_{AE}$ and L$_{AX}$, and the following relationship exists [15]:

$$L_{eq}(8\,h) = L_{AX} - 44.6\,dB(A)$$

A less commonly used parameter is the L_{90}, which is the dB(A) level exceeded for 90 per cent of the time, and is mainly of interest when it is necessary to estimate the background level of noise in a community where the ambient noise fluctuates throughout the day. The impact of proposed intrusions, such as new roads or noisy industries, can then be assessed [16]. Similarly L_{10}, the dB(A) level exceeded for 10 per cent of the time, is an estimation of peak noise levels during the same period.

Sound level meter (SLM)

The basic components of a SLM are a microphone, an amplifier, weighting networks, a rectifier and a meter calibrated in dB to display the reading. All sound level meters measure the rms value, but some have the added capability of measuring peak levels, so that single events, or a series of impacts, can be evaluated.

The International Electrotechnical Commission (IEC) has produced a standard, IEC 651 (1979) 'Sound Level Meters', which specifies four grades of meters. The

Figure 7.12 Sound level meter (SLM) in use (courtesy of Bruel and Kjaer)

British Standards Institution's 'Specification for Sound Level Meters' BS 5969:1981 follows the IEC grading:

Type 0 Laboratory grade
Type 1 Precision grade (*Figure 7.12*)
Type 2 Industrial grade
Type 3 Survey meter

Only types 1, 2 and 3 are suitable for field measurements.

To minimize the effects of the body as a sound insulator and absorber, all SLMs should be held at arms length, or mounted on a tripod (as shown in *Figure 7.12*), when making measurements.

Calibrator

All types of sound level meter must be calibrated before and after use by adjusting the displayed level to that of a reference tone of fixed frequency and intensity.

Reverberation time

See BS 5363:1976 (1986) for details of measurement. In an enclosed space, the time taken for the sound to decay after a sound source has been switched off is an indication of the sound-absorbing properties of the room surfaces. The reverberation time (R_T) is measured by tracing the sound level of a pistol shot, or other loud broad-band sound on a graphic level recorder as it decays through 60 dB (*Figure 7.13*). Industrial background noise is usually too high to allow a full 60 dB fall in level, therefore it is normal practice to extrapolate. Since R_T varies with frequency, the measurement is repeated for each octave band. Alternatively the sound can be captured on a tape recorder, and replayed to obtain the decay slopes.

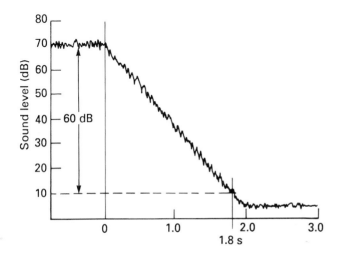

Figure 7.13 Reverberation time (courtesy of Sound Research Laboratories)

Narrow band analyser

A narrow band analyser is desirable for positively identifying pure tones and their harmonics, and relating these, for example, to blade passage frequency, pumping frequency, gear-mesh frequency or shaft rotation frequency when investigating noise sources. When connected to the output of the SLM, the analyser's tunable filter sweeps across the frequency spectrum, continuously measuring the sound level within a narrow band, which may be either 1, 2 or 4 per cent band width. The analyser drives a connected graphic level recorder, producing a continuous spectrogram (compare in *Figure 7.14*).

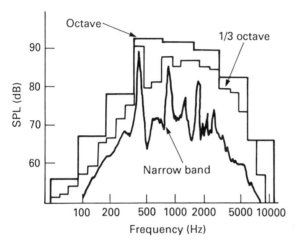

Figure 7.14 Octave, one-third octave, and narrow band spectra compared (courtesy of Morgan-Grampian)

Real time analyser

This instrument contains a bank of parallel octave-band filters, covering the whole audiofrequency range. The received signal is passed through all of the filters simultaneously, and the instantaneous octave-band analysis is displayed on a small screen by means of light emitting diodes (LEDs); a light-transparent recording chart can be placed over the screen to make a permanent record.

Personal noise dosimeter or sound exposure meter

Where a person remains in the same position throughout the working day, and the noise to which he is exposed remains constant, then the $L_{eq}(8\,h)$ is the sound level in dB(A) at his ear. This rarely occurs in industry and the L_{eq} can be obtained only by means of a personal monitor.

In the dosimeter, a microphone is clipped to the helmet or lapel, where it monitors the sound energy received by the ear (*Figure 7.15*). The instrument can be worn throughout the work-shift, or for a representative period, during which the signal is 'A' weighted and integrated. With a dynamic range of 60 dB, and no applied time constant, the dosimeter responds equally well to continuous, intermittent and impulsive sound.

Figure 7.15 Noise dosimeter in use (courtesy of Bruel and Kjaer)

On most instruments, the reading displayed is the percentage of the daily maximum permitted noise dose received by the wearer. This value can be converted to L_{eq} using the slide rule supplied with the instrument. A warning light indicates when the maximum permitted instantaneous level has been exceeded.

Some dosimeters are designed to give a reading in pascals-squared-hours ($Pa^2.h$). Since this value is directly proportional to the 'A' weighted energy, the instrument will need no modification when the allowable limit is reduced (by contrast with those based on the limit being 100 per cent noise dose). In fact it will be simpler, since $1\,Pa^2.h \equiv 85\,dB(A)$ for 8 h (*see Table 7.3*).

The major factors which determine the degree of hearing loss are the sound intensity and its duration whose product is the 'noise dose'. It follows that there is a tradeoff between noise level and time, as shown in *Table 7.3* where the 100 per cent noise dose is equivalent to an $L_{eq}(8\,h)$ of 90 dB(A). It can be seen that for each 3 dB increase in the L_{eq}, the exposure time must be halved. This is known as the '3 dB rule' and the 'equal energy principle' and applies in the United Kingdom. A few countries, notably the United States, use the '5 dB rule', where for a 5 dB increase in L_{eq} the exposure must be halved. This relationship makes an allowance for the recovery of temporary threshold shift (TTS) during the extended quiet period enjoyed when exposure time is reduced.

Frequent calibration is important for the dosimeter, since it tends to drift more than the SLM. An electronic dosimeter calibrator can be used to calibrate the SLM, but has the additional capability of delivering the reference tone for a fixed period representing 100 per cent noise dose (for example, 116 dB at 1 kHz for 65.5 s). Note that the 'A' weighting is zero at 1 kHz.

Noise control

Where noise is found to be at a potentially damaging level, it must be controlled; the procedure is to consider control methods in the following order:
1. Control at source.
2. Increase the distance between source and receiver.
3. Reduce the time of exposure.
4. Place barriers between the source and exposed persons.

Control at source

Noise is produced by vibration of parts of a machine, the materials being worked, and ancillary items. A frequency analysis may indicate which of these is responsible for the highest noise levels in the sound spectrum (*Figure 7.16*). For example, in the case of a fan or a pump, the offending pure tone may be associated with the fan blade passage frequency (*f*), where

$$f \text{ (Hz)} = (\text{fan motor rpm} \times \text{number of blades})/60$$

or the pumping frequency where

$$f \text{ (Hz)} = (\text{pump motor rpm} \times \text{number of pistons})/60$$

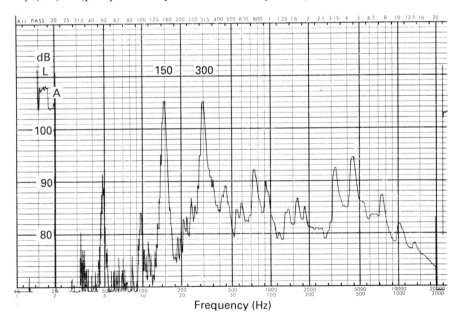

Figure 7.16 Sound spectrogram with pure tones

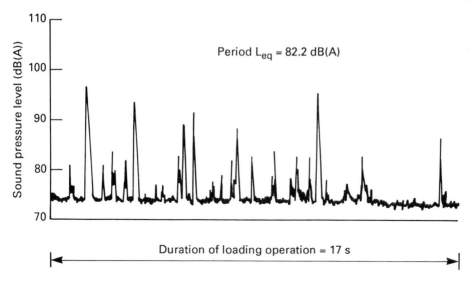

Duration of loading operation = 17 s

Figure 7.17 Time history of impacts (courtesy of PERA)

Alternatively, where the work cycle produces a series of impacts, a chart-recorded time-history of the noise will reveal which events relate to the major peaks (*Figure 7.17*). The noisiest parts or events must be dealt with in order of decreasing magnitude. Treatment could include any, or all the following remedies, some of which involve minimal effort or cost.

Planned maintenance

The first line of attack is to check that regular maintenance is carried out. Lubrication, replacement of worn bearings and tightening loose bolts can reduce the sound level by as much as 10 dB.

Modify speed

Where the noise is a low frequency tone, reducing the speed of the moving part will reduce the frequency either to below the audible range, or to a region where the ear is less sensitive. For example, halving the speed of a fan halves the frequency produced, and the sound level at that frequency may be reduced by as much as 17 dB. A high frequency problem may be resolved by increasing the speed to raise the frequency above the audible range.

Reduce impact noise

Use resilient materials such as heavy duty rubber or neoprene, brush strip, cork, between the impacting surfaces. Where manufactured metal objects or scrap metals fall into a metal bin, line the bin with rubber, or replace it with a perforated or woven metal container, reduce the drop height or break the fall. These remedies only apply where the impact is incidental to the process.

Redesign the process

Where the impact is a necessary part of the process, this should be examined to see if it can be effected by alternative (quieter) means, such as fixing with bolts or screws rather than nails, applying gradual rather than instantaneous force when cutting or working metal.

Stiffen panels

A thin metal panel, forced into vibration, either by being worked upon, or because it is a machine casing, may radiate high noise levels, particularly when forced to vibrate at its natural frequency. Stiffening the panel with struts at intervals will raise its natural frequency, and thus avoid resonance at the forcing frequency. With broad-band excitation, adding a damping layer of viscoelastic material to the panel, vibrational energy will be converted to heat in deforming it, and noise at resonant frequencies will be drastically reduced.

Replace equipment

When there is a proposal to buy new machines or plant, the opportunity should be taken to include a low noise level in the specification.

Insert an attenuator/silencer

An attenuator is a noise-reducing device, of which there are three main types, (a) dissipative, (b) reactive, and (c) active. They are inserted into duct-work or pipe-work associated with fans, pumps or compressors to reduce the transfer of noise energy, while allowing unhindered movement of the gas or liquid. The same principles can be applied to both media of sound propagation. Attenuators are usually positioned directly adjacent to, and on both sides of the sound source, with flexible connections to the duct-work.

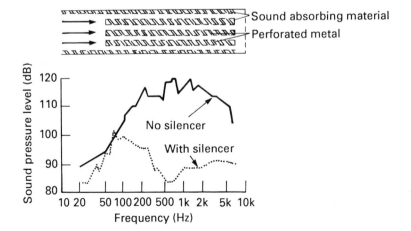

Figure 7.18 Dissipative silencer showing attenuation provided (courtesy of Sound Research Laboratories)

DISSIPATIVE SILENCERS

These are designed to absorb acoustic energy by means of friction. They can be either cylindrical or rectangular, and are lined with absorbent material. To improve efficiency, the internal surface area is increased by adding absorbent-lined splitters (*Figure 7.18*). This provides broad-band attenuation, with maximum performance at middle to high frequencies. The lining material must be faced with a thin plastic membrane, or with perforated metal, to prevent two-way contamination between absorbent and gas. These protective surfaces have the added advantage of improving the low frequency attenuation. The drawback with this type of silencer is the resulting pressure drop, therefore it should only be used where this is acceptable.

REACTIVE SILENCERS

These may be in the form of an expansion chamber, a tuned side-branch resonator, or a Helmholtz resonator (*Figure 7.19*). They are designed for high performance at a specific frequency, and so are ineffective either side of a very narrow band. They are used to attenuate pure tones produced by fixed-speed machinery. No pressure loss results from insertion of the last two types of silencer.

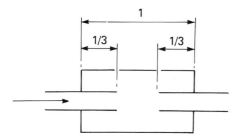

Figure 7.19 Reactive silencer (courtesy of Sound Research Laboratories)

ACTIVE ATTENUATORS

These are used for constant low frequency noise in duct-work, or exhaust systems. They comprise a high quality microphone and loudspeaker system, connected through an amplifier with a phase-shifting network. The sound is received, and played back at the same point in antiphase, thus in theory cancelling the noise, regardless of its frequency content (*Figure 7.20*). In practice, the noise is reduced by up to 40 dB in the 15–500 Hz region. By combining this with a dissipative silencer, the size of the latter can be reduced, and attenuation extended to all frequencies.

PNEUMATIC EXHAUST SILENCERS

These can be classed as dissipative. In many automatic processes, the exhaust ports of pneumatic control valves produce intermittent noise at levels usually well over 110 dB. Pneumatic silencers screw directly onto the outlet, allowing the air to escape by expanding through a metal or plastic porous body. Since the pores may become clogged with oil, these silencers must be inspected regularly, and cleaned

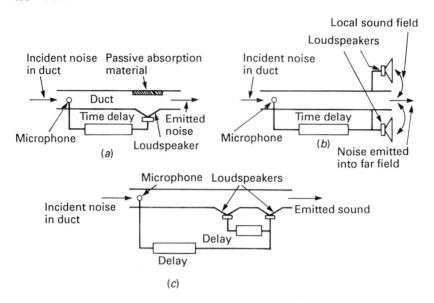

Figure 7.20 Three examples of active attenuation (courtesy of Morgan-Grampian)

or replaced when necessary. A noise reduction of up to 25 dB can be achieved by this means with little effort or outlay.

Increase distance

Where quiet and noisy operations are carried out in close proximity, if they are part of the same process, and must be in the same building, the quiet jobs can all be relocated to one end of the building, and the noisy ones to the other. In a noisy area, such as a machine shop, or press shop, there will be areas between machines where the sound is at a lower level. By painting the floor a different colour where the noise drops below 90 dB(A), these areas can be readily identified, and the operators should be encouraged to move into them whenever their duties allow. This is particularly pertinent to the person minding several automatic machines.

Sound intensity is inversely proportional to the square of the distance from the sound source (inverse square law). In decibel terms, this means that doubling the distance from the source, reduces the sound level by 6 dB, that is:

$10\log_{10}(I_1/I_2) = 10\log_{10}(2^2) = 6\,\mathrm{dB}$ where $I = Q/4\pi r^2$
I_1 and I_2 are intensities at distances r and $2r$ (respectively)

This law applies only in free-field conditions, because in an enclosed space sound is reflected back into the room from the room surfaces. This raises the general level of sound by an amount which depends on the distance from the reflecting surface. To reverse this degradation of the effect of distance, the room surfaces can be lined with sound absorbent material, the two main types being porous (open-cell plastic form), and fibrous (glass fibre or mineral wool).

Sound absorption

For a sound source operating in an enclosed space, the total sound field comprises the sound direct from the source plus the reverberant (reflected) sound (*Figure 7.21*). The combined formula is:

$$SPL = SWL + 10\log_{10}(Q/4\pi r^2 + 4/R)$$

where R is the room constant $S\alpha/(1 - \alpha)$ defined below.

The relative proportion of each component received by the ear of a person working within the room is governed by the position of the source and the receiver with respect to the room surfaces.

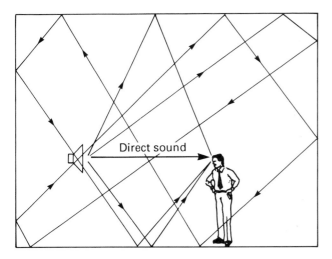

Figure 7.21 Direct and reverberant sound fields in an enclosed space (courtesy of Blackwell Scientific Publications Ltd)

The level of the reverberant component depends upon the room volume and the nature of the room surfaces. In a small room with highly reflecting surfaces, little sound passes through the walls. Most of the incident sound energy is reflected back into the room, and the remainder is absorbed by the walls. Multiple reflections within the room space will result in a build-up of sound throughout the space, so that the inverse square law no longer applies to the reduction of sound level at distances from the source, because at any point in the room, except within 1 m from the source, the reverberant component will be greater than the direct component and the sound level measured in any part of the room will show little variation (*Figure 7.22*) (live room).

In a room with large dimensions and surfaces which absorb sound (dead room), the direct sound component predominates since the level of reflected sound will be minimized. Because of the greater distances involved, the direct sound will have fallen to a much lower level at the room boundaries, where its energy will be dissipated on contact with the absorptive surfaces. The consequent low levels of reflected sound will be diminished further where the listener is some distance from the walls.

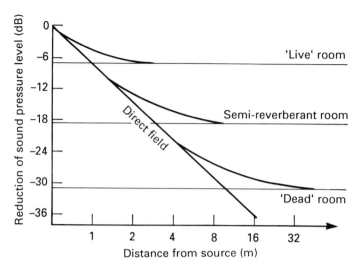

Figure 7.22 Combination of direct and reverberant sound fields (courtesy of Bruel and Kjaer)

Most industrial environments fall between these two extremes (semireverberant room), but they serve to illustrate two effective means of reducing noise exposure for an individual. Increasing the distance between the sound source and the listener reduces the direct sound level at the listener's ear by approximately 6 dB per doubling of distance, and increasing the quantity or quality of the absorptive surfaces reduces the reverberant level, more so if the listener can be distanced from the surfaces which reflect most sound.

The energy balance when sound meets a surface is $\alpha + r + \tau = 1$

where α = fraction absorbed   (absorption coefficient)
            r = fraction reflected   (reflection coefficient)
            τ = fraction transmitted   (transmission coefficient)

Since τ is relatively small, the equation approximates to:

$$r = 1 - \alpha \quad \text{as} \quad \alpha \to 1, r \to 0$$

therefore the reverberant sound level can be reduced by increasing the amount of absorption in the room. A reduction of 3 dB is achieved each time the absorption is doubled, so the first increase is the most cost-effective, particularly where there is little absorption initially. Values of α at each octave band for most commonly-used materials have been tabulated [17]. Improvements are monitored by re-measuring reverberation time.

Reduce exposure time

The 'equal energy principle' allows a tradeoff between noise level and time of exposure. As shown in *Table 7.3*, if the sound level is above 90 dB(A), the $L_{eq}(8\,h)$ can be kept below 90 dB(A) by reducing exposure time. This can be achieved by job rotation.

Exposure to 93 dB(A) for 4 h or even 99 dB(A) for 1 h is permissible, provided that the remainder of the working day, *and the non-working hours*, are spent in a quiet environment of <80 dB(A). However, since the aim is to keep the L_{eq} as far below 90 dB(A) as feasible, and not just to regard the limit as a dividing line between safe and hazardous, reducing exposure time should be combined with other methods of controlling noise.

Acoustic barriers

If other methods fail to achieve the required reduction in sound level, an acoustic barrier must be placed between the sound source and the exposed person(s). This can take various forms, and would be designed to suit the particular set of circumstances, but for maximum efficiency it should be as close as possible to either the source, or the receiver.

Barriers close to the source may be either acoustic cladding of the source, a complete enclosure built around the source, or where people work on only one side of the source, a partial enclosure may suffice, particularly if overheating would otherwise occur. An enclosure around the receiver is known as an acoustic haven.

Hearing protective devices such as earmuffs or earplugs are in the latter category (barriers close to the receiver), but should only be used as a temporary measure, or as a last resort if all other attempts at noise control fail.

Sound insulation

Desirable characteristics for sound-insulating materials are high mass per unit area, high inherent damping and low stiffness, and therefore the ideal material is lead sheet. Insulation is always combined with an absorptive layer on the side facing the noise source.

The measure of acoustic insulation is the ratio of the total incident sound energy to the energy transmitted. In decibel terms

Sound reduction index (SRI) = $10\log_{10}(1/\tau)$ dB
where τ = transmission coefficient

As shown in the energy balance equation, for the acoustic energy transmitted through the barrier to be low, the amount reflected must be high, which means that the barrier must be capable of resisting movement when sound pressure is applied. The properties which determine such resistance are the mass (inertia) of the barrier, and the frequency of the incident sound.

This combination of effects is represented in the 'mass law of sound insulation', since it has been shown that the transmitted sound (τ) is inversely proportional to the square of the mass of the barrier and the frequency of the incident sound:

$$SRI = 10\log_{10}(1/\tau) = 20\log_{10}(\pi f M/420) = 20\log_{10}(Mf) - 43 \text{ dB}$$
where M = surface density of barrier (kg/m^2)
         f = frequency of sound (Hz)
         420 = impedance of air (mks Rayls)

This is true where sound impinges on a dense, limp material at normal incidence. Usually sound energy approaches from all directions (random incidence), particularly in a reverberant room. For field measurements, the relationship was found to be:

$$SRI = 20\log_{10}(Mf) - 48 \text{ dB}$$

These formulae predict an increase of 6 dB on doubling the mass or the frequency, but in practice the improvement is nearer 5 dB.

If the barrier is stiff, its insulating capacity will be degraded at low frequencies

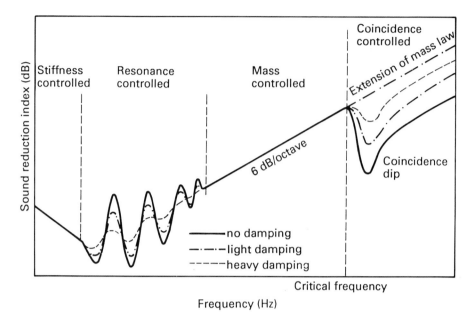

Figure 7.23 Ideal sound reduction index curve and the effect of damping (courtesy of Bruel and Kjaer)

due to resonances. At high frequencies, coincidence between the bending wave in the barrier and the incident sound wave results in a modification of the mass law as illustrated in *Figure 7.23*. Thus the mass law applies only at frequencies between these two extremes.

Damping

Both resonance and coincidence effects can be reduced by adding damping (*see Figure 7.23*) to one side of the barrier in the form of a mastic layer, bonded to the barrier material, so that sound energy is converted into heat energy on deforming the mastic. This is known as unconstrained layer damping.

Metals such as steels, brasses, and aluminium have very low damping capacities, so a more durable method of applying damping is to sandwich the damping layer between two sheets of the metal. This is called constrained layer damping.

Composite barriers

The degree of insulation provided by a barrier or enclosure depends upon its homogeneity. Doors, windows and holes for entry of services provide opportunities for sound to circumvent the barrier. Furthermore since doors and windows usually have a lower SRI than walls, the overall SRI will be reduced in relation to the area they occupy. The average absorption coefficient is then

$$\tau_{\text{average}} = (\tau_m S_m + \tau_w S_w + \tau_d S_d + \tau_h S_h + \ldots)/S_{\text{total}}$$

where m = masonry, w = window, d = door, h = hole etc., S = surface area (m^2).

This value is then used to calculate SRI. Should it be necessary to upgrade the sound insulation, windows can be double-glazed with an airspace not less than 150 mm. Panes of different thicknesses should be mounted in separate frames set in rubber mouldings, and the reveals lined with absorptive material. Gaps around doors can be eliminated by fitting an airtight seal.

The following formula is used to predict the sound level in a room adjacent to that housing a sound source of known SWL:

$$\text{SRI} = \text{SPL}_1 - \text{SPL}_2 + 10\log_{10}(S/A_2)\,\text{dB}$$

where   1 and 2 relate to source room and receiving room (respectively)

S = area of dividing wall (m^2)
A_2 = total absorption in room ($S_2\overline{\alpha}_2$) in sabins

This calculation assumes that all of the sound in the receiving room has been transmitted through the dividing wall. In practice, flanking transmission occurs where the sound can bypass the wall via the ventilation system or the suspended ceiling. These pathways must be blocked if better insulation is required. It may then be worthwhile to construct a cavity wall.

Transmission loss is an alternative term for SRI, and values at each octave band of frequencies for a range of sound insulating materials have been measured and tabulated [2].

Personal protection

Ideally there are only four conditions under which hearing protection should be used.

1. To give immediate protection where a noise problem has been identified, and while solutions are being devised.
2. As a last resort, if attempts at control have failed to reduce noise exposure below the recommended limit of 90 dB(A) L_{eq}(8 h).
3. In an area of high noise level, such as a plant room, entered infrequently, or only for short periods, for example, for maintenance work.
4. Where noise exposure exceeds 85 dB(A) L_{eq}(8 h), hearing protectors must be made available when the EEC Directive on Noise has been implemented in the United Kingdom [18].

Hearing protection devices (HPDs)

The two main types of hearing protectors are earmuffs and earplugs, collectively known as hearing protection devices (HPDs).

Earmuffs

These are designed completely to cover the external ear and a foam-filled or fluid-filled cushion seal ensures a close fit to the head. The ear seals and the headband tension must be regularly checked, since their efficiency diminishes rapidly with constant use.

Earplugs

These are of more varied design than earmuffs. The three main types are (1) glass down; (2) soft, solid plastic or silicone rubber inserts; and (3) plastic foam plugs.

All types fit into the ear canal and, with the exception of glass down which is disposable, can be worn many times, and washed when necessary. Glass down and foam plugs are user-moulded before insertion and expand to fit the ear canal.

Plastic or silicone rubber plugs are premoulded and made in a range of up to seven sizes, depending on the adaptability of the plug. For a good seal, a size slightly larger than the width of the ear canal is essential. This may mean a different size for each ear.

A person who cannot find a comfortable fit among the ready-made plugs may need custom-made plugs, which are individually moulded into the ear and allowed to set. These then last indefinitely.

The solid and foam plugs are available with an attached neckcord, for use where it is important that they do not fall into the product. In addition, foam plugs attached to a lightweight, flexible headband can be obtained, thus combining the advantages of plugs with the ease of donning and removing normally associated with muffs.

There are specially designed muffs and plugs for protecting against impulse noise (such as gunfire or explosions), without affecting speech communication.

Advantages and disadvantages

The advantages and disadvantages of muffs and plugs are listed below.

1. Muffs are more visible than plugs, making it easier to check if they are being worn.
2. Because of their size, muffs are not so easily lost.
3. People with ear infections may be able to wear muffs, whereas they should not wear plugs without consulting a doctor.
4. Muffs come in one size which fits most heads. Down and foam plugs are also one size, but the solid type of plug is not so adaptable, and comes in several sizes, or may be custom-made.
5. Dirt and toxic matter may be transferred to the sensitive skin in the ear canal if plugs are inserted with dirty hands, or if they have been placed on a contaminated surface.
6. Plugs are much cheaper than muffs, a few pence compared with a few pounds. Even allowing for more frequent replacement of plugs, the overall cost would be lower.
7. Plugs are more comfortable than muffs in hot environments.
8. Glasses (safety or prescription), long hair or facial hair do not interfere with the fit of plugs, but prevent a good seal between muffs and head.
9. Plugs are more suitable where the head must be manoeuvred in a confined space.

Degree of protection afforded by HPDs

The acoustic attenuation of HPDs must be evaluated by a recognized standard test. Attenuation data are supplied in either tabular, or graphic form, giving the mean and the standard deviation for each band of frequencies. The Code of Practice for Reducing the Exposure of Employed Persons to Noise defines 'assumed protection' as the mean minus one standard deviation, being the minimum attenuation provided by a device for 84 per cent of users. Where a higher level of protection is demanded, for 97 per cent of users the minimum attenuation would be the mean minus twice the standard deviation.

The standard deviation is a measure of the adaptability of the hearing protector to the wide range of sizes and shapes of heads, ears and ear canals, and the consistency with which the test is carried out. It must be emphasized that the values given for assumed protection will only be achieved in practice if the device is fitted properly, and worn continuously.

It has been shown [19] that removing the HPD for only 15 min of an 8 h continuous exposure will reduce the protection provided by a high performance HPD from 30 dB to 15 dB overall, and from 20 dB to 14 dB for a less efficient protector (*Figure 7.24*). For example, a device selected to reduce the wearer's exposure from 115 dB(A) L_{eq} to 85 dB(A) L_{eq} will only achieve 100 dB(A) L_{eq} under these conditions, leaving the wearer grossly underprotected. The application of the 'equal energy principle' (described earlier, and illustrated in *Table 7.3*) to wearing time, indicates that if the HPD is only worn for 50 per cent of the exposure time, the protection afforded is limited to a maximum of 3 dB, regardless of the quality of the protector. In practice, up to 5 dB protection was achieved.

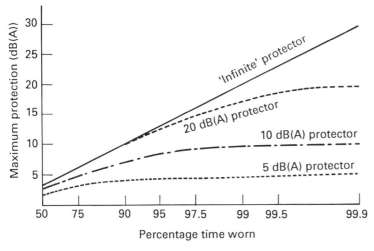

Figure 7.24 Maximum hearing protection as a function of wearing time (courtesy of BNF Metals)

Since the main reasons for removing HPDs are physical discomfort and a feeling of isolation, comfort assumes equal importance to attenuation, and overprotection can be as detrimental as underprotection. Furthermore, the maximum attenuation attainable by any HPD varies with frequency between 40 dB and 55 dB [20], the difference between the hearing threshold by bone conduction and by air conduction. The bone conduction route via the skull to the inner ear bypasses both types of HPD (*Figure 7.25*) [21] and even if the whole head could be encased in a soundproof helmet, sound would be transmitted via the chest and neck to the skull.

The examples given in *Table 7.4* show that, in general, all HPDs give better protection against high frequency sound than against low frequencies. However, most plugs provide better attenuation than muffs at low frequencies (below 500 Hz), while high frequency attenuation (above 1 kHz) is similar for both, and more than adequate for most acoustic environments.

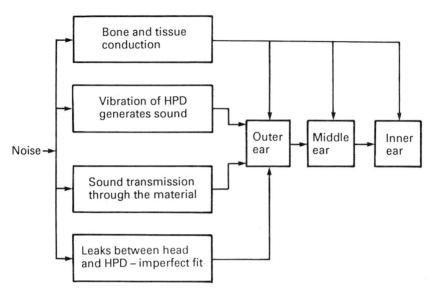

Figure 7.25 Noise pathways through HPDs (after A.M. Martin, Proceedings of BNF Metals' Seminar *Industrial Noise and its Reduction* [16]

Table 7.4 Attenuation data for various types of hearing protectors (evaluated in dB as recommended in BS 5108:1974)[22]

Protector	Statistic	Octave band centre frequency (Hz)							
		63	125	250	500	1 k	2 k	4 k	8 k
Glass down	mean	13.7	15.4	16.8	18.8	19.9	25.0	31.7	33.8
	SD	6.2	6.1	6.2	6.8	6.6	4.8	6.8	5.2
AP 84 per cent			7.5	9.3	10.6	12.0	13.3	20.2	24.9	28.6
Plastic plug	mean	21.7	22.0	21.3	20.3	23.0	29.8	44.9	43.3
	SD	4.4	4.9	5.8	5.1	4.6	5.5	4.2	5.2
AP 84 per cent		17.3	17.1	15.5	15.2	18.4	24.3	40.7	38.1
Foam plug	mean	24.8	26.1	26.7	28.9	30.4	32.8	43.6	44.4
	SD	7.3	7.8	7.4	7.0	5.9	4.9	5.0	5.5
AP 84 per cent		17.5	18.3	19.3	21.9	24.5	27.9	38.6	38.9
Earmuff (light)	mean		7.2	8.5	16.3	25.2	31.4	36.7	34.6
(foam seal 130 g)	SD		4.3	3.5	3.0	3.8	5.5	4.0	6.1
AP 84 per cent			2.9	5.0	13.3	21.4	25.9	32.7	28.5
Earmuff (medium)	mean		19.6	25.0	27.2	36.9	34.8	37.4	35.0
(foam seal 260 g)	SD		3.1	3.1	3.7	3.4	3.6	3.7	5.5
AP 84 per cent			16.5	21.9	23.5	33.5	31.2	33.7	29.5
Earmuff (heavy)	mean	13.6	15.9	21.8	28.2	36.3	35.3	35.1	25.0
(liquid seal 373 g)	SD	5.4	6.3	4.5	5.6	5.1	3.8	5.4	5.8
AP 84 per cent		8.2	9.6	17.3	22.6	31.2	31.5	29.7	19.2

SD = standard deviation
AP 84 per cent = assumed protection for 84 per cent of wearers

The three muffs are listed in *Table 7.4* in increasing order of weight and quality, and are from different manufacturers. Each of the last two claims to be for use in extreme conditions, and particularly effective at lower frequencies. Although for any specific model, liquid-filled seals give better attenuation than foam-filled, *Table 7.4* shows that this rule cannot be applied when comparing different makes, since other factors which maximize performance are the density of the earcup material, its enclosed volume, and the headband pressure.

Threshold shift methods of measuring the performance of HPDs

Table 7.5 illustrates the importance of comparing data obtained by the same test method. The two most widely used standards are the UK BS 5108 and the US ANSI S3 19-1974 (ASA STD1-1975) [22,23]. The British Standard test [22] is carried out in anechoic conditions using an array of loudspeakers to produce a diffuse sound field. The acoustic attenuation of the hearing protector under test is the difference in dB between the normal hearing threshold and the raised threshold when wearing the muffs or plugs.

Table 7.5 Attenuation data for earplugs measured by UK and US standard test methods (BS 5108:1974 and ANSI S3 19-1974)[22,23]

Standard	Centre frequency (Hz)									
	63	125	250	500	1k	2k	3.15k	4k	6.3k	8k
BS 5108 (mean)	19.8	19.9	20.0	22.2	24.1	30.7	38.8	41.4	41.5	40.8
(SD)	7.5	7.8	6.4	4.9	3.5	4.3	4.5	4.7	4.5	5.9
AP 84 per cent	12.3	12.1	13.6	17.3	20.6	26.4	34.3	36.7	37.0	34.9
ANSI S3 (mean)		22	25	29	34	37	45	46	41	38
(SD)		3.5	2.3	2.5	3.1	2.4	3.4	4.2	4.4	3.9
AP 84 per cent		18.5	22.7	26.5	30.9	34.6	41.6	41.8	36.1	34.1

SD = standard deviation
AP 84 per cent = assumed protection for 84 per cent of wearers

The original American Standard test, ANSI Z24.22-1957, obtained the hearing thresholds in an anechoic chamber with one loudspeaker facing the subject and producing pure tones at nine octave band centre frequencies between 125 Hz and 8 kHz. This was later thought to be untypical of industrial noise exposure, and the 1974 Standard [23] rectified this by specifying nine one-third octave bands of noise in a diffuse sound field. The British method, which in general produces lower mean attenuation values than the American test, is accepted as more closely resembling typical industrial conditions.

Estimation of noise level received by the protected ear

Ideally the calculations outlined in *Table 7.6* should always be carried out when selecting suitable HPDs for a specific acoustic environment. This is particularly important where the noise contains significant pure tone components, or is predominantly low frequency. Allowance should be made for the fact that in use the attenuation measured under the BS test conditions may not be achieved, because for the test the device is new and carefully fitted, the subjects are motivated, and only required to wear the device for 15 min. In the industrial setting, earplugs may gradually work loose due to jaw movements, and muffs may

Table 7.6 Estimation of noise level received by the ear when using lightweight muffs in two different acoustic environments

	Octave band centre frequency (Hz)								
	63	*125*	*250*	*500*	*1 k*	*2 k*	*4 k*	*8 k*	*Overall*
Shake-out (dB)	77	82	87	89	90	97	96	102	105 (dB)
A-weighting	−26	−16	−9	−3	0	+1	+1	−1	
SPL (dBA)	51	66	78	86	90	98	97	101	104 (dBA)
AP 84 per cent	0	3	5	13	21	26	33	29	
Received level	51	63	73	73	69	72	64	72	79 (dBA)
Furnace (dB)	96	102	94	82	80	77	77	74	103 (dB)
A-weighting	−26	−16	−9	−3	0	+1	+1	−1	
SPL (dBA)	70	86	85	79	80	78	78	73	90 (dBA)
AP 84 per cent	0	3	5	13	21	26	33	29	
Received level	70	83	80	66	59	52	45	44	85 (dBA)

SD = standard deviation
AP 84 per cent = assumed protection for 84 per cent of wearers

be displaced during work activities, or long hair and glasses may prevent a good seal with the head.

A frequency analysis of the noise, in dB(linear) octave bands, is made at the ear of the exposed person. If the noise level fluctuates, an integrating sound level meter may be required to obtain L_{eq} measurements. The 'A' weighting corrections are applied, and at this stage the resulting values can be combined (as in *Table 7.2*) to give an overall dB(A) level, if this were not measured directly. The 'assumed protection' data are then subtracted for each octave band, and the resulting values are the levels received by the protected ear. These dB(A) values are combined to give the overall exposure level in dB(A), which should be below 85 dB(A).

Table 7.6 illustrates the importance of obtaining more information than just the overall dB or dB(A) value. The lightweight muffs provide 25 dB attenuation in sound which is mainly high frequency, but only a barely adequate 5 dB attenuation in the furnace noise, because of their poor performance at low frequencies. The hearing mechanism itself provides considerable protection (represented by the 'A' weighting) at low frequencies, which accounts for the large difference between the linear and 'A' weighted overall values for the furnace, by contrast with the minimal difference for the shakeout noise.

In very high noise levels, even the most efficient HPD may not provide the degree of protection required. An extra 5 or 6 dB attenuation may be gained by wearing earplugs and earmuffs.

Physical properties of HPDs

BS 6344 Parts 1 and 2 concern the physical rather than the acoustic properties of ear muffs and plugs [24]. Part 3, dealing with more specialized devices, will follow.

Protection from infrasound and ultrasound

Attenuation data supplied with HPDs is limited to the octave bands 63 Hz to 8 kHz, as produced by the standard test. A special study [25] of the attenuation provided by both muffs and plugs in the infrasonic and ultrasonic frequency ranges, showed that earmuffs do not attenuate infrasound, and may even amplify it, whereas well-fitting earplugs give adequate protection. At ultrasonic frequencies, not only does the hearing sensitivity decrease by about 100 dB/octave, but the performance of all types of HPDs is excellent.

Statutory provisions

Until the proposed noise regulations [18] become law in 1989, compliance with the 1972 Code of Practice [26] must be regarded as an essential requirement of the Health and Safety at Work Act 1974 Section 2 [27].

In the latest DHSS leaflet NI.2, April 1987 listing prescribed diseases, 'occupational deafness' is No. A10. Detailed information is given in leaflet NI.207 March 1984 *Occupational Deafness.*

References

1. *CRC Handbook of Chemistry and Physics* 66th edn (1985–86)
2. Sharland, I. *Woods Practical Guide to Noise Control*, 4th impression, Woods, Colchester (1986)
3. Acton, I. and Grime, R.P. Legal liability for occupational deafness, *New Law Cassettes*, Butterworths, London (1979)
4. Kemp, D.T. Simulated acoustic emissions from within the human auditory system. *Journal of the Acoustic Society of America*, **64**, 1386–1391 (1978)
5. Pickles, J.O. *An Introduction to the Physiology of Hearing.* Academic Press, London (1982)
6. Beales, P.H. *Noise, Hearing and Deafness.* Michael Joseph (1965)
7. BS 3383:1961 and ISO R226(1961) *Normal Equal-Loudness Contours for Pure Tones and Normal Threshold of Hearing under Free-Field Listening Conditions*
8. Robinson, D.W. and Dadson, R.S. A re-determination of the equal loudness relation for pure tones, *British Journal of Applied Physics*, **7**, 166 (1956)
9. BS 3593:1963 and ISO R266(1975) *Recommendations on Preferred Frequencies for Acoustical Measurements* (BSI confirmed 1986)
10. Beranek, L.L. Criteria for office quietening based on questionnaire rating studies, *Journal of the Acoustical Society of America*, **28**, 833 (1956)
11. Kosten, C.W. and van Os, G.J. *Community reaction criteria for external noises,* Proceedings of Conference on Control of Noise. HMSO, London (1982)
12. Webster, J.C. Speech communication as limited by ambient noise. *Journal of the Acoustic Society of America*, **37**, 692–699 (1965)
13. ISO R3352(1974) *Assessment of Noise with Respect to its Effect on the Intelligibility of Speech*
14. Burns, W. *Noise and Man*, 2nd edn. John Murray, London (1973)
15. HSC Consultative Document *Protection of Hearing at Work 1981. Proposed regulations, draft Code of Practice and Guidance Note*
16. BS 4142:1967 (amended 1982) *Method of Rating Industrial Noise affecting Mixed Residential and Industrial Areas.*
17. Evans, E.J. and Bazley, E.N. *Sound Absorbing Materials* 2nd edn. National Physical Laboratory (1978)
18. *Prevention of Damage to Hearing from Noise at Work*, Draft Proposals for Regulations and Guidance, HSC Consultative Document, HMSO (1987)
19. Else, D. A note on the protection afforded by hearing protectors—implications of the energy principle. *Annals of Ocupational Hygiene*, **16** (1973)
20. Zwislocki, J. In search of the bone conduction threshold in a free sound field. *Journal of the Acoustic Society of America*, **29**, (1957)
21. Martin, A.M. *Industrial Noise and its Reduction. Proceedings of BNF Metals' Seminar* (1975)

22. BS 5108:1974 *Method of Measurement of the Attenuation of Hearing Protectors at Threshold* (revised 1983)
23. American National Standards Institute (ANSI) S3 19-1974 *Method for the Measurement of Real-Ear Protection of Hearing Protectors and Physical Attenuation of Earmuffs.*
24. BD 6344 Part 1: 1984 *Specification for Ear Muffs.* Part 2: 1988 *Specification for Ear Plugs*
25. Berger, E.H. Attenuation of hearing protectors at the frequency extremes. *11th International Congress on Acoustics, Paris, Vol. 3* (1983)
26. *Code of Practice for reducing the exposure of employed persons to noise,* Health and Safety Executive, HMSO, London
27. Health and Safety at Work etc. Act 1974. HMSO, London

Chapter 8

Ionizing radiation

R.C. Hack

Introduction

Certain wavelengths of electromagnetic waves (say, less than 20 nm) and energetic particles are capable of ionizing matter through which they pass. Such agents are collectively referred to as 'ionizing radiation'. Ionization results in atoms becoming electrically charged by either losing or gaining electrons, the resultant atoms being known as ions. The resultant deposition of energy causes chemical changes and hence biological damage on a microscopic scale. Biological effects (nearly always harmful to the healthy being) may be divided into somatic (that is, affecting the person exposed to the radiation) and hereditary (affecting the descendants of the exposed person). They may also be subdivided into stochastic and non-stochastic effects. A stochastic effect is one in which the probability of the effect occurring is proportional in some way to the exposure with no threshold; the severity of the effect, if it does occur, is independent of the degree of exposure. A non-stochastic effect is one in which the severity of the effect is proportional to the degree of exposure above a threshold value; the effect is certain to occur at exposures above the threshold. Both stochastic and non-stochastic effects caused by low level chronic exposures are usually associated with long latent periods (years to tens of years).

This chapter outlines the basic physical concepts, defines the quantities involved, considers the various sources of radiation both natural and man-made, describes the various effects and considers the risk factors associated with stochastic effects, looks at the various limits recommended for both occupational exposure and general populations, and finally outlines the basic control methods of radiological protection.

Basic physical concepts

Structure of the atom

The atom (radius approximately 10^{-8} cm) is the smallest part of matter which can take part in a chemical reaction, there being 6×10^{23} atoms in the gram atomic weight (6×10^{23} is termed Avogadro's number).

Atoms may be considered to consist of negatively charged electrons in orbits or shells at different radial distances from a nucleus. The total number of electrons, Z, in a neutral atom characterizes the element; Z is termed the atomic number.

The transition of electrons from one orbit to another is associated with the emission or absorption of electromagnetic waves (to inner shells X-rays are emitted, to other shells ultraviolet or visible light is emitted). Electron shells are labelled K, L, M, N. K is nearest to the nucleus.

Structure of the nucleus

The nucleus (radius approximately 10^{-12} cm) consists of positively charged protons (mass 1836 times that of an electron) and of neutral neutrons (mass same as protons). The number of protons = the number of electrons (Z) = more or less number of neutrons (N).

Definitions:

The mass number (A) is the number of nucleons (protons or neutrons) in the nucleus. It is approximately equal to the atomic weight.

The atomic number (Z) is the number of protons in the nucleus. It is also the number of units of charge and hence the number of electrons in a neutral atom.

The neutron number (N) is the number of neutrons in the nucleus.

The isotopic number (I) is $N-Z$.

A nuclide is a nucleus containing a specific number of protons and neutrons, written:

$$^A_Z X \qquad \text{(for example, } ^{235}_{92}U\ ^{12}_6C)$$

Since the atomic number (Z) is known for a given element this is often omitted:

$$^A X \qquad \text{(for example, } ^{60}Co,\ ^{24}Na)$$

Sometimes it is written:

$$_Z X^A \quad \text{(for example, } _{38}Sr^{90}) \quad \textit{or} \quad X^A \quad \text{(for example, } S^{35})$$

Isotopes are nuclides with the same number of protons (Z) and therefore the same nuclear charge, the same arrangement of extranuclear electrons and hence the same chemical properties; they differ because of different numbers of neutrons. Elements with even Z usually have several stable isotopes, whereas the majority of those with odd Z have one or at most two.

Mass, energy and charge

Mass

The atomic mass unit (amu) is defined so that the mass of the ^{12}C atom = 12.000:

1 amu	$= 1.66 \times 10^{-24}$ g
Electron mass	$= 0.00055$ amu
Proton mass	$= 1.0073$ amu
Neutron mass	$= 1.0097$ amu

Energy

The electron volt (eV) is the energy acquired by an electron in dropping through a potential of 1 volt.

$1\,\mathrm{eV} = 1.6 \times 10^{-12}\,\mathrm{ergs}.$

From the equivalence of mass and energy formula ($\mathrm{E} = mc^2$),

$1\,\mathrm{amu} = 931\,\mathrm{MeV}$

Charge

The indivisible unit of charge is that of the electron or proton ($4.8 \times 10^{-10}\,\mathrm{esu}$).

Radioactivity

Some nuclides are stable because they have a correct balance of protons and neutrons, others are unstable (and many can be made unstable) because they have relatively too many protons or neutrons, and thus too much energy. The spontaneous disintegration of nuclei is generally known as radioactive decay.

Modes of nuclear disintegration

ALPHA EMISSION

Alpha particles are helium nuclei (two protons, two neutrons) emitted mostly by heavy nuclei (A > 208, for example, polonium, radium, uranium). The mass number of the parent nucleus is reduced by 4.

for example, $^{239}\mathrm{Pu} \rightarrow {}^{235}\mathrm{U} + {}^{4}\alpha + \text{energy}$

BETA EMISSION

This is the creation and ejection of a positive electron (positron) or negative electron (negatron) together with a virtually undetectable neutrino or antineutrino.

(1) If the nucleus is unstable because of neutron excess a neutron transforms into a proton and a negatron.

$n \rightarrow p + \beta^-$

and the resultant nucleus is transmuted into an element of one higher atomic number

for example, $^{14}\mathrm{C} \rightarrow \mathrm{N}^{14} + \beta^- + \text{energy}.$

(2) If the nucleus is unstable because of neutron deficiency, a proton transforms into a neutron and a positron,

for example, $p \rightarrow n + \beta^+$

and the resultant nucleus is transmuted into an element one less in atomic number

for example, $^{22}\mathrm{Na} \rightarrow {}^{22}\mathrm{Ne} + \beta^+ + \text{energy}.$

Beta particles are emitted from a nuclide in a broad spectrum of energies, the mean of which is usually taken as one-third of the maximum.

GAMMA EMISSION

A nucleus may be formed in an excited state (or may gain energy by collision) and on its return to the ground state loses this excess energy by emitting gamma rays.

This usually occurs very rapidly (within 10^{-13} s) but sometimes the time before emission is a few hours and this excited state is then called a metastable state. Gamma emission usually follows alpha or beta emission.

for example, following β^- decay of ^{137}Cs:

$$^{137}_{56}\text{Ba} \rightarrow {}^{137}_{56}\text{Ba} + \gamma$$

ELECTRON CAPTURE (K CAPTURE)

This is a form of beta decay. Instead of a nucleus emitting a positron it may capture an electron from the K shell which transforms with a proton into a neutron,

$$p + e^- \rightarrow n$$

The atomic electron then drops to fill the inner shell vacancy emitting X-rays.

for example, $^{51}_{24}\text{Cr} + e^- \rightarrow {}^{51}_{23}\text{V} + \text{X-ray}$

INTERNAL CONVERSION

The excited nucleus loses its excess energy to a K shell electron which is ejected. Note that these electrons are monoenergetic unlike those from beta decay which have a continuous spectrum.

for example, $^{119}_{50}\text{Sn} \rightarrow {}^{119}_{50}\text{Sn} + e^- + \text{X-ray}$

FISSION

A heavy nucleus (A approximately 240) splits into two roughly equal fragments with mass numbers around 90 and 140. This is rare in nature but is the basic mechanism in nuclear reactors. Almost at the instant of fission, fast neutrons are emitted with average energies around 1 MeV.

for example, $^{235}_{92}\text{U} \rightarrow {}_{38}\text{Sr} + {}_{54}\text{Xe} + 2\text{--}3 \text{ neutrons} + \text{energy}$

The fission products have a neutron excess and are negative beta emitters.

Production of radioisotopes

Artificial radioisotopes are produced by bombarding the stable element by either neutrons (from a nuclear reactor) or charged particles (from an accelerator):

BY THERMAL NEUTRONS

(n, γ) Reactions produce most beta-emitting radioisotopes in the dense thermal neutron flux in a reactor.

for example, $^{35}_{17}\text{Cl} + {}^{1}_{0}n \rightarrow {}^{36}_{17}\text{Cl} + \gamma$

The emitted gamma ray is termed a 'capture gamma'.
[Note the method of summarizing a nuclear reaction which in general can be written:

Target atom + bombarding particle
$\rightarrow$ resultant atom + ejected particle

or more briefly,

Target atom (bombarding particle, emitted particle) resultant atom]

BY FAST NEUTRONS

(n, p) reactions

for example, $^{35}_{17}Cl + ^{1}_{0}n \rightarrow ^{35}_{16}S + ^{1}_{1}H$

(n, α) reactions

for example, $^{35}_{17}Cl + ^{1}_{0}n \rightarrow ^{32}_{15}P + ^{4}_{2}He$

BY CHARGED PARTICLES

By charged particles in an accelerator (usually a cyclotron).

for example, $^{12}C(p, 3p3n)\ ^{7}Be$
$\qquad\quad ^{24}Mg(d,\alpha)\ ^{22}Na$
$\qquad\quad ^{52}Cr(\alpha, 4n)\ ^{52}Fe$

Radioactive decay

For a given isotope every nucleus has a definite probability of decaying in unit time (called the decay constant λ) which is independent of the chemical or physical state of the element and independent of environmental factors such as temperature and pressure. The rate of decay (called the activity) at any instant is directly proportional to the number of radioactive atoms present at that instant (N):

that is, $\qquad \dfrac{dN}{dt} = -\lambda N$

Integrating this equation we have

$N = N_0 e^{-\lambda t}$

where $\qquad N_0 = N$ at $t = 0$

Radioactive decay is therefore exponential, the actual decay rate being determined by the decay constant and the number of active nuclei present. The reciprocal of the decay constant is called the mean or average life (t_m),

$t_m = 1/\lambda$

However, the most widely used index of decay is the half-life ($t_{1/2}$) which is the time required for the number of active nuclei (or for their activity) to decay to half its initial value. (Note that this is independent of the amount of the radioisotope present.) It can easily be shown that,

$t_{1/2} = {}^{0.693}/\lambda$

that is, the half-life is inversely proportional to the decay constant.

Since the activity (A), that is dN/dt, is proportional to the number of active atoms (N) we also have,

$A = A_0 e^{-\lambda t}$

and combining this with the half-life formula we have,

$$A = A_0 e^{-0.693} t/t_{1/2}$$

which is the most widely used equation.

Problems concerned with decay are easily solved with a pocket calculator with exponential functions or graphically (a linear plot of time versus a log plot of activity yields a straight line).

Units

At present we are in a transitional state between the 'old' traditional units and the 'new' SI units and so it is necessary to consider both.

Absorbed dose

This is a measure of energy deposition in a medium, that is, the energy deposited per unit mass. It is necessary to quote the medium concerned, soft tissue, muscle or bone, for example. In radiation protection it is usual to consider energy deposition in soft tissue which is for all practical purposes equivalent to water.

The rad (old unit) is defined as an energy deposition of 100 ergs/g.

The gray (new unit) is defined as an energy deposition of 1 J/kg; 1 gray (Gy) = 100 rads.

Dose equivalent

Since the same absorbed dose of different types of radiation does not necessarily give the same biological effect, the absorbed dose must be multiplied by a factor (called the quality factor QF in radiation protection) to give an index whereby the effects of different types of radiation may be added simply (linearly). The quantity obtained when the absorbed dose is multiplied by the quality factor is called the dose equivalent.

The rem (old unit) = QF × rads.
The sievert (new unit) = QF × grays (Gy)
1 sievert (Sv) = 100 rems.

The quality factor depends on the linear density of ionization caused by the radiation and varies from 1 (X-rays) to 20 (alpha particles).

Activity

This is the rate of decay of a radionuclide.

The curie (Ci) (old unit) was originally defined as the activity of 1 g of radium being formally standardized as 3.7×10^{10} disintegrations per second.
The becquerel (Bq) (new unit) is defined as 1 disintegration per second; it is a very small unit.

Table 8.1 summarizes the relationships between the old and the new units.

Table 8.1 Summary of units

Quantity	New (SI) unit	Old unit	Conversion factor
Absorbed dose	Gray (Gy)	Rad	$1\,Gy = 100\,rad$
Dose equivalent	Sievert (Sv)	Rem	$1\,Sv = 100\,rem$
Activity	Becquerel (Bq)	Curie (Ci)	$1\,Bq = 2.7 \times 10^{-11}\,Ci$

Sources of radiation

Naturally occurring sources

Man is, and always has been subjected to a certain amount of radiation exposure from his surroundings (the so-called 'natural background'). This background radiation comes from several sources.

Cosmic rays

These are a complex mixture of high energy radiations originating in outer space but producing a wide variety of secondary radiations when the primary radiation meets the earth's atmosphere. The primary rays are mostly high energy protons producing a complex mixture of secondaries—protons, neutrons, electrons, photons and mesons (the latter are particles of many types intermediate in mass between those of the electron and proton). The intensity of cosmic radiation varies with latitude and elevation above sea-level. The latter is of importance in high altitude supersonic air transport.

The earth's crust

This is mostly due to the presence of the components of the uranium and thorium decay series and, to a lesser extent, from ^{14}C and ^{40}K. Individual doses are heavily dependent on local conditions such as the composition of the surface layer of the earth's crust and the composition of buildings. There is considerable global and local variation in the dose received via this route.

Radioactivity in air

This is mainly due to radon and thoron which diffuses from the ground and buildings. The modern tendency towards central heating and double glazing has resulted in increased doses from this cause due to the lack of effective ventilation in modern dwellings.

Food and drink

Ingested ^{40}K is the major component of dose from diet apart from radon decay products.

In the United Kingdom 87 per cent of the average exposure comes from natural causes made up as follows [1]

from cosmic rays	14 per cent
from the ground and buildings	19 per cent
from the air	37 per cent
from food and drink	17 per cent

This results on average individual dose of $1870\,\mu Sv$ (187 mrad). The remaining 13 per cent comes from man-made sources as dealt with in the next section.

Man-made sources

Outside the nuclear power industry and specialized research laboratories, the most likely sources of ionizing radiation are radioactive sources and high voltage devices which emit X-rays (by design or otherwise).

Radioactive sources

These can be classified in various ways.

ALPHA SOURCES

Because of the short range in material of alpha particles the containment is a compromise between emitting a useful proportion of the particles available and containing the (usually) highly toxic isotope—a realistic maximum window thickness is about $1\,mg/cm^2$. Radiation damage of window and/or the containment becomes significant for surface activity levels more than about $0.1\,Ci/cm^2$.

BETA SOURCES

These have similar but less severe (dependent on energy) problems of containment and absorption. There are three main types: foil, ceramic and gas, dependent on their construction.

GAMMA SOURCES

By comparison with alpha and beta sources, their containment (and self-absorption) presents less difficulty and large effective sources can be made, but for energies below about $100\,keV$ self absorption becomes comparable with beta sources. The chemical form of the radioactive material in the source should be inert and massive rather than finely powdered.

NEUTRON SOURCES

These are of relatively low output, the three main types being:

Spontaneously fissile material (for example, Cf^{252})
(α,n) for example, $Be^9 + \alpha^4 \rightarrow C^{12} + n^1$
(γ,n) for example, Be/Sb^{124}

The two latter consist of an intimate mixture of the target material (usually Be) and the radioactive isotope.

Source manufacture

There are two main methods.

IRRADIATION IN A REACTOR OR ACCELERATOR AFTER ENCAPSULATION

The need here is to consider possible longlived induced activity in cladding material (Al alloys, nickel and platinum—no iridium—are often used) and/or soldering and brazing alloys, stainless steel is practicable for long irradiations. Reactivation of a 'spent' source appears attractive but is seldom used.

ENCAPSULATING RADIOACTIVE MATERIAL

Compared with irradiation after encapsulation, there is a wider choice of capsule material but work has to be performed remotely in a shielded and/or contained enclosure ('cave'). Usually stainless steel, nickel–copper alloys or platinum alloys are preferred, the latter particularly for clinical use. The base radioactive material should be chemically stable, of high melting point and low volatility, and must not react with the capsule material.

Medical use of sources

For interstitial radiotherapy, three techniques are usually employed:

Needle (^{60}Co, ^{137}Cs or radium)
Permanent implantation for shortlived nuclides (for example, ^{198}Au 'seeds')
Wire (^{192}Ir) via a plastic tube.

For intracavity radiotherapy stronger sources than for interstitial techniques are used with modern emphasis on 'after-loading' techniques.

For surface treatment (such as of skin and eye) ^{90}Sr plates or ^{32}P or ^{90}Sr dispersed in polyethylene sheet are used.

Radioactive chemicals as pharmaceuticals have a wide variety of uses:

Diagnostic: determination of body composition; physical tracing; isotope tracing; scanning techniques.
Therapeutic: which depends on the preferential concentration of the radioactive material in the desired organ or tissue.

Industrial use of sources

THICKNESS AND DENSITY GAUGING

There are two main methods, direct transmission gauging and a back-scattering technique. Beta, *bremsstrahlung* or gamma sources are used.

XRF TECHNIQUES (X-RAY FLUOROSCOPY)

Low energy gamma, X-ray or *bremsstrahlung* radiation are used to excite characteristic X-radiation from the target material and can be used to measure coating thickness and for making a rapid non-destructive chemical analysis.

GAMMA RADIOGRAPHY

These sources offer greater penetration than that achieved by kilovolt (kV) X-ray machines; ^{60}Co, ^{137}Cs, ^{192}Ir and ^{170}Tm are commonly used isotopes (decreasing order of energy).

THERMOELECTRIC GENERATORS

These utilize heat produced during radioactive decay to produce electricity via thermocouples.

INDUSTRIAL IRRADIATION

For initiating chemical reactions, altering physical properties, killing bacteria, its main use is for sterilization of medical supplies; ^{60}Co, ^{137}Cs, sources or spent fuel elements are the main sources used.

X-ray emitting devices

PRODUCTION

X-rays are produced when high speed electrons suddenly change their kinetic energy being produced deliberately in X-ray tubes by stopping the electrons in a target. The essentials of an X-ray tube are:

A source of electrons (the cathode): usually a hot filament of tungsten
An accelerating field: from various types of high voltage generator. The applied voltage is usually specified in terms of the peak value of the voltage waveform (kVp).
A target: of high melting point material (the anode).
A tube (evacuated): to contain the above with a window to allow X-rays to pass through.

'Unwanted' or adventitious X-rays may also be produced in any device where electrons are accelerated through a few kV or more, by, for example, high voltage rectifying valves or electron beam equipment.

The emitted X-ray spectrum is continuous, superimposed with characteristic X-radiation dependent on the target material.

X-RAY 'QUALITY'

X-rays of low energy (long wavelength) are called 'soft', high energy (short wavelength) 'hard'. The degree of hardness (related to the ability to penetrate matter) is called the 'quality'. The half-value thickness (hvt) in a given material (for example, copper, aluminium or beryllium) is often quoted as a measure of quality. The effective energy (defined as that energy of monoenergetic radiation which has the same half-value thickness as the beam in question) is another index of quality.

FACTORS AFFECTING X-RAY OUTPUT

The intensity (I) is directly proportional to the tube current (i) and roughly proportional to the square of the voltage (V), that is

$$I \propto iV^2$$

For thin targets $I \propto Z^2$, for thick targets $I \propto Z$, where Z is the atomic number of the tube target material.

X-RAY APPLICATIONS

Medical   Diagnosis (radiography and fluoroscopy); therapy (chiefly malignant disease).

Industrial and research   Radiography and fluoroscopy of inanimate objects; crystallography; thickness gauges; particle accelerators.

Other devices   Other devices which produce X-rays adventitiously are: cathode ray tubes; electron microscopes; magnetrons; klystrons; valve rectifiers.

AVERAGE DOSE TO THE POPULATION

The average annual dose in the United Kingdom from radiation of man-made origin accounts for about 13 per cent of the total exposure made up as follows [1]:

Medical irradiation	11.5 per cent
Fallout from atomic weapons	0.5 per cent
Luminous watches, air travel	0.5 per cent
Occupational exposure	0.4 per cent
Release from the nuclear industry	0.1 per cent

This results in an average individual annual dose of $282\,\mu Sv$ (28.2 mrad) giving a combined total from all sources (natural and man-made) of $2152\,\mu Sv$ (215.2 mrad).

Radiation effects

Ionizing radiation effects

An ionizing event in tissue has typically an energy of about $100\,eV$ and since the average chemical bond only requires about $3\,eV$ to break, it is clear that ionizing radiation can exert a powerful molecular action resulting in effects such as:

1. Cell death
2. Mutations
3. Chromosome breaks
4. Inhibition of cell division
5. Formation of giant cells
6. Reduction of survival
7. Action on macromolecules (DNA, for example)
8. Reduction of metabolism and protein synthesis.

The deposition of energy on a microscopic scale from ionizing radiation should be contrasted with the macroscopic deposition of energy from non-ionizing radiation which usually results in energy being transmitted to complete molecules, the resultant effects being due to heat generated.

Effects of whole-body irradiation

The acute radiation syndrome

If the entire body is uniformly irradiated in a short time the result is the killing of cells of various radiosensitive organs. If the dose is high enough, death results with a threshold of around 5 Gy (500 rads). At the lower end of the dose range, the usual cause of death is damage to the blood-forming organs, resulting in infection, haemorrhage and anaemia; at somewhat higher doses, damage to the intestine results in death from diarrhoea, dehydration, loss of body salts and massive bacterial invasion from the lining of the bowel. At even higher doses, death results rapidly from damage to the CNS.

Radiation-induced cancer

Carcinogenesis is considered to be the chief risk of irradiation at low doses. The effect is stochastic, that is, the probability of the effect occurring but not the severity of the effect is a function of exposure (dose) without a threshold figure. Although not all radiation-induced cancers are fatal, it is usual to consider the risk (or probability) of the induction of fatal cancers because the consideration of non-fatal risks is full of problems of definition.

SOURCES OF HUMAN EXPOSURE DATA

Data on the effects of low-level irradiation on human beings is available from a large number of sources including the following:

1. The Japanese atomic bomb survivors at Hiroshima and Nagasaki: this source provides the largest sample to date of exposed individuals over a large range of exposures, and has resulted in a truly enormous amount of information for the risk outcome of a 'single-shot' exposure.
2. Children exposed prenatally from abdominal X-ray examinations of the mother during pregnancy: since the irradiations occurred at the most radiation-sensitive time during an individual's lifespan the results almost certainly give an upper limit to the risk.
3. Adults undergoing radiotherapy for ankylosing spondylitis: this involved irradiation of the spine resulting in substantial exposure to the bone marrow with smaller but significant exposures of other organs. This source is significant in that it represents a major use of radiotherapy for a non-malignant condition with (very nearly) whole-body irradiation.

Of the above, the atomic bomb survivors and the patients with ankylosing spondylitis have probably yielded the most clearcut risk data so far, and we will consider these two sources of information in greater detail.

EXPOSURE OF PATIENTS WITH ANKYLOSING SPONDYLITIS

In Great Britain between 1935 and 1954, patients with ankylosing spondylitis were treated by radiotherapy using wide field X-ray exposures to the whole of the spine with doses ranging from 1.2 to greater than 300 Gy, in from one to eight courses of treatment. About 13 000 patients were so treated and were followed from the date of the first treatment to 1960 [2] and the deaths from different types of neoplasms compared with those expected from national mortality rates.

Cancer deaths other than leukaemia have been divided into two categories depending upon the amount of radiation which the primary site is likely to have received. The results are reasonably clearcut. The number of leukaemia deaths are grossly increased—the maximum relative mortality (giving an observed:expected ratio of nearly 13) occurring between 3 and 8 years after the first irradiation. Cancer of the lightly irradiated sites shows no significant increase in the total number of deaths ($P \cong 0.4$), but that of the heavily irradiated sites shows a highly significant increase ($P < 0.000001$).

It is possible, however, that the above data might overestimate the risk slightly— some association between leukaemia and rheumatic diseases has been observed and most of the patients involved in the study had been exposed to drugs known to depress bone marrow cellular activity.

JAPANESE ATOMIC BOMB SURVIVORS

The surviving populations of Hiroshima and Nagasaki in 1950 contained about 60 000 people who received some significant radiation exposure, including about 7000 who experienced the major symptoms of the acute radiation syndrome. A tremendous effort has been made to study this population and a lifespan study is still being made of some initial 82 000 survivors all told. Based on theoretical and experimental information concerning the air dose from the weapons, the location and shielding history of the individual, and the attenuation factors for the particular shielding, a dose assignment has been made for about 79 000 of those in the study; this dose information (often considered one of the weak links of the study) is still continuously being updated and refined.

Observed:expected mortality ratios using the lowest dose category (0–0.09 Gy) as controls have been given by Goss [3] for leukaemia and all other cancers respectively during the period 1950–70. For both categories an increase in mortality with increasing dose is clearly demonstrated with a ratio of over 18 for leukaemias and 1.5 for all other cancers at doses in excess of 2 Gy. Data from the atomic bomb survivors and patients with ankylosing spondylitis demonstrate the ease with which an increase of a comparatively rare disease (leukaemia) can be demonstrated compared with a common disease (all cancers) when the absolute numbers in each case are of the same order. The time distribution of mortalities indicates that the majority of radiation-induced leukaemias have already occurred. Apart from the doubts in absolute dose figures mentioned previously many of the exposures were extremely non-uniform. The validity of the various control populations, made in the study so far, has been the subject of considerable debate—the approach by Goss to use the lightest exposed group as the control group disposes of many of the criticisms about the cost of a lower statistical accuracy.

THE RISK AT LOW DOSES

The data from the various studies detailed previously are remarkably consistent in their estimates of risks for doses of (say) 0.5 Gy upwards.

For leukaemia: 3×10^{-3} per Gy (3×10^{-5} per rad)
For all cancers: 10^{-2} per Gy (10^{-4} per rad).

For low doses of the order of 0.01 Gy per year, the above risk estimates will be the risk per year at equilibrium if a linear dose–effect relationship is assumed to hold.

Theory and cellular experiments indicate that the dose–effect relationship is probably a square law; thus assuming a linear relationship for risk coefficients obtained at high doses will overestimate the risk at low doses. Current radiation protection standards assume a linear relationship between dose and risk and therefore almost certainly represent an upper limit of risk. It is clearly highly desirable to be able to demonstrate that one's risk estimates are realistic (or at any rate erring on the safe side) and it can be demonstrated [4] that for the risk coefficients quoted above only a national or international survey is capable of demonstrating their validity in a reasonable period of time. Such a survey has been started in the United Kingdom by the National Radiological Protection Board (NRPB) who eventually hope to include every occupationally exposed person in the United Kingdom on their register.

Their estimates of minimum detectable risks for various observation times and staff turnover rates for 200 000 in-service and ex-workers show that a large survey on occupational exposure would not be expected to produce any first conclusions for at least 20 years. However, if total exposures are much less than 1000 man Gy/ year, or if the risk is less than 10^{-2} per Gy the time required to prove a positive effect of radiation on the incidence of deaths from cancer becomes very high, and with little prospect of making statistically valid intermediate statements.

Effects of partial body irradiation

Irradiation of single organs can induce (stochastically) latent cancers in those organs as in whole-body irradiation. However, many single organs exhibit specific non-stochastic effects such as:

1. Erythema of the skin
2. Cataract formation in the lens of the eye
3. Production of temporary or permanent sterility (applied to both sexes).

In general the doses required to produce such effects are high and are only likely to occur following accidental exposure; for the three effects mentioned above the threshold absorbed doses are approximately 20 Gy (2000 rad) for erythema, 15 Gy (1500 rad) for cataracts, and 7 Gy (700 rad) for permanent sterility in man. The fetus is relatively sensitive to radiation and radiography of the lower abdomen of women during pregnancy must be avoided unless there are overriding clinical considerations. In fact, any woman should be treated as though she might be pregnant and, in general, radiographic (or nuclear medicine) investigations should be limited to a 10 day interval from the onset of the last menstrual period—the so-called '10 day rule'.

Genetic effects

Damage to the chromosomes (caused by faulty recombination following ionization) can produce genetic effects. Thus another aim of radiation protection procedures is to reduce the genetically significant dose to large populations, for example, by minimizing as far as possible the dose to the gonads in radiographic examinations. For general populations, the average risk of serious hereditary defects in the first two generations is about 4×10^{-3} per sievert (4×10^{-5} per rem). The hereditary risk, therefore, is considerably smaller than that for cancer induction.

Principles of control of radiation exposure

Basic requirements

The International Commission for Radiological Protection (ICRP) has recommended [5] a system of dose limitation based on three fundamental concepts:

(1) No practice shall be adopted unless its introduction produces a positive net benefit. It is usually impossible, in a given instance, to quantify either the benefit or the detriment. The applications to medicine and power production are usually deemed to have an obvious net benefit but not everyone would agree.

(2) All exposures shall be kept as low as reasonably achievable (ALARA), economic and social factors being taken into account. This ALARA principle again usually involves unquantifiable judgements, and the 'economic' aspect implies some form of cost-benefit analysis, which immediately involves the ethical problem of putting a monetary value on either detriment (for example, disease or death) or the risk of detriment.

(3) The dose equivalent to individuals shall not exceed the limits recommended for the appropriate circumstances by the Commission. It is this third concept that usually receives the most attention. It is well received by lawyers, managers and trades unions alike since it usually represents criteria of compliance or otherwise. However, it must be emphasized, first, that the dose limit does not represent a fine dividing line between 'safe' and 'unsafe' conditions and, second, that the dose limit is only a 'long stop' to the general principle of ALARA.

Exposure limits

National and international regulatory bodies set limits (usually annual) of exposure to individuals and to groups (large and small). These are usually broadly based on the ICRP recommendations detailed in their publication number 26 [5] which in turn are based on the risk estimates given previously. Note that background ('natural') radiation and that from medical investigations are not included in the recommended limits since the former is, in general, not controllable and the benefit from the latter is automatically assumed to outweigh any risk. Limits are quoted for both stochastic and non-stochastic effects and both for occupational exposure and the adventitious exposure of the general population. Limits for the latter are lower because of the inclusion of potentially sensitive categories such as the very young, the very old, the sick and the pregnant, and also to minimize the genetic risk associated with exposing large populations. In general, the limits are intended to prevent non-stochastic effects altogether and to limit the occurrence of stochastic effects to an 'acceptable' level. Most of the discussion which occurs about dose limits is concerned with defining 'acceptable'.

Basic limits

Those quoted here are for occupational exposure; for the general population one-tenth of these figures are recommended. For the prevention of non-stochastic effects the recommended annual dose-equivalent limit is,

0.5 Sv (50 rem)

except for the lens of the eye which is,

0.15 Sv (15 rem)

For the limiting of stochastic effects the recommended annual dose-equivalent limit when the whole body is irradiated uniformly is:

0.05 Sv (5 rem)

Since different organs exhibit different degrees of sensitivity weighting factors are applied if the irradiation of the body is not uniform such that the overall estimate of risk should be the same irrespective of the number of organs irradiated. Mathematically this may be expressed thus:

$$\sum_T W_T H_T \leqslant H_{wb}, L$$

where
W_T is a weighting factor representing the proportion of the stochastic risk resulting from tissue T to the total risk, when the whole body is irradiated uniformly
H_T is the annual dose equivalent in tissue T, and
H_{wb}, L is the recommended annual dose equivalent limit for uniform irradiation of the whole body (0.05 Sv).

Values of W_T are given in *Table 8.2*.

Table 8.2 Organ weighting factors

Tissue	W_T
Gonads	0.25
Breast	0.15
Red bone marrow	0.12
Lung	0.12
Thyroid	0.03
Bone surfaces	0.03
Remaining five most sensitive tissues (5 × 0.06)	0.03

Note that if the annual dose limit for a single organ is computed from $0.05/W_T$ and the figure exceeds the non-stochastic limit (0.5 Sv) the latter figure applies.

Doses from internally deposited nuclides

The limits given above for both the whole body and for individual organs refer to the total doses received from both external and internal radiation. In practice the dose from external radiation is usually comparatively easy to measure (from personal dosimeters) that from internal emitters being much more difficult to estimate.

Application of the limits

As stated previously the recommended limits are absolute upper limits with a requirement to keep individual and collective doses as low as possible. Continual exposure for a given individual at or near the limit year after year would only be considered acceptable in exceptional circumstances. Experience has shown that in largish occupation groups operating to an upper exposure limit, the frequency-dose distribution tends to fit a log-normal function with an arithmetic mean of about 10 per cent of the limit and very few values even approaching the limit.

Control of the external hazard

The external radiation hazard results from radiation sources outside the body of sufficient energy to penetrate the (dead) outer layers of the skin. Accepting that the source of radiation used is the smallest which meets the requirements of the specific use or experiment, the methods of control are:

(1) Time
(2) Distance
(3) Shielding

or any combination of these.

Control by time

It is assumed that over the range usually encountered in radiation protection work biological effects are independent of dose-rate, that is

Dose (for a given effect) = dose-rate × exposure time.

Thus exposure can be kept below a given figure by adjusting the working time, but note that control by this method becomes increasingly difficult as the dose-rate increases. Practice on a non-active mock-up can often reduce the time required to do a given job.

Control by distance

The intensity of radiation from a source decreases with distance and may often be used to reduce the radiation exposure. The actual fall-off law depends on the physical size and shape of the source relative to the distance of interest.

FOR A POINT SOURCE

The dimensions of the source are small compared to the distance being considered), the dose-rate is inversely proportional to the square of the distance from the source; this knowledge of the dose-rate at a known distance enables the dose-rate at any other distance to be calculated (assuming air attenuation to be negligible); thus

$$D_1 r_1^2 = D_2 r_2^2$$

where D_1 is the dose-rate at distance r_1 from the source
D_2 is the dose-rate at distance r_2 from the source.

A simple direct consequence of the inverse square law is the dramatic decrease in hand exposure when using (even short) tongs or tweezers when handling sources—which should never be picked up with either bare or gloved hands.

FOR A LINE SOURCE

For a line source of infinite length (long compared with the distance of interest) the dose-rate is inversely proportional to the distance (assuming no air attenuation); thus:

$$D_1 r_1 = D_2 r_2$$

In principle, the fall-off law can be calculated for any size and shape of source although often the algebra becomes tedious. (For the softer radiations the fall-off is more rapid than the above formulae indicates due to air absorption—this is strictly an example of shielding.)

Control by shielding

This may be achieved by putting materials between the source and the exposed person to absorb all or part of the radiation. Roughly speaking, directly ionizing particles (such as alpha particles, beta particles or protons) can be 'stopped', but indirectly ionizing radiation (X-rays, gamma-rays or neutrons) can only be attenuated exponentially.

ALPHA PARTICLES

Since the range in air of even the most energetic alpha particles is only a few centimetres and they are stopped by a sheet of paper or the (dead) outer layer of the skin, they are not normally considered to pose an external radiation hazard. (They do, however, post a very serious *internal* hazard.)

BETA PARTICLES

Although beta radiation is more penetrating than alpha particles, in the energy range produced by most isotopes (up to a few MeV) it can be stopped by using moderate thicknesses of material (say 1 cm of perspex or equivalent). Note that a beta source emits a continuous spectrum of electrons from zero up to a maximum energy (E_{max}). The mean energy for radiation protection purposes is usually taken as $E_{max}/3$.

The intensity of *bremsstrahlung* radiation (secondary X-rays produced by the rapid slowing down of electrons in the source itself and the shielding material) is proportional to the atomic number (Z), of the stopping material, so that beta shielding should be of light materials (perspex or aluminium, for example) to minimize *bremsstrahlung* production. In extreme cases it may be necessary to employ a double shield—an inner layer of light materials to stop the betas and an outer layer of more dense material such as lead to attenuate the *bremsstrahlung* produced in the inner layer.

GAMMA-RAYS AND X-RAYS

These are attenuated exponentially through material:

$$D_t = D_0 e^{-\mu t}$$

where D_t is the dose-rate emerging from a shield of thickness t with incident dose-rate D_0.

The linear absorption coefficient is μ (which has the dimension of inverse length), and is a function both of the energy of the incident radiation and of the material of the shield.

In practice the linear absorption coefficient is only one of various indices used, the others being:

1. *Mean free path (λ) = $1/\mu$*
2. *Half-value layer (HLV) = 0.693λ = $0.693/\mu$*
3. *Tenth value layer (TVL) = 2.303λ = $2.303/\mu$*

Note the similarity (mathematically) between shielding attenuation and radioactive decay.

The above only applies to 'good geometry' where scattered radiation does not reach the point of interest. Under conditions of 'poor geometry', where scattered radiation can reach the point of interest, the required shield thickness is underestimated and may be compensated for by a build-up factor (B) available in tables as a function of radiation energy and shield material and thickness and

$$D_t = BD_0 e^{-\lambda t}$$

NEUTRONS

Fast neutrons are usually shielded by hydrogenous materials such as water, wax or polythene which slow them to thermal energies (less dangerous *per se*) and absorbing the thermal neutrons with (say) cadmium. One may then have to shield the cadmium by lead to attenuate the prompt gammas from the (n,γ) reaction.

Calculating external dose rates

Crude estimates of dose rates from a given source activity are often useful to estimate working times or shielding thickness, and also to estimate doses following an inadvertent exposure.

POINT SOURCE OF GAMMA RADIATION

The following formula gives the dose-rate over the energy range 0.3–4 MeV to within about 20 per cent:

$$D = \frac{143\ BE}{r^2}$$

where D = dose-rate in μ Gy/h.
  B = source strength (GBq).
  E = source energy (MeV).
  r = distance (m).

Note that E is the total energy per disintegration taking into account any branching ratio (that is, the ratio of the particle (or photon) emission rate to the source disintegration rate).

POINT SOURCE OF BETA RADIATION

$$D = \frac{0.81\ B}{r^2}$$

where D = dose-rate in Gy/h.
   B = source strength (GBq).
   r = distance (units of 10 cm).

Note that dose-rate is independent of energy and that the formula neglects air absorption.

POINT SOURCE OF NEUTRON RADIATION

The strength of a neutron source is usually given by the neutron emission rate Q—the number of neutrons emitted per second. The flux ϕ at a distance r from the source is then given by

$$\phi = \frac{Q}{4\pi\ r^2}$$

The dose equivalent-rate is then computed from tables or graphs of neutron flux versus DE rate (this is a function of neutron energy) [6]. Very roughly,

For thermal and slow neutrons: 7×10^6 n/m^2/s $\cong 25\mu$ Sv/h.
For fast (1–5 MeV) neutrons: 2×10^5 n/m^2/s $\cong 25\mu$ Sv/h.

OUTPUT OF X-RAY MACHINES

This can usually be estimated from graphs or tables [6].

Check of efficiency of control measures

The results of any calculated exposure pattern should be subsequently checked by:

1. Monitoring the working environment with a suitable instrument. Care must be taken to ensure that the instrument used is appropriate for the radiation or mixture of radiations from the source. Particular care must be taken when dealing with very low energy sources and with pulsed radiation sources. In both instances the advice of the instrument manufacturer should be sought.
2. Monitoring the individual with appropriate personal dosimeters. The film dosimeter (the 'film badge') is cheap, light and reasonably accurate for a wide energy range of beta and gamma rays and of dose levels. It consists of a piece of fast X-ray film in a plastic holder containing a selection of filters which enable an assessment to be made of both the type, quantity and quality (that is, the energy) of the incident radiation to be made from densitometry of the blackened film. Other available passive dosimeters are based on a thermoluminescent material such as lithium fluoride which when heated after exposure emits visible light—the quantity of light emitted being proportional to the integrated dose incident on the dosimeter. Both the above types of dosimeter require to be processed after exposure to obtain a dose estimate; there are available many active dosimeters which can, in some instances, give an alarm at preset levels.

Such instruments are, however, expensive and require regular maintenance and calibration.

Control of the internal radiation hazard

The internal radiation hazard results when the body is contaminated (either internally or externally) with radioactive material. The presence of radioactive material in or on the body usually presents a more serious problem than exposure to external radiation for two reasons:

1. The radioactive material is in intimate contact with the body tissues and organs resulting in a high local dose-rate (remember the inverse square law).
2. Whereas an external source can be removed or shielded, you cannot do either of these things with an internal source which irradiates the body for a full 168 h/week.

Routes of entry

These are (as for any toxic material):

1. *Inhalation* of radioactive dust or gas.
2. *Ingestion* of active material via drink, food or tactile transfer.
3. *Absorption* of active material through the intact skin.
4. *Direct entry* of active material through a puncture or wound.

The fate of the material once in the body depends on the chemical and physical nature of the material (and to some extent on the person). The dose to a particular organ (other things being equal) depends on the biological half-life (t_b) as well as the radioactive half-life (t_r). These may be combined to produce an effective half-life (t_{eff}) thus:

$$\frac{1}{t_{eff}} = \frac{1}{T_r} + \frac{1}{t_b}$$

Standards

For each radioactive nuclide the ICRP [7] specifies an 'annual limit on intake' (ALI) for both ingestion and inhalation and for different chemical forms (that is, for different solubilities in the body fluids). The ALI is defined as the activity of a radionuclide which would give, in a year, a harm commitment to the irradiated organs equivalent to that assumed for the whole body irradiation specified by the ICRP as an annual limit (currently 50 mSv (5 rem)). The ALI is converted into derived working limits (DWLs) thus:

1. *Derived air concentrations (DAC):* the inhalation ALI of a radionuclide divided by the volume of air inhaled by ICRP reference man [8] in a working year (2400 m^3).
2. *Drinking water concentration (DWC):* the ingestion ALI of a radionuclide divided by the volume of liquid ingested annually by ICRP reference man (0.8 m^3).

3. *Surface contamination*: the DAC is converted to a DWL from a resuspension factor f_r defined as

$$f_r = \frac{\text{Air concentration}}{\text{Surface contamination}}$$

In the absence of known specific values the resuspension factor is usually assumed to be 10^{-6}.

Methods of control

Apart from using the minimum amount of material possible, these, by and large, follow the usual methods for any toxic material:

1. Containment of source.
2. 'Containment' of personnel.
3. Correct procedures.

CONTROL OF THE SOURCE: CONTAINMENT

As many levels as possible (but at least two) should be used; for example (working outwards from source):

1. Immediate containment of the source itself.
2. Splash tray.
3. Ventilated hood (fume cupboard) or glove box.
4. Controlled area with change barriers and contamination monitors.

CONTROL OF PERSONNEL: PROTECTIVE CLOTHING

The use of any protective clothing (laboratory coats, coveralls, caps, gloves, shoes or overshoes) should be confined to the active area involved to avoid the spread of contamination. To be most effective the clothing should be designed so that the wearer can remove it easily, without transferring any contamination either to himself/herself or his/her environment.

CONTROL OF PERSONNEL: RESPIRATORY PROTECTION

The usual rules apply—it is important to emphasize that any respiratory protection device may only be used for those hazards for which they are designed; filter type respirators do not provide protection against radioactive gases. Whenever possible a supplied air mask or hood should be used since as the pressure in the breathing zone is higher than atmospheric, leakage will be from the inside outwards.

CORRECT PROCEDURES

These can be summed up as cleanliness (good housekeeping) and good personal discipline not only technical but personal—no smoking or eating. The design and operational methods of the facility should be such as to encourage the observance of good procedures.

Monitoring procedures

SURFACE CONTAMINATION

This can often be measured directly using a suitably calibrated probe but in some cases, in a high radiation background level or when looking for loose contamination or an active object, for example, a smear (or swab) survey is taken. A filter paper is wiped over a known area (usually about a foot square; $0.09\,m^2$) and then counted using a detector of known efficiency for the nuclide in question; the level of contamination can then be calculated:

$$\text{Surface contamination level } (\text{Bq/m}^2) = \frac{C_c}{E_c A\ E_f}$$

where C_c = counting rate (corrected for background—and dead-time if a geiger detector is used) in cps
E_c = fractional efficiency of counting system
A = area smeared (in m^2)
E_f = fraction of loose contamination wiped off onto paper.

Since E_f is not usually known, it is often taken as 0.1 (10 per cent).

AIR MONITORING

Particulate activity is determined in the same way as other hygiene measurements of airborne levels; a known volume of air is drawn through a filter paper which is then counted on a detector of known efficiency for the nuclide in question:

$$\text{Airborne concentration } (\text{Bq/m}^3) = \frac{C_c}{E_c V}$$

where C_c = corrected counting rate in cps
E_c = fractional efficiency of counting system
V = volume of air sampled in m^3.

Gaseous activity is usually measured by passing the filtered air through a detection chamber directly and the gas to be monitored becomes the gas filling of the detector.

Checking efficiency of control measures

This is performed by:

1. Surveying the area for loose contamination and airborne activity, using methods appropriate for the particular radioactive material involved.
2. Personal monitoring: the intake of radioactivity can be estimated from:

   (a) personal air samplers for particulate activity;
   (b) biological monitoring of urine, faeces and blood; to estimate the retained activity (and hence the dose incurred) knowledge is required of the time of the intake and the effective half-life of the nuclide in the body;
   (c) whole body counting, the measurement of external dose-rate measurements.

Calculation of doses to specific organs

This can often be calculated using the following formula for the dose-rate in an infinite medium containing a uniformly distributed known nuclide or known mixture of nuclides:

$$D = 0.58\,TE$$

where D = dose-rate in μGy/h
T = specific activity in kBq/kg
E = energy per disintegration in MeV

This formula is applicable if the dimensions of the organ are large compared with the range of the particles (or tenth value length in the case of gamma-rays and X-rays).

References

1. National Radiological Protection Board. *Living with Radiation*, HMSO, London (1986)
2. Court Brown W.R. and Doll, R. *British Medical Journal*, **2**, 1327 (1965)
3. Goss, S.G. *NRPB Report R20*, HMSO, London (1974)
4. Reissland, J.A. *et al.* Observation and analysis of cancer deaths among classified radiation workers. *Physics in Medicine and Biology*, **21 (6)**, (1976)
5. ICRP Publication 26. *Recommendations of the International Commission on Radiological Protection.* Pergamon Press, Oxford (1977)
6. ICRP Publication 21. *Data for Protection against Ionising Radiation from External Sources.* Pergamon Press, Oxford (1971)
7. ICRP Publication 30. *Limits for Intakes of Radionuclides by Workers.* Pergamon Press, Oxford (1978)
8. ICRP Publication 23. *Report of the Task Group on Reference Man.* Pergamon Press, Oxford (1975)

Chapter 9

Thermal stresses in occupations

J.R. Allan

Introduction

Man, like other mammals, has the ability to maintain his body temperature within narrow limits. This characteristic is known as homeothermy and conveys important evolutionary advantages by permitting high levels of body activity which are substantially independent of environmental temperatures over a wide range. A disadvantage, however, is that where environmental or other circumstances are such as to exceed the capacity of man's thermoregulatory processes to control his body temperature within a degree or two of 37 °C, then the consequences may be serious. Such consequences range from the mere loss of comfort and contentment to the impairment of physical or mental performance or even, under extreme circumstances, to thermal illnesses or injuries. All these consequences may be seen from time to time in a variety of industries and the occupational physician must be able to understand and recognize them and to give practical advice on control techniques.

The evaluation and control of occupational thermal stress requires knowledge of its origins and of the way in which the physical components of the environment (temperature or humidity, for example) influence the exchange of heat between man and his surroundings. It is also important to understand the highly significant effects of clothing upon these thermal exchanges and the various techniques available for measuring the thermal environment and physiological responses to it. These topics form the subject matter of this chapter; it is not intended to give a detailed account of the physiology of human thermoregulation for which the reader may refer to one of the texts given under the further reading section at the end of this chapter. A brief summary follows as an *aide memoire*.

Thermal physiology

A comfortable person has a deep body temperature close to 37 °C and a mean skin temperature of 33 °C. Skin temperatures over the trunk will usually be 3–4 °C higher than those over the limbs. Heat generated by metabolic processes in the organs and muscles is transferred to the skin surface via the blood supply. In comfortable conditions the metabolic heat load is lost at the skin surface primarily by convection and radiation. In hot conditions heat loss is increased first by vasodilatation which increases the flow of blood to the skin and raises skin temperatures. If this is insufficient body temperature will rise further and sweating commences, thus increasing heat loss by evaporation. In cold conditions heat loss is restricted by vasoconstriction which reduces blood flow to the skin and lowers skin temperature. If this is insufficient to control body temperature then shivering will commence so as to increase metabolic heat production. Repeated exposure to heat leads to modified responses in the sweating mechanism and cardiovascular system which are referred to as heat acclimatization. Sweat losses may increase by up to 100 per cent and body temperature and heart rate during work will be lower than before acclimatization. Repeated exposure to cold does not lead to comparable spectacular changes. Cold acclimatization is a controversial topic and probably without significant practical importance save that subjective tolerance may be improved.

Origins of thermal stress

There are three main sources of thermal stress in industry.

First, there is exposure to hot or cold environmental extremes which may either be those that occur naturally such as in the desert, the jungle or the arctic; or those that are man-made such as are found in the steel, glass and mining industries or in cold stores.

Second, there is the level of physical work involved in a task. An environment which may not be stressful to an individual undertaking sedentary or light work may be highly stressful to an individual undertaking heavy physical work. The converse may be equally true in that the performance of workers in sedentary tasks may be adversely affected if environmental conditions are too cool.

Third, there is clothing. Normally individuals will adjust the clothing they wear to suit the environmental conditions but sometimes this is not possible. In many industries workers are required to wear specific protective clothing to guard against hazards or injuries from chemicals, radiation or other physical dangers in the environment. Such clothing is frequently made from materials impermeable to water vapour which inhibit the evaporation of sweat and thus cause thermal stress as a side-effect.

In practice, of course, all three of these sources of thermal stress may exist together.

Environmental components and their measurement

The important components of the thermal environment are: air temperature, radiant heat, humidity, and air movement or wind.

Air temperature

A measurement of air temperature, sometimes referred to as shade temperature or dry bulb temperature, is the obvious starting point for thermal environment assessments. It is important that techniques for measuring air temperature exclude the possible effects of radiant heating or cooling on the sensor. A common example of errors from this source is when a mercury thermometer is used for measuring air temperature and readings may be erroneously high due to direct radiant heating of the thermometer.

Traditionally, air temperatures are measured with mercury-in-glass thermometers shielded from radiation by being enclosed in some form of screen. It is also common to measure air temperature in conjunction with wet bulb temperature using an instrument known as a psychrometer (see below). Because mercury freezes at about $-50\,°C$, where extremely low temperatures are anticipated it is advisable to use an alcohol thermometer. A disadvantage of mercury or alcohol thermometers, apart from their fragility, is that they do not lend themselves to automatic measurement or recording. It has therefore become commonplace to use various electrical temperature sensors such as thermocouples or thermistors. The latter are now widely available and consist of small beads of semiconductor material the electrical resistance of which varies with temperature in a predictable manner. In conjunction with appropriate electronic circuitry they can be used to measure temperature over a very wide range and with accuracies up to $\pm 0.05\,°C$ or 1 per cent of the range. Readings can be taken either directly from a suitable meter or stored in a solid state data logger or recorded on a chart.

Radiant heat

Radiant heat is given out by all bodies at all temperatures. Radiant exchange between two bodies varies as the difference in the fourth power of the absolute temperature of their surfaces and is influenced by the geometry of the surfaces and by a characteristic known as emissivity. Matt black surfaces emit and absorb radiant heat in greater quantities than white surfaces.

The wavelength of radiant heat emissions depends upon the temperatures of the radiant heat source. In most industrial settings the source temperatures are not particularly high and the radiant heating is in the infrared range. By contrast, solar radiation comes from a source at very high temperatures and includes a considerable short-wave element in the visible band of the spectrum. In a number of practical situations the distinction can be of some importance. For example in work areas surrounded by glass or other transparencies, such as greenhouses or vehicles, the short-wave component of sunlight passes freely through the glass or transparency and warms up those structures on which it falls. The heated structures then re-radiate but the radiation is now in the infrared region of the spectrum and is not readily transmitted by glass or other transparencies. Thus the heat becomes 'trapped' within the glass enclosure—a phenomenon that is known as the 'greenhouse effect'.

The mean radiant temperature of the surroundings can be calculated with reasonable accuracy from measured surface temperatures and corresponding angle factors between the source and the individual exposed. This is a complicated procedure and is described in detail by Fanger [1]. A much simpler technique is to measure the globe temperature, T_g, using a black globe thermometer. In its original form this consists of a hollow copper sphere, 150 mm in diameter and painted matt black. The sphere has a thermometer, either mercury-in-glass or a thermistor, placed at its centre. Because of the inconvenience of the traditional globe thermometer there has been a tendency in recent times to use smaller globe thermometers with a diameter of 50 mm. Mean radiant temperature may be calculated from globe temperature, air temperature and air movement using the following equation:

$$\overline{T}_r = T_g + k.V^{0.5} (T_g - T_a)$$

where $k = 2.2$ for T in °C, V in m/s air movement

T_g is for a 150 mm diameter globe.

If a smaller globe is used then the constant k in the above equation should be substituted by k_d which may be calculated from:

$$k_d = k \left(\frac{0.15}{d} \right)^{0.4}$$

where d is the diameter in metres of the globe used.

Most heat stress indices (see below) which incorporate T_g use a 150 mm globe and it may be necessary to calculate this from readings of the smaller globe and wind speed [2].

In occupational health practice it is not always necessary to go to the complication of measuring mean radiant temperatures because the measurements of globe temperature may be used directly in a range of heat stress indices for predicting and controlling thermal stress. Under very special circumstances where a detailed geography of the radiant surroundings is required, the technique of infrared thermography may be used for which there are a number of highly effective, if expensive, modern instruments.

Humidity

Humidity is the concentration of water vapour in the air and is usually expressed as relative humidity. This is the ratio of the actual amount of water vapour in the air to the amount that would be present if the air were saturated at the same temperature—expressed as a percentage. Occasionally humidity is expressed in absolute terms as mass per unit mass of air (kg/kg) or as a partial pressure (mm of mercury (mmHg) or kPa). Relative humidity does not directly convey the amount of water vapour in the air because this depends on temperature. For example air at 40 °C with 50 per cent relative humidity contains very much greater quantities of water vapour than air at 20 °C and 50 per cent relative humidity. An advantage of expressing humidity as a partial pressure is that it gives an immediate indication as to the likely effect on sweat evaporation. For example, we know that sweating skin at a temperature of say 38 °C has a water vapour pressure at the surface of 5.6 kPa. If the environment also has water vapour pressure of 5.6 kPa, there is clearly no vapour pressure gradient available for sweat evaporation.

Classically, humidity is measured by measuring the wet bulb depression. If the bulb of a thermometer (or a thermistor for that matter) is covered with a wick and kept wet with distilled water and placed in an air stream, the evaporation of the water cools the thermometer which therefore reads below a dry bulb thermometer in the same situation. These measurements are usually obtained by using an instrument known as a psychrometer which includes both the dry bulb and wet bulb sensors. Air flow over the thermometers is induced either by physically swinging them through the air, as in a whirling hygrometer, or by mounting the thermometers in tubes through which air is drawn by an electrically driven fan as in the Assman psychrometer. The readings of dry bulb and wet bulb temperature obtained with a psychrometer are used to enter a psychrometric chart to obtain relative or absolute humidity as shown in *Figure 9.1*.

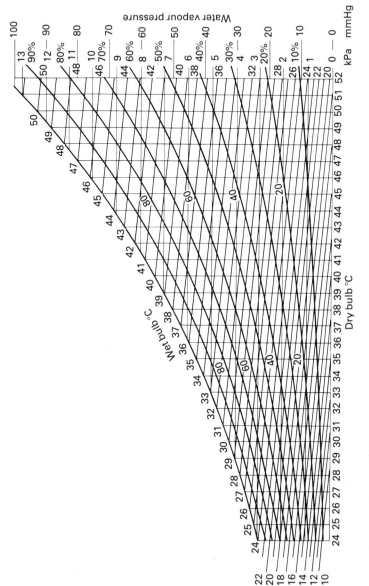

Figure 9.1 A psychrometric chart: the chart is entered with wet and dry bulb temperatures. The point of intersection can be read across to the right-hand scale for absolute humidity. Relative humidity is obtained by interpolation between the lines of equal relative humidity given in 10 per cent steps

In recent years the use of psychrometers for the measurement of humidity has been less common due to the advent of a number of solid state humidity sensors. Modern 'solid state' humidity sensors employ a thin film polymer capacitor element specially designed to absorb atmospheric water vapour and whose capacitance varies with ambient relative humidity. They are unsuitable at low water vapour concentrations and extremes of temperature, and must be protected from contamination by pollutant particles especially those containing sulphur. A typical accuracy of ±2 per cent relative humidity can be maintained by regular recalibration. These sensors can lend themselves conveniently to automatic data logging techniques.

Air movement

Air movement has an extremely important effect on human heat exchange both at high temperatures, where it affects convection and sweat evaporation, and at low temperatures where it produces the well-known windchill effect (see below). In locations out of doors air movement is usually unidirectional and can be measured using traditional instruments such as vane or cup anemometers. In these instruments the speed of rotation of the cups or vanes is calibrated in terms of wind speed. The speed of rotation is either measured by mechanical means or, in the more modern instruments, by electronic pulse counters. In industrial settings air movement is frequently omnidirectional and in these circumstances the traditional vane and cup anemometers are liable to errors. Omnidirectional air flow can be measured using devices based on a hot wire anemometer principle which measure the cooling effect of air flow over a heated wire. In modern instruments the heated wire has been replaced by a heated thermistor but otherwise the principle is the same.

A range of environmental measuring instruments is shown in *Figure 9.2*.

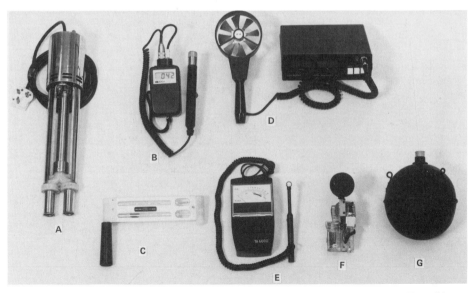

Figure 9.2 Environmental measuring instruments. *Key*: A. Assmann psychrometer; B. solid state humidity sensor; C. whirling psychrometer; D. vane anemometer; E. hot wire anemometer; F. combined 50 mm globe thermometer and psychrometer; G. 150 mm black globe thermometer

Human heat exchange

A naked man exchanges heat with his environment via one or more of four physical pathways, conduction (K), convection (C), radiation (R) and evaporation (E). If body temperature is to remain constant the net result of his heat exchanges with his environment must be a loss to the environment of an amount of heat equal to his metabolic heat production (H). If this balance is not maintained then the excess heat lost or gained must alter the total body heat content (Q) and lead to a change in body temperature. Man's conversion of food energy into physical work is not an efficient process and for most activities something like 80 per cent of the metabolic energy production appears as heat. Thus for a man in heat balance with his environment we may write an equation as follows:

$H = M - W = E + R + C + K + S$

where H = metabloic heat production
M = metabolic energy production
W = rate of external work
E, R, C and K are the rates of heat loss by evaporation, radiation, convection and conduction respectively
S is the rate of storage of heat in the body

All quantities may be expressed in watts per square metre of body surface area (W/m^2). A gain of heat by any channel is a negative loss of heat.

For a given individual any imbalance in his heat exchange with his environment which results in heat being stored or lost in his body will lead to changes in the total body heat content (Q) which are related to his body mass (M), the mean specific heat of human tissues $(c = 3.48 \times 10^3 \text{ J/kg/°C})$ and changes in mean body temperature, ΔT_b by using the following equation:

$Q = c/M \, (\Delta \overline{T}_b)$

Thus, for a given imbalance in heat exchange with the environment a heavy man's body temperature will change less than that of a light man.

It is a preoccupation of human thermal physiologists and of those interested in developing mathematical models of human thermoregulation, to endeavour to produce mathematical relationships by which heat exchange by conduction, convection, radiation and evaporation may be exactly calculated. The accuracy of these calculations is somewhat controversial and, based as they usually are on naked persons, great difficulties arise in predicting the complicated and varied influence of clothing upon these relationships. Nevertheless, in a majority of practical circumstances in industry the occupational physician will be concerned with clothed persons, indeed some of them clothed in complicated protective assemblies. However, the value of a short account of the mathematical description of human heat exchange lies in showing the way in which the various environmental components affect each exchange pathway. Beyond that the reader is advised to adopt a healthy scepticism as to the precision and usefulness of these expressions when applied to individuals, although they can be useful when mean responses of large groups are required.

Conduction

Heat is exchanged by conduction between objects at different temperatures in contact with each other. Conduction is relatively unimportant in human heat exchange except in the rather special circumstances of cold water immersion. Conductive heat transfer (K) depends on the temperature difference between the objects in contact and the conductance between them as follows:

$$K = k\,(T_1 - T_2)$$

The conductance (K) is the property of a material or interface between objects which determines the rate of heat transfer/unit area/unit temperature difference and the normal units are W/m² °C. It may be convenient to think of conductance as the inverse of resistance.

Convection

When a fluid (liquid or gas) at one temperature flows over a surface at a different temperature heat is gained or lost by convection. In the case of a naked person in air, the rate and direction of convective exchange depends on the temperature difference between the air and the skin surface. The situation is more complicated in a clothed individual and this is dealt with below. Convective heat transfer may be calculated as follows:

$$C = h_c(\overline{T}_{sk} - T_a)$$

where C = the heat exchange by convection (W/m²)
h_c = convective heat transfer coefficient, (W/m² °C)
$\overline{T}_{sk}$ = mean skin temperature (°C),
$\overline{T}_a$ = dry bulb temperature (°C).

The convective heat exchange coefficient, h_c, is not constant under all environmental conditions and depends upon the rate of air movement. It may be calculated from:

$$h_c = 8.3\ V^{0.5}\ (\text{W/m}^2\,°\text{C})$$

where V is the air movement in m/s.

The important point to note in this relationship is that heat exchange by convection depends both on the temperature difference between the body and the surrounding air or water and on the rate of movement of the air. It does not matter whether the latter is naturally or artificially generated.

Evaporation

When water evaporates from a surface, energy is absorbed during transition from the liquid to the gaseous state. This energy is termed the latent heat of vaporization, and in the case of sweat evaporation a figure of 2500×10^3 J/kg is approximately correct. This figure emphasizes the extraordinary power of the human sweating mechanism as a heat loss pathway.

In environments where the air temperature is the same or higher than skin temperature then sweating is the sole means available for dissipating the metabolic heat production. Under such conditions anything that limits evaporation, such as high ambient humidity or impermeable clothing, will rapidly lead to heat storage and a rise in body temperature.

Heat loss by evaporation (E) depends upon the vapour pressure gradient between the skin surface and the ambient air; it may be calculated from:

$$E = h_e(P_{sk} - P_a)$$

where E = evaporative heat loss (W/m^2)
           h_e = the coefficient of evaporative heat exchange W/m^2 kPa
           p_{sk} = water vapour pressure at the skin surface (kPa)
           p_a = the ambient water vapour pressure (kPa)

The coefficient for evaporative heat exchange (h_e) incorporates the latent heat of vaporization of sweat and the highly important effect of air movement on evaporation. For practical purposes h_e may be calculated from:

$$h_e = 124 \, V^{0.5} \text{ W/m}^2 \text{ kPa}$$

where V is the air movement in m/s

It is possible to show [3] that for a given air movement there is a constant relationship between the heat exchange coefficients for evaporation and convection, the ratio h_e:h_c being approximately 15. This relationship holds good for a human subject of given size in air of fixed properties.

As with convection the important point to note from the above is that sweat evaporation is determined by the vapour pressure gradient between the skin and the ambient air and the rate of air flow across the skin.

Radiation

Exchange of heat by radiation between two surfaces depends upon the difference in the fourth powers of the absolute temperatures of the two surfaces. In practice it is often acceptable to use a first power relationship as follows:

$$R = h_r \, (\overline{T}_{sk} - \overline{T}_r)$$

where R = the heat exchange by radiation, (W/m^2)
           h_r = the first power combined radiation coefficient
           T_{sk} = mean skin temperature, (°C)
           T_r = mean radiant temperature of the surroundings, (°C)

The coefficient h_r depends upon the temperature of the two surfaces, the geometrical relation between them and such characteristics of the surfaces as emittance and reflectance. Thus it may be seen that heat transfer by radiation depends principally on the temperature of the two surfaces concerned and is unaffected by air movement or the distance between the surfaces. It can take place across a vacuum.

In most industrial settings hot surfaces in the surroundings radiate in the long infrared. A notable exception to this is the radiation received from the sun in the open air. Here the radiation is substantially in the visible range (0.4–0.7 μm). In the desert the heat energy derived from solar radiation can exceed 1000 W/m^2 body surface area. For practical purposes the value of h_r may be taken as 5.2 W/m^2 °C.

Operative temperature

It will be helpful here to introduce the concept of operative temperature which has some useful practical applications. It will have been noticed above that the processes of convection and radiation have in common that each depends on the difference between skin temperature and an environmental temperature either T_a or T_r. Operative temperature is the result of efforts to combine the air temperature and the mean radiant temperature into a single figure and is defined as the uniform temperature of a radiantly black enclosure in which an occupant would exchange the same amount of heat by radiation plus convection as in the actual non-uniform environment. Operative temperature is numerically the average, weighted by respective heat transfer coefficients (h_c, h_r), of the air and mean radiant temperatures. Thus:

$$T_o = (h_c \, T_a + h_r \, \overline{T}_r)/ (h_c + h_r).$$

At air speeds of 0.4 m/s or less and $T_r < 50\,°C$, operative temperature is approximately the simple average of the air and mean radiant temperatures. It can also be shown that the operative heat exchange coefficient, h_o, is equal to the sum of the heat exchange coefficients for convection and radiation, $h_c + h_r$. *Table 9.1* [4] gives the values for all the heat exchange coefficients at various wind speeds and will facilitate calculations of likely heat balance under a variety of conditions.

Table 9.1 Heat exchange coefficients at various wind speeds h_c and h_e are calculated as $8.3 \cdot V^{0.5}$ and $124 \cdot V^{0.5}$ respectively; h_r is assumed to be 5.2 W/m² °C; $h_o = h_c + h_r$

V m/s	h_c W/m² °C	h_e W/m² °C	h_o W/m² °C
0.1	2.6	39	7.8
0.2	3.7	55	8.9
0.3	4.5	68	9.7
0.4	5.2	78	10.4
0.5	5.9	88	11.1
0.6	6.4	96	11.6
0.7	6.9	104	12.1
0.8	7.4	111	12.6
0.9	7.9	118	13.1
1.0	8.3	124	13.5
1.2	9.1	136	14.3
1.4	9.8	147	15.0
1.6	10.5	157	15.7
1.8	11.1	166	16.3
2.0	11.7	175	16.9
2.5	13.1	196	18.3
3.0	14.4	215	19.6
3.5	15.5	232	20.7
4.0	16.6	248	21.8
4.5	17.6	263	22.8
5.0	18.6	277	23.8

The thermal effects of clothing

The descriptions of human heat exchange given above are related to unclothed subjects. However, workers are normally clothed and sometimes wear special protective clothing for a variety of reasons. The effects of clothing on human heat exchange are extremely complicated and difficult to describe in exact mathematical terms. The reader who wishes to pursue this difficult topic in greater detail is referred to Kerslake [4].

Such detailed knowledge, however, is unnecessary for the practising occupational physician but he should understand some of the general principles involved so that he will have a basis on which to formulate practical advice.

In general, clothing impairs the loss of heat to the environment. While this is a considerable advantage under cold conditions it may be a serious disadvantage under warm conditions. The impairment of heat loss effects all four physical pathways—conduction, convection, radiation, and evaporation. Impairment through the conductive pathway is of some importance in industrial workplaces where it provides the basis for protecting workers from injuries due to contact with very hot or cold surfaces. This topic is dealt with in greater detail in Chapter 21 on protective clothing.

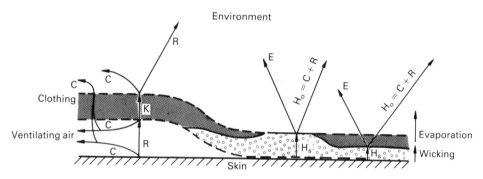

Figure 9.3 Diagram of heat exchange through clothing by convection, radiation and evaporation (after Kerslake [4]. *Key:* C = heat exchange by convection; R = heat exchange by radiation; E = heat exchange by evaporation; H_s = net heat loss from the skin; H_0 = sensible heat loss = $C + R$

The effect of clothing on heat loss by convection and radiation is to decrease conductance between the skin surface and the surrounding air. The general situation is illustrated in *Figure 9.3*. Sensible heat from the skin must first pass through the barrier represented by the clothing insulation, I_c. At the clothing surface the heat transfer coefficients for convection and radiation, h_c and h_r, may be combined into the operative heat transfer coefficient, h_0, as described above. The reciprocal of h_0 is referred to as the insulation of the environment, $I_a = 1/h_0$. The total resistance or insulation between the skin and the environment in a clothed subject is the sum of the clothing insulation, I_c, and the insulation of the environment, I_a, thus:

$$I = I_c + I_a.$$

This describes the situation in its simplest form represented by a sedentary individual in still air. However, if the individual is working, or there is a wind, then things may change dramatically. In outdoor conditions where there is a wind, or even in some indoor conditions where there is significant air movement induced by ventilation installations, the effective insulation of a clothing assembly is reduced by reductions both in I_a and I_c. The reduction in I_c will depend largely on the degree of wind penetration into the clothing system. Highly permeable open weave garments will lose their effective insulation significantly whereas impermeable garments or garments with impermeable outer layers will not do so, although they will be subject to decreases in I_a due to disturbance of the boundary air layers by the wind.

The effect of physical work on clothing insulation arises through two mechanisms. First, physical work such as walking may induce air penetration of the clothing layers in much the same way as wind. Equally important is a phenomenon known as bellows action or pumping. This comes about when the exercise induces exchanges of the air beneath the clothing with ambient air either directly through the clothing in permeable fabrics or through openings at the wrist, ankles or neck. Bellows action is a useful mechanism for promoting heat loss during exercise in workers dressed for cold conditions at rest.

With the increasing use of immersion suits or survival suits for the protection of workers in the offshore oil industry, the fishing industry or the merchant marine, the importance of clothing insulation as protection in the event of immersion in cold water has been more widely recognized in recent years. The prime function of the immersion suits or survival suits is to preserve clothing insulation by keeping it dry.

The insulation provided by clothing is largely determined by the thickness of the trapped air layer within the clothing fabrics themselves or between layers of fabrics. Very little is contributed by the physical characteristics of the fibres themselves. For this reason it is possible to estimate a mean figure for insulation by measuring the thickness of the clothing including the air layers when the insulation may be calculated as $0.25\,°C\ m^2/W$ per 10 mm of thickness. If the air contained within the clothing is displaced by water leakage during immersion or exposure to rain or spray, then its insulation is dramatically decreased, as shown in *Figure 9.4*. A wet suit constructed from closed cell neoprene foam can also provide effective protection in water because the air contained within the cells is not displaced by water during immersion and the suit limits convective heat loss at the skin surface. However, it is interesting to note in passing that a loose-fitting wet suit may exhibit a significant loss of effective insulation due to the water equivalent of the bellows action described for air above. This is known as flushing and the exchange of water between the layer beneath the suit and the open sea can effectively halve the insulation provided by the wet suit [5].

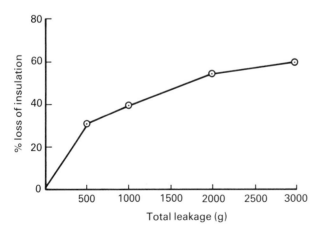

Figure 9.4 Figure showing percentage loss of insulation plotted against water leakage into the insulation worn beneath an immersion suit

Clothing insulation may be measured by a number of techniques. Traditionally this is done by measuring the insulation of samples of the clothing fabric on a device known as a guarded hotplate [6]. The difficulty with this technique is computing the overall insulation of a multilayered clothing assembly from the individual figures for each layer since these do not allow for the variable air layers trapped between the layers of fabric or for the rather complicated boundary effects. This may, however, be done approximately by multiplying the sum of the individual insulations by 0.82 [7]. Many groups of workers have used heated metal manikins of various kinds for the measurement of clothing insulation and these have the advantage of measuring the overall insulation for whole clothing assemblies rather than for individual items [8]. Modern manikins have the capability of leg and arm movements thus realistically reproducing the pumping effects of activity [9]. Immersible manikins may also be used for measuring the effective insulation of immersion protective clothing when they have the additional advantages of assessing effects of leakage of water into the insulation and of the loss of insulation due to hydrostatic compression on immersion [10].

Clothing and radiation

The effect of clothing on radiation heat exchange depends significantly on the wavelength of the radiation. In respect of radiation in the long infrared, clothing of any colour acts more or less as a black body and absorbs the radiation. Nevertheless in the case of severe infrared radiation the clothing may still provide protection against skin burns. In the case of ultraviolet radiation this may be significantly reflected by white clothing which therefore conveys an advantage upon the wearer. This consideration, however, is usually outweighed by the advantages of loose-fitting cotton garments on ventilation and sweat evaporation, and this is why the Bedouin in the desert are not persuaded to exchange their black loose-fitting clothing for white.

Clothing and evaporation

Clothing materials impede heat transfer by evaporation more than sensible heat transfer. This restriction to evaporative heat transfer varies over a very wide range, from being virtually total in the case of impermeable fabrics such as neoprene-coated nylon, to being insignificant in the case of thin cotton fabrics.

In the case of impermeable fabrics sweat evaporating from the skin will rapidly saturate the microenvironment between the skin and the garment and evaporation will then cease. In practice this is usually not a complete process because openings at the wrist, ankles or the neck permit exchange of the saturated air beneath the clothing with the environment and this allows some evaporation to proceed. Nevertheless the wearing of impermeable garments is almost always associated with severe loss of comfort and a significant risk of heat stress due to lack of sweat evaporation.

In recent years a number of new fabrics have appeared on the market which have the property of being permeable to water vapour while being waterproof in the sense that droplet water will not pass through. Somewhat exaggerated claims are sometimes made on behalf of these fabrics but they do appear to reduce the disadvantages of impermeable materials in terms of comfort and sweat evaporation. The fabrics are either constructed as laminations, with a middle layer of microporous polytetrafluoroethylene (PTFE), or are coated with a new modified polyurethane which transports water by a chemical pathway. They are dealt with in more detail in Chapter 21 on protective clothing.

When clothing is permeable both to water vapour and to sweat, evaporation may occur at any level from the skin surface to the clothing surface (see *Figure 9.3*). Thus with loose-fitting, thin, cotton clothing and substantial ventilation of the microclimate, evaporation directly from the skin may predominate. With less permeable fabrics or multilayered clothing assemblies the sweat may be wicked through the clothing and evaporate either at some intermediate layer within the clothing or from the surface. Clearly when this occurs the cooling effect upon the person will be less effective. In order to develop a method for describing the effect of clothing on evaporation a number of indices have been described, for example the permeability index [11] or the permeation efficiency factor [12].

Woodcock's permeability index, I_m, is based on the ratio of the conductance of a clothing system for water vapour, k_e, to that for sensible heat, k_s. The ratio k_e/k_s is compared with a standard for air for which the corresponding ratio is h_e/h_c. The dimensionless permeability index is thus calculated from $k_e/k_s/h_e/h_c$ and ranges from 0 in the case of impermeable fabrics to unity in the case of air.

Evaporative heat transfer through clothing, where sensible heat and water vapour transfer are at rates well in excess of those possible by molecular diffusion, can be determined from:

$$E = 16.5 \frac{P_a - P_{sk}}{I}$$

where

E = evaporative heat transfer (W/m^2)

$P_a - P_{sk}$ = vapour pressure difference between the skin and the air (kPa)

I = insulation (m^2 °C/W)
(the constant has the dimensions °C/kPa)

Woodcock [11] suggested expanding the above equation so as to allow for the water vapour transfer characteristics of different clothing by including his dimensionless permeability index (i_m) to give

$$E = 16.5 \, i_m \frac{P_a - P_{sk}}{I}$$

From this it may be seen that the ratio $i_m{:}I$ is highly significant in determining the evaporative heat transfer through a clothing system and it is widely used. It represents the fraction of the maximum evaporation cooling possible in a given environment without wind (insulation (I) being a 'still air' determination).

Indices of thermal stress

It will be appreciated that there are a great many factors which contribute to the overall stress of thermal environments and the resulting thermal strain in individuals exposed to them. These factors include the environmental components—air temperature, humidity, air movement and radiant temperature; physical properties of the body such as shape, size, movement and skin colour; the physical characteristics of the clothing assembly worn; the physiological character-istics of the individual and his work rate. This long list of variables can prove rather alarming to the occupational physician faced with giving simple, succinct advice for the control of thermal stress in the workplace.

Many attempts have been made to combine the above factors into a simple numerical description or index, capable of predicting the likely level of thermal stress with some degree of accuracy. The fruits of these endeavours may be divided into three main groups.

Group 1

Indices in this group are based on an analysis of heat exchange and inevitably involve complex mathematical calculations. Examples are the heat stress index of Belding and Hatch [13] (HS1) and the index of thermal stress of Givoni [14], ITS. On the cold side this group includes the well-known windchill index [15].

Group 2

Indices in this group are based empirically on physiological observations. Examples are the predicted 4 h sweat rate (P_4SR) [16], the wet bulb globe temperature index (WBGT) [17], and the wet dry index (WD), also known as the Oxford index [18].

Group 3

Includes indices based on immediate subjective sensations of warmth on entering an environment. The most common example is the effective temperature scale (ET) [19] and a derivative allowing for radiant heat which is known as the corrected effective temperature (CET) [20].

For detailed descriptions of these indices the reader is referred to the references given for each. In practice, however, many of them have received minimal practical application and they may be regarded as the toys of environmental physiologists. Notable exceptions, however, are the effective temperature scales, the wet bulb globe temperature index and the windchill index. These will therefore be described in greater detail.

Effective temperature (ET)

Effective temperature scales were developed by Yaglou and his associates in the 1920s [19]. They are based on the immediate subjective impression of warmth gained when subjects walk from one environment to another. The scales take the form of nomograms, the normal scale being for subjects wearing normal indoor clothing and the basic scale for subjects stripped to the waist. *Figure 9.5* gives the nomogram for the effective temperature normal scale. It can be seen that the index allows for air temperature, wet bulb temperature and air movement. The index was originally criticized for not making allowance for radiant heating and a number of modifications have been proposed to make good this deficiency. The simplest of these is merely to enter the scale using globe temperature in place of air temperature and the resulting index is then referred to as the corrected effective temperature (CET) [20].

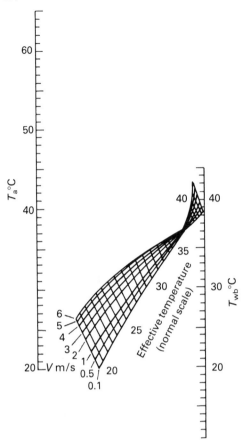

Figure 9.5 Nomogram for effective temperature (normal scale). A line is drawn between the measured dry bulb and wet bulb temperatures. ET is read at the point of intersection of this line with the appropriate curve for the measured wind speed

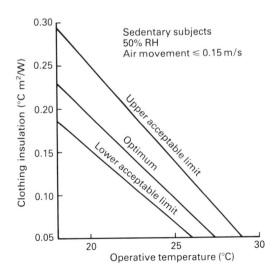

Figure 9.6 Clothing insulation necessary for comfort in sedentary subjects at various operative temperatures (after ASHRAE Standard 55-81[7])

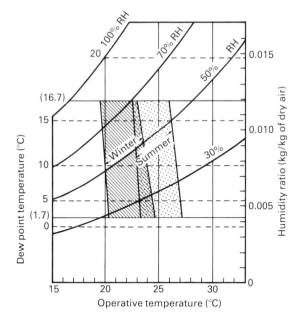

Figure 9.7 Acceptable ranges of operative temperature and humidity for comfort in persons clothed in typical summer and winter clothing undertaking light activities (after ASHRAE Standard 55-81[7])

Since the effective temperature scale was devised on the basis of subjective sensation, it is hardly surprising that it has proved inaccurate as a predictor of physiological response. It makes no allowance for work rate. Specifically the effective temperature scales tend to overestimate the effect of high humidity in comfortable conditions and underestimate the harmful effects of low wind speed and high humidity in the heat. Its main application, therefore, has been within the comfort range and since in this range the effect of humidity is relatively small, several authorities have abandoned effective temperature scales in favour of the simpler operative temperature [7]. Examples of the use of operative temperature to define both the level of clothing insulation required for comfort, and acceptable ranges of operative temperature and humidity for individuals clothed in typical summer and winter clothing are given in *Figures 9.6* and *9.7* respectively. For industrial applications involving moderate or high work rate it is recommended that the WBGT index be used as described below.

The wet bulb globe temperature index (WBGT)

The WBGT was originally developed by Yaglou and Minard [17] as a simple index for use in controlling heat casualties in the United States Marine Corps during work in the open desert. As originally described it was calculated from measurements of a standard 150 mm black globe thermometer, air temperature and a wet bulb thermometer which was not artifically ventilated as in a psychrometer but merely exposed to the ambient air movement and radiation. The index is calculated as follows:

$$\text{WBGT} = 0.7\,T_{\text{wb}} + 0.2\,T_{\text{g}} + 0.1\,T_{\text{a}}$$

An alternative formula has been proposed for use with a psychometric wet bulb temperature (obtained using an artificially ventilated wet bulb) as follows:

$$WBGT = 0.7\ T_{wb} + 0.3\ T_g$$

A derivative of the WBGT known as the wet, dry or Oxford index may be used where radiant heat loads are absent and is calculated from:

$$WD = 0.85\ T_{wb} + 0.15\ T_{db}\ °C.$$

The WBGT has the great advantage of simplicity. It is also capable of being measured directly, with appropriate electronic instrumentation doing the calculation 'online'. As a predictor of thermal strain it is most accurate under the conditions for which it was originally devised, that is in the open air with wind movement and a radiant heat load. However, WBGT does not allow *per se* for work rate or clothing, and if used to establish limits for environmental control purposes it must be clearly understood that the limits will be related to specific rates of working and specific clothing assemblies. In most industrial settings work rates and clothing are known quantities and recommendations for WBGT limits can be made with this knowledge. The detailed application of the WBGT index in the control of industrial thermal stress is dealt with below.

Windchill index and equivalent chill temperature

The index of windchill was originally developed by Siple [15] in 1945 from experiments on the time taken to freeze a cylinder of water in a range of temperatures and windspeeds. A windchill chart is shown in *Figure 9.8* from which

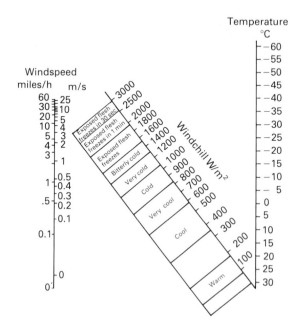

Figure 9.8 A windchill chart: a line is drawn between the measured air temperature on the right-hand scale and the measured windspeed on the left-hand scale. Windchill is read from the centre scale in W/m² or descriptive terms

Wind speed		Air temperature (°C)																	
mph	m/s	−2	−1	−4	−7	−9	−12	−15	−18	−21	−23	−26	−29	−32	−34	−37	−40	−43	−46
1–5	0.5–2.3	−1	−4	−7	−9	−12	−15	−18	−21	−23	−26	−29	−32	−34	−37	−40	−43	−46	−48
6–10	2.6–4.4	−7	−9	−12	−15	−18	−23	−26	−29	−32	−37	−40	−43	−46	−51	−54	−57	−59	−65
11–15	5.0–6.7	−9	−12	−18	−21	−23	−29	−32	−34	−40	−43	−46	−51	−54	−57	−62	−65	−68	−73
16–20	7.2–9.0	−12	−15	−18	−23	−29	−32	−34	−40	−43	−48	−51	−54	−59	−62	−68	−71	−76	−79
21–25	9.5–11.3	−12	−18	−21	−26	−29	−34	−37	−43	−46	−51	−54	−59	−62	−68	−71	−76	−79	−84
26–30	11.7–13.5	−15	−18	−23	−29	−32	−37	−40	−46	−48	−54	−57	−62	−65	−71	−73	−79	−82	−87
30+	>13.5	−15	−21	−23	−29	−34	−37	−40	−46	−51	−54	−59	−62	−68	−73	−76	−82	−84	−90

Little danger or frostbite	Increasing danger of frostbite	Great danger of frostbite

Figure 9.9 Equivalent chill temperature chart. The chart is entered with local air temperature and windspeed. The intersection gives the equivalent chill temperature in °C (see text)

it may be seen that the chart is entered with inputs of air temperature and wind speed. The centre windchill scale may be read in physical terms as W/m^2 or in subjective terms as indicated.

The windchill index is useful in predicting frostbite for exposed dry skin, but is less satisfactory as a basis for deciding suitable clothing for the protection of personnel working outdoors. For this purpose the equivalent chill temperature (or equivalent still air temperature) is more useful. The equivalent chill temperature is the still air temperature that would produce the same cooling as the temperature and windspeed actually measured. An equivalent chill temperature chart is given in *Figure 9.9*.

Evaluation and control of occupational thermal stress

In the final section of this chapter it is proposed to show how the information given above may be applied directly to the evaluation and control of specific thermal stresses in industry. This will be done in three sections. The first will outline the objectives of such thermal evaluations and control techniques. This will be followed by a section dealing with the techniques for evaluating industrial thermal stresses. Finally the range of control techniques will be described. The general approach to evaluation and control of industrial thermal stress is illustrated by the flow chart in *Figure 9.10*.

Objectives

The first objective of any evaluation and control technique is the prevention of thermal illness and injuries. This includes not only the prevention of classical thermal illnesses such as heat exhaustion, heat stroke or hypothermia but also thermally induced injuries such as contact or thermal burns, frostbite or non-freezing cold injuries.

In relation to the prevention of thermal illness and injuries this consideration should be extended to include the exacerbation of non-thermal illnesses by exposure to uncomfortable or stressful environments. A number of conditions can be exacerbated by thermally stressful environments and these include a range of skin conditions and cardiovascular problems.

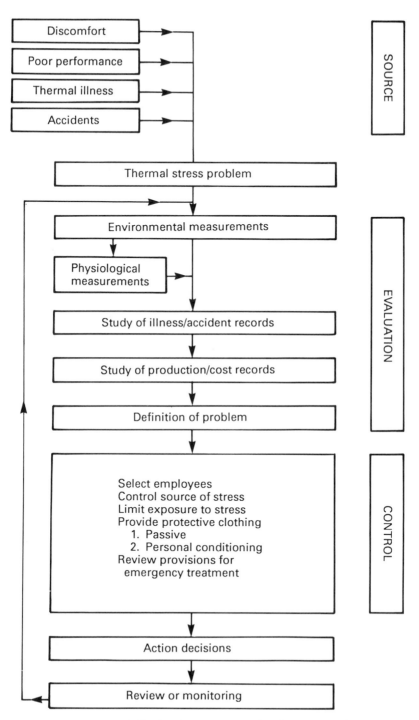

Figure 9.10 A flowchart indicating the steps to be taken in the evaluation and control of industrial thermal stress

In the unlikely case where managements are not persuaded to install adequate control techniques by humanitarian or economic arguments then the provisions of the Health and Safety at Work Act (1974) may be invoked to persuade them to do so. Section 2 (1) of that Act imposes a broad and all-embracing duty on employers to ensure, so far as is reasonably practical, the health, safety and welfare at work of all employees. Then there are the more specific requirements in Section 2 (2)e of the Act under the heading 'Safe Working Environment' requiring employers to provide and maintain a working environment for employees that is, so far as is reasonably practical, safe without risks to health and adequate as regards facilities and arrangements for their welfare at work.

A second objective in the control of occupational thermal stress is the maintenance of adequate performance whether the task be a purely physical, unskilled one or a skilled or mental task. Where thermal conditions overtly affect the performance of workers in industry the economic consequences are usually obvious and the occupational physician will have no difficulty in persuading management to install adequate control measures. For example, if workers in a cold store have to be removed to a suitable place to warm up for 20 min in every hour this will result in a 30 per cent increase in labour costs. The provision of adequate thermal protective clothing may enable full shifts to be worked without breaks and the economic argument for such measures can be persuasive. Equally,

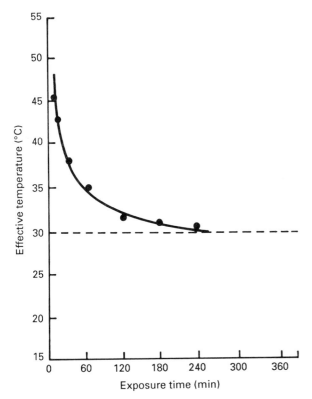

Figure 9.11 Graph showing upper limits for unimpaired mental performance in terms of effective temperature and exposure time (after Wing[23])

workers in a sedentary task such as typists can show marked decrements in performance if their working environment is uncomfortably cold.

Generally speaking conditions of environment and work rate that lead to sustained deep body temperatures in excess of 38 °C are likely to lead to decrements in performance particularly where the task has a high level of skill or mental activity [21, 22]. *Figure 9.11* demonstrates how the duration of work for which performance is unimpaired decreases with increments in effective temperature [23].

A final objective of control measures for thermal stress is the maintenance of contentment among a labour force. Thermal discomfort, which may fall well short of the levels of stress at which thermal illnesses or impaired performance may be seen, can nevertheless lead to general disgruntlement among a work force and to irritability among its members. This in turn can lead to a loss of job satisfaction coupled with poor performance at supervisory or management skills.

Evaluation

The cornerstone of a thermal evaluation is measurement of the environmental thermal components, the air temperature, mean radiant temperature (or globe temperature), humidity and air movement, using the techniques described earlier. These measurements should be made at sites that are relevant to the positions at which individuals actually work [24].

If a preliminary check indicates marked thermal gradients then measurements should be made 0.1, 1.1 and 1.7 m above the ground for standing workers, or 0.1, 0.6 and 1.1 m for sitting workers, corresponding to the height of the head, abdomen and ankles. The mean WBGT is then calculated from

$$\text{WBGT} = \frac{\text{WBGT head} + (2 \times \text{WBGT abdomen}) + \text{WBGT ankles}}{4}$$

Occasionally it may be adequate to make simple spot measurements here and there, but more often it will be appropriate to arrange for the continuous recording of the variables throughout a full working shift under a range of seasonal conditions. For this purpose modern recorders and data logging equipment are extremely useful. The manner in which these measurements are subsequently used to institute control procedures is described below.

The measurement of environmental components alone is not always sufficient to make an adequate assessment of thermal stress. This particularly applies when complicated multilayered or impermeable protective clothing is worn or when work rates are highly variable. Under these conditions environmental measurements should be supplemented by measurements of physiological variables such as deep body and skin temperature, heart rate and sweat loss. Where a thermal stress situation is so severe as to warrant the installation of personal conditioning systems, such as air-ventilated or water-cooled suits, then it will almost certainly be necessary to make a physiological appraisal as part of the evaluation procedure.

The final part of a comprehensive thermal evaluation will include a study of the illness and accident records of the place of employment to determine whether there is any specific evidence of heat or cold-induced illness on injuries, or an above-average incidence of accidents that may be unexplained.

Finally, but not necessarily of least importance, it is worth enquiring into the production/cost records of the workshop or factory. Occasionally these may show evidence of seasonal variations in relation to external temperature conditions or periods of poor performance associated with weather extremes. In studying this aspect the occupational physician must be on his guard not necessarily to interpret as cause and effect relationships between thermal conditions and production which may have other explanations. A simple example of such erroneous attribution is that warm summer weather is frequently associated with an increase in the number of staff on holiday or in absenteeism.

Control techniques

Control techniques available for occupational thermal stress fall into four categories—employee selection, controlling the sources of thermal stress, limiting exposure to uncontrolled stresses, or the provision of specific thermal protective clothing.

Selection of employees

There are no very effective methods for preselecting employees suitable for work at either temperature extreme and there are no exact age or sex limits. The process is more one of excluding employees who are specifically at risk and there are one or two obvious categories that justify this decision. For example, individuals with recent experience of non-freezing cold injuries, possibly from some circumstance not related directly to their work, would be best excluded from work involving significant cold stresses. The same would apply to groups with peripheral vascular disease, such as Raynaud's phenomenon and related variants or obliterative vascular disease of any kind. Similarly employees with cardiac or vascular disease should be excluded from significant heat stress which inevitably involves increasing the strain on the cardiovascular system. Some skin conditions such as severe acne or seborrhoeic dermatitis are liable to exacerbation under heat stress causing sweating, and such cases should also be excluded.

Controlling the source of thermal stress

From the measurements made during the evaluation stage of the control process it should be possible clearly to identify the main sources of thermal stress. In hot conditions it will be important to come to some conclusion as to the relative contributions of air temperature, humidity, poor air movement or radiant heat. Occasionally the main difficulty may simply be the lack of adequate air movement and this may be solved by the installation of fans without necessarily resorting to any more complicated forms of air conditioning. Similarly where the source of stress is almost entirely due to a very high radiant source in the environment such as the front of a furnace or other very hot structure or material, then the installation of simple shielding may be all that is required.

Where no single main source of stress is amenable to obvious control, and the problem is just that conditions are generally too hot and humid, it is likely that some form of air-conditioning system will be required. A good approach to this decision is to compute the WBGT index from the individual measurement made during evaluation or from direct measurements of WBGT with special instrumentation.

The International Standards Organization, standard number ISO 7243 [23] suggests threshold limit values (TLV) for WBGT in relation to work rate for heat acclimatized and unacclimatized individuals normally clothed ($I_{cl} = 0.09$ m²°C/W). These are based on a maximum permissible rectal temperature of 38 °C and are summarized in *Table 9.2*. *Figure 9.12*, reproduced from ISO 7243, shows how WBGT threshold limit values should be modified for different work/rest cycles. It has also been proposed that the WBGT threshold limit values should be further reduced if the body is partially covered by impermeable clothing (by 2 °C WBGT)

Table 9.2 Upper limits for WBGT at various work rates for fit individuals, normally clothed (I_{cl} 0.09 m²°C/W) Adapted from ISO 7243 (1982) [24]

Metabolic rate (for a mean skin surface area of 1.8 m²) W	Upper limit for WBGT			
	Acclimatized subjects °C		Unacclimatized subjects °C	
< 117	33		32	
117–234	30		29	
234–360	28		26	
	No air movement	Air movement	No air movement	Air movement
360–468	25	26	22	23
> 468	23	25	18	20

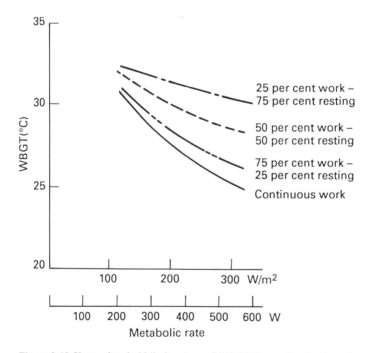

Figure 9.12 Shows threshold limit values of WBGT for acclimatized workers at various workrates and work/rest cycles (from Annex B of ISO Standard Reference number 7243 (1982)[24]

or fully covered by impermeable clothing (by 4 °C WBGT), but these are tentative limits and it remains preferable to make individual physiological measurements for such cases.

If the measured conditions exceed the WBGT limits suggested above then either the pattern or duration of work must be changed or air conditioning introduced to lower the WBGT values to acceptable limits for appropriate work/rest regimes.

For office staff or other individuals undertaking sedentary tasks the approach put forward by the American Society of Heating, Refrigerating and Air Conditioning Engineers (ASHRAE) in their Draft Standard 55-74R is recommended. *Figure 9.7* is slightly modified from the ASHRAE Standard and indicates the range of conditions in terms of operative temperature and humidity which will be perceived as comfortable by approximately 80 per cent of the persons present when wearing typical summer or winter clothing.

As an alternative it is possible to use the corrected effective temperature scales for establishing environmental limits for sedentary or office work. In 1969 the World Health Organization recommended an upper environmental limit of 30 °C CET for sedentary work but many would consider this too high, particularly for skilled work which has been shown to deteriorate significantly above an upper limit of 26.7 °C CET. For practical purposes lower and upper limits of 16 °C and 22 °C CET respectively are recommended for sedentary work. Heating or air conditioning systems should therefore be designed to produce environments within those limits.

In cold conditions, either in industry or in the open, there is usually little scope for controlling the source of cold stress since this has either been generated deliberately for the specific industrial purpose or, in the case of most external cold environments, is incapable of significant modification. Nevertheless the reader is referred to the windchill chart given in *Figure 9.8* from which the obvious advantages of reducing wind exposure may be seen. For some external working conditions it is possible to provide wind breaks of various kinds and these may be highly effective in reducing windchill. It can prove cost-effective, for example in the building industry, to enclose the work site totally in a polythene structure.

Limiting exposure to uncontrolled thermal stress

Where it is impossible to exercise control over the source of thermal stress, whether this is for economic or other reasons, it will be necessary to introduce working schedules that will limit the duration of exposure of individual workers to the uncontrolled conditions. It has been found that unacclimatised men working at 90 W metabolic heat production can endure saturated environments of 39.5 °C (equivalent to a WBGT of 39.5 °C if there is no radiant load) for about 60 min. At higher work rates the WBGT limit obviously will decrease so that at 210 W it reduces to 36.5 °C at 315 W to 35.5 °C and at 420 W to about 31.5 °C. These figures should be regarded as only approximate indications of limits for a particular case. It is essential that where conditions are so severe and uncontrolled that exposures to them are limited to relative short durations then the durations and the rest periods should be determined on the basis of a detailed physiological assessment of each particular job. The reason for this recommendation is that when conditions are severe, minor differences between one industrial situation and another in terms of work rate, type of clothing or protective clothing worn, the exposure to radiant heat loads and the physical characteristics of the labour force can have a dramatic effect on physiological responses. It is therefore most unwise to give generalized advice for these fringe conditions. Very often in such situations it will be more satisfactory to provide specific thermal protective clothing as detailed below.

Providing specific thermal protective clothing

When working environments are outside the limits indicated above it will often prove more economical to provide specific thermal protective clothing rather than to institute costly, time-limited exposure schedules. In hot conditions the choice is between a range of air-ventilated clothing or the more modern liquid-cooled systems. There are also a number of cooling garments on the market which incorporate ice as a heat sink. These of course must be refrozen at intervals during use. For cold conditions the choice is between insulating garments and those which provide direct heating by electrical means. All these specific thermal protective garments are described in greater detail in Chapter 21 on protective clothing.

Emergency treatment

In most industrial conditions the risk of serious heat illness or cold injury is very small indeed and no specific provisions for emergency treatment are required other than the general recommendations contained in Chapter 24. However, where a significant risk does exist then special provisions should be made as follows.

Heat illness

In severe hot environments, exceeding a WBGT of 30 °C, and especially where these conditions are combined with very high rates of physical work, then the risk of heat hyperpyrexia or heat stroke must be considered. This quite commonly applies in the naturally hot parts of the world but may also be found in some industrial settings in otherwise temperate climates. Particular examples may be found in industries involving the use of furnaces, in mining (especially deep mining) and in a variety of special or emergency circumstances such as apply to the maintenance of furnaces, mine rescue teams, conditioning system failures and fires. The essential requirements for the emergency first aid treatment of heat stroke are a means for accurately measuring deep body temperature and for the immediate institution of effective cooling. A rectal thermometer with a range extending at least to 43 °C should be available. For cooling purposes an open-weave or net bed or stretcher together with a wet sheet and high capacity fans to generate a high air flow over the patient will generally be sufficient. This technique is usually at least as effective as some of the sophisticated water spray beds available and will be infinitely cheaper to provide. Covering the patient with a wet sheet during transit to hospital can also be highly effective as a method of maintaining cooling.

It is beyond the scope of this chapter to provide specific details of the clinical diagnosis and management of heat illnesses which are covered in the further reading section. Most occupational physicians will have very little direct experience of serious heat illnesses and it is essential therefore that, where their parish includes cover for severe heat stress, they should rehearse the basic details of the clinical management of heat illness. It is also important that all supervisory and nursing staff are made fully aware of the risks of serious heat illness and the action to be taken in the event that it should occur.

Hypothermia and cold injuries

As with heat illness it is only in those jobs in which there is a significant risk of hypothermia or cold injury that specific provisions are required. For example, there will always be a risk of hypothermia from cold water immersion for workers in the offshore oil industry in the North Sea, both in the event of rig accidents or from helicopter ditching. Risks of immersion accidents may also be significant in the fishing industry and the merchant marine. Accidental hypothermia may also occur in any form of work involving the use of closed refrigerated stores of various sizes.

For most instances of immersion hypothermia of relatively short duration the recommended treatment consists of rapid rewarming in a bath of water at approximately 41 °C. The necessity for provision of a facility for such rewarming at a workplace rather depends on the availability of other medical facilities close by and of the means of transport thereto. If there is any doubt at all as to the rapidity with which an immersion victim can be transferred to skilled medical care then the provision of onsite rewarming facilities should be considered. Again it is essential to have available a suitable rectal thermometer capable of reading body core temperatures down to at least 30 °C. For the details of the clinical management of accidental hypothermia or immersion victims or cold injury the reader is referred to the text listed under the further reading section.

References

1. Fanger, P.O., *Thermal Comfort*. McGraw-Hill, New York (1972)
2. Hey, E.N. Small globe thermometers. *Journal of Scientific Instrumentation*, **1**, 955–957 and corrigendum, 1260 (1968)
3. Brebner, D.E. Kerslake,D.McK. and Waddell, J.L. The relationship between coefficients for heat exchange by convection and by evaporation in man. *Journal of Physiology London*, **141**, 164–168 (1958)
4. Kerslake, D.McK. *The Stress of Hot Environments*. Cambridge University Press, London (1972)
5. Wolff, A.H., Coleshaw, S.R.K., Newstead, C.G. and Keatinge, W.R. Heat exchanges in wet suits. *Journal of Applied Physiology*, **58**, 770–777 (1985)
6. British Standards Institution. BS 4745 *Thermal Resistance of Textiles* (1971)
7. American Society of Heating, Refrigerating and Air Conditioning Engineers. *Thermal Environmental Conditions for Human Occupancy*. ASHRAE Standard No. 55–81 (1981)
8. Kerslake, D.McK. *A Heated Manikin for Studies on Air Ventilated Clothing*. Flying Personnel Research Committee. Ministry of Defence, London, Memorandum No. 214 (1964)
9. Clesen, B.W., Shiwinska, E. Madsen, T.L. and Fanger, P.O. *Effect of Body Posture and Activity on the Thermal Insulation of Clothing: Measurements by a Movable Thermal Manikin*. ASHRAE. Trans. 32 (1982) 791–805 (1982)
10. Allan J.R., Higenbottam, C. and Redman, P.J. The effect of leakage on the insulation provided by immersion-protection clothing. *Aviation and Space Environmental Medicine*, **56**, 1107–1109 (1985)
11. Woodcock, A.H. Moisture transfer in textile systems. *Textile Research Journal*, **32**, 628–633 (1962)
12. Nishi, Y. and Gagge, A.P. Moisture permeation of clothing—a factor governing thermal equilibrium and comfort. *Transactions of the American Society of Heating and Refrigeration Air-Conditioning Engineers*, **76**, 1 (1970)

13. Belding, H.S. and Hatch, T.F. Index for evaluating heat stress in terms of the resulting physiological strain. *Heat Pipe and Air Conditioning*, **27**, 129–136 (1955)
14. Givoni, B. A new method for evaluating industrial heat exposure and maximum permissible work load. *International Journal Bioclimatology and Biometeorology*. **8**, 115–124 (1964)
15. Siple, P.A. and Passel, C.F. Measurements of dry atmospheric cooling in subfreezing temperatures. *Proceedings of the American Philosophical Society*, **89**, 177–199 (1945)
16. McArdle, B., Dunham, W., Holling, H.E., Ladell, W.S.S., Scott, J.W., Thomson, M.L. and Weiner, J.S. *The Prediction of the Physiological Effects of Warm and Hot Environments*. Report Number RNP 47/391 Medical Research Council, London (1947)
17. Yaglou, C.P. and Minard, D. Control of heat casualties at military training centres. *American Medical Association Archives Industrial Health*, **16**, 302–316 (1957)
18. Leithead, C.S. and Lind, A.R. *Heat Stress and Heat Disorders*. Cassell, London (1964)
19. Yaglou, C.P. Temperature humidity and air movement in industries: the effective temperature index. *Journal of Industrial Hygiene*, **9**, 297–309 (1927)
20. Bedford, T. *Basic Principles of Ventilation and Heating*, 2nd edn, Lewis, London (1964)
21. Allan, J.R., Gibson, T.M. and Green, R.G. Effect of induced cyclic changes of deep body temperature on task performance. *Aviation and Space, Environmental Medicine*, **50(6)**, 585–589 (1979)
22. Allan, J.R. and Gibson, T.M. Separation of the effects of raised skin and core temperature on performance of a pursuit rotor task. *Aviation and Space, Environmental Medicine*, **50(7)**, 678–682 (1979)
23. Wing, J.F. Upper thermal tolerance limits for unimpaired mental performance. *Aerospace Medicine*, **36**, 960–964 (1965)
24. International Organization for Standardization. Standard No. ISO 7243 (E) *Hot environments—estimation of the heat stress on working men, based on the WBGT-index (wet bulb globe temperature)* (1982)

Further reading

Burton, A.C. and Edholm, O.G. *Man in a Cold Environment.* Edward Arnold, London (1955)
Cena, K. and Clark, J.A. (eds) *Bioengineering, Thermal Physiology and Comfort.* Elsevier Scientific Publishing Company, Amsterdam (1981)
Keatinge, W.R. *Survival in Cold Water.* Blackwell, Oxford (1969)
Kerslake, D.McK. *The Stress of Hot Environments.* Cambridge University Press, London (1972)
Khogali, M. and Weiner, J.S. Heat-stroke: report on 18 cases. *The Lancet, **2**, 8189, 276–278 (1980)
Leithead, C.S. and Lind, A.R. *Heat Stress and Heat Disorders*, Cassell, London (1964)

Chapter 10

Biological monitoring of exposure to industrial chemicals

A. Bernard and R. Lauwerys

Introduction: definition and role of biological monitoring of exposure

The objective of biological monitoring is to prevent excessive exposure to chemicals which may cause acute or chronic adverse health effects. In this approach, the health risk is assessed by comparing the value of the measured parameter with its currently estimated maximum permissible value in the medium analysed (so-called biological limit value). Biological monitoring of exposure, like ambient monitoring, is essentially a preventive activity and, in this respect, it must be clearly distinguished from biological monitoring of effects (also called medical screening or health surveillance), which, by means of sensitive biological markers, aims to detect early signs of toxicity [1,2]. Biological monitoring directly assesses the amount of chemical effectively absorbed by the organism, that is, the internal dose. Depending on the characteristics of the selected biological parameter (particularly its biological half-life) and the conditions under which it is measured, the term internal dose may have different meanings such as the total amount or a fraction (for instance, the biologically active dose) of a chemical recently absorbed (recent exposure), the amount stored in one or several body compartments (total integrated exposure or specific organ dose), or the amount bound to the target molecules (target dose).

Biological monitoring of exposure is usually reserved for chemicals which penetrate into the organism and exert systemic effects. Very few biological tests have been proposed for the identification or the monitoring of chemicals present at the interface between the environment and the organism (the skin, gastrointestinal mucosa or respiratory tract mucosa). The analysis of nickel in the nasal mucosa and the counting of asbestos bodies in sputum could be considered as examples of such tests. For systemically active chemicals, biological monitoring of exposure represents the most effective approach for assessing the potential health risk since a biological index of internal dose is necessarily more closely related to a systemic effect than any environmental measurement.

Biological monitoring of exposure integrates the chemical absorption by all routes (pulmonary, oral, cutaneous) and from all possible sources (occupational, environmental, dietary, and so on). This is particularly useful when assessing the overall exposure to widely dispersed pollutants. Even for elements present in the environment under different chemical forms with different toxicities (for example, inorganic arsenic in water or in the industrial setting and organic arsenic in marine organisms), it may still be possible correctly to estimate the health risk by speciation of the element in the biological medium analysed. Biological monitoring of exposure takes into account the various individual factors which influence the uptake or the absorption of the chemical such as sex, age, physical activity, hygiene and nutritional status.

In general, the proper application of a biological test for determining the internal dose of a chemical requires the collection of relevant information on its metabolism (absorption, distribution, excretion), its toxicity and on the relationships between internal dose, external exposure and adverse effects. The knowledge of the latter permits the direct (from the internal dose-adverse effect relationship) or indirect (from the threshold limit value and the internal dose-external exposure relationship) estimation of the maximum permissible internal dose (biological limit value) [1, 2]. Unfortunately, for many industrial chemicals, one or all of the preceding conditions are not fulfilled, which limits the possibilities of biological monitoring. As mentioned above, biological monitoring is usually not applicable to substances acting locally nor is it useful either for detecting peak exposures to rapidly acting substances. Excessive exposure to these chemicals should rely mainly on the continuous monitoring of the pollutant concentration in the environment.

Methods in biological monitoring of exposure

Determination of the chemical or its metabolites in body fluids

The great majority of the tests currently available for biological monitoring of exposure to industrial chemicals (*Table 10.1*) rely on the determination of the chemical or its metabolites in biological media. In practice, the biological samples most commonly used for analysis are urine, blood and alveolar air. The analysis of other biological materials such as milk, fat, saliva, hair, nails, teeth, placenta is less frequently performed. As a general rule, urine is used for inorganic chemicals and for organic substances which are rapidly biotransformed to more water-soluble compounds; blood is used for most inorganic chemicals and for organic substances poorly biotransformed; alveolar air analysis is reserved for volatile compounds.

Tests measuring the chemical or its metabolites in biological media can be classified into two broad categories: selective tests and non-selective tests.

Selective tests

This category includes the majority of tests currently used in occupational medicine. The unchanged chemical is measured in biological media when the substance is not biotransformed (which is the case for nearly all inorganic chemicals); when it is poorly biotransformed (for example, some solvents such as methylchloroform or tetrachloroethylene); when the exposure is too low for a significant amount of metabolite to be produced (such as in slight exposure to benzene); or when a high degree of specificity is required. The determination of the unchanged chemical may indeed have a greater specificity than that of a metabolite that may be common to several substances.

Most organic chemicals are rapidly metabolized in the organism to more water-soluble compounds that are easily excreted via the urine or bile. Exposure to these chemicals is generally monitored by measuring specific urinary metabolites. These tests are more readily accepted by the workers because they do not require blood collection. Furthermore, they offer the advantage that when urine is collected at the appropriate time (at the end of a shift or the morning of the next day), the concentration of the metabolite in urine is much less influenced by very recent exposure than that of the unchanged chemical in blood or in alveolar air.

Non-selective tests

These tests are used as non-specific indicators of exposure to a group of chemicals. Examples of non-selective exposure tests currently available are given below.

DETERMINATION OF DIAZOPOSITIVE METABOLITES IN URINE

This test has been proposed to monitor exposure to aromatic amines.

ANALYSIS OF THIOETHERS IN URINE

The urinary excretion of thioethers increases following exposure to electrophilic substances and has therefore been proposed to monitor occupational exposure to carcinogenic or mutagenic substances. The specificity of this test is, however, limited: the urinary excretion of thioethers may be increased by several compounds that are not carcinogenic or mutagenic (toluene, o-xylene or biphenyl, for example); various endogenous substances are also eliminated in urine as thioethers; and finally, smoking is an important confounding factor.

DETERMINATION OF THE MUTAGENIC ACTIVITY IN URINE

An enhanced mutagenic activity has been observed in the urine of rubber workers, coke plant workers, workers exposed to epichlorhydrin, anaesthetists and nurses handling cytostatic drugs. As with the thioether test, smoking is an important confounding factor. In smokers, the mutagenic activity of the urine is increased proportionally to the daily cigarette consumption.

Because of their lack of specificity and the existence of a large individual variability, non-selective tests cannot be used to monitor exposure on an individual basis. However, when an adequate control group is used as a reference, they may be useful qualitatively to identify groups at risk.

Recently, non-invasive methods have been developed for measuring the *in vivo* metal content of selected tissues. These methods, which are usually based on neutron activation or on X-ray fluorescence techniques, have been applied for the determination of cadmium in kidney or liver, of lead in bones and of mercury in the central nervous system and bones [3,4]. They enable the monitoring of the long-term retention of heavy metals in the organism and, in some cases, the target organ dose (cadmium in the kidney, for example).

Determination of a non-adverse biological effect related to the internal dose

A biological effect is considered as non-adverse if the functional or physical integrity of the organism is not diminished, if the ability of the organism to face an additional stress (homeostasis) is not decreased or if these impairments are not

Table 10.1 Biological methods for monitoring exposure to industrial chemicals (after Lauwerys [1])

Chemical agent	Biological parameter	Biological material	Normal value	Tentative maximum permissible value*	Remarks
A. Inorganic and organometallic substances					
Aluminium	Al	Serum	< 1 µg/100 ml		
	Al	Urine	< 50 µg/g creatinine	< 150 µg/g creatinine	
Antimony	Sb	Urine	< 2 µg/g creatinine		
Arsenic	Total As	Urine	< 40 µg/g creatinine		
	Total As	Blood			Interference from fish consumption
	Total As	Hair	< 1 µg/g		
	Inorganic As and its metabolites	Urine	< 20 µg/g creatinine	< 130 µg/g creatinine (TLV: 50 µg/m³) < 50 µg/g creatinine (TLV: 10 µg/m³)	No interference from fish consumption
Barium	Ba	Urine	< 2 µg/g creatinine		
Beryllium	Be	Urine	< 2 µg/g creatinine		Non-smokers
Cadmium	Cd	Urine		10 µg/g creatinine	
	Cd	Blood	< 0.5 µg/100 ml	1 µg/100 ml	
	Metallothionein	Urine			
Chromium (soluble compounds)	Cr	Urine	< 5 µg/g creatinine	30 µg/g creatinine	
Cobalt	Co	Urine	< 2 µg/g creatinine	30 µg/g creatinine	
Fluoride	F	Urine	< 0.4 mg/g creatinine	3–4 mg/g creatinine	Difference between preshift and postshift values
Lead	Pb	Blood	< 35 µg/100 ml	60 µg/100 ml / 40 µg/100 ml	Adult males / Young females
	Pb	Urine	< 50 µg/g creatinine	150 µg/g creatinine	
	Pb (after 1 g EDTA IV)	Urine	< 600 µg/24 h	1000 µg/24 h	
	Free porphyrins	Erythrocytes	< 75 µg/100 ml RBC	300 µg/100 ml RBC	
	Zinc protoporphyrin	Blood	< 40 µg/100 ml	150 µg/100 ml	
			< 2.5 µg/g haemoglobin	12.5 µg/g haemoglobin	
	δ-Aminolaevulinic acid	Urine	< 4.5 mg/g creatinine	10 mg/g creatinine	
	Coproporphyrins	Urine	< 100 µg/g creatinine	250 µg/g creatinine	
	dehydratase	Erythrocytes			
Lead tetraethyl	Pb	Urine	< 50 µg/g creatinine	100 µg/g creatinine	

Chemical agent	Biological parameter	Biological material	Normal value	Tentative maximum permissible value*	Remarks
Manganese	Mn	Urine	< 3 µg/g creatinine		
	Mn	Blood	< 1 µg/100 ml		
Mercury (inorganic)	Total Hg	Urine	< 5 µg/g creatinine	50 µg/g creatinine	
	Total Hg	Blood	< 2 µg/100 ml	3 µg/100 ml	
	Total Hg	Saliva			
Mercury (methyl)	HgCH$_3$	Blood	< 2 µg/100 ml	10 µg/100 ml	
	Hg	Hair			
Nickel (soluble compounds)	Ni	Urine	< 5 µg/g creatinine	70 µg/g creatinine	
	Ni	Plasma	< 1 µg/100 ml	1 µg/100 ml	
Nickel carbonyl	Ni	Urine			
Nitrogen protoxyde	N$_2$O	Urine			
	N$_2$O	Expired air			
Selenium	Se	Urine	< 25 µg/g creatinine	50 µg/g creatinine	
Thallium	Tl	Urine	< 1 µg/g creatinine		
Uranium	U	Urine	< 0.3 µg/g creatinine		
Vanadium	V	Urine	< 1 µg/g creatinine		

B. Organic chemicals

Non-substituted aliphatic and alicyclic hydrocarbons

Chemical agent	Biological parameter	Biological material	Normal value	Tentative maximum permissible value*	Remarks
Cyclohexane	Cyclohexanol	Urine		3.2 mg/g creatinine	
	Cyclohexane	Blood		45 µg/100 ml	During exposure
	Cyclohexane	Expired air		220 ppm	During exposure
n-Hexane	2-Hexanol	Urine		0.2 mg/g creatinine	
	2,5-Hexanedione	Urine		5.3 mg/g creatinine	
	n-Hexane	Blood		15 µg/100 ml	During exposure
	n-Hexane	Expired air		50 ppm	During exposure
2-Methyl-pentane	2-Methyl-2-pentanol	Urine			
	2-Methyl-pentane	Expired air			
3-Methyl-pentane	3-Methyl-2-pentanol	Urine			
	3-Methyl-pentane	Expired air			

Non-substituted aromatic hydrocarbons

Chemical agent	Biological parameter	Biological material	Normal value	Tentative maximum permissible value*	Remarks
α-Methylstyrene	Atrolactic acid	Urine			
Benzene	Phenol	Urine	< 20 mg/g creatinine	45 mg/g creatinine	If TLV = 10 ppm
				< 20 mg/g creatinine	If TLV < 10 ppm
				< 0.022 ppm	If TLV < 1 ppm
	Benzene	Expired air		5 µg/100 ml	If TLV = 1 ppm
	Benzene	Blood			

Chemical agent	Biological parameter	Biological material	Normal value	Tentative maximum permissible value*	Remarks
Cumene (isopropylbenzene)	2-Phenylpropanol	Urine		200 mg/g creatinine	
	Cumene	Expired air			
	Cumene	Blood			
Diphenyl	4-Hydroxydiphenyl	Urine			
Ethylbenzene	Mandelic acid	Urine		2 g/g creatinine	
	Ethylphenol	Urine			
	Ethylbenzene	Blood		0.15 mg/100 ml	During exposure
Mesitylene	3,5-Dimethylhippuric acid	Urine			
Styrene	Mandelic acid	Urine		1 g/g creatinine	
	Phenylglyoxylic acid	Urine		350 mg/g creatinine	
	Styrene	Blood			
	Styrene	Expired air			
Toluene	Hippuric acid	Urine	< 1.5 g/g creatinine	2.5 g/g creatinine	
	o-Cresol	Urine	< 0.3 mg/g creatinine	1 mg/g creatinine	
	Toluene	Expired air		20 ppm	During exposure
	Toluene	Blood		0.1 mg/100 ml	During exposure
Xylene	Methylhippuric acid	Urine		1.5 g/g creatinine	
	Xylene	Expired air			
	Xylene	Blood		0.3 mg/100 ml	During exposure
Halogenated hydrocarbons					
Dichloromethane (methylene chloride)	Carboxyhaemoglobin	Blood	< 1 per cent	5 per cent	Non-smokers
	Dichloromethane	Blood		0.08 mg/100 ml	
	Dichloromethane	Expired air		35 ppm	
Halothane	Trifluoroacetic acid	Urine		10 mg/g creatinine	After 5-day exposure
	Trifluoroacetic acid	Blood		0.25 mg/100 ml	After 5-day exposure
	Halothane	Urine		250 µg/g creatinine	
Monobromomethane (methylbromide)	S-Methylcysteine	Urine			
	Bromide	Blood			
Monochloromethane (methylchloride)	S-Methylcysteine	Urine			
Monochlorobenzene	4-Chlorocatechol	Urine			
p-Dichlorobenzene	2,5-Dichlorophenol	Urine			
Polychlorinated biphenyl	PCB	Serum	< 0.3 µg/100 ml		
Tetrachloroethylene	Tetrachloroethylene	Expired air		60 ppm	During exposure
	Tetrachloroethylene	Blood		4 ppm	After exposure (16 h)

Chemical agent	Biological parameter	Biological material	Normal value	Tentative maximum permissible value*	Remarks
1,1,1-Trichloroethane	Trichloroethanol + trichloroacetic acid	Urine		50 mg/g creatinine	During exposure
	Trichloroethanol	Urine		30 mg/g creatinine	After 5-day exposure
	Trichloroethane	Blood			
	Trichloroethane	Expired air		50 ppm	After 5-day exposure
Trichloroethylene	Trichloroethanol	Urine		125 mg/g creatinine	After 5-day exposure
	Trichloroacetic acid	Urine		75 mg/g creatinine	After 5-day exposure
	Trichloroethanol	Plasma		0.23 mg/100 ml	After exposure (16 h)
	Trichloroethylene	Expired air		< 0.5 ppm	During exposure
				12 ppm	After 5-day exposure
	Trichloroacetic acid	Plasma		5 mg/100 ml	During exposure
	Trichloroethylene	Blood		0.6 mg/100 ml	
Vinylchloride	Thiodiglycolic acid	Urine	< 2 mg/g creatinine		
Amino-derivatives and nitro-derivatives					
Aniline	p-Aminophenol	Urine		10 mg/g creatinine	
	Methaemoglobin	Blood	< 2 per cent	5 per cent	
Benzidine-derived azo-compounds	Benzidine	Urine			
Ethyleneglycol dinitrate	Ethyleneglycol dinitrate	Urine			
	Ethyleneglycol dinitrate	Blood			
4,4'-Methylene *bis* (2-chloroaniline) or MOCA	MOCA	Urine		80 µg/g creatinine	
Monoacetylbenzidine derived azo-compounds	Monoacetylbenzidine	Urine			
Nitroglycerine	Nitroglycerine	Blood			
Nitrobenzene	p-Nitrophenol	Urine		5 mg/g creatinine	
	Methaemoglobin	Blood	< 2 per cent	5 per cent	
Several aromatic amino-compounds and nitro-compounds (aniline, nitrobenzene, dinitrobenzene etc.)	Methaemoglobin	Blood	< 2 per cent	5 per cent	
	Diazo-positive metabolites	Urine			
	Parent compound (for example, benzidine, beta-naphthylamine)				
Trinitrotoluene	2,4 and 2,6 Dinitro-aminotoluene	Urine			
2,4-Dinitrotoluene	2,4 Dinitrobenzoic acid	Urine			

Chemical agent	Biological parameter	Biological material	Normal value	Tentative maximum permissible value*	Remarks
Alcohols, glycols and derivatives					
Dioxane	Beta-hydroxyethoxy-acetic acid	Urine			
Ethyleneglycol	Oxalic acid	Urine	< 100 mg/24 h		
	Ethyleneglycol	Serum			
Ethyleneglycol monoethyl ether	Ethoxyacetic acid	Urine		150 mg/g creatinine	
Furfurylalcohol	Furoic acid	Urine	< 65 mg/g creatinine		
Methanol	Methanol	Urine	< 2.5 mg/g creatinine	7 mg/g creatinine	
	Formic acid	Blood			
	Formic acid	Urine			
Methylcellosolve	Methoxyacetic acid	Urine			
Ethylene glycol mono-butylether	Butoxyacetic acid	Urine			
2-Propylene-glycol monomethylether (1-Methoxy-2-propanol)	Propyleneglycol	Urine			
Cetones					
Acetone	Acetone	Urine	< 2 mg/g creatinine	30 mg/g creatinine	
	Acetone	Blood	< 0.2 mg/100 ml	5 mg/100 ml	
	Acetone	Expired air			
Methylethylketone	Methylethylketone	Urine		2.6 mg/g creatinin	
Methyl-*n*-butylketone	2,5-Hexanedione	Urine			
Aldehydes					
Formaldehyde	Formic acid	Urine			
Furfural	Furoic acid	Urine	< 65 mg/g creatinine	200 mg/g creatinine	
Amides					
Dimethylformamide	N-methylformamide†	Urine		40 mg/g creatinine	
	Dimethylformamide	Blood		0.15 mg/100 ml	
	N-methylformamide†	Blood		0.1 mg/100 ml	
	Dimethylformamide	Expired air		1 ppm	During exposure
Dimethylacetamide	N-methylacetamide	Urine			
Phenols					
Phenol	Phenol	Urine	< 20 mg/g creatinine	300 mg/g creatinine	
p-Tert-butylphenol	p-Tert-butylphenol	Urine		2 mg/g creatinine	

Chemical agent	Biological parameter	Biological material	Normal value	Tentative maximum permissible value*	Remarks
Asphyxiants					
Acrylonitrile	Acrylonitrile	Urine			
	Thiocyanate	Urine	< 2.5 mg/g creatinine	5 per cent	Non-smoker
Carbon monoxide	Carboxyhaemoglobin	Blood	< 1 per cent	10 ml/100 ml	Non-smoker
	Carbon monoxide	Blood	< 0.15 ml/100 ml	18 ppm	Non-smoker
	Carbon monoxide	Expired air	< 2 ppm		Non-smoker
Cyanide and aliphatic	Thiocyanate	Urine	< 2.5 mg/g creatinine	6 mg/24 h	Non-smoker
nitriles	Thiocyanate	Plasma	< 0.6 mg/100 ml		Non-smoker
	Cyanide	Blood			
	Ratio between thiocyanate in urine (mg/g creatine) and carboxyhaemoglobin (per cent)	urine + blood		3	
Methaemoglobin-forming agents	Methaemoglobin	Blood	< 2 per cent	5 per cent	
Pesticides					
Baygon	2-Isopropoxyphenol	Urine			
Carbaryl	1-Naphthol	Urine		10 mg/g creatinine	
Chlorophenoxyacetic acid derivatives	2,4 D	Urine			
	2,4,5 T	Urine			
	MCPA	Urine			
DDT	DDT	Serum			
	DDT+DDE+DDD	Blood			
	DDA	Urine			
Dieldrin	Dieldrin	Blood		15 µg/100 ml	
	Dieldrin	Urine			
Dinitro-orthocresol	Dinitro-orthocresol	Blood		1 mg/100 ml	
	amino-4-nitro-orthocresol	Urine			
Endrin	Endrin	Blood		5 µg/100 ml	
	Anti-12 Hydroxyendrin	Urine		0.13 µg/g creatine	
Hexachlorobenzene	Hexachlorobenzene	Blood		0.03 mg/100 ml	
	2,4,5-Trichlorophenol	Urine			
	Pentachlorophenol	Urine			

Chemical agent	Biological parameter	Biological material	Normal value	Tentative maximum permissible value*	Remarks
Carbamates insecticides	Cholinesterase	Erythrocytes		30 per cent inhibition	
		Plasma		50 per cent inhibition	
		Whole blood		30 per cent inhibition	
Lindane	Lindane	Blood		2 µg/100 ml	
Organophosphorus esters	Cholinesterase	Erythrocytes		30 per cent inhibition	
		Plasma		50 per cent inhibition	
		Whole blood		30 per cent inhibition	
	Dialkylphosphates	Urine			
Parathion	p-Nitrophenol	Urine		0.5 mg/g creatinine	
Pentachlorophenol	Pentachlorophenol	Urine		1 mg/g creatinine	
Mutagenic and carcinogenic substances					
	-mutagenicity	Urine			
	-thioethers	Urine			
	-protein or DNA adducts	Blood, urine			
	-cytogenetic alterations (chromosomal aberrations, sister chromatid exchanges, micronuclei)	Lymphocytes, exfoliated cells			
Others					
Carbon disulfide	Iodine azide test	Urine		> 6.5 (Vasak index)	
	2-Thiothiazolidine 4-carboxylic acid	Urine		5 mg/g creatinine	
Diethylstilboestrol	Diethylstilboestrol	Urine	30 mg/g creatine		24 hours urine
Ethylene oxide	Ethylene oxide	Expired air	0.5 mg/m^3		During exposure
		Blood	0.8 µg/100 ml		During exposure

*Except when indicated otherwise, the values correspond to analyses performed on samples collected at the end of the exposure period.
†The in vivo metabolite is the N-hydroxymethyl-N-methylformamide which is detected as N-methylformamide by gas chromatography.

likely to occur in the near future (delayed toxicity). The advantage of methods measuring a non-adverse biological effect is that they may provide information on the amount of chemical liable to react with the target sites. The application of such methods is limited, however, and only a few examples can be given [1]:

1. The inhibition of pseudocholinesterase by organophosphorus pesticides.
2. The inhibition of the erythrocyte enzyme δ-aminolaevulinic acid dehydratase by lead.
3. The formation of adducts with blood proteins (alkylated haemoglobin [5], aniline–haemoglobin conjugate [6]).
4. The increased urinary excretion of β-hydroxycortisol as an index of exposure to chemicals inducing microsomal enzymes (for example, polychlorinated biphenyl—PCB).
5. The increased urinary excretion of metallothionein following exposure to some heavy metals (cadmium, for example) [7].

The non-adverse biological effect, however, may have no more predictive value than the mere determination of the chemical itself. For instance, in the biological monitoring of exposure to cadmium, the analysis of metallothionein in the urine seems to offer no advantage over that of cadmium except for being insensitive to external contamination [7].

Determination of the amount of chemical bound to target molecules

The most useful biological monitoring methods are those which directly measure the amount of active chemical bound to the target molecules (the target dose). When feasible, that is when the target site is readily accessible, these methods may assess the health risk more accurately than any other monitoring test. The carboxyhaemoglobin test used in industry for several decades belongs to this category. Progress in this approach is to be expected in the field of genetic toxicology where the determination of nucleic acids adducts (that is, molecular dosimetry) is considered to be one of the most promising methods for monitoring human exposure to carcinogenic or mutagenic chemicals. Examples of DNA adducts for which immunoassays have been described are aflatoxin B_1-DNA, benzo(a)pyrene-DNA and 2-acetylaminofluorene-DNA [8]. The DNA adducts are measured either in hydrolysates of DNA-molecules (sampled in blood cells, for example) or in degradation products of DNA released in body fluids such as urine. However, these techniques are currently at the developmental stage and have seldom been applied to groups of workers exposed to industrial carcinogens and mutagens.

Criteria for selecting biological tests

In practice, only a few biological tests can be used routinely to monitor exposure to industrial chemicals. Before a biological monitoring programme is implemented, the most appropriate parameter (or parameters) must be selected by taking into account:

1. Its specificity.
2. Its sensitivity; there should be a strong relationship between the parameter and

external exposure, and this relationship must exist at exposure levels below those associated with adverse effects.
3.   The analytical and biological variability of the test.
4.   The applicability of the test including cost and possible discomfort to the subject.
5.   The capability of the test to evaluate a risk to health.

In these considerations, the existence of a biological limit value is an important element which must be taken into account when selecting a biological monitoring test.

Tests which can estimate the target dose or the target organ dose are better to assess the risk to health. When the chemical is not itself toxic but must be metabolically activated before reaching the target site, the determination of the reactive chemical may be more relevant than that of the parent compound or of any other metabolite not directly related to the toxic effects. For example, to assess the risk linked with exposure to *n*-hexane or methyl-*n*-butylketone, it might be more relevant to measure 2,5-hexanedione in blood or in urine than to determine these solvents directly in blood or expired air.

Interpretation of the results

Results can be interpreted on an individual basis. However, this is possible only if the intraindividual variability of the parameter is small and its specificity is high. The results may also be interpreted on a group basis by considering their distribution. If all the observed values are below the biological limit value, the working conditions are satisfactory. If all, or the majority of the results are above the biological limit value, the overall exposure conditions must certainly be corrected. A third condition may also occur: the majority of the workers may have values below the biological limit level but a few have abnormally high values (the distribution is bimodal or polymodal). Two interpretations can be put forward:

1.   Either the subjects with the high values perform activities exposing them to higher levels of the pollutant, in which case the biological monitoring programme has identified job categories for which work conditions need to be improved.
2.   These workers do not perform different activities and, in this case, their higher internal dose must result from different hygiene habits.

When interpreting the results, it must be kept in mind that the metabolism of xenobiotics may be influenced by various endogenous or exogenous factors. Endogenous factors may be genetic or pathophysiological such as age, sex and diseases. For instance, hepatic insufficiency may be associated with a decreased biotransformation of xenobiotics, whereas renal diseases may impair their elimination in the urine. Alcohol consumption is a frequent exogenous confounding factor. In the body, ethanol is transformed to acetate by two successive oxidations, one catalysed by alcohol dehydrogenase and the other by aldehyde dehydrogenase. This metabolic pathway is not specific to ethanol and many other organic compounds including some solvents may be oxidized by the same enzymes. When these substances are absorbed concomitantly with ethanol, metabolic interferences may occur. In man, ethanol has been shown to inhibit the oxidation of methanol, trichloroethylene, xylene, toluene and styrene.

This inhibition of the biotransformation of a solvent following the ingestion of a large dose of alcohol may also result in a rise of the blood level of the solvent. But when alcohol is regularly consumed the opposite may be observed. In workers exposed to toluene, Waldron et al. [9] found that blood toluene concentrations were lower in those who drank regularly. Presumably, this results from the induction by alcohol of the microsomal oxidizing system of the liver. Many other substances, for example barbiturates, polycyclic hydrocarbons and the organochlorine pesticides, are inducers of microsomal enzyme activity, and may therefore interfere with the biotransformation of xenobiotics.

Smoking may also be a confounding factor. Dossing et al. [10] reported that smoking may increase the urinary excretion of orthocresol up to four times in heavy smokers probably because cigarettes contain orthocresol. Smoking must also be taken into account when monitoring the mutagenic activity of urine or its thioether concentration. Finally, confounding may also arise from the diet, the consumption of drugs, or the exposure to several industrial chemicals, such as with mixed exposure to phenol and benzene.

Analytical and ethical aspects

In any monitoring programme, the analytical aspect is of paramount importance. The parameter selected must be sufficiently stable to allow the transportation of the sample and possibly its storage for a few days. Therefore, before a new programme of biological monitoring is implemented, preliminary investigations must be conducted to test its stability under the conditions of sampling (including the type of container, stability in the biological fluid and the effect of physical factors). The laboratory responsible for the analysis must adopt good laboratory practices, which implies the use of a well-standardized method and the implementation of regular internal and external quality control programmes.

Finally, it must be kept in mind that in biological monitoring, humans are used as an integrator of exposure. The ethical aspects must receive a great deal of attention. In particular, the monitoring procedure itself must be without health risks. Sufficient information must be given to the subjects before and after monitoring and the individual results must remain confidential.

References

1. Lauwerys, R. *Industrial Chemical Exposure: Guidelines for Biological Monitoring*. Biomedical Publications, Davis, California (1983)
2. Bernard, A. and Lauwerys, R. General principles of biological monitoring of exposure to organic chemicals. In *Biological Monitoring of Exposure to Chemicals*, Vol. 1: *Organic Compounds* (eds M.H. Ho and H.K. Dillon) pp. 1–16 John Wiley and Sons, New York (1986)
3. Lauwerys, R. *In vivo* tests to monitor body burdens of toxic metals in man. In: *Chemical Toxicology and Clinical Chemistry of Metals* (eds S. Brown and J. Savory) Academic Press, New York, p.113 (1983)
4. Bloch, P. and Shapiro, I.M. An X-ray fluorescence technique to measure *in situ* the heavy metal burdens of persons exposed to these elements in the workplace. *Journal of Occupational Medicine* **28**, 609–614 (1986)

5. Farmer, P.B., Bailey, E. and Campbell J.B. Use of alkylated proteins in the monitoring of exposure to alkylating agents. In *Monitoring Human Exposure to Carcinogenic and Mutagenic Agents*. IARC, Lyon, pp. 189–198 (1984)

6. Lewalter, J. and Korrallus, U. Blood protein conjugates and acetylation of aromatic amines. New findings on biological monitoring. *International Archives of Occupational and Environmental Health*, **56**, 179–196 (1985)

7. Roels, H., Lauwerys, R., Buchet, J.P. *et al.* Significance of urinary metallothionein in workers exposed to cadmium. *International Archives of Occupational and Environmental Health*, **52**, 159–166 (1983)

8. Adamkiewicz, J., Nehls, P. and Rajewsky, M.F. Immunological methods for detection of carcinogen—DNA adducts. In *Monitoring Human Exposure to Carcinogenic and Mutagenic Agents*. IARC, Lyon, pp. 199–216 (1984)

9. Waldron, H.A., Cherry, N. and Johnston, J.D. Effects of ethanol on blood toluene concentrations. *International Archives of Occupational and Environmental Health*, **51**, 365–369 (1983)

10. Dossing, M., Bachum, J., Hansen, S.H., Lundqvist, G.R. and Anderson, N.T. Urinary hippuric acid and ortho-cresol excretion in man during experimental exposure to toluene. *British Journal of Industrial Medicine*, **40**, 470–473 (1983)

Chapter 11

Pulmonary function and assessment

R.B. Douglas

Introduction

The lung is the external organ of respiration with the objective of maintaining blood gases between closely prescribed limits. The structure and function of the lung in health are such as to be able to maintain the required partial pressures of arterial oxygen (O_2) and carbon dioxide (CO_2) over large variations in demand. The lung divides dichotomously for about 23 generations until it reaches the alveolar sacs which number approximately 300 million and cover some $70\,m^2$. Distributed over this relatively large surface (approximately the size of two tennis courts) is 80 ml of blood (two-thirds of a wine glass) in the alveolar capillaries through which CO_2 is given up and O_2 absorbed.

The conducting airways are surrounded by smooth muscle and are innervated. They are also lined with specialized cells, some of which produce mucus, others which carry cilia. Together they form an escalator carrying mucus and deposited inhaled material upwards to maintain the lung in a sterile condition. These features combine to determine the airflow resistance of the lung. The alveolar surface is covered with a surfactant lining (perhaps 0.5 nm thick) which determines the normal compliance (distensibility) of the parenchyma.

Objective measurements of lung function are necessary for assessment of individuals and populations for whom reference values are available.

Structure and function

An important difference between fishes and warm-blooded mammals such as whales and man is the large 'gill' area required to sustain the higher metabolic rate of mammals. In man this alveolar surface is sufficient to cover two tennis courts and if totally exposed would present major problems in terms of water loss by evaporation. Successful evolution has led to a lung which is protected by being folded inside the thorax and ventilated through a system of dichotomously branching conducting airways. This is where the problems can begin. In order to maximize alveolar ventilation the conducting airways need to be as narrow as possible—thus reducing the respiratory 'dead space'. However, the smooth flow of gases through tubes (*Figure 11.1*) is governed by Poiseuille's equation

$$\dot{v} = \frac{\pi pa^4}{8\eta l}$$

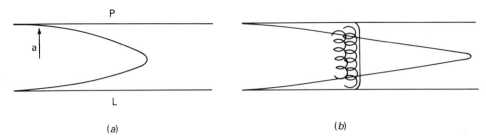

Figure 11.1 (a) The smooth flow of gas down a tube produces a parabolic profile. The flow is adequately described by Poiseuille's equation (see text). (b) At high flow rates the highly ordered laminar flow breaks down into disordered or turbulent flow

where $\dot{v}$ is the flow
        p is the pressure difference
        a is the radius of the tube
        l its length
        η the viscosity of the gas

The resistance to flow is critically dependent on the radius a which in the above equation is 4th order. For example halving the radius (or the diameter) produces a 1600 per cent (that is, 2^4) increase in resistance, and to a first approximation a 1 per cent change in diameter causes a 4 per cent change in resistance. Thus, the airways should be as wide as possible to minimize airflow resistance. Clearly, this is not compatible with the desire to minimize the diameter in order to reduce the dead space. In practice a functional, dynamic compromise is reached but which is responsive to varying situations.

The lung is a branching system. It starts off asymmetrically when at the bottom of the trachea the right main bronchus continues sharply downwards, deviating only slightly from the vertical, while the left main bronchus branches sharply to the left. There are three lobes on the right side of the chest and two on the left reflecting the accommodation of the heart within the thoracic cavity. Within the lobes branching is dichotomous with a mean branching angle of 37° from the

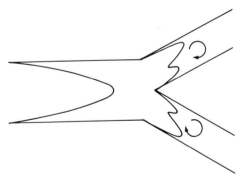

Figure 11.2 There is an intermediate smooth flow associated with dichotomous branching. This is known as secondary flow and the flow profile becomes 'double bumped' with a rotation. This carries implications for inhaled particle deposition

'straight on' position. Aeronautical engineers have demonstrated that this is the optimum branching angle for minimum disturbance of flow. At high flow rates the highly ordered laminar flow of *Figure 11.1* breaks down into disordered or turbulent flow. However, there is an intermediate ordered, smooth flow associated with dichotomous branching. This is known as secondary flow (*Figure 11.2*) in which the flow profile is 'double bumped' and has a rotation. This carries implications for inhaled particle deposition. The conducting airways are lined with cilia which propel mucus upwards towards the trachea, and there are approximately 23 generations before the respiratory bronchioles give way to the alveolar ducts. These are surrounded by the alveolar sacs totalling approximately 300 million. For such a large number of generations branching at 37° the lung cannot be entirely symmetrical as outer branches would curl up on themselves and this lack of symmetry imposes constraints on models of respiratory function.

Like the gut, the lung is topologically an external surface, but unlike the gut, the lung maintains itself in a sterile condition. Phagocytic cells, the alveolar macrophages, scour the alveolated surface scavenging any available material. This, generally, is transported to the ciliated escalator or to lymph tissue for indefinite storage (as may be seen particularly well in sections from the lungs of coal-miners at autopsy).

Surfactant

Mechanically, the alveolar surface is intrinsically unstable. In the classical experiment of producing and then joining together two soap bubbles of different radii (*Figure 11.3*) they do not equilibrate but the smaller empties into the larger. How can this be? Consider an 'equator' drawn around a bubble. The contractile forces due to the surface tension T per unit length are those acting on the line, $2\pi RT$. For two surfaces, inside and outside, we have $4\pi RT$. For a closed bubble, these balance the expansive forces due to pressure P per unit area over the curved internal surface. Resolving the contribution at all surface points into those perpendicular and horizontal with respect to the opposing contractile tension at the 'equator' gives an effective area equal to the equatorial area of cross-section resulting in a net expansive force of $\pi R^2 P$.

Equating the expansive and contractile forces we have

$$\pi R^2 P = 4\pi\, RT$$
$$\therefore\ P = \frac{4T}{R}$$

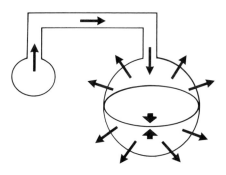

Figure 11.3 The pressure inside a soap bubble is greater inside a smaller bubble. If two bubbles are connected the smaller bubble empties into the larger one. How can this inherent instability be resolved in the lung? (*see text*)

In other words, the smaller the radius the greater the pressure and the small bubble does indeed empty into the larger. Similarly, motor car tyres may be at $165\,kN/m^2$ (24 lb per sq inch) and bicycle tyres $689\,kN/m^2$ (100 lb per sq inch). The mature, healthy lung is saved from the relentless process of one alveolus emptying into another by being lined with surfactant. The surface tension of the surfactant lining is small and is also variable. The first person to grasp the significance of this was Pattle [1–3] who had been examining experimental lung washings. According to a personal communication, by considering the bubbles in his glass of beer he came to the conclusion that the lung washings contain a surface-active material the surface tension of which could become increasingly smaller as the film contracted, thus defeating the natural instability. Many of the physical properties were soon estimated, but due to the small volume and the limitations of contemporary analytical techniques the precise chemical nature of 'Pattle's Peculiar Protein' remains elusive. Recent reviews [4, 5] cover advances in our understanding of the cellular origin and nature of this important material and the creation of aerosols of artificial surfactant for treatment of respiratory distress syndrome in the newborn.

Innervation

The lung is heavily innervated as is the respiratory smooth muscle which surrounds the airways, the intercostal muscles and the diaphragm. The function of the nerves is primarily concerned with the efficient control of ventilation under varying conditions. Protection of the lung in terms of cough, bronchoconstriction and mucus secretion are all neurologically mediated. The most relevant nerve in this respect is the Xth cranial nerve, the vagus. The vagal afferent systems which mediate respiratory reflexes in mammals are as follows.

Pulmonary stretch receptors

These are slowly adapting myelinated fibres found in the smooth muscle of the airways extending from the trachea down to the bronchioles which Adrian [6] concluded were responsible for the Hering–Breuer inflation reflex. In man this reflex is very weak and bilateral vagotomy does not cause slow, deep breathing as observed in other mammals. Experimental evidence by Guz *et al.* [7], and Guz and Widdicombe [8] showed that bilateral anaesthetization of the vagus nerves caused no change in the pattern of quiet breathing or in end-tidal $P\text{CO}_2$.

Work on animals has shown that increased activity is produced by pulmonary congestion and atelectasis. Also, stimulation of these receptors causes a reflex relaxation of tracheobronchial smooth muscle [9]. However, they are insensitive to pathological changes such as microembolism, mild bronchoconstriction and inhalation of irritants and dust.

Type-J receptors

First reported by Paintal [10] type-J receptors lie in the alveolar wall and have non-myelinated vagal afferent fibres. They are stimulated by microembolism, congestion and oedema and also by inhalation of irritant gases and of halothane. The reflex action is to cause rapid shallow breathing, hypotension and bradycardia.

Cough receptors

The cough receptors are concentrated at the carina and at bronchial bifurcations and decrease in number in smaller bronchi. They are found superficially between the epithelial cells and are relatively insensitive to chemical irritants. However, they are very sensitive to mechanical stimulation, for example by inhalation of carbon dust [11]. At low concentration this produces bronchoconstriction but at higher concentrations it also elicits coughing.

Lung irritant receptors

These are found in the epithelial layer of the intrapulmonary airways from the trachea to the large bronchioles. They have myelinated fibres in the vagus nerves which produce reflex bronchoconstriction and hyperpnoea. They are more sensitive to chemical than to mechanical irritation. The receptors may be stimulated by inhalation of irritant gases, pulmonary microembolism, cigarette smoke, carbon dust, intravenous histamine and also histamine aerosol. In man they are thought to contribute to the sensation of breathlessness [12]. They are also stimulated by bronchoconstriction and hyperpnoea thus providing a reinforcing positive feedback which may prolong any response. The bronchoconstriction can be abolished or prevented by isoproterenol, indicating that the effect is due to contraction of airway smooth muscle. Atropine also blocks the bronchoconstriction suggesting that the effect is mediated via postganglionic cholinergic pathways.

Different subjects show greater and lesser sensitivities to stimulation of irritant receptors. *Figure 11.4* (after Nadel [13]) depicts the mechanisms by which reflex bronchoconstriction is thought to occur.

Lung volume measurements

The first quantitative measurements of lung volumes were obtained by Hutchinson [14]. Using a water-filled spirometer he recorded vital capacities (the amount which can be exhaled following a full inspiration) and corresponding anthropometric measurements. The results, expressed numerically in litres, are what physicians had known intuitively about the necessity for adequate alveolar ventilation. More rudimentary, non-validated measures such as chest expansion with a tape measure continue to be used to this day in some instances such as recruitment of firemen although this should soon be replaced.

Following a full exhalation starting from total lung capacity (TLC), the amount remaining is known as the residual volume (RV). *Figure 11.5* shows a standard normal spirogram with the various subdivisions labelled. The volume which may be exhaled following a full inspiration is known as the vital capacity (VC). The volume remaining in the lung (not measurable directly) is known as the residual volume (RV), (see also *Figure 11.17*). The sum of these two volumes is known as the total lung capacity (TLC). In health, the RV may constitute 20–25 per cent of TLC but rises with age. Normal breathing is at a volume approximately 60 per cent of the VC below TLC and requires a tidal volume (V_t) of approximately 600 ml. The volume of gas in the lung at end expiration is known as the functional residual capacity (FRC). The volume which may be inhaled from FRC is known as the inspiratory capacity (IC) and the volume which may be exhaled below FRC is

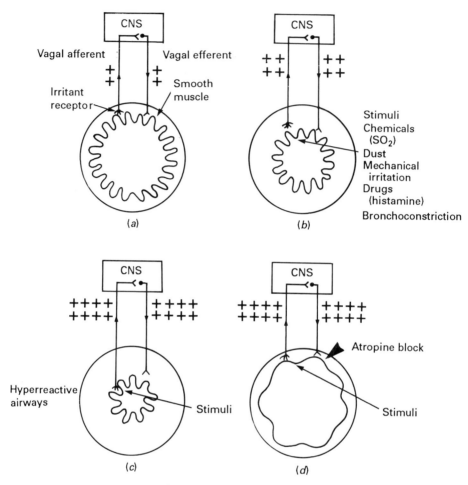

Figure 11.4 (*a*) The airways contain smooth muscle in which there is normal vagally mediated tone. (*b*) Stimulation of irritant receptors which ramify between the cells of the epithelium produces contraction of smooth muscle and narrowing of the lumen. Note the addition of bronchoconstriction to the list of possible stimuli. (*c*) Hyperreactive 'irritable' airways produce an exaggerated response. (*d*) The mechanism may be blocked by atropine

known as the expiratory reserve volume (ERV). In obstruction the RV is typically elevated at the expense of the VC and the ratio of RV to TLC may be as high as 60 per cent. In restrictive conditions the RV, VC and TLC are all reduced but the RV/TLC ratio may be elevated, normal or reduced depending largely on the degree of severity and the possibility of concomitant obstruction, especially in cigarette smokers.

Environmental temperature, barometric pressure and humidity are subject to considerable local variation with corresponding effects on spirometric measurements. To overcome this it is usual practice to correct measurements to body temperature and pressure saturated with water vapour (BTPS).

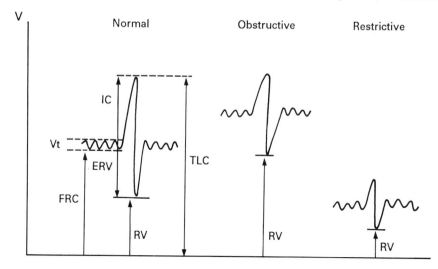

Figure 11.5 Lung volumes and sub-divisions may conveniently be recorded using a water-filled spirometer (*see text*)

The pathophysiology of the lung encompasses two broad categories: obstructive, in which ventilation is impaired; and restrictive in which gas transfer at the alveolar surface is reduced.

Adjacent to the normal spirogram are examples of abnormal spirograms. The first, which is characterized by a somewhat reduced vital capacity, elevated functional residual capacity (FRC) and elevated residual volume, is representative of the classification known as obstruction and includes asthma, bronchitis and emphysema. The second is characterized by a reduced vital capacity, residual volume and total lung capacity. It is representative of the classification known as restriction and includes diseases such as fibrosing alveolitis and, classically, asbestosis. In the absence of other information it is not possible to discriminate between the spirograms of a tall man with asbestosis and a small physically fit woman because both would have small, fast-emptying lungs.

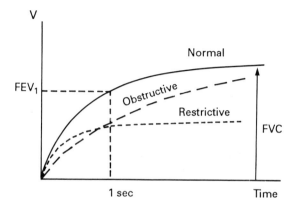

Figure 11.6 The vital capacity delivered at maxium rate is called the forced vital capacity (FVC). The volume expelled in the first second is known as the forced expiratory volume in 1 s (FEV_1)

A major advance in measurement took place when Tiffeneau [15] recorded the rate at which the vital capacity could be delivered. He called this the forced vital capacity (FVC) (*Figure 11.6*) and he simultaneously recorded the amount which could be delivered in the first 1 s. This is called the forced expiratory volume in 1 s (FEV_1) and may be standardized for volume by quoting as a ratio FEV_1/FVC expressed as a percentage. Approximately three-quarters of the vital capacity can normally be expelled in the first second, the precise ratio depending predominantly upon age.

Peak flow

A useful and simple measure of airflow is the Wright peak flowmeter (*Figure 11.7*). This is a rotating vane instrument that records the peak flow which can be sustained over 100 ms during a short, sharp exhalation (a maximum puff). As well as being simple, it requires neither water nor electricity. Nonetheless, for a while it appeared to be giving way to direct writing spirometers providing accurate measurements of FEV_1 and FVC. However, more recently, the introduction of very cheap mini peak flowmeters has allowed patients to carry these instruments in their pockets or handbags and make serial measurements of their own respiratory function throughout the 24 h. This is particularly useful in some drug

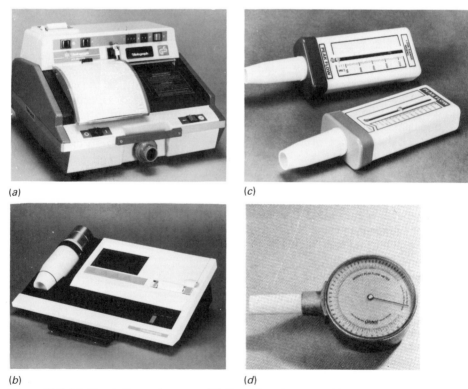

(a)   (c)

(b)   (d)

Figure 11.7 (*a*) Dry bellow spirometer. (*b*) Portable spirometer. (*c*) Portable peak flow meter (*d*) Traditional peak flow meter

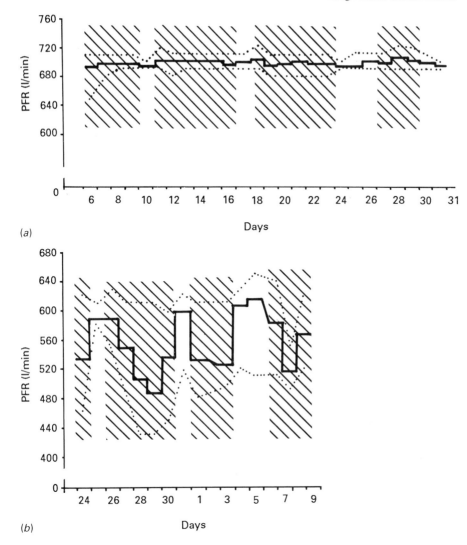

Figure 11.8 (*a*) In normal subjects there is a small circadian rhythm but a stable mean value (solid line indicates mean value, dashed line maximum and minimum values). (*b*) In occupational asthma there may be a steady deterioration in mean throughout the working week with large swings between maximum and minimum. Weekends and holidays may show a marked improvement

trials and in evaluating cases of occupational asthma. These measurements should be seen as complementary to the more formal spirometric measurements.

Respiratory function exhibits a circadian rhythm being minimal in the early hours and maximal in the afternoon, apart from which it should be reasonably reproducible. In an occupational investigation [16], a variation in peak flow of greater than 20 per cent has been taken as evidence of asthma. If it improves in days away from the workplace (weekends and holidays) (*Figure 11.8*) this has been taken to indicate that the asthma is occupational.

Flow volume curves and early detection of abnormality

It used to be believed, quite reasonably, that the small airways ($\leqslant 2\,\text{mm}$ internal diameter) were the main source of airflow resistance. This is especially plausible in view of the fourth power of the diameter law for laminar flow through tubes. However, morphological measurements have since shown that the increase in cross-sectional area as one progresses down the lung is not only large but may better be considered as explosive. In America this is described as the 'thumb tack' model (*Figure 11.9*), and the increase in available area of cross-section more than

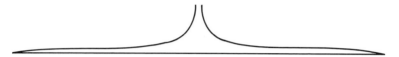

Figure 11.9 The 'thumb tack model' frequently referred to in the literature emphasizes the explosive increase in area of the cross-section with dichotomous branching in the lung

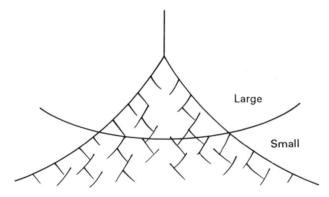

Figure 11.10 The 'small' airways are those equal to and less than 2 mm internal diameter. Together they form a large cross-sectional area and in health contribute only 10–20 per cent of the total airflow resistance

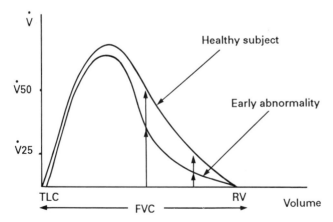

Figure 11.11 A flow volume curve which is responsive to early changes in the small airways. Flow is usually recorded with 25 per cent of the vital capacity remaining ($\dot{V}_{25}$) and also with 50 per cent remaining ($\dot{V}_{50}$)

offsets the increase in resistance of individual units. Thus it transpires that in health 80–90 per cent of resistance is in the large airways and only 10–20 per cent in the small airways—the so-called 'silent zone' of the lung (*Figure 11.10*).

This is perhaps unfortunate in measurement terms because it is in this area that the disease process often begins, especially if related to cigarette smoking. For a robust measure such as the FEV_1 which is responsive to all the airways *in toto* there is little chance of detecting early changes in the small airways and a special, specific test is required.

The numerical value of the peak flow measurement is determined by sex, age, height and ethnic group. It is also determined by the lung volume at which the measurement is made. Normally this is made close to total lung capacity following a full inspiration. Starting from the normal resting position (functional residual capacity, FRC) produces a lower value and at residual volume the expiratory flow drops to zero. By using an instrument which measures flow and volume continuously it is possible to draw a flow-volume curve as shown in *Figure 11.11*. The last half of the blow is independent of effort and reflects the nature of the small airways. A suitable instrument may be obtained comprising a flow measuring device providing *via* a transducer an electrical signal which is electronically integrated to provide volume. A pneumotachograph is a tube containing either a screen or a bundle of short tubes in parallel. As the exhaled air passes through the low resistance screen a small pressure drop is created proportional to the flow. This pressure drop is transmitted to the transducer to give an electrical output proportional to the flow.

Alternatively, a volume signal from a low inertia rolling seal dry spirometer may be differentiated to give flow which is simultaneously displayed against volume.

Both the above systems may be designed to record FEV_1 and FVC while performing the flow volume test and both have their advantages and disadvantages of which a user should be aware.

For instance, pneumotachographs being gas viscosity instruments are sensitive to variations in gas composition (for example, CO_2 content) throughout the blow whereas the calibration is usually performed with room air and which, unlike exhaled air, will not normally be saturated with water vapour. In addition, electronic integrators are liable to 'drift' of derived volume.

Dry spirometers, on the other hand, provide reliable volume measurements (after correction to BTPS), but even though the back pressure may be very small, the inertia may be sufficient to question the value of flow obtained by differentiation. However, in the part of the curve of interest for flow-volume loops, the error is not likely to be very large.

Theory

Simultaneous recordings of flow, volume and transpulmonary pressure (see below under lung compliance) may be recorded for different degrees of effort and represented as in *Figure 11.12*. Taking a vertical section A–A through the family of flow volume curves at a given volume and plotting the flows against the appropriate corresponding pressures produces an isovolume pressure flow curve (α), the slope of which with respect to the ordinate is resistance. Repeating the procedure at B–B produces a second curve (β) of slightly increased resistance. At a lower volume C–C the curves all merge together. The corresponding pressure flow curve (γ) now exhibits a plateau. This is interpreted as increasing effort being met by increasing resistance and the flow is now effort independent.

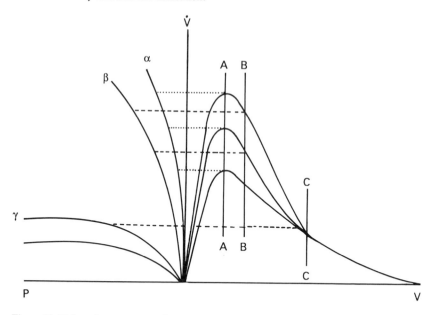

Figure 11.12 Isovolume pressure flow curves (*see text*)

The mechanism may be understood by considering *Figure 11.13*. This represents conducting airways supplied by an alveolated region inside a thoracic cavity. At functional residual capacity the elastic recoil P_{el} is typically 5 cm water (cmH$_2$0; 0.5 kPa). At rest this is balanced by an equal but negative pressure in the intrapleural space, P_{pl}. The net alveolar pressure is zero with respect to the mouth, and flow is also zero. At inspiration muscular effort is applied to make the intrapleural pressure more negative, for instance -15 cm water. Against the elastic recoil of $+5$ cm water this leaves a net alveolar pressure of -10 cm water and inspiration follows. At expiration the intrapleural pressure may now become positive, for example $+15$ cm water. Adding to the elastic recoil pressure produces an alveolar pressure of $+20$ cm water with respect to the mouth, and expiration follows. The pressure falls progressively along the conducting airway from the alveolus until at the equal pressure point the pressure inside the tube equals the pressure outside the tube. Beyond this point the pressure outside exceeds the pressure inside and there is a tendency to squeeze down the conducting airway. Air which does get through to the mouth is that originating from the alveolar generator above the equal pressure point. This flow is limited by the resistance of the small airways (R_{sa}) between the alveolar region and the equal pressure point. It is also proportional to the pressure difference between the alveoli (P_A = in this example) and the equal pressure point (EPP, equals P_{pl} in this example). The difference of 5 cm water is the elastic recoil pressure.

Thus $\dot{V}_{max} = \dfrac{P_{el}}{R_{sa}}$

The pleural pressure, reflecting muscular effort is eliminated from the equation and the maximum flow is seen to be independent of effort. Increase in resistance

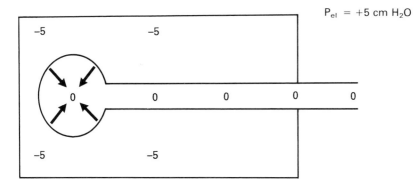

P_{el} = +5 cm H₂O

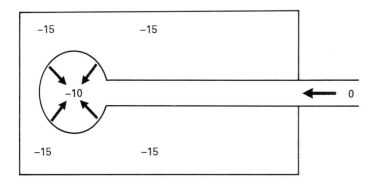

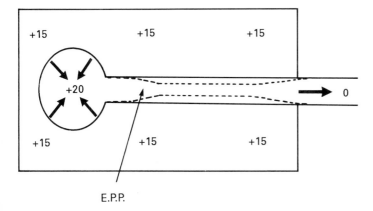

E.P.P.

Figure 11.13 Stylized representation of conducting airways inside a thoracic cavity (*see text*)

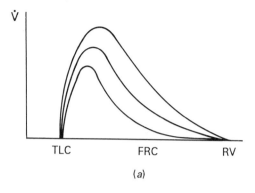

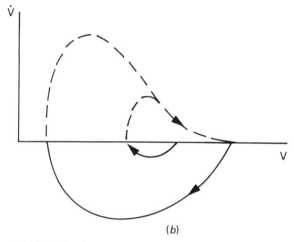

Figure 11.14(*a*) The flow volume curve and the effects of chronic airflow limitation (CAL) as exemplified by chronic bronchitis and by emphysema. (*b*) Partial expiratory flow curves can produce increased sensitivity, especially in studying relatively acute effects

of the small airways R_{sa} as in chronic bronchitis produces a reduction in expiratory flow in the above equation.

In emphysema P_{el} is reduced and this produces a further reduction in flow (*Figure 11.14a*). Current semantic arguments about definition and diagnosis of chronic bronchitis (precisely, but inadequately, defined in terms of sputum production for 3 months for 2 or more consecutive years) and emphysema (parenchymal spaces of equal or greater than 1 mm) during life have produced the term chronic airflow limitation (CAL) which includes both conditions.

As the reduction in flow becomes progressively worse in the region of the functional residual capacity (FRC), tidal breathing—especially during exercise—becomes progressively more limited. The only escape, mechanically, is to move the FRC to the left on the volume axis, that is, to a progressively higher lung volume. For clarity, the three curves in *Figure 11.14a* are shown with the same vital capacity and residual volume, but in practice the residual volume increases, essentially at the expense of the vital capacity and the whole curve moves to the left, further increasing the pressure on the FRC. The reader may care to imagine the situation in restrictive disease. Here the vital capacity and residual volume both reduce and the curve shifts to the right.

The undoubted sensitivity of the flow volume curve has gained in popularity in some epidemiological surveys including byssinosis. Some of the sensitivity can be lost due to the taking of a full inspiration prior to exhalation (see also under body plethysmography) and this has been approached by using partial expiratory flow volume curves (*Figure 11.14b*). Here the subject inhales slightly above FRC and then exhales forcefully to residual volume. He then inhales to TLC to define the vital capacity and allow location of the 50 per cent and 25 per cent points.

An ingenious extension of the flow volume curve is the helium flow volume curve. Turbulent flow is limited by density and not by viscosity and at high lung volumes, the expiratory flow of a subject breathing 80 per cent He, 20 per cent O_2 would be elevated (*Figure 11.15*). As the equal pressure point becomes

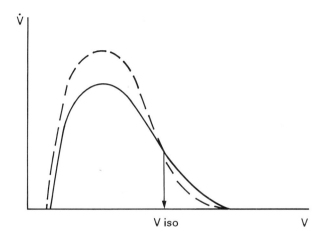

Figure 11.15 Helium flow volume curves (*see text*)

intrathoracic and intrapulmonary the Reynold's number in the distal airways drops and the flow becomes laminar. Laminar flow is dependent on viscosity and not upon density (Poiseuille's equation). Helium is less dense but more viscous compared with air and the flow is reduced. Where the helium curve crosses the flow volume curve for breathing air is called the volume of isoflow. Changes in this volume may correlate with severity of emphysema during life [17].

Mid-maximum expiratory flow

The mid-maximum expiratory flow (MMEF) is obtained (*Figure 11.16*) by drawing a chord from the volume at 75 per cent lung inflation to 25 per cent inflation and dividing by the time. The mid-maximum expiratory flow is thought to reflect flow in the medium and small airways. A convenient test, readily obtainable from a trace of forced expiration, it is related to flow at 50 per cent of vital capacity ($\dot{V}_{50}$)—normally obtained from a flow volume curve—but which can be seen to be the tangent to the volume time curve at the midpoint (*see Figure 11.16*). Assuming a

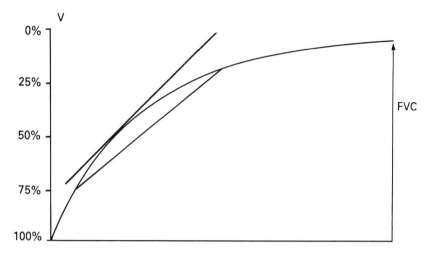

Figure 11.16 The mid maximum expiratory flow (MMEF) is obtained by drawing the chord across the middle half of the vital capacity and dividing the volume by the time. It may be related to the tangent to the curve at 50 per cent point which is $\dot{V}_{50}$ (*see text*)

simple (monoexponential) curve it has been possible to demonstrate the relationship [18] between MMEF and $\dot{V}_{50}$ as:

$$\frac{\text{MMEF}}{\dot{V}_{50}} = -\ln 3, \text{ which is } 0.92 \text{ approx.}$$

Departures from this ratio are a measure of non-linearity on the descending flow volume curve and provide a measure of abnormality of the curve. Since the ratio describes the intrinsic shape of the curve it is sensibly independent of height. For investigators with access to prediction equations for either MMEF or $\dot{V}_{50}$ but not the other, and non-corresponding measurements, it does provide a method of conversion to a first approximation.

Residual volume

The residual volume measurement is important for correctly positioning other lung volumes on a spirogram. It is also important in its own right in discriminating obstructive conditions in particular and also restrictive defects. There are three main methods of assessment.

Helium rebreathing

In this technique an 'anatomical cast' of the lungs at functional residual capacity is obtained by allowing the subject to rebreathe a known quantity of helium from a spirometer in closed circuit (*Figure 11.17*).

It is important to allow the subject a little time on the mouthpiece with the valve turned to room air to allow breathing to settle down. By observing the rise and fall of the chest it is possible to turn the subject at the appropriate moment to the spirometer when he is at functional residual capacity. Carbon dioxide is removed

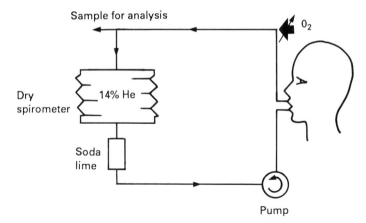

Sample for analysis

O_2

Dry
spirometer

14% He

Soda
lime

Pump

Figure 11.17 Residual volume measurement by the technique of helium rebreathing (*see text*)

by soda lime and oxygen admitted via a needle valve to make up that absorbed by the subject.

The initial concentration of helium in the spirometer is recorded (about 14 per cent) and most subjects come to equilibrium at a lower concentration within about 3 min. Since the mass of helium present is constant the drop in concentration is due to the added lung volume (the FRC) in the closed circuit.

Thus: $C_1V_1 = C_2V_2$

where the subscripted letters represent initial and final concentration and volumes.

However $V_2 = V_1 + FRC$

$$\therefore FRC = V_1 (C_1 - C_2)/C_2$$

Residual volume is then calculated by subtraction of the expiratory reserve volume (ERV) this being the volume which can be exhaled below the FRC (*see Figure 11.5*).

Body plethysmography

The whole body plethysmograph (*Figure 11.18*) is an instrument resembling an airtight telephone booth with a volume of approximately 600 l. By measurement of pressure variations in the mouth and the box while panting—at zero flow— against a closed shutter and then by application of Boyle's law it is possible to calculate the volume of gas in the lung. After subtraction of the ERV, this yields the residual volume.

In healthy subjects there is good agreement with the method of helium dilution. However, in patients with poorly communicating spaces in the parenchyma (for example, in emphysema) the helium method will underestimate the residual volume. However, the plethysmograph, using a different principle includes such 'trapped air' and the difference between the two readings provides an estimate of the volume of the emphysematous bullae.

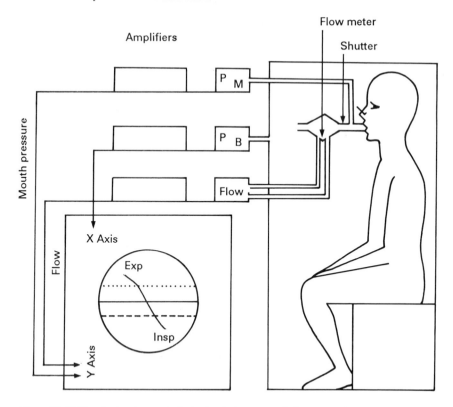

Figure 11.18 Constant volume whole body plethysmograph for measuring lung volumes and airways resistance (*see text*)

Planimetry of X-radiographs

This method is most suitable if the subject is to have a chest radiograph taken as part of his assessment. However, although convenient, it does have a disadvantage in that two films, an anterioposterior and a lateral, are required. By tracing round the outline of the shadows, a very quick estimation of total lung capacity is obtained, and by subtraction of vital capacity the residual volume is calculated.

Airways resistance by body plethysmography

The body plethysmograph (*see Figure 11.18*) may be used to measure airways resistance. It is exquisitely sensitive at measuring small changes over time following a challenge with a bronchoconstrictor or bronchodilator. For instance, from smoking one cigarette it is possible to demonstrate a 20 per cent bronchoconstriction. This is perhaps not too surprising considering the great sensitivity of flow to airway diameter (*see above*). However, such changes are not readily discernible using a normal spirometer. One reason for this is that plethysmographic measurements are made at functional residual capacity using a panting manoeuvre. If by comparison one measures an FEV_1 this is preceded by a full inspiration to total lung capacity. This activates pulmonary stretch receptors which tend to reverse any vagally mediated reflex brochoconstriction. Another advantage of the plethysmograph is that the panting manoeuvre abducts the vocal chords and by reducing the contribution due to the larynx allows the investigator to look more closely at change in the lower airways which are of interest.

Resistance is pressure divided by flow. The flow through the airways is readily measured by a pneumotachograph. However, the pressure difference required is that between alveolus and mouth. To avoid using an oesophageal balloon (see below) the subject carries out a panting manoeuvre inside the closed box while flow and box pressure are recorded simultaneously. At end expiration, a shutter is brought down to occlude the mouthpiece while the subject continues to pant, but with zero flow. Under this condition, the pressure at the mouth is the same as that in the alveolus. The corresponding compressions and rarefactions in the box continue to be recorded. By eliminating box pressure from the calculations it is possible to obtain the relationship between pressure and flow, that is, the airways resistance (R_{AW}). In practice it is more common to quote the reciprocal of this, conductance (G_{AW}), and to correct this for variation with lung volume by dividing by the volume at which the measurement is made to give specific conductance of the airways (SG_{AW}).

Measurements are usually made at inspiration rather than expiration. This is particularly appropriate in some patients with obstruction where collapse of airways during exhalation produces a greater variability in replicate measurements.

In normal subjects the body plethysmograph has been particularly useful in producing dose–response curves for irritant gases and corresponding thresholds as an adjunct to the problem of setting hygiene standards for workroom air [19,20].

Transfer factor

Some patients may complain of dyspnoea even in the absence of airways obstruction. In such cases it may well be that although the lung is sufficiently well-ventilated, nonetheless, the ability of the lung to transfer oxygen into the blood is compromised. Classically, interstitial fibrosis such as is found in asbestosis would produce this condition. One way of assessing this condition would be by inhaling oxygen, breath-holding and then seeing at exhalation how much has been retained. However, this would necessitate a knowledge of the back tension in the blood which involves an extra procedure. An alternative method, generally accepted, is to inhale a low concentration of carbon monoxide (0.28 per cent) and assume the back tension to be very low if the subject has not been smoking. Carbon monoxide has an affinity for blood approximately 200 times that of oxygen. After exhaling to residual volume, the subject inhales to total lung capacity from a bag containing the gas in low concentration. After a corrected breath-hold time of approximately 10 s, the subject exhales (*Figure 11.19*). The first proportion of the exhalate is recorded but voided. A middle portion is collected for analysis as representing an alveolar sample. The only problem is that the diffusion of the carbon monoxide across the alveolar membrane is preceded by dilution due to the (unknown) residual volume. To overcome this, the inhaled gas also contains 14 per cent of helium. Although this is diluted in an equal manner to the carbon monoxide, it does not diffuse across appreciably into the blood. Thus the relative difference for the two gases is attributable to transfer of the carbon monoxide into the blood. Additionally, the method produces a single breath estimate of alveolar volume

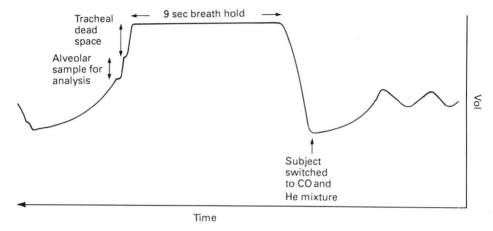

Tracheal dead space

Alveolar sample for analysis

9 sec breath hold

Subject switched to CO and He mixture

Vol

Time

Figure 11.19 Carbon monoxide transfer test. An actual trace (which goes from right to left) from a patient is shown (*see text*)

from the helium measurements on inhalation and exhalation. This may be compared with that obtained from the helium rebreathing method (*see above*) and the ratio is a measure of inhomogeneity of gas mixing in the lung.

Lung compliance

An important lung parameter is the compliance which represents the distensibility of the lung. It is defined as the change in lung volume for unit change in pressure $\Delta V/\Delta P$. The compliance varies somewhat throughout the vital capacity and is best

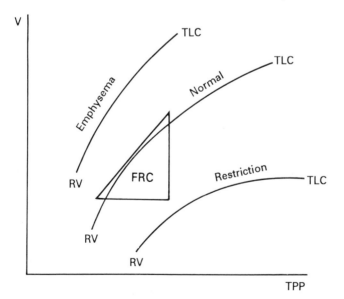

V

TLC

Emphysema

Normal

TLC

RV

FRC

Restriction

TLC

RV

RV

TPP

Figure 11.20 Pressure volume curves for estimation of lung compliance. In emphysema the pressures are reduced and the volumes elevated, shifting the curve up and to the left. In asbestosis the reverse occurs producing a stiff lung of low compliance measured at a stated volume such as functional residual capacity (FRC) (*Figure 11.20*).

Volume changes in the lung are readily measured spirometrically. However, the required corresponding pressures are transpulmonary pressure differences. The lung is intrinsically springy and if removed from the thorax will deflate under its own elastic recoil which exerts a positive deflationary pressure of about 5 cm water at FRC. The normally inflated lung adheres to the pleura lining the thorax due to a negative pressure (actively maintained by the osmotic pull of the plasma proteins). At rest, this negative pressure in the intrapleural potential space is, say, minus 5 cm water (cancelling the positive elastic recoil pressure and resulting in net

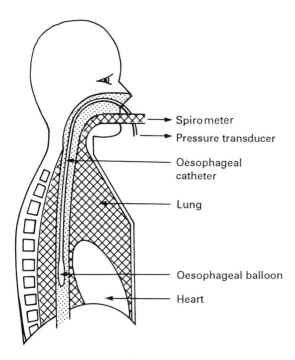

Figure 11.21 Oesophageal balloon technique for measuring lung compliance

alveolar pressure of zero). The changes in the negative pressure on the outside of the lung with volume change are estimated by measuring corresponding pressure changes in the oesophagus (which has similar—if not identical—pressure variations). The pressure in the oesophagus is measured with a small oesophageal balloon on the end of a rigid catheter which is connected to a pressure transducer (*Figure 11.21*). To avoid the gag reflex the catheter and balloon are introduced *via* the nares.

Patients with a restrictive condition, such as asbestosis, show a reduced compliance, i.e. a stiff lung, and those with emphysema show a high compliance, i.e. a soft, spongy lung lacking elasticity.

Dynamic compliance

Lung compliance may also be measured dynamically (C_{dyn}). In this method, transpulmonary pressure, flow and volume are recorded continuously on a three-channel recorder. A spirometer would be unable to keep up adequately with breathing at high frequency and so volume is derived from electronic integration of flow. From the recorder output points of zero flow may be identified. Adjacent zero flow points correspond to end of inspiration and end of expiration, and the corresponding volume and pressure changes are used to calculate the dynamic compliance at that frequency of breathing. Normal subjects show little variation with frequency but with small airways disease, the measured value of C_{dyn} will tend to fall with increased frequency of breathing. In emphysema, the measured value is high at low frequencies but falls with increasing frequency eventually falling below normal.

Time constant

This is an important concept for understanding lung emptying. If compliance is multiplied by resistance the product is found to have the dimensions of time and is called the time constant.

$$\frac{V}{P} \times \frac{P}{\dot{V}} = T$$

The time constant is the time for a monoexponential system to drop to $1/e$ of its original value. This is easily related by a simple transformation to the half-life with which many readers will be familiar from their studies of the exponential decay of radioactivity and also of the washout of anaesthetic gases.

The reader should now be in a position to justify the adjectives fast and slow in relation to restrictive and obstructive conditions. In asbestosis the compliance is reduced, as also is the resistance, as the fibrosis tends to hold open the airways delaying the onset of collapse. Thus the product of compliance and resistance is reduced giving a small time constant and a fast emptying lung. In bronchitis the resistance is elevated producing a slowing of expiration. In emphysema the compliance is also elevated giving an even larger time constant and further prolongation of expiration.

Blood gas measurements and pH

Blood gases and pH may rapidly and reliably be measured with special electrodes using blood from arterial puncture. Alternatively, arterialized venous blood from a finger or an ear lobe made hyperaemic with cantharides ointment may be used—especially for successive estimations—but there may be some loss of accuracy.

During exercise there may be a widening of the alveolar arterial oxygen gradient, or there may be alveolar underventilation. Due to the sigmoid shape of the oxygen dissociation curve (*Figure 11.22*) the reduction in PaO_2 occurs before a reduction in saturation. However, unsaturation, when it does occur, may be measured continuously and non-invasively by an ear lobe or finger oxymeter. These instruments work by measuring transmission of light through the tissue to a detector on the far side worn rather like a clip-on earring and are finding popularity not only in exercise physiology but in the important area of sleep studies. Nocturnal

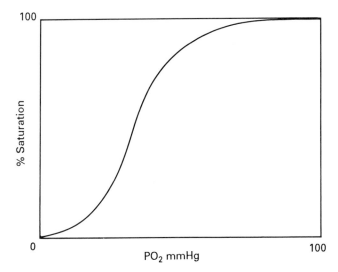

Figure 11.22 The oxygen dissociation curve

unsaturation may be more widespread than previously thought and left untreated can lead to pulmonary hypertension.

As a guide to the adequacy of ventilation the arterial P_{aCO_2} may be more useful than the P_{aO_2} and is readily measured by a rebreathing technique [21]. This technique may be employed just as readily on a factory premises as at the bedside. The procedure requires the subject to rebreathe from a bag containing 1 l of oxygen for 1½ min. The subject then breathes fresh air for 2 min and then comes back to the bag for five breaths or 20 s, whichever is the longer. The CO_2 concentration in the bag now represents mixed venous CO_2. Subtraction of 6 mmHg from the mixed venous yields the arterial P_{CO_2}.

P_{aCO_2} in asthma is usually normal or even reduced testifying to the success of the increased ventilatory drive (*Figure 11.23*). In this respect they resemble the emphysematous 'pink puffer' and contrast with the CO_2 retaining 'blue bloater' of chronic obstructive bronchitis. However, if CO_2 retention does develop in the asthmatic the drop in pH is more marked than in the chronic bronchitic in whom the development of a metabolic alkalosis serves to ameliorate the fall.

Reference values

Lung function varies between individuals. Prime sources of variation are gender, body size, age and ethnic group. Body size is usually expressed in terms of standing height (seated height has occasionally been mooted but has not found wide acceptance). Weight is not usually a significant factor as for a given height it is difficult to attribute how much weight is due to build and how much is due to fat.

The most popular method of producing reference values of respiratory function, for example, the FEV_1 is to produce equations based on classical multiple regression:

$$FEV_1 = a + b \times \text{height} + c \times \text{age}.$$

	Asthma	Bronchitis	Emphysema
PaO_2	↓	↓ ↓	↓
P_AO_2	↑	↓	—
$PaCO_2$	—/↓	↑	—
Ventilatory response to CO_2	↓	↓ ↓	↓
Neurogenic drive to breathing	↑ ↑	? ↑	↑
Δ FEV_1/FVC %	↑	—/↑ *	—/↓ **
T_ICO	—/↑	—/↓	↓ ↓
C	—	—	↑
C dyn	↓	↓	↑/↓

Figure 11.23 Summary of differences in measurements between asthma, bronchitis and emphysema. ↑ = Elevated or improved; ↓ = reduced; ↓ ↓ = much reduced; — = no change. FEV_1, FVC not included as they are reduced in all three conditions. *In so-called chronic irreversible bronchitis there is no improvement following inhalation of salbutamol. However, recent studies in Arizona estimate that 30 per cent of such cases are 'hidden asthmatics'. Following intensive steroid therapy they can become labile to bronchodilators. **In emphysema the effect of a bronchodilator is often more evident in the FVC than in the FEV_1. Although both may be improved, the ratio often drops in consequence

The majority of such equations are based on cross-sectional studies [22]. Early studies tended to mix smokers and non-smokers together but the data are more satisfactory if based on lifelong non-smokers in spite of the epidemiological bias which may be introduced.

Similar equations are used for the FEV_1 and the peak flow. Since the FEV_1 and FVC both contain height, the FEV_1/FVC ratio is commonly taken to be a function of age only (but *see below*).

Cole [23], using secondary data from other studies, assumed a linear function for the regression on age and by an elegant use of mathematical support functions set out to determine the best power fit (α) for the height (*ht*) used multiplicatively:

$$FEV_1 = ht^\alpha (C + D \times \text{age})$$

He produced a value of α=2.05 which can conveniently be written as 2.00 with slight adjustments to *C*, *D*. For this work [23] he received the annual prize of the Royal Statistical Society and the work does have the merit that the FEV_1 and *ht*— both of which are measured directly—can be taken together opposite a linear function of age:

$$\frac{FEV_1}{ht^2} = C + D \times \text{age}$$

The age regression coefficient for adults is usually negative, but for children it is positive. Somewhere it must be zero (if continuous). Knudson *et al.* [24] arbitrarily divided their data at age 25 years and fitted separate linear regression lines for the data below and above 25. Unfortunately, they neglected to constrain the regression lines to meet, so for age 25 there are two predicted values.

Schoenberg *et al.* [25], recognizing that in general terms respiratory function rises with age, plateaus in early adult life and then begins a slow decline (*Figure 11.24*) fitted a complex polynomial to the data. This had the definite advantage of reducing the heteroscadicity of the data but had the disadvantage of not allowing ready comparison from one study to another—especially in terms of age coefficients which are usually of prime interest.

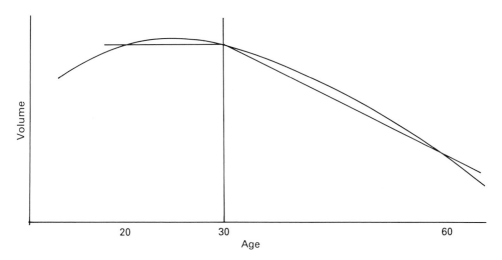

Figure 11.24 Throughout life FEV and FVC grow, plateau and then decline. Regression lines may be fitted in many, and sometimes complex ways, but first order (linear) functions of age have advantages for ease of understanding and group comparisons

Nonetheless, their feeling that the decline in adult life is curvilinear is echoed elsewhere [26]. Now, if persons of below age 20 are separated and analysed separately, linear regressions for adults over 20 may be biased and underestimate the rate of decline due to the plateau in early life. A suitable alternative may be to split adult data at about age 30 and fit linear regression lines constrained to have a common solution (*see Figure 11.24*). For older adults (>30) this should produce more appropriate linear regression coefficients.

If young adults (say 20–30) are to be taken as the benchmark for measurement—perhaps for pre-employment medicals—then combining with Cole's approach of linear age function and higher order height function may have definite advantages.

Due to the plateau (for age <30) the numerical value of D will be small and prone to variation. Thus if D is arbitrarily set equal to zero we have:

$$\frac{FEV_1}{ht^2} = \text{constant}$$

and all the values for young adults can be pooled to give a single, unique value for that population.

However, the plateau may not always occur in the 20–30 age group. In a recent careful study of London firemen [27] for three groups (approximately 500 in each) of non-smokers, ex-smokers and smokers linear regressions were found to be the most appropriate for each group declining steadily (at different rates) from the 20–24 age group and the plateau must, presumably, have occurred earlier—possibly reflecting earlier maturation. Furthermore, although it is generally regarded that since FEV_1 and FVC are each determined by height, the ratio FEV_1/FVC is independent of height. This is often a reasonable assumption but for tall, fit men with large lungs, 1 s is not a very long time in which to deliver 70 or 75 per cent of the vital capacity. In the analysis of the population of firemen [27], taller men were found to have a somewhat smaller FEV_1/FVC ratio. Crosbie [28] found that North Sea divers had vital capacities approximately 20 per cent greater than predicted for their age and height and this further served to reduce the FEV_1/FVC ratio. Thus a tall, fit man could have an actual ratio considerably below that predicted and care must be exercised [29] in making judgements on fitness based on such ratios.

Standardization of lung function testing and acceptable reference values has been reported on by a working party for the Commission of the European Communities [30] and also, more recently, the American view on standardization of techniques has been reported [31].

Assessment of individuals for clinical management or for financial compensation employs many of the techniques described above [32]. Impairment of function may lead to disability and this in turn may, depending very much on circumstances, be productive of handicap in employment. It is this last item which is susceptible to reasonably equitable assessment for compensation [33].

However, for a patient, more acute than the social anxieties associated with employment can be the subjective sensation of dyspnoea, and research is continuing into both assessment [34] and pharmacological manipulation [35] of this distressing condition. Where possible subjective opinions or scales should be related to objective measures of respiratory function and changes in these variables.

References

1. Pattle, R.E. Properties, function, and origin of the alveolar lining layer. *Nature (London)*, **175**, 1125–1126 (1955)
2. Pattle, R.E. Properties, function, and origin of the alveolar lining layer. *Proceedings of the Royal Society of London*, **148**, 217–240 (1958)
3. Pattle, R.E. The cause of the stability of bubbles derived from lung. *Physical and Medical Biology*, **5**, 11–26 (1960)

4. Robertson, B., Van Golde, L.M.G., Batenburg, J.J. *Pulmonary Surfactant*, Elsevier, Amsterdam (1984)
5. Jobe, A. and Ikegami, M. State of art: surfactant for the treatment of respiratory distress syndrome. *American Review of Respiratory Diseases*, **136**, 1256–1275 (1987)
6. Adrian, E.D. Afferent impulses in the vagus and their effect on respiration. *Journal of Physiology*, **79**, 332–358 (1933)
7. Guz, A., Noble, M.I. and Widdicombe, J.G. The role of vagal and glossopharyngeal afferent nerves in respiratory sensation, control of breathing and arterial pressure regulation in conscious man. *Clinical Science*, **30**, 161–170 (1966)
8. Guz, A. and Widdicombe, J.G. Reflexes respiratoires provenant des poumons de l'homme. *Poumon Coeur*, **24**, 1033–1057 (1968)
9. Widdicombe, J.G. and Sterling, G.M. The autonomic nervous system and breathing. *Archives of Internal Medicine*, **126**, 311–315 (1970)
10. Paintal, S. *The Mechanism of Excitation of Type J Receptors and the J Reflex. Hering-Breuer Centenary Symposium, CIBA*. Churchill, Edinburgh (1970)
11. Widdicombe, J.G., Kent, D.C. and Nadel, J.A. Mechanism of bronchoconstriction during inhalation of dust. *Journal of Applied Physiology*, **17**, 613–616 (1962)
12. Sellick, H. and Widdicombe, J.G. Stimulation of lung irritant receptors by cigarette smoke, carbon dust and histamine aerosol. *Journal of Applied Physiology*, **31**, 15–19 (1971)
13. Nadel, J.S. Aerosol effects on smooth muscle and airways visualization technique. *Archives of Internal Medicine*, **131**, 83–87 (1973)
14. Hutchinson, J. *Transactions of the Medical and Chirurgical Society of London*, **29**, 137–252 (1846)
15. Tiffeneau, R. and Pinelli, A. Air circulant et air captif dans l' exploration da la fonction ventilatrica pulmonair. *Paris, Medical*, **133**, 614–628 (1947)
16. Finnegan, M.J., Pickering, C.A.C., Burge, P.S., Goffe, T.R.P., Austwick, P.K.C. and Davies, P.S. Occupational asthma in a fibre glass works. *Journal of the Society of Occupational Medicine*, **35**, 121–127 (1985)
17. Cosio, M., Ghezzo, H., Hogg, J.C., Corbin, R., Loveland, M., Dosman, J. and Macklem, P.T. The relations between structural changes in small airways and pulmonary-function tests. *New England Journal of Medicine*, **298**, 1277–1281 (1978)
18. Douglas, R.B. The midmaximum expiratory flow. *Bulletin Européen de Physiopathologie Respiratoire*, **16(5)**, 283–286 (1980)
19. Douglas, R.B. PhD Thesis. *Human reflex bronchoconstriction as an adjunct to conjunctival sensitivity in defining the threshold limit values of irritant gases and vapours.* London (1975)
20. Douglas, R.B. Inhalation of irritant gases and aerosols. In *Respiratory Pharmacology* (ed. J. Widdicombe), Chapter 15, p. 297–333. Pergamon Press, Oxford (1981)
21. Campbell, E.J.M. and Howell, J.B.L. Rebreathing method for measurement of mixed venous PCO_2. *British Medical Journal*, **2**, 630–633 (1962)
22. Cotes, J.E. *Lung Function*. 4th edn. Blackwell Scientific Publications, London (1979)
23. Cole, T.J. Linear and proportional regression models in the prediction of ventilatory function. *Journal of the Royal Statistical Society A*, **138**, 297–325 (1975)
24. Knudson, R.J., Slatin, R.C., Lebowitz, M.D. and Burrows, B. (1976) The maximal expiratory flow volume curve. Normal standards, variability and effects of age. *American Review of Respiratory Diseases*, **113**, 587–600 (1976)
25. Schoenberg, J.B., Beck, G.J. and Bouhuys, A. Growth and decay of pulmonary function. *American Review of Respiratory Diseases*, **113**, 90 (1976)
26. Smith, D.J. and Searing, C.S.M. Letters to the editor. *Journal of the Royal Naval Medical Service*, **72**, 114–117 (1986)
27. Douglas, R.B. Report to Home Office: A six year follow-up study of respiratory morbidity in London firemen. Research Report No. 25, *Pulmonary function of firemen: a follow-up study* (1985)
28. Crosbie, W.A., Clarke, M.B., Cox R.A.F, *et al.* Physical characteristics and ventilatory function of 404 commercial divers working in the North Sea. *British Journal of Industrial Medicine*, **35**, 104–108 (1977)
29. Douglas, R.B. Detection of early ventilatory impairment in RN submarines. *Journal of the Royal Naval Medical Services*, **72**, 111–115 (1986)

30. Report: Standarized lung function testing, *Journal of the Societas Europaea Physiologiae Clinicae Respiratoriae*, Suppl. 5 (1983)
31. Standardization of Spirometry 1987 update. *American Review of Respiratory Diseases*, **136**, 1285–1298 (1987)
32. Douglas, R.B. Occupational lung disease and aerosols. In *Aerosols and the Lung: Clinical and Experimental Aspects* (eds S.W. Clarke and D. Pavia). Butterworths, London (1984)
33. Ward, F.G. Industrial benefits and respiratory diseases, *Thorax*, **41**, 257–260 (1986)
34. Killian, K.J. Assessment of dyspnoea. *European Respiratory Journal*, **1**, 195–197 (1988)
35. Stark, R.D. Dyspnoea: assessment and pharmacological manipulation. *European Respiratory Journal*, **1**, 280–287 (1988)

Chapter 12

Audiometry

B. Malerbi

Introduction: structure and function of the ear

Hearing can be damaged in many different ways, and by a number of agents. Audiometry provides a means of investigating the extent and the nature of the damage, since the graph produced, known as an audiogram, is a plot of the degree of hearing loss versus frequency. The particular pattern recorded is indicative of the site of the damage, and thus of possible causes. The characteristic pattern for noise-induced hearing loss (NIHL) is readily identifiable where noise is the only damaging agent, but confusion may arise where other types of hearing damage coexist. It is therefore useful to be aware of, and to recognize, other causes of impaired hearing.

Hearing mechanism

The hearing mechanism comprises three distinct parts (*Figure 12.1a*).

The outer ear

This comprises the visible part, known as the auricle (pinna), which collects and funnels sound into the auditory canal (meatus). This is about 2.5 cm in length and one of its functions is to direct sound pressure pulses onto the eardrum (tympanic membrane), causing it to vibrate.

The middle ear

This is also air-filled, and contains the ossicles. These are three tiny, articulated bones known as the hammer (malleus), the anvil (incus) and the stirrup (stapes), which transmit vibrations from the eardrum to the oval window (fenestra ovalis), a membrane separating the middle ear from the fluid-filled inner ear.

The ossicles provide a three to one mechanical advantage, and the effective area of the eardrum is about 13 times greater than that of the oval window. When combined with other (attenuating) factors, the overall effect is that the pressure exerted on the oval window is about 20 times that on the eardrum, thus gaining the increased pressure required to vibrate the fluid.

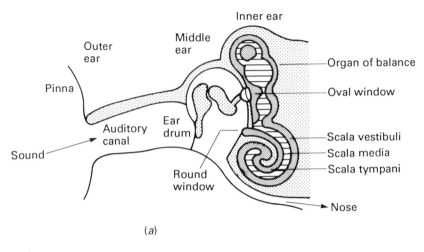

Key:

▨ Bone

▨ Perilymph

▭ Endolymph

▭ Air

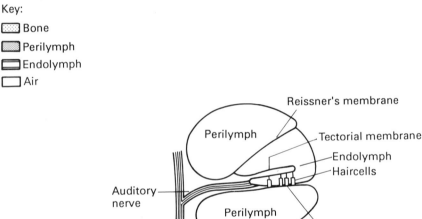

Figure 12.1 (*a*) Simplified diagram of the ear. (*b*) Cross-sectional view of the cochlea showing organ of Corti

THE EUSTACHIAN TUBE

The Eustachian tube connects the middle ear to the nasal cavity (nasopharynx), allowing pressure to equilibrate on both sides of the eardrum as atmospheric pressure fluctuates.

The inner ear

This contains the semicircular canals which constitute the organ of balance, and the cochlea which is the hearing organ. The cochlea is basically a tube coiled around a central stem, resembling a snail shell, hence its name, and is encased in bone. The tube is divided longitudinally into three fluid-filled canals, two of which, the scala vestibuli and the scala tympani, are connected at the apex by a narrow passage, the helicotrema, and contain the liquid perilymph. Between these lies the scala media. It has a triangular cross-section (*Figure 12.1b*) and contains the organ of Corti, comprising about 24 000 special hair cells, with hair-like processes called stereocilia, which are embedded at the upper end in the overhanging tectorial membrane. Thus pressure changes in the surrounding endolymph create a shearing force on the cilia.

The perilymph and endolymph are of different chemical composition, the major cation in the former being Na^+, and in the latter K^+, with Ca^{2+} playing an important role. The canals containing them are continuous, via the vestibule, with those surrounding the organ of balance, as illustrated in *Figure 12.1a*. Reissner's membrane separates the scala vestibuli from the scala media.

The hair cells are connected to the nerve endings of the auditory (VIIIth cranial) nerve, and lie along the basilar membrane, which responds to, and transmits vibrations from the perilymph in the scala tympani, to the endolymph, thus exciting the nerve endings via the movement of the stereocilia. There are two separate rows of hair cells; the outer row is three or four cells deep, whereas the inner row is a single line of cells. An enclosed channel between the two rows of hair cells (the tunnel of Corti), is filled with cortilymph, whose chemical composition is more like perilymph than endolymph.

If uncoiled, the cochlea would be about 35 mm long, and although its diameter decreases towards the apex, the width of the basilar membrane increases, from 0.04 mm at the base to 0.5 mm at the apex, which explains why its response to high frequency sound begins at the base of the cochlea, moving towards the apex for the lower frequencies.

The basilar membrane's function as a frequency analyser was shown by von Békésy about 45 years ago. This was later disproved, but more recently confirmed again, by a succession of researchers using increasingly sophisticated techniques. The properties and design of this membrane are very different from those of the adjacent membranes. It is under differing degrees of tension throughout its length, being stiffer at the basal end, and becoming more elastic towards the apex.

Sound pressure pulses transferred to the perilymph create a travelling wave, which disperses at a point of maximum displacement of the basilar membrane, the location of which is specific for a particular frequency. Thus for high frequencies, only the basal end is stimulated, whereas low frequencies excite most of its length before the travelling wave dies away.

The round window (fenestra rotunda) is a relief valve, allowing the near-incompressible fluid in the cochlea to move when the ossicles vibrate the oval window. The inner ear can be described as a transducer, converting vibratory energy to electrical impulses, which are then transmitted via the auditory nerve to the brain, where they are perceived as sound.

If any part of this complicated pathway is damaged, or its function impaired, there will be a hearing loss of some kind.

Categories of hearing loss

There are two main categories of hearing loss: conductive and perceptive (sensorineural).

Conductive hearing loss

This type of loss relates to disorders in the outer or middle ear, and can occur in the following ways:

1. Either impacted wax, or a foreign body in the ear canal, can reduce the amplitude of sound entering the ear, both by creating a sound barrier, and by obstructing the eardrum's vibratory movement. This is a temporary condition, and is easily corrected.
2. The eardrum may be ruptured by an explosion, or a blow on the head, or it may be perforated by disease, such as measles. It has remarkable powers of self-healing, but in severe cases it can be repaired by plastic surgery. In one case where one-third of the surface of the eardrum had been destroyed by an explosion, it had healed completely within 9 months.
3. If the Eustachian tube is blocked due to discharge or swelling, the middle ear cannot adjust to atmospheric pressure. The eardrum will thus be under tension, and will not respond efficiently to sound. Although neglect of this condition can have serious consequences, it is amenable to medical treatment, or operation.
4. The ossicles may be dislocated by the blast from an explosion, or a blow on the head. Their movement will be restricted if disease causes fluid to be formed in the middle ear (otitis media), or the articulated joints, usually the stapes, to become fixed (otosclerosis). These conditions are now all capable of being corrected by surgery.

Perceptive (sensorineural) hearing loss

This describes damage to the inner ear, involving the hair cells, the auditory nerve itself, or the hearing centre of the brain, and as with conductive deafness, can have a variety of causes, the main categories of which are described below. The inaccessibility of the inner ear, and the non-mechanical nature of its function, preclude the possibility of successful operation or investigation.

Congenital deafness

This can be associated with diseases such as rubella and influenza suffered by the mother during early pregnancy, or with medication taken at that time.

Accidents at birth

Accidents at the time of birth, or certain diseases of the newborn, can cause perceptive deafness, as can some childhood diseases and the drugs used to treat them. Hearing loss related to viral diseases, such as measles, is usually bilateral, although that associated with mumps is nearly always unilateral and total.

Ototoxicity

This is one of the side-effects of some commonly used drugs, and their action is either on the hair cells, or the auditory nerve. Examples include some antibiotics, antirheumatic and anticancer drugs, some diuretics, also quinine, nicotine, alcohol, contraceptive pills, eraldin and aspirin if taken regularly for long periods. In the case of streptomycin, very small doses [1] have been known to cause severe hearing loss.

Fracture of the base of the skull

Fracture of the base of the skull, whereby the delicate structures of the inner ear are damaged, is a common cause of deafness.

Acoustic neuroma

If detected early, this can now be successfully removed by surgery. The hearing loss is usually unilateral.

Nerve deafness

The worst condition is where the auditory nerve is disrupted or destroyed by disease, resulting in the person becoming 'stone deaf'. This type of nerve deafness is an irreversible, complete loss of hearing, by contrast with other forms of hearing loss which exhibit differing degrees of impairment over the audiofrequency range.

The two most widespread forms of sensorineural hearing loss are presbyacusis (age-related hearing loss), and noise-induced hearing loss (NIHL).

Presbyacusis

Presbyacusis begins at an early age. The hair cells at the base of the cochlea which respond to the higher frequencies are affected first. This selective erosion of

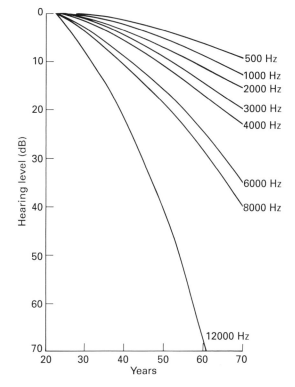

Figure 12.2 Hearing loss with age after Hinchcliffe (courtesy of Sound Research Laboratories)

hearing is thought to be because the hair cells in this region (adjacent to the oval window) bear the brunt of the pressure changes in the surrounding fluid, as described earlier. Hardening of the arteries is ruled out because if this were the cause, the deterioration would be expected to occur equally at all frequencies. Furthermore, no such vascular changes have been observed [1, 2].

In various studies of hearing loss in the different age groups of healthy populations, some high frequency impairment has been detectable by audiometry in the early teens. However, it is only in later life, when the deterioration has reached the middle frequencies, and speech starts to become unintelligible, that the loss is actually perceived.

Figure 12.2 illustrates the average presbyacusis loss for the whole population, but it is generally found that the loss is more pronounced in men. This is thought to be related to their different lifestyle, because other studies comparing groups of elderly people showed that those living in the town had more advanced presbyacusis than those in the same age group living in the country. Even more surprising were studies [3] of the hearing acuity of the Mabaan tribe, who live in a remote part of Africa, where their environment is both noise-free and stress-free. The hearing of the men at 70–79 years was slightly better than that of 30–39-year-old men tested in the United States, and the Mabaan women's hearing was similar.

These findings are said to indicate that presbyacusis is exacerbated by the general hubbub of modern life to which the ear is always exposed, although it is just as likely that the common factor in the various studies is the degree of stress.

Noise-induced hearing loss (NIHL)

Noise-induced hearing loss is also a gradual and irreversible process. It is distinct from damage caused by an explosion, which may rupture the eardrum, and dislocate the ossicles, and result in conductive failure. This would protect the inner ear from the full force of the blast, and the damaged mechanical parts could be repaired. The more common type of injury; resulting from long-term exposure to excessive levels of broad band noise (defined in Chapter 7 on noise), is degeneration of the hair cells centred around the 4–6 kHz region. Experiments reveal a link between these frequencies and ear canal resonance. The site of damage caused by continuous pure tone or narrow band noise appears to depend upon the frequency of the noise.

The micrographs in *Figure 12.3a and b* compare normal hair cells with an example of noise-damaged cells. Stereocilia and hair cells in the outer row are always damaged or destroyed before the inner row begins to be affected. If noise exposure continues, the organ of Corti in the affected region will eventually be completely destroyed, and the associated nerve fibres will gradually disappear.

Although NIHL is not a progressive condition like presbyacusis, since if exposure to noise ceases the hearing loss will be arrested, the superimposed advance of presbyacusis will eventually magnify any hearing loss at 4–6 kHz. It is at this stage of the working life, usually towards the middle or late 50s, that a noise-exposed person who has hitherto been unaware of any hearing damage will complain that he is 'going deaf'.

A change in a person's hearing acuity, whether temporary or permanent, is known as a *threshold shift*.

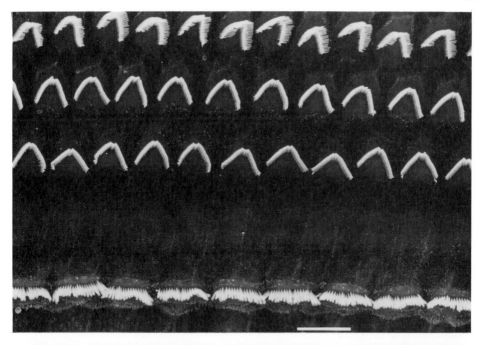

Figure 12.3 (*a*) A scanning electron micrograph showing the pattern of normal hair cells—three rows of outer cells and a single row of inner cells with their normal stereociliary arrangements (in white) Magnification bar = 10 μm. (*b*) A scanning electron micrograph showing noise damaged hair cells— many outer hair cells have completely degenerated, while on others remnants of stereocilia can be seen. Most of the inner hair cells are still intact, but some show disarray of stereocilia. Magnification bar = 10 μm (courtesy of Dr Ade Pye, Institute of Laryngology and Otology, London)

Development of noise-induced hearing loss

Temporary threshold shift (TTS)

Initially the effect of excessive noise on hearing is temporary. The time taken for recovery depends upon the period of exposure and the sound level, but it can be as much as 48 h, or more in extreme cases. If overtime is normally worked, full recovery of hearing may not be achieved until the weekend, and then only if non-working periods are spent in quiet pursuits. TTS is a manifestation of the body's response to a threat of injury. It is thought to be metabolically induced, and is associated with the following protective mechanisms.

The first is the aural reflex. Should the ear be subjected to noise above 90 dB for more than 10 ms, the muscles controlling the tension of the eardrum and the action of the stapes will tighten and reduce the movement of the mechanical parts of the ear, attenuating the sound by 12–14 dB. The 10 ms delay means that this protection does not extend to unexpected loud noises, such as explosions. The ear has not yet evolved to deal efficiently with man-made noise.

The second protective device is activated by noise levels above 140 dB, and changes the normal to-and-fro oscillation of the ossicles to a sideways rocking motion, reducing the pressure transferred to the cochlea.

TTS occurs at the frequency of the noise for low level exposures to narrow band noise, and at half an octave above for high levels. It increases in proportion to the logarithm of the exposure time.

Permanent threshold shift (PTS)

Incomplete recovery of TTS before further noise exposure allows the residual threshold shift to become permanent, and where this pattern is repeated over the working life, PTS accumulates. The most rapid increase occurs during the first 10 years of continuous exposure. To estimate the extent of NIHL, the average presbyacusis loss for a person of the same age and sex is obtained from standard graphs [4, 5] or from tables (*Table 12.1*) and subtracted from the measured PTS.

Table 12.1 Average correction (in dB) for presbyacusis. For comprehensive values see [6]

Age (year)	500	1000	2000	3000	4000	6000	8000	12 000
Males								
20	0	0	0	0	0	0	0	0
30	1	1	1	2	2	3	3	5
40	2	2	3	6	8	9	11	14
50	4	4	7	12	16	18	23	42
60	6	7	12	20	28	32	39	64
70	10	11	19	31	43	49	60	>70
Females								
20	0	0	0	0	0	0	0	0
30	1	1	1	1	1	2	2	5
40	2	2	3	4	4	6	7	14
50	4	4	6	8	9	12	15	42
60	6	7	11	13	16	21	27	64
70	10	11	16	20	24	32	41	>70

Column header: Test frequencies (Hz)

*Hinchcliffe's values [4] at 12 000 Hz throughout.
NB. These values have been corrected to the nearest whole number as recommended in the NPL Report [6], which should be referred to where intermediate ages are of interest.

Occupational deafness

Occupational deafness is the term used to describe hearing loss which results from exposure to noise at work. Because it would be impossible to quantify added effects due to non-occupational noise or other unrelated causes, tables have been prepared [7] and incorporated in a British Standard [8] giving an indication, in terms of a person's age, and the level and duration of their noise exposure, of the likelihood that the mean of the hearing levels at 1, 2 and 3 kHz would exceed 30 dB. This is the level at which hearing handicap is said to begin.

Hearing handicap is a term used to describe the degree of impairment which would affect a person's normal social or domestic activities. This becomes a *disability* when it affects a person's employment prospects.

Estimation of the percentage of an exposed population becoming handicapped in this way is based on their noise exposure over a period of years, and is expressed as the noise immission level (NIL) which is the cumulative $dB(A)L_{eq}(8h)$ derived as follows:

$$NIL = L_{eq} + 10 \log_{10} (T/T_0)$$
where T = duration in years
          T_0 = one year

Using this formula to estimate NIL at the end of each successive decade in a working life, it can be seen that after 10 years, NIL has increased by 10 dB, whereas after subsequent 10 year periods, the additional increase drops to 3, 2, 1...dB.

Although it is agreed that high noise exposure damages the hair cells in the inner ear, opinions differ as to the mechanism involved. Some workers [9] disagree with the long-held theory that loud noise directly damages the hair cells by continued excessive movement of the hairs. They propose that hearing loss is caused by the long-term effects of the body's reaction to stress, loud noise being a form of stress.

The stress reaction involves the release of hormones, which induce redistribution of the blood supply to prepare the body for 'fright, fight or flight'. In the case of noise stress, they believe that the hormones protect the inner ear by reducing its blood supply, and when the noise has ceased, the hormone levels slowly return to normal, thus restoring normal blood circulation. This could be the mechanism for TTS occurring, and its return to normal when exposure ceases. When noise is continuous, and repeated day after day, the hormone levels never completely subside, and there is a slow build-up, causing a permanent restriction in the blood supply to the inner ear, resulting in hair cell death. This effect was studied in rats and humans, and the suggestion was made that blood hormone levels would give an earlier warning of impairment than an audiometric test.

Other investigators into 'urban stress' [10] found that when a person is subjected to loud noise, if he believes he has direct control over its termination, even if he does not exercise that control, its effect is less detrimental than if he has no control. Where he has only indirect control, through access to another person in power, even though that person does not guarantee assistance, he feels in control, and the stressful consequences are ameliorated.

The indications are that a person's attitude towards the noise in his working environment governs the degree of hearing impairment suffered, and probably accounts for individual differences in susceptibility to NIHL. A detailed study [11] of three industrial workers, comparing their attitudes to the noise in their workplace with the associated degree of hearing loss, supported this conclusion.

Measurement of hearing loss

Primitive methods

A century ago, before audiometers were available, a Scottish physician named Thomas Barr made the first objective measurements of occupational deafness. Previously only subjective assessments had been recorded. He carried out a survey of hearing acuity among three large groups of workers; foundry-workers and boiler-makers, with 100 postmen as the control group. Their hearing levels were measured in 'inches of hearing', being the distance away from the ear at which the subject could just hear a pocket-watch ticking [12].

Another method (still in use) was based on the distance in feet at which the subject could correctly repeat the series of words whispered by the examiner, using as a reference a standard distance regarded as 'normal'.

Derivation of audiometric zero

Hearing loss or impairment is now measured on an audiometer over the range of frequencies necessary for normal conversation to be heard and understood. It is the difference between the resulting raised hearing threshold level, and 'normal' hearing. This assumes that it is possible to judge what is normal.

Normal threshold

A 'normal' threshold of hearing was established in 1952 after more than 1200 otologically normal 18–25-year-old subjects were tested in two independent studies carried out at the National Physical Laboratory (NPL) and the Central Medical Establishment of the Royal Air Force (CME). The results were in general agreement, and the modal values were used to formulate the original British standard [13] which was published in 1954.

From the NPL test results, Burns [14] subsequently calculated the following standard deviations about the mean hearing threshold:

Pure tone frequency (Hz) 125 250 500 1k 2k 4k 6k 8k
Standard deviation (dB)   6.8 7.3 6.5 5.7 6.1 6.9 9.1 8.7

Since the spread of hearing thresholds for these subjects extends over three standard deviations (99.7 per cent), ± 20 dB about the mean encompasses the 'normal' range. For this reason, audiometers measure down to −10 dB HL (hearing level), and a hearing loss of <20 dB is not considered to be significant.

An 0 dB reference level for audiometers had been established a year earlier, in 1951, by the American Standards Association (ASA) using information obtained in the 1930s by Sivian and White [15]. This had set the normal hearing threshold at a level which was 10 dB above the British findings. When the British standard threshold was confirmed by similar studies in other countries, the International Organization for Standardization (ISO) responded to the need for a universal standard by publishing the recommended reference zero [16] in 1964. This was arrived at by averaging the results from 15 studies of suitable subjects aged between 18 and 30 years in five different countries.

The American National Standards Institute (ANSI) adopted this standard in 1969. Since there might still be audiometers in use complying with ASA-1951, and old records of tests carried out on such instruments may be unearthed to compare with recent tests, the 10 dB adjustment must be made in such circumstances.

The current British standard BS 2497:1988 Part 5 [17], a revised version of BS 2497:1968, Parts 1 and 2, is equivalent to ISO-389, which was similarly revised in 1975 and 1985. The need for revisions can be understood, considering the different methods employed over the years, and in different countries, to measure normal hearing.

Parts 6 and 7 of BS 2497 have also been published [18–20], providing RETSPL values (*defined below*) for a more extensive range of calibration equipment, and BS 6950:1988 (*ISO* 7566:1987) gives [18] standard reference equivalent threshold vibratory force levels for a standard bone vibrator used in bone conduction tests (*see below*).

BS 2497 defines an 'otologically normal subject' as 'a person in a normal state of health, who is free from all signs or symptoms of ear disease, and from obstructing wax in the ear canal, and who has no history of undue exposure to noise'.

Calibration

The auditory thresholds of the large groups of otologically normal young people used to derive normal hearing, were obtained in the following manner. The sound pressure levels at the entrance to the auditory canal when each of a series of pure tones played through earphones was just audible, were measured indirectly (as rms voltages applied to the earphones). These values are called 'equivalent threshold sound pressure levels' (ETSPLs), but because they differ depending upon the electrical impedance and clamping force of the earphones used, these parameters must be specified and adhered to consistently.

The modal values of the ETSPLs at each specified frequency are known as 'reference equivalent threshold sound pressure levels' (RETSPLs). They have been accepted as the universal standard, and incorporated in audiometers as *audiometric zero*, which is 0 dB HL at all frequencies used. It must be emphasized that this is not the same as *acoustical zero* which is used as the reference level for sound measurement, in other words 0 dB SPL (sound pressure level), as described in Chapter 7 on noise (p.118). The comparison is illustrated in *Figure 7.5*, the difference being that 0 dB HL is an average hearing threshold, whilst 0 dB SPL is the best of the better-than-average thresholds. However, unlike the equal loudness contours (ELC), which are not parallel, the dB HL scale increases in parallel 5 dB increments.

Having established audiometric zero as the baseline, it was necessary to devise a means of calibrating the audiometer during manufacture, and at regular intervals when in use. This is the reverse of the procedure for obtaining the RETSPLs. An artificial ear is designed to allow the microphone of a sound level meter to be inserted at the lower end of a cavity, representing the eardrum and hearing mechanism at the end of the ear canal. A similar device, known as a coupler, serves the same purpose. The audiometer is calibrated by clamping each earphone in turn over the open end of the 'canal', and while the pen is set at 70 dB hearing level, adjusting the potentiometers in the audiometer until the sound levels measured on the meter at each frequency coincide with the calibration values supplied (*Figure 12.4*).

The internationally accepted reference zero, to which all audiometers should now be calibrated, corresponds to values of RETSPLs which depend upon the

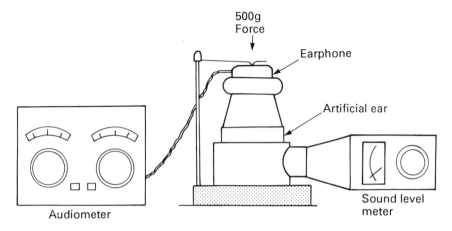

Figure 12.4 Calibration of the output from an audiometer using a standard artificial ear and a sound level meter (courtesy of M.E. Bryan and W. Tempest)

standard calibration equipment favoured in various countries. In the United Kingdom, the values recommended in BS 2497 are for the coupler/ear specified [19, 20] in BS 4668:1971 and BS 4669:1971 [21, 22]. Audiometers are therefore calibrated so that each of these values corresponds to 0 dB HL at the associated frequency. *Figure 12.4* illustrates the calibration instrumentation in use.

Simple check

The calibration procedure described above should be carried out on an annual basis. A simple daily or weekly check may be made by recording the hearing thresholds of one or two people with normal hearing who work in a quiet environment. Their audiograms should not vary from week to week by more than 5 dB at any frequency.

Audiometric test

For industrial screening, the aim is to identify NIHL, therefore the frequencies of interest are 4–6 kHz. The audiometer used generates pure tones at 0.5, 1, 2, 3, 4, 6 and 8 kHz, which are presented in sequence to the subject through earphones. Each ear is tested separately, starting with the left, the intensity being increased in 5 dB steps until the subject responds, then decreased until the tone is inaudible. This is repeated several times for each frequency, until the threshold of audibility is established. The results are recorded, either manually or automatically, on a graph (audiogram).

The tones can be either pulsed, as normally used to measure auditory threshold, or continuous for calibration or diagnostic purposes. Hearing levels from −10 dB to 90 dB or more can be recorded.

Preparation

Before the test can take place, there are three important aspects to consider.

Selecting an audiometer

It must be borne in mind that for industrial use speed is essential, since the cost of lost production must be added to other costs incurred when presenting the case for audiometry to an employer. The method of test must therefore be simple to explain to each subject, and the test must be of short duration.

Of the audiometers available, many are 'diagnostic', providing a battery of tests to identify positively the site of the hearing loss. This is not our purpose; audiometric screening allows the occupational health nurse, or the occupational hygienist, to separate individuals with no sign of NIHL from those who need further investigation. However, if the former show signs of unrelated hearing problems, they must be referred to a hearing specialist for diagnosis.

Three instruments suitable for industrial use are described below together with their advantages and disadvantages in relation to individual circumstances.

Selecting the test room

Since the subject is listening for sounds at the hearing threshold, it is essential that these should not be masked by ambient noise. Before investing in a soundproof booth, other possible locations should be investigated. The medical department may contain a quiet room; an interview room or conference room is unlikely to be in use continuously.

Having found such a quiet room, its suitability as a test room must be ascertained by measuring the octave band sound levels, and comparing them with the 'ambient noise limits for industrial audiometry' tabulated in the HSE Discussion Document [21] *Audiometry in Industry*. Noise levels inside a soundproof booth should be similarly checked.

Table 12.2 Limits for ambient noise in an audiometric test-room

| | | | | Columns | | | |
| | | | * | * | | * | |
1 (Hz)	2 (dB)	3 (dB)	4 (dB)	5 (dB)	6 (dB)	7 (dB)	8 (dB)
31.5	76	0	76	86	–	76	75
63	61	1	62	72	1	62	60
125	46	2	48	58	9	55	50←
250	31	5	36	46	13	44	40←
500	7	7	14	24	24	31	21←
1000	1	15	16	26	30	31	21←
2000	4	25	29	39	39	43	20
4000	6	31	37	47	44	50	19
8000	9	23	32	42	35	44	18

Column
1 Octave band (OB) centre frequencies
2 Allowable OB SPL discounting headset attenuation
3 Attenuation of earphones with standard MX41/AR cushions
4 Allowable OB SPL with standard earphones (columns 2 plus 3)
5 Allowable OB SPL for measurement down to 0 dBHL (column 4 plus 10dB)
6 Attenuation of typical noise-excluding headset
7 Allowable OB SPL with noise-excluding headset (columns 2 plus 6)
8 Ambient OB SPL measured in test-room
Columns 4, 5 and 7 (marked with an asterisk) are each compared with the values in column 8 to assess the suitability of the test-room under the three conditions. For measurements down to –10 dB HL using the standard earphones, the values marked with an arrow in column 8, exceed the allowable SPLs in column 4, but for measurements down to 0 dBHL (column 5) the room would be suitable. An alternative would be to use the noise-excluding headset (column 7).

Should the measured levels exceed the limits, the attenuation of the earphones can be improved by fitting a noise-excluding headset over the earphones. A further allowance can be made if measurements of hearing level down to 0 dB (as indicated in Annex II of the EEC Directive [22], rather than to −10 dB, are acceptable. The limiting values can then be increased by 10 dB as depicted in *Table 12.2*.

Preparing the subject

On the day preceding the test, both ears should be examined for visible abnormalities, and excessive wax on the drum should be removed. It is inadvisable to carry out the test while the subject has a heavy cold.

If the intention is to measure PTS, the subject must not have been exposed to levels of noise likely to induce TTS. The best time for a test is before work commences on a Monday morning, provided that the weekend has been spent in a noise-free environment. An alternative is to ensure that the subject wears adequate hearing protection during the period preceding the test.

Records of personal details and a medical questionnaire should be completed before the test, so that hearing losses due to other causes, as listed on p.248 can be identified. Typical questionnaires can be found in the 1978 Discussion Document [21] and in other publications devoted to the subject [5]. The test procedure is then explained to the subject.

The earphones must be fitted correctly, the convention being that the red one covers the right ear, and the blue on the left. Spectacles and earrings should be removed, and hair pushed away from the ears so that the earphones form an intact seal with the head or ears.

Industrial screening audiometers

There are two main types of audiometer suitable for industrial screening, and various test procedures have been recommended over the years. All measure the 'pure-tone air-conduction hearing threshold', and the convention is to test the left ear first. Where hearing thresholds have been determined by alternative methods, they have been comparable, with no greater differences than are found when repeating the same test on an individual subject, where a difference of 5–10 dB at any frequency is not uncommon. Typical examples of procedures used with each type of instrument are:

1. Manual, using the Hughson–Westlake procedure.
2. Automatic self-recording:
   (a) Békésy method
   (b) Hughson–Westlake

Manual audiometry

Frequency and intensity levels are set by the operator on the manual audiometer. The Hughson–Westlake procedure starts at 40 dB and 1 kHz, a tone which is audible for most people, and the intensity is reduced in 10 dB steps until the subject responds by raising a finger when the tone becomes inaudible (descending threshold). The operator then increases the intensity in 5 dB steps until the tone becomes just audible (ascending threshold). The descending threshold is frequently

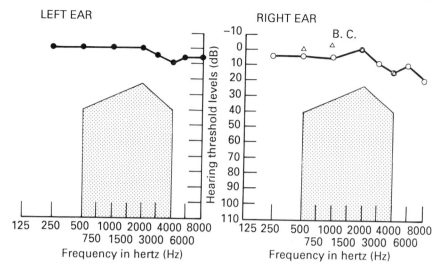

Figure 12.5 Audiogram from 60-year-old woman, obtained on a manual audiometer. Shaded area indicates speech frequencies and levels

5 dB lower, in which case the two thresholds are averaged. It is usually necessary to check the threshold level several times, after which the average value is recorded on a standard chart. This is repeated for 2, 3, 4, 6 and 8 kHz, then 0.5 and again at 1 kHz as a check. The whole procedure is repeated for the right ear, thus producing an audiogram (*Figure 12.5*).

The normal test takes 15 minutes to complete, during which time the operator is fully occupied, either altering the settings, or recording the results, but this is only a disadvantage where there are large numbers of tests to complete. The advantage lies in the relatively low cost of the instrument (£400–£500).

Automatic self-recording audiometry

With a Békésy instrument, the audiogram is automatically traced on a chart. The operator starts the test after ensuring that the recorder pen is on the baseline of the chart (usually at −10 dB HL), but the subject is in control thereafter. Starting at the lowest frequency (usually 500 Hz), the intensity automatically increases until the subject responds by pressing a hand-switch, and holding it until the sound fades away, when he releases it, thus allowing the intensity to increase again. This is repeated for 30 s before the next frequency is presented. When all frequencies have been tested, the procedure is repeated for the other ear. The pen follows a zigzag course across the threshold of hearing for each frequency (*Figure 12.6*). There are built-in checks to detect malingering.

The operator has only a supervisory role during the test, and therefore could be in charge of more than one audiometer. Although the test itself only takes 8 min, the mean thresholds must be estimated later, and plotted on a graph, or listed. The high cost of £2000–£4000, depending upon the model, can only be justified where speed is essential, and the instrument is to be used regularly. Some audiometers have a built-in computer interface (or this may be an optional extra), whereby the required information can be stored, and any necessary calculations programmed,

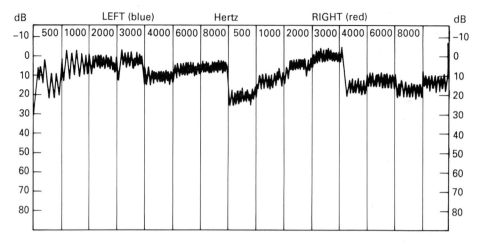

Figure 12.6 Békésy audiogram from the same subject in a room with low frequency ambient noise causing raised thresholds at 500 Hz

for example, mean threshold, hearing loss category or any increase in hearing threshold level since the last test (*Figure 12.7*).

The more recently introduced audiometers with a built-in microprocessor, automatically search for the subject's hearing threshold using the Hughson–Westlake procedure, and printing out the search pattern as it progresses. If a threshold cannot be satisfactorily estabished, this fact is indicated on the printout,

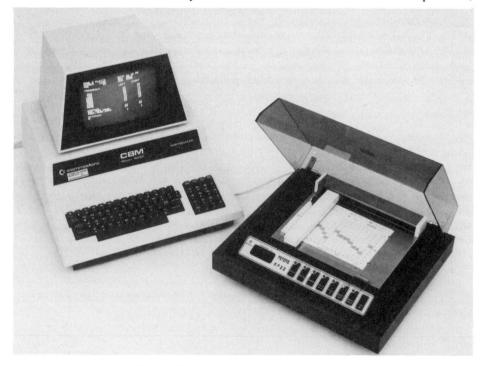

Figure 12.7 Békésy-type audiometer linked to a computer (courtesy of Alfred Peters and Sons Ltd)

SUBJECT: *alice M.*

X.............................

DATE: 14 05 86
JOB NO.:....................
NOISE EXP.:..............
PROTECTOR: *None*
BIRTH DATE: 27.9.1896
SEX: *F*.....................

L 1K   40-60+50-55+45-
       50-55+
L .5K 65+55-60+50-55+
       45-50-55+
L 1K   45-65+55+45-50-
       55+45-50-55+
L 2K   65+55-60-65+55-
       60+50-55-60-65+
L 3K   75+65+55-60+50-
       55+45-50-55-60+
L 4K   70+60-65-70+60-
       65-70+
L 6K   80-95+85-90+80-
       85+75-80-85+
L 8K   95+85+75-80-85+
       75-80-85+
R .5K 40-60+50+40-45+
       35-40-45-50+40-
       45-50+
R 1K   60+50+40-45+35-
       40-45+
R 2K   55-75+65+55-60+
       50-55-60+
R 3K   70+60+50+40-45-
       50-55-60+50-55-
       60+
R 4K   70+60-65+55-60-
       65-70+60-65+
R 6K   75+65-70-75-80-
       85-90-95-95-
R 8K   95+85-90-95+85-
       90-95+
R 6K   40-60-80-95-95-

              AUDIOGRAM

FREQ.   L dB   R dB

  500HZ   55   50
 1000HZ   55   45
 2000HZ   65   60
 3000HZ   60   60
 4000HZ   70   65
 6000HZ   85   NR
 8000HZ   85   95

   THRESHOLD AVERAGE

.5-1-2K   58   52
 1-2-3K   60   55
 2-3-4K   65   62
 3-4-6K   72   --
 4-6-8K   80   --
.5-1-2-3K   59   54

   INCOMPLETE TEST

NR-NO RESPONSE

MAICO MA728 SN 30165
CALIBRATED 7-85
ANSI S3.6-1969,R1973

EXAMINER: *Bhn*.

Figure 12.8 Hughson–Westlake printout from microprocessor audiometer

and the test is labelled 'incomplete' (*Figure 12.8*). It may then be necessary to resort to a manual test.

Much time is saved by this means, since the test is completed in 6 min, and the results are printed-out, together with average hearing levels for various combinations of frequencies as required for categorization. The price is in the same range as for the Békésy types.

Interpretation of audiograms

It is important to note that an audiogram is presented with the horizontal line near the top of the chart representing audiometric zero at specified frequencies, and the dB HL scale on the vertical axis increasing in 5 dB steps towards the bottom of the chart. Therefore, unlike most graphs, a horizontal line high up on the chart indicates normal hearing, and a downward deviation shows at which frequencies hearing loss has occurred.

Identifying NIHL as distinct from other causes

A person exposed to excessive levels of industrial noise will almost invariably display the same effect in both ears, which will be a dip or notch in the audiogram centred around the 4 or 6 kHz region, with normal hearing levels retained on either side of the dip. This pattern will become more pronounced over the years if noise exposure continues. *Figure 12.9* is a typical example of the development of NIHL, and is distinguishable from presbyacusis (as in *Figure 12.10*) by the upward slope at high frequencies. The superimposition of presbyacusis as the individual nears the end of his working life will make this distinction less apparent.

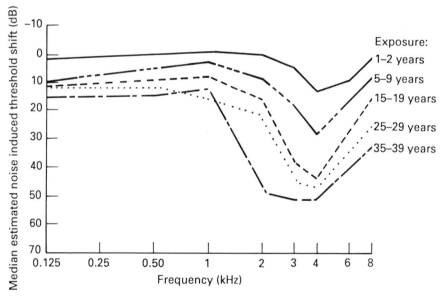

Figure 12.9 Audiograms showing development of noise-induced hearing loss (courtesy of Bruel and Kjaer)

A flat audiogram which shows a similar hearing loss at all frequencies as in *Figure 12.11*, or one where the loss is greater at the low frequencies, with a gradual improvement towards the high frequencies, is indicative of conductive loss. This supposition can be confirmed by a bone conduction test, whereby a standard electromagnetic vibrator is connected to the audiometer, and applied to the mastoid bone behind the ear.

NAME ....   DATE ....TIME ....

ID No. .... AGE ....   OPERATOR ....

REMARKS ....   LOCATION ....

....   AUDIOMETER ....

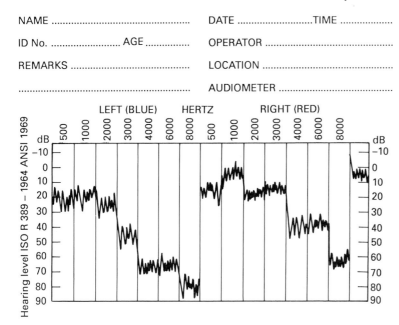

Figure 12.10 Advanced presbyacusis (courtesy of Bruel and Kjaer)

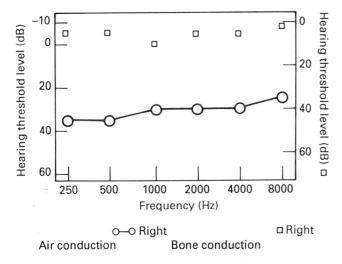

o—o Right   □ Right
Air conduction   Bone conduction

Figure 12.11 Conductive hearing loss as a result of damage to ossicles. Normal bone conduction confirms the diagnosis

The sound is transmitted to the inner ear via the skull, thus bypassing the middle ear. Since by bone conduction the sound is transmitted equally efficiently to both ears, in most cases it is necessary to mask the ear not being tested. If the audiogram is normal by bone conduction, and abnormal by air conduction, as illustrated in *Figure 12.11*, occupational deafness can be discounted.

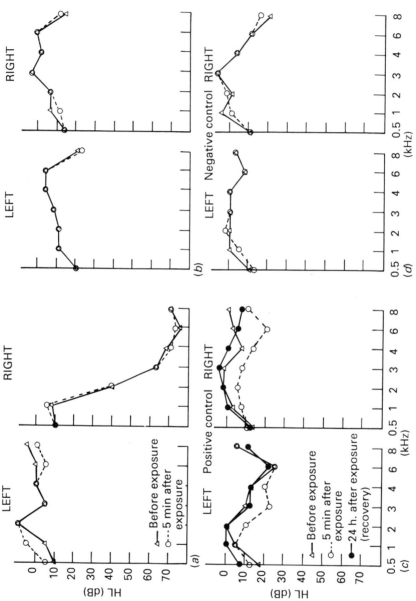

Figure 12.12 (*a*) Audiograms before and after exposure for 15 min to 110dBA while wearing earmuffs. Severe deafness in right ear following attack of mumps as a child. (*b*) Audiograms from subject with normal hearing under the same conditions as for (*a*). (*c*) Audiograms before and after exposure for 10 min to 110dBA without hearing protection. Shows complete recovery of TTS after 24 h. (*d*) Repeat audiograms for non-noise-exposed person show normal variations of 5 dB

Where a history of high noise exposure is coupled with an audiogram of the characteristic shape for both ears, this is sufficient indication that the hearing loss is noise-induced. However, should the loss be greater than expected for the intensity and level of exposure experienced [8], the answers to the questionnaire may throw some light on the discrepancy.

Most of the other causes of sensorineural hearing loss listed above may produce an audiogram similar to that for NIHL. For example, hearing loss due to rifle-shooting, being noise-induced, has the characteristic dip, though somewhat steeper, but only in one ear—the one *not* protected by the gun stock.

The audiogram in *Figure 12.12a* was surprising for a 25-year-old scientist, until the medical history revealed that the subject had mumps as a child, and this is typical of the resulting severe unilateral hearing loss. This example introduces another factor which must be considered. Unlike an eye test, where one eye can be closed while the other is tested, there is a limit to the attenuation which can be achieved by blocking the good ear, because where the difference in hearing levels for the two ears approaches 60 dB, the high intensity test tones required for the 'deaf' ear will be heard by the good ear before they have reached the threshold level of the ear under test. The sound crosses the head by bone conduction and is attenuated by at least 40 dB en route, therefore this effect is not noticeable where hearing loss is bilateral. The problem is solved by introducing masking noise into the good ear while testing the opposite one.

Assessment of audiograms

Three categories of hearing levels have already been discussed:

1. ≤20 dB HL at any frequency is within 'normal' limits and therefore does not necessarily indicate hearing impairment.
2. ≥30 dB HL averaged over 1, 2 and 3 kHz is the definition of hearing handicap in BS 5330:1976 used in Common Law.
3. 50 dB HL averaged over 1, 2 and 3 kHz is the minimum level for claiming disablement benefit and represents 20 per cent disability.

Items 2 and 3 are aimed at preserving the ability to hear speech, but as illustrated in *Figure 12.9*, NIHL does not significantly affect the speech frequencies until the loss at 4 kHz is well advanced, and where NIHL centres around 6 kHz, as is more often the case, it takes much longer to spread to the speech frequencies.

The HSE Discussion Document on Audiometry [21] attempted to redress the balance by taking into account the hearing levels at 0.5, 1 and 2 kHz, as well as those at 3, 4 and 6 kHz, also by including age in the categorization. Section 5 of that document describes the action that should be taken for each of five categories. The hearing levels in each group of frequencies are summed for each ear, and compared with the warning levels and referral levels in *Table 12.3*, according to age grouping. The appropriate action for categories (1) to (3) must be decided by the medical officer. Individuals in category (4) should be warned that their hearing must be protected, and category (5) requires no action. Category (1) is for audiograms where the summed hearing levels exceed those in the previous test by at least 30 dB, or 45 dB where 3 years has elapsed between tests. Category (2) is for differences between the two ears of 45 dB for low frequencies and 60 dB for high frequencies using summed hearing levels.

Table 12.3 Categorization of hearing levels (in dB). From [21], reproduced with permission of the Controller of Her Majesty's Stationery Office

	0.5, 1, 2 kHz		3, 4, 6 kHz	
Age group (years)	Warning	Referral	Warning	Referral
20–24	45	60	45	75
25–29	45	66	45	87
30–34	45	72	45	99
35–39	48	78	54	111
40–44	51	84	60	123
45–49	54	90	66	135
50–54	57	90	75	144
55–59	60	90	87	144
60–64	65	90	100	144
65+	70	90	115	144

Other uses of audiometry

Once an audiometer has been purchased for routine tests, it can be used as an educational tool to demonstrate, for example, the extent of TTS during even short periods of high noise exposure, the importance of allowing time for complete recovery after such exposure, and the adequacy of the hearing protectors prescribed.

The subject whose audiogram is featured in *Figure 12.12a* took part in an exercise to illustrate these points. He had asked for an assurance that the hearing protection he wore while developing a process which generated noise levels up to 110 dB(A) was giving adequate protection. The audiograms obtained before, and 5 min after exposure at that level for 15 min, are shown superimposed in *Figure 12.12a and b*, the latter relating to his assistant who was similarly exposed, since the equipment was located in a small highly reverberant room.

Bearing in mind that 110 dB(A) for only 4 min represents an $L_{eq}(8 h)$ of 90 db(A), if they had not used hearing protection they would have received four times the limit recommended in the Code of Practice [23], and the 'second action level' proposed for the Noise Regulations [24] to be introduced in 1989.

The positive control (*Figure 12.12c*), was their supervisor, who never wore hearing protectors, but was only infrequently exposed to noise. He remained in the room with them for 10 min, with his right ear closer to the noise source. The negative control (*Figure 12.12d*), spent 15 min in a quiet office without hearing protection.

Figure 12.12d shows that even for a non-noise exposed subject, audiograms repeated within a short period can differ by as much as 5 dB at some frequencies, therefore the 'before and after' differences in *Figure 12.12a and b* are not significant, particularly when compared with the extent of the effects displayed in *Figure 12.12c* when no protection was worn. After only 10 min, the threshold shift was 10 dB at most frequencies, and 20 dB at 6 kHz for the ear receiving higher noise levels. The third trace in *Figure 12.12c*, obtained 24 h after noise exposure had ceased, served to demonstrate that the effect was temporary if given sufficient time to recover between periods of exposure. The dip at 6 kHz in the left ear was a permanent shift and, being unilateral, was likely to be associated with previous military service.

This set of audiograms confirmed that the theoretical attenuation claimed for the hearing protectors used was being achieved in practice.

Conclusions

Audiometry is not a substitute for noise control, which must always have priority. However, it can be complementary to other activities aimed at protecting hearing. Opinions are divided about its usefulness, but are rarely unbiased. Biological monitoring is regarded by some as 'shutting the stable door after the horse has bolted', and using the employee to monitor the harmfulness of the working environment. They insist that all efforts should be directed towards control of the acoustic environment, so that audiometry is unnecessary. It is also feared that a pre-employment or preplacement audiometric test may be used to reject applicants who have hearing deficiencies, and regular testing of existing employees may result in some being removed from a noisy job against their wishes.

An alternative view is that even in a controlled environment, because of individual differences, a small percentage of those exposed will not be adequately protected, therefore audiometry is needed to identify them. The fact remains that the audiogram is used as legal evidence to prove or deny the existence of NIHL, and to quantify it. A prudent employer will regularly test those at risk, taking preventive action at the first sign of NIHL, thus ensuring that the condition never develops.

References

1. Beales, P.H. *Noise, Hearing and Deafness*. Michael Joseph Ltd, London (1965)
2. Sataloff, R.T. and Sataloff, J. *Occupational Hearing Loss*. Marcel Dekker Inc., New York (1987)
3. Rosen, S., Plester, D. *et al.* Presbycusis study of a relatively noise-free population. *Annals of Otology, Rhinology and Laryngology*, **September** (1962)
4. Hinchcliffe, R. The threshold of hearing as a function of age. *Acustica*, **9**, 303–308 (1959)
5. Bryan M.E. and Tempest, W. *Industrial Audiometry*. Published by the authors, Tel. 0253 712550 (1983)
6. Shipton, M.S. Tables relating pure-tone audiometric threshold with age. *NPL Acoustics Report Ac 94*. London (1979)
7. Robinson, D.W. and Shipton, M.S. Tables for the estimation of noise-induced hearing loss, *NPL Acoustics Report Ac 61* (1973)
8. BS 5330:1976. *Method of Test for Estimating the Risk of Hearing Handicap due to Noise Exposure*. BSI, London (1976)
9. Bennett, T.E. Penn State researches say hearing loss caused by stress. *Noise News*, **7**, 20 (1978)
10. Glass, D.C. and Singer, J.E. *Urban Stress*, Academic Press, London (1972)
11. Malerbi B. *Paper presented at the Institute of Acoustics Autumn Conference, Aston University* (1978)
12. Barr, T. Enquiry into the effects of loud sounds upon the hearing of boilermakers and others who work amid noisy surroundings. *Proceedings of the Philosophical Society of Glasgow* (1886)
13. BS 2497:1954. *The Normal Threshold of Hearing for Pure Tones by Earphone Listening* (revised as in 16 below) BSI, London (1954)
14. Burns, W. *Noise and Man*, 2nd edn. John Murray, London (1973)
15. Sivian, L.J. and White, S.D. On minimum audible sound fields. *Journal of the Acoustic Society of America*, **4**, 288 (1933)

16. ISO-R389 *Standard Reference Zeros for Pure Tone Audiometers*. International Organization for Standardization (ISO), London (1964)
17. BS 2497:1988 *Standard Reference Zeros for the Calibration of Pure Tone Air Conduction Audiometers,*
    Part 5 (ISO 389–1985) Specification for a Standard Reference Zero using an Acoustic Coupler complying with BS 4668.
    Part 6 (ISO 389–1985/Add. 1–1983) Specification for a Standard Reference Zero using an Artificial Ear complying with BS 4669.
    Part 7 (ISO 389–1985/Add. 2-1986) Specification for a Standard Reference Zero at Frequencies Intermediate between those given in Parts 5 and 6 of BS 2497
18. BS 6950:1988 (ISO 7566–1987) *A Standard Reference Zero for the Calibration of Pure Tone Bone Conduction Audiometers.*
19. BS 4668:1971. Specification for an acoustic coupler (IEC reference type) for calibration of earphones used in audiometry.
20. BS4669:1971. Specification for an artificial ear of the wide band type for the calibration of earphones used in audiometry.
21. HSE Discussion Document. *Audiometry in Industry*. HMSO, London (1978)
22. EEC Council Directive (12 May 1986) on the protection of workers from the risks related to exposure to noise at work. No. L 137/28. *Official Journal of the European Communities*, **24 May** (1986)
23. *Code of Practise for Reducing the Exposure of Employed persons to Noise*, HMSO, London (1972)
24. *Prevention of Damage to Hearing from Noise at Work*, Draft Proposals for Regulations and Guidance. Consultative Document, HSC (1987)

Chapter 13

Vaccination policies

J.A. Lunn

Introduction

Vaccination plays a significant, but not total, role in providing protection against infectious diseases. It is important to acknowledge the additional support which good and safe working practices play in controlling or eliminating risks from infection. Education and training of occupational groups must always be seen to be an essential supplement to a vaccination programme. In discussing the practical details of such a programme it is not realistic to attempt to produce definitive lists of workers at risk from any given infection and broad guidelines only will be given. The risks of acquiring infectious diseases vary significantly not only with occupation but also with the varying incidence and prevalence of infection in the surrounding population and environment. It is also worth emphasizing that no infectious disease is exclusively occupational in origin. It will therefore be necessary for individual occupational physicians and nurses to identify in detail the exposure risks of the different working groups within the organizations for which they are medically responsible. Increasing numbers of employees, especially in multinational companies, are required to travel abroad in the course of their work. They may be exposed to new infectious diseases against which additional vaccinations will be necessary. More people are going abroad for holidays. It is not always easy, or profitable, to differentiate between infectious risks from occupation as opposed to leisure. It may well seem prudent for an industrial organization to provide protection from infection regardless of its origin.

It is important that clear and precise guidelines on the indications for giving vaccines and the methods for using them are available for reference by each member of the occupational staff involved in carrying out the programme. Details of all vaccination procedures must be recorded in individual medical records in a way which is accessible and easily read. These details must include the type of vaccine, the batch number and manufacturer, method of giving, and dose and date given. A personal record card should be completed for retention by the employee. Call-forward systems should be established to ensure further attendances which may be necessary to complete a vaccine course. Simple card index or computer-based systems can be used depending on the number of employees and the financial resources available. Although computer-based recording systems have some sophisticated benefits over humbler and simpler card index systems, the latter are reliable and are immune from technical tantrums.

A medicolegal point which is applicable in the United Kingdom and in some other countries arises from the legal requirement for vaccines to be prescribed. In the majority of occupational health units the vaccination programmes are carried out by occupational health nurses. In most instances vaccinations are given routinely under an overall departmental policy and not on an individually prescribed basis. In such circumstances it is necessary for the doctor with overall responsibility for the programme to issue a signed document to each member of the nursing staff, authorizing each to carry out routine vaccinations. The document must state which vaccines are to be used routinely. It is also recommended that these documents make reference to the need to note and follow any written departmental policies on vaccination.

Vaccines

Immunity to infection can be induced by the use of appropriate antigens which will stimulate the immune system to produce antibodies against the disease-producing organism. Alternatively, preformed antibodies may be introduced into the body. In the former circumstances the immunity which is induced is called 'active' and in the latter, 'passive'. In practice, active immunity is induced by the use of vaccines prepared either from bacteria or viruses which have been killed, or from living organisms which have been subcultured to produce an attenuated and non-virulent form. The latter are referred to as 'live' vaccines. Whooping cough, typhoid and cholera are examples of killed or inactivated vaccines. Poliomyelitis, rubella, yellow fever and BCG (bacillus Calmette-Guérin) are examples of live vaccines. A few vaccines, tetanus and diphtheria in particular, rely for their immunizing effect on the presence of a toxoid prepared from bacterial toxins detoxified by treatment with formaldehyde. It is important to identify vaccines which are live because there are some restrictions and contraindications to their use. Live viral vaccines should not usually be given during pregnancy, to anyone who is immunosuppressed or who is receiving corticosteroids, or where there is evidence of a disease process of the reticuloendothelial system. More than one live vaccine should not normally be given at the same time but separated by an interval of at least 3 weeks. If it is necessary to give more than one live viral vaccine at the same time they must be given simultaneously and at different sites and not in the same limb. It is also recommended that at least 3 weeks should elapse between giving live viral vaccines and BCG or immunoglobulin. The latter may suppress the development of antibodies to the vaccine, although it is doubtful whether in practice this is a significant issue. Passive immunity is induced by injecting immunoglobulin derived from human plasma. By contrast with active immunity, passive immunity is shortlived, lasting usually for less than 6 months.

Methods of administration

Oral

This route is usually limited to poliomyelitis vaccine although in some eastern European countries BCG is given orally to infants. Vaccine may be placed on a sugar lump, spoon or given directly into the mouth.

Deep subcutaneous or intramuscular

This is the preferred route for most vaccines. A 21 G 40 mm needle should be used, or for thin subjects a 23 G 32 mm size will suffice.

Intradermal

BCG must only be given by this method. Typhoid, cholera and rabies may be given in this way. This has the advantage that a smaller volume of vaccine is necessary with a consequent reduction in adverse reactions and cost. These latter vaccines should not routinely be given intradermally, however, unless it is known that the operators possess the necessary skill to give intradermal injections properly. Inexperienced operators may give the injection too deeply with a consequently diminished immune response. Adequate training and experience in giving intradermal injections is essential for anyone intending to use this method. An intradermal needle should always be used. These are short bevelled 25 G, 10 mm.

General precautions before giving vaccines

Anaphylaxis from vaccines is extremely rare. It is essential, however, to be prepared for such a possibility and ampoules of adrenaline 1 in 1000 should be readily available. It is also inadvisable to undertake vaccination procedures unless a couch is available for use in the event of a hypotensive or anaphylactic reaction occurring. A final, but elementary, point: *never neglect to read the manufacturer's information sheet accompanying the vaccine.*

Types of vaccine

Cholera

This vaccine is prepared from heat-killed cholera organisms. The primary course consists of two injections: the first of 0.5 ml and the second of 1 ml given 1 months later. If time is limited, the second injection can be given within 7 days of the first. A booster injection should be given every 6 months where there is a continuing risk of exposure. The injections should be given subcutaneously or intramuscularly. If there is a history of a severe reaction to this vaccine a smaller quantity of 0.1 ml and 0.2 ml respectively can be given intradermally, resulting in a less severe or minimal reaction.

Travellers to most hot climates should be given cholera vaccine. It is, however, probably the least effective of available vaccines and good hygiene is essential if the risk of infection is to be avoided.

Diphtheria

This is a toxoid preparation. It must only be given to non-immune subjects as demonstrated by the Schick test. This test consists of an intradermal injection of 0.2 ml of Schick test toxin given into the anterior surface of the left forearm and 0.2 ml of Schick test control solution into the equivalent site on the right arm. Readings should be made at 24–48 h and at 5–7 days. No reaction at either site indicates immunity. An erythematous response at the test site, but not the control, indicates susceptibility. An equal response at both test and control sites indicates a false-positive response and therefore immunity. If there is a reaction at both sites, but the one on the test arm is significantly larger, a true positive response has occurred in addition to a false one, indicating susceptibility. The primary immunizing course consists of three injections of 0.5 ml of adsorbed diphtheria vaccine given subcutaneously or intramuscularly, the second 1 month after the first and the third 6 months later. A booster injection of 0.5 ml is sufficient to restore immunity after a primary course has been given.

Diphtheria vaccine is recommended for subjects liable to be exposed to cases of diphtheria.

Hepatitis B

This is an inactivated viral vaccine. Until 1986 the only hepatitis B vaccines available were derived from the surface antigen of the hepatitis B virus. This antigen was isolated from the plasma of human carriers of the virus. Genetically engineered vaccines from yeast sources are now becoming available and early reports of their efficacy are encouraging.

The primary course consists of three injections each of 1 ml. There should be an interval of 1 month between the first and second injections and 5 months between the second and third. An initial blood test to assess immunity before the vaccine is given is not routinely necessary. For a given individual whose history strongly suggests immunity it may be worthwhile looking for antibodies in the expectation that the expensive vaccine may not be necessary. When large numbers of people are being immunized, antibody assessment is not medically necessary or cost-effective. The vaccine must only be given intramuscularly. It should not be given subcutaneously or intradermally as its effectiveness when given by these routes has not been proved. It must not be given into the buttock as absorption from this site is poor.

ASSESSMENT OF POSTVACCINATION IMMUNITY

It is important to carry out postimmunization antibody assessment since in some subjects the primary vaccine course does not produce a satisfactory antibody response. Postvaccination screening should be done 2–3 months after the third injection and not sooner than 4 weeks. Those who fail to produce a satisfactory antibody response should be given a further 1 ml booster injection and antibody estimations repeated after the appropriate interval. In all subjects a further antibody assessment should be made after an interval of 2 years and a booster injection given if necessary. Further assessments may then be left for 5 years. The World Health Organization recommends that antibody levels of 10 international units (IU) per litre or more indicate satisfactory immunity.

POSTINOCULATION INJURY RISKS

These are as follows:

1. If a person at risk has been immunized and has had a satisfactory antibody level assessed within 2 years, no further action is necessary. When the assessment has been longer than 2 years a 1 ml booster injection of vaccine should be given.

2. Where there is no history of previous vaccination or of antibody level assessment specific immunoglobulin should be given for immediate passive protection. Antibody level assessment may take up to 2 weeks for a result and is therefore not a practical help in deciding whether or not to give immediate protection to someone at risk. A course of hepatitis B vaccine may be started at the same time to initiate the development of active immunization. When this is done it is not necessary to give the second injection of specific immunoglobulin.

Health care workers—especially in drug dependency and sexually transmitted disease clinics, infectious disease wards, mental handicap units and laboratories— should be offered protection. In addition, dentists, ambulance personnel, prison warders and some members of the police force may also need protecting.

CONTRAINDICATIONS

There are no known contraindications. It is not recommended, however, in pregnancy because it has not yet been licensed for this use in the United Kingdom. In the United States the vaccine is offered in pregnancy.

Immunoglobulin

NON-SPECIFIC

This is prepared from the pooled plasma of normal healthy adults. It contains sufficient antibodies to afford some immediate protection for susceptible contacts of measles, rubella and especially for those at risk from infectious hepatitis. For the latter condition where the exposure risk is not longer than 3 months, 250 mg of immunoglobulin should be given intramuscularly into the buttock. When the exposure risk is likely to be longer than 3 months, 500 mg should be given. This will extend the prophylaxis to approximately 5 months.

SPECIFIC IMMUNOGLOBULIN AGAINST HEPATITIS B

This is prepared from the plasma of subjects who have had hepatitis B and have developed appropriate antibodies. It is expensive and in short supply and should only be given when there has been a defined risk of infection from hepatitis B. It should be given as soon as possible after the risk of infection and not later than 7 days. The normal protective course consists of a first injection of 500 mg of immunoglobulin followed by a further 500 mg 1 months later. Hepatitis B vaccine may be started at the same time as the first injection, in which case the second injection a month later is not necessary.

Injections of immunoglobulin must always be given intramuscularly into the buttock because of the volume (5 ml) of each injection. It is also essential to keep the recipient under observation for 20 min because of the possibility of anaphylaxis to the injected protein.

Influenza

This is a killed viral vaccine. Several strains of influenza viruses are included in these vaccines, which vary according to the anticipated epidemic for any given year.

A single dose of 0.5 mg intramuscularly or deep subcutaneously is usually sufficient to give a degree of immunity for about 5 months. Mass inoculation is not now generally recommended, but sufferers from chronic heart and chest conditions should be offered the vaccine.

The virus is grown on egg albumin or chick embryos and the vaccine should not therefore be given to subjects known to be sensitive to eggs or poultry proteins.

Poliomyelitis

This is a live attenuated viral vaccine. There is a theoretical risk that unprotected contacts of those immunized could develop vaccine-related paralysis and therefore they should be protected at the same time. In practice this is difficult to effect. It is not now considered an essential procedure not only because of the practical difficulties but also because of the remote nature of the risk. The primary course consists of three doses given orally at intervals of 4–6 weeks. Each dose consists of three drops of vaccine. A booster dose of three drops should be given every 5 years but an interval of 10 years will still suffice to stimulate immunity. In the event of an individual being exposed to a known case of poliomyelitis a booster dose is recommended even if the primary course was given less than 5 years previously.

The vaccine should not be given in the event of pregnancy, acute febrile illnesses, enteritis and impaired immune responses. The general restrictions previously indicated for live viral vaccines are also applicable. There is a minute quantity of penicillin in the vaccine which is a theoretical reason why it should not be given to penicillin-sensitive subjects. In practice the amount is insufficient to constitute a contraindication to the use of the vaccine in these circumstances. It is recommended that all age groups are immunized.

Rabies

This is an inactivated vaccine. The primary course consists of two injections each of 1 ml separated by a month, given intramuscularly or deep subcutaneously. A booster injection of 1 ml should be given 6–12 months later. In the event of an individual being bitten by a dog suspected of having rabies, a further booster should be given without delay. The vaccine can be given intradermally using 0.1 ml. This obviously reduces the cost of protection. A note of caution must be given, however, where there is the possibility of an inadequate intradermal method being used. If 0.1 ml of the vaccine is given by a poor intradermal technique—that is, too deeply—there will be an inadequate antibody response.

All travellers to risk areas abroad are recommended to have protection against rabies.

TREATMENT OF SUSPECTED RABIES

The vaccine affords some protection during the incubation period of rabies for subjects who have not previously been immunized. The dosage and monitoring is a matter for particular experience and in the United Kingdom the Central Public Health Laboratory, Colindale (telephone: 01 205 7041) should be contacted. Whenever there is the remotest doubt about a risk of rabies it is *essential* to seek further advice.

Rubella

This is a live attenuated vaccine. One injection of 0.5 ml given subcutaneously will give protection and subsequent antibody assessment is usually not necessary. It should normally be given only to seronegative subjects and it is therefore necessary before giving the vaccine to check for the presence of an adequate level of protecting antibody. Women of child-bearing age should be immunized and all health care workers, whether male or female, who are liable to transmit the rubella virus to pregnant women. This is especially important for staff working in antenatal units. The vaccine must not be given during the first 3 months of pregnancy. It is important to ensure that recipients are not pregnant at the time of the injection and that they understand the need to avoid pregnancy for 3 months after receiving the vaccine. It is recommended that before having the vaccine all females should sign a document confirming that they understand the need to comply with these important requirements.

From time to time reactions to the vaccine occur 10 days later and the subject should be warned of the possibility. Pyrexia, maculopapular rashes and glandular enlargement may occur and, occasionally, painful arthropathy. These side-effects are usually self-limiting but may require the subject to rest. Occasionally the arthropathy can be troublesome and require treatment.

Tetanus

The vaccine is prepared from tetanus toxin. The primary course consists of three injections each of 0.5 ml given deep subcutaneously. There should be an interval of 6 weeks between the first and second injections and 6 months between the second and third. Booster injections of 0.5 ml should be given at intervals of approximately 10 years. Too frequent boosters are unnecessary and may give rise to allergic reactions and should therefore be avoided. It is also unnecessary and undesirable to repeat a primary course as a booster injection will stimulate an antibody response much longer than 10 years after a primary course. In the absence of written evidence and where there is uncertainty whether a primary course has been given previously, it will be necessary to repeat it. It is worth noting that anyone who has served in the armed services of most countries will almost certainly have had a primary course of tetanus toxoid. In the case of a suspected reaction to a previous tetanus injection, a test dose intradermally is not recommended since a thickened and nodular skin reaction may occur at the test site. When there is a history of a reaction to a previous tetanus injection it must be a clinical decision whether or not to give a further injection.

It is recommended that the general population is protected, particularly those whose occupations put them at greater risk. Maintenance and sewage workers, engineers, farmworkers and gardeners and workers whose occupations entail the risk of sustaining breaches of the skin must be included in the higher risk category.

Tuberculosis

This is a live attenuated vaccine containing the organism bacillus Calmette-Guérin, usually abbreviated to the well-known initials, BCG. It is important to ensure proper selection of subjects who are to be given this vaccine. If it is given to subjects incorrectly classified as tuberculin-negative or who already have immunity to the tubercle bacillus from previous BCG, severe accelerated reactions to the vaccine will occur with the probability of ulceration. Subjects with a negative tuberculin response but who have evidence of a satisfactory BCG immunization previously should not be given further BCG. A BCG scar 4 mm or more in diameter should be considered satisfactory evidence of the existence of immunity. All BCG sites should be inspected 4–6 weeks after inoculation to confirm a satisfactory response and for adverse reactions to be identified. Immunity develops approximately 6 weeks after the initial injection provided there has been a satisfactory response. All staff at risk from tuberculosis who are tuberculin-negative and who have not had a successful BCG immunization should be kept from risk until protected. Post-BCG tuberculin testing is not necessary in the presence of a satisfactory reaction.

Freshly prepared vaccine (0.1 ml) should be given intradermally at the site of the insertion of the deltoid muscle of the left arm. If it is given at other sites where there are underlying tendons or bone, poor scars will result. Approximately 90 per cent of BCG inoculations given over the acromioclavicular joint result in unsightly keloid scars. This site must be avoided. The vaccine should be given by a doctor well experienced in intradermal techniques. Poor administration and poor selection of subjects are invariably the reasons for adverse reactions to this vaccine.

Protection should be given to health care workers who are exposed to infectious patients or materials and to the general population in areas where the prevalence of pulmonary tuberculosis is high. The vaccine should not be given in the presence of tuberculin positivity, previous satisfactory BCG, immunosuppression, local eczema or infection.

INADVERTENT NEEDLE INOCULATION AFTER BCG

Occasionally a doctor or nurse may sustain an inoculation injury to a finger, often the terminal pulp, from a needle which has been used to give BCG. A cold abscess may occur in these circumstances and isoniazid resistance has been reported where treatment has been delayed. It is recommended that isoniazid should be given immediately in cases of BCG inoculation injuries.

TUBERCULIN TESTING

No discussion on BCG would be complete without considering details of tuberculin testing. There are a variety of methods for doing this, the Mantoux and Heaf techniques being the commonest ones in the United Kingdom. Disposable tests, the Tine in particular, are not recommended as they have a high false-negative rate. Whichever test is used it is important to appreciate that under certain circumstances a positive response may be suppressed. Many viral infections, especially glandular fever, will do so. Testing should not be done during any illness, especially an upper respiratory infection. It should not be done within 3 weeks of a live viral vaccine being given. Skin sensitivity and thickness vary and the test should always be given at a standard site on the flexor surface of the left forearm at the junction of the upper third and lower two-thirds. Induration and not erythema should be measured. It is the normal convention to measure transverse diameters and to record in millimetres, taking the lower mark when the measurement falls between two divisions on the scale.

Mantoux test   Tuberculin (0.1 ml of 1:1000) is given intradermally. Tuberculin may be absorbed onto the material of disposable syringes and testing must be done as soon as possible after loading the syringe. The test should be read at 48 or 72 h. Occasionally a reaction will occur as long as a week after the test; 5 mm of induration is regarded as a positive response. Reactions of 15 mm or more are regarded as strong reactions and follow-up chest X-rays may be indicated to confirm or otherwise the possibility of active pulmonary disease.

Heaf test   A concentrated tuberculin solution, 100 000 units/ml, is placed on the forearm and the Heaf gun applied and 'fired'. Six needles penetrate the skin introducing small quantities of tuberculin. The needle length can be adjusted to 2 mm or 1 mm for use for adults or children respectively. The test should be read after 7 days. Responses are graded as follows:

Grade   0   no response
Grade   I   discrete induration at four or more of the needle sites
Grade   II   merging of the individual reactions to form a ring
Grade   III   induration filling the centre of the ring
Grade   IV   breakdown of the induration with vesicles and/or some ulceration or induration of more than 10 mm.

Interpretation: Grades 0 and 1 are negative, grade 2 positive and grades 3 and 4 strongly positive.

Space does not allow full details of the techniques and problems of tuberculin testing and giving BCG (see further reading section).

Typhoid

This is a killed bacterial vaccine. The monovalent form should be used. The primary course consists of two injections each of 0.5 ml given subcutaneously or intramuscularly at an interval of 4–6 weeks. A booster injection of 0.5 ml may be given every 3 years but even if given after a much longer interval it will still be effective. Where reactions to the primary course have been severe, subsequent injections of 0.1 ml intradermally may be given to reduce adverse reactions. The vaccine should not be given where there is a history of extreme hypersensitivity to a previous injection of monovalent vaccine.

Drainage and sewage workers, plumbers and microbiological laboratory staff should be offered the vaccine. Hospital laundry workers may also be given protection although the risk of these workers contracting typhoid is remote. Travel abroad to certain countries may also indicate the need for the vaccine.

Acknowledgement

I would like to record my gratitude to Diane Domagala for her invaluable help in preparing the typescript.

Further reading

Immunisation against Infectious Disease. Department of Health and Social Security, Welsh Office, Scottish Home and Health Department. HMSO, London (1988)

Protection of NHS Workers against Tuberculosis. Notes for Guidance. Society of Occupational
    Medicine, London (1986)
Geddes, A.M. and Lane, P.J.L. Immunization. *Medicine International*, **51**, 2082–2089 (1988)

Chapter 14

Principles of occupational epidemiology

O. Axelson

Introduction: historical background

Epidemiology probably started in England with John Graunt, a haberdasher, publishing *Natural and Political Observations...made upon the Bills of Mortality* in 1662. A new interest in the quantitative aspects of the causes of death arose in the earlier part of the nineteenth century, including occupational aspects, when William Farr, responsible for the mortality statistics in England, made observations on Cornish tin miners. Some years later Augustus Guy, also in England, took an interest in 'pulmonary consumption' and physical exertion in letterpress printers and other occupational groups [1, 2].

It was not until the 1940s and 1950s, however, that modern epidemiology began and led to the present boom of studies starting in the 1970s and catalysed by methodological progress. This development is reflected in the appearance of several textbooks on epidemiology (including glossaries) [3–12] and there have also been ethical debates on the implications of epidemiological results for disease preventions [13, 14].

This chapter will hopefully serve as an introduction to occupational health epidemiology with emphasis on some particular features of the area.

General concepts

Population character and disease occurrence

The general population of a country or a region, with a turnover of individuals due to births, deaths and migration, is *open* or *dynamic* in character [15]. A group of once-identified individuals without turnover is a *closed* or *static* population, usually referred to as a cohort (originally a unit in the Roman army). Individuals in a cohort have something in common, such as working in a certain factory, a similar exposure although in different factories, or in a chemical accident, etc. A defining event could also be birth in a particular year, inclusion in a particular birth cohort, or having been selected by some criterion and enrolled into a particular register or study.

The open or dynamic population, as considered in an epidemiological study, is a circumscribed part of the general population and defined by living in a certain area or belonging to a particular group, occupational or other. To take another perspective, holding a particular state for some time might be used as a group-defining event, transferring the individuals of an open population into a closed one and establishing a cohort.

A time aspect is always involved in epidemiology, in other words, a population–time segment forms the *base* for any observation of mortality or morbidity, irrespective of whether the population is open or closed [15]. The population in a base may be referred to as the base population or study population, being considered during the *study period*.

For a closed population, the measure of morbidity or mortality is the *cumulative incidence* (*rate*), which is the fraction or percentage of affected individuals and requires a specification of the time-span involved [16]. Since the numerator is part of the denominator, this means that the closed population formally retains its cases. By contrast, the open population by definition omits the cases or deaths in the turnover process. So for open populations the number of cases or deaths has to be related to a denominator of subject–time of healthy persons, that is, the candidates for falling ill or dying. This point is particularly clear for mortality, since those already dead are not contributing any subject-time at risk of dying. For example, an open population (say, a small town) with an average of 1000 individuals followed for 5 years (producing, say, ten cases) would result in 5000 person-years of observation. The measure to be applied is the *incidence rate* or *incidence density*, here ten/5000 or two per 1000 person-years [16]. However, cancer registration usually provides cases per mean population per year so that the surviving cases are 'incorrectly' contributing to the person-years in the denominator of a cancer rate. In recurrent disease studies cases would also have to generate person-years.

The time period of the cumulative incidence can also be an age-span; for example, the cumulative incidence may consider cases or deaths from 50 to 59 years. The cumulative incidence corresponds well to the concept of *risk*, that is, the probability of attracting a certain morbidity or mortality during a time period or in an age-span. 'Risk' is also commonly used as a general term, synonymous with rate.

The incidence rate is also applicable to closed populations by observing cases in relation to the accumulation of person-years as used in occupational cohort studies and discussed below. The incidence rate of a closed population takes care of favourable or poor survival (or health), as the person-years under observation are obtained from surviving (or healthy) individuals only, whereas the cumulative incidence is insensitive in this respect. As an extreme example, the mortality from all causes over a long follow-up period would lead to a cumulative incidence approaching 100 per cent, irrespective of whether many died young or survived until old age. However, poor survival would reduce the number of person-years observed, which makes the incidence rate more sensitive, especially if competing mortality risks are operating.

Whether the population is closed or open, a *cross-sectional* study can be undertaken with regard to the occurrence of a disease at a particular point in time, in other words, the study period is reduced to zero. The number of individuals with the disease, out of all individuals observed at the particular point in time, is the *prevalence rate*.

To be meaningful, rates usually have to be considered for both sexes and different age classes, and some relationships between rates are shown in Appendix A [9]. The rates may be thought of as absolute measures of the occurrence of disease or death, but tend to be somewhat abstract and difficult to appreciate. Therefore, the rates are usually compared between populations as a quotient or a difference, that is, by the *rate ratio* (or risk ratio, when the cumulative incidence is involved) and *rate difference* (or correspondingly, risk difference). These parameters may be taken as relative measures of morbidity or mortality among compared populations, but would also serve as absolute measures of the effect of an exposure when exposed individuals are contrasted against a non-exposed reference population.

Determinants of disease

Most disease rates are dependent on age and sex and so is mortality. Therefore, age and sex are determinants of the morbidity or mortality, but age, for example, would not necessarily increase the risk of disease, since the probability for some conditions developing also decreases with age. This makes 'determinant' a more general term than 'risk factor' as applicable to preventive factors. A 'determinant' can also be merely descriptive in character and is not necessarily a true cause or preventive factor. For example, male sex is a determinant of prostatic cancer but is certainly not its cause.

In occupational health epidemiology, the determinants are usually various chemical or physical exposures but workload and various stress factors may also be considered. In addition, general determinants or background variables like age and sex, smoking, alcohol use and perhaps eating habits need attention. There may also be an interaction with an industrial factor, for example, smoking increasing the risk of lung cancer among asbestos workers.

Industrial exposures

Occupational epidemiology should aim to be specific in relating health hazards to agents or work processes rather than to job titles, since prevention cannot be directed towards occupations. Great difficulty is involved in assessing exposure, not only for obtaining exposure information but also conceptually in accounting for both intensity and duration [17, 18]. Usually, the time integrals of the concentration or the intensity have been used as measures of exposure, for example, working level-months for radon daughter exposure in mines, fibre-years in the context of asbestos studies etc. Long-term exposure to a low concentration is not the same as short-term high exposure, however, although indistinguishable from it.

In carcinogenesis at least, recent exposure may play a lesser role than earlier exposure, suggesting a *latency time* requirement by disregarding exposure for the past few years or about a decade before diagnosis. Previous knowledge may sometimes support discarding both remote and late exposure, accepting exposure to be effective only in a 'time window'. There are also suggestions in the literature of more sophisticated methods for a time-weighing of the exposure [17, 18], although this is rarely practised, presumably because of the conceptual complexity involved. Often just the amount of years spent in an occupation are taken as the measure of exposure, but hygienic standard setting also requires information on exposure intensity.

For the practical assessment of exposure, so-called *job-exposure matrices* have been elaborated, but can only suggest a specific exposure for a worker on a national or regional probability basis with regard to the job title [19, 20]. Direct information about exposure is preferable. Also data from *biological monitoring* [21] of exposed workers, say determinations of lead in blood or mandelic acid in urine as a styrene metabolite may sometimes be a useful basis for a study, especially when a new effect of the exposure is feared.

Epidemiological study designs

Descriptive versus aetiological studies

Descriptive

For practical and economic reasons, the occurrence of disease in a population (country or region), may be investigated through a representative sample to allow any inference about the total population. When no aetiological questions are involved, the study is descriptive and the information obtained is specifically anchored in time and place.

Aetiological

Aetiological (or analytical) epidemiology, on the other hand, concerns a general, scientific question, for instance, whether or not a particular exposure may cause a disease [9]. No representation of the population of a country or region is required (but any study is obviously bound in time and place and may be thought of as representing a kind of hypothetical 'superpopulation' with just the characteristics of the study population). Instead comparison of the health outcome between individuals with and without the exposure is the key issue—for example, as in the experiment where animals are exposed and the outcome evaluated without need for any representation of a particular animal population. Instead the important aspect is the comparability between the exposed and the non-exposed animals through a random distribution of exposure to individuals. In the absence of any randomization of the exposure in epidemiology, the demand instead is to select a suitable population for study in relation to the relevant determinant, so as to have comparable individuals with and without exposure.

Where an occupational disease is a clearly recognizable and unique effect of an exposure, for example, silicosis due to inhalation of silica dust, it may be sufficient to demonstrate a *dose–response* pattern in comparable worker groups—that is an increasing disease rate with degree of exposure. However, the disease(s) of interest in current occupational epidemiology are of all kinds, some with known or suspected causes, among industrial exposures. With a proper reference population it would be possible to distinguish an effect of the exposure, by an altered rate of the disease compared with the spontaneous occurrence of that disease in the reference population. Any indication of a dose–response phenomenon would also be of interest.

Aetiological study designs

The most straightforward type of study would be to follow a *closed population* with both exposed and non-exposed individuals for a period of time, comparing their disease outcome. This design would be a *cohort study*, with the non-exposed as a reference cohort. Similar to the exposed population (or *index population*), the reference population is closed, and the defining event may be taken as just the enrolment of individuals in the study at a given point in time.

In occupational epidemiology it is common to use a variant of the cohort study without any reference, and rely on a comparison with an expected number of cases derived from the rates in the general population [22]. This study design may be thought of as involving the construction of a hypothetical, non-exposed reference cohort, which is similar to the index population. Although a common approach in occupational epidemiology, it is applicable only to studies of mortality and cancer incidence, since other rates cannot usually be obtained from the general health statistics.

For an aetiological study in an *open population*, all the cases are ascertained, but usually only a sample of healthy subjects (with regard to the disease studied) is drawn from the base to account for the distribution of the exposure and other determinants of interest. The cases are then considered in relation to the individuals of the sample with regard to (various categories of) the exposure at issue. This approach, involving only a sample of the base population, is known as a *case-referent* or *case-control study*. If a sample is drawn from a closed population, that is, within a cohort, then a so-called *nested* case-referent study is obtained. The nested case-referent study is usually to elucidate particular aspects, when cohort data constitute the main study.

A drawback with the case-referent approach is that the case:referent ratio in the exposed and non-exposed population sectors only express a sort of relative rate, allowing only an estimate of the rate ratio (*Figure 14.1*). There are indirect ways to some other estimates given that the overall rate for the study population is known or can be estimated from some extraneous source [16]. It may be seen in this context that if the number of referents is increased, they may finally encompass all healthy individuals in the study base, so there is no longer a sample study but rather a census of the entire base population.

With a continuous census over time in an open study base, it would be possible to compare the incidence rates among workers in industries with different degrees of exposure; or a comparison could even be made with regard to incidence rates

Taking D_1 as the exposure of interest one obtains

	D_1+	D_1-
Cases	13	15
Base	4u	12u
Sample/ referents	4s	12s

Figure 14.1 Study base with two independently (additively) operating determinants, D_1 and D_2. Cases are symbolized with dots in relation to population units (u; the squares), taken as either individuals (including cases, that is, cumulative incidence) or person-years of healthy individuals as also reflected in a sample (s) over time of (healthy) referents in a case-referent study. The rate ratio, *RR*, with regard to D_1 is $RR = (13/4u)/(15/12u)$, or for case-referent data, the odds ratio, *OR*, is $OR = (13 \times 12s)/(15 \times 4s) = (13/4s)/(15/12s) = RR$, where $OR = RR$ becomes an approximation for nested case-referent data involving the cumulative incidence

in regions with a more or less concentrated representation of the type of industry under consideration. Although some studies of this type have been presented, this design cannot be recommended because of the dilution with non-exposed individuals. The design is more useful in environmental epidemiology, for example, studying health effects of air pollution. It may be referred to as a *correlation study*, especially when the rates in several areas are correlated with some particular exposure.

The fact that studies can involve only a sample of the base population as in case-referent studies, or complete information as in cohort and correlation studies, has lead to an alternative, more clear view and more precise terminology [9, 15]. Hence, one might think of open and closed populations and then take the view that studies involve either a census or a sample of the study base. This is illustrated in *Table 14.1*, where the *aetiological fraction* is also mentioned as explained in Appendix B. This gives information about the fraction of cases that may be due to the exposure (*preventive* fraction is the corresponding measure in preventive situations [23]).

Table 14.1 Type of approach and epidemiological study designs in reference to character of population along with various measures of morbidity or mortality

Type of approach	Character of population		Obtainable measures
	Open*	Closed†	
Census studies	Correlation study	Cohort study	Rate*†, rate ratio and difference; aetiological fraction
Sample studies	Case-referent study	Nested case-referent study	Rate ratio; aetiological fraction

*Incidence rate, prevalence rate and their derivatives.
†Cumulative incidence, incidence rate; prevalence rate and their derivatives (the odds ratio only approximating the risk ratio in nested case-referent studies).

Comparability and standardization

Ideally the distribution of compared populations should be equal with regard to various determinants of the disorder at issue. *Figure 14.1* shows a situation with two determinants, D_1 and D_2, operating symmetrically—that is, the area is assumed to correspond to the study population in terms of individuals or person-years and the dots to the number of cases appearing in each sector or unit of the study base. Whichever determinant is considered as the exposure, this base has comparability between the sectors with or without the determinant at consideration—the exposure.

By contrast with *Figure 14.1*, consider the populations in *Table 14.2* with a different distribution over age (but any determinant could have been considered). The overall, unadjusted or *crude* rates differ, but the age-specific rates are the same for two of the populations (II and III). For comparison it is necessary to weigh together the age-specific rates from each population in a similar way, either against a particular standard population (for example, as defined by WHO as 'World population', 'European population', etc.) or against a subpopulation within the study (usually the non-exposed) [24]. This adjustment procedure, shown in the table, is called *direct* standardization and the corresponding rate ratio is called the *standardized rate ratio* (SRR).

There is also an *indirect* standardization, resulting in a rate ratio referred to as the *standard mortality* (or morbidity) *ratio* (SMR); also SIR (standard incidence ratio) may be used when dealing with morbidity. This standardization means that the rates of the standard population (usually that of the country) are weighted by

Table 14.2 Example of standardizations. Hypothetical populations, of which II and III have the same age-specific rates but a different age structure. 'Denominator' refers to either individuals, person-years or referents

Age	Population I		Population II		Population III	
	Cases	Denominators	Cases	Denominators	Cases	Denominators
40–49	12	1200	8	800	2	200
50–59	16	800	20	500	20	500
60–69	20	400	30	200	120	800
40–69	48	2400	58	1500	142	1500
Crude rate	20	1000	39	1000	95	1000
Crude rate ratio	(1.0)		2.0		4.8	
Standardization rate*	20	1000	43	1000	43	1000
SRR**	(1.0)		2.2		2.2	
SMR***	(1.0)		2.1		2.7	

*Direct standardization (with population I as the standard), that is, rates from population II applied to population I: $(1200 \times 8/800 + 800 \times 20/500 + 400 \times 30/200)/(1200 + 800 + 400) = 104/2400 = 43/1000$. Similarly for population III.
**SRR = (standardized rate II)/(standardized rate I) = $(43/1000)/(20/1000) = 2.2$; the reference is indicated by parentheses in the table.
***Rates from population I is applied to populations II and III, which gives the expected number; then SMR = (observed)/(expected): population II gives $58/[(12/1200) \times 800 + (16/800) \times 500 + (20/400) \times 200] = 58/[(8 + 10 + 10)] = 58/28 = 2.1$ and population III gives $142/[(12/1200) \times 200 + (16/800) \times 500 + (20/400) \times 800)] = 142/[(2 + 10 + 40)] = 142/52 = 2.7$.

the number of person-years or individuals in the different age groups of the study population. This leads to an expected rate (or number of cases) for the index population, and the quotient between the observed and the expected in the index population is the SMR (the procedure is like calculating the expected number of cases or deaths in an occupational cohort with the national rates as the reference, Appendix C).

Indirect standardization does not provide for comparability between more than two populations at a time, since the standards (that is, the index populations), will otherwise vary in age composition, which influences the numerical value of the SMR as seen in *Table 14.2*. Taking the referents as the denominator, the same standardization procedures may be applied to case-referent data.

Confounding, modification and inference

Two different study bases are shown in *Figure 14.2a and b* with asymmetrical relations between the two determinants for the disease, D_1 and D_2. It is assumed that D_1 increases the background rate threefold and that D_2 doubles it, and jointly the action is sixfold, that is, a multiplicative interaction [25]. If the crude rate ratio for the presence versus the absence of D_1 is calculated, one gets 3.5 and 2.5 respectively instead of the 'real' rate ratio of 3.0, because D_2 exerts *confounding*. Hence, a confounding factor is a determinant of the disease, not necessarily causal but reflecting some underlying causal mechanism, as appearing in different frequencies in the populations compared. It may be positively or negatively associated with the exposure, thereby exaggerating or masking the effect of the exposure (D_1 in this case) [26, 27]. Standardization, either by the SMR or the SRR, brings back the threefold increase in risk.

If there is independent activity of the two determinants (*Figure 14.2c*), the rate ratio appears as decreased even after standardization, simply because the stratum-specific rate ratio is only 2.0 in the D_2 stratum due to the elevated background among the non-exposed (those without D_1). This phenomenon is usually referred to as a modification of the effect, and is of formal rather than biological character

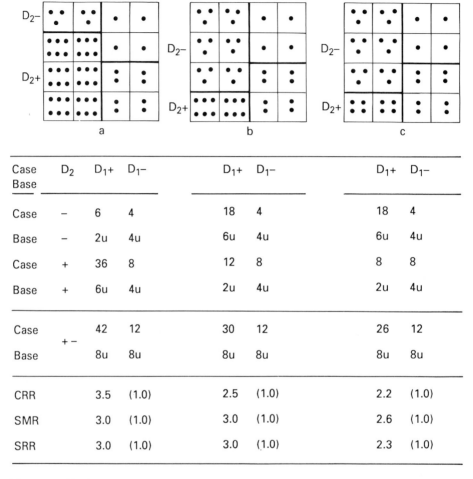

Case Base	D_2	D_1+	D_1-	D_1+	D_1-	D_1+	D_1-
Case	–	6	4	18	4	18	4
Base	–	2u	4u	6u	4u	6u	4u
Case	+	36	8	12	8	8	8
Base	+	6u	4u	2u	4u	2u	4u
Case	+–	42	12	30	12	26	12
Base		8u	8u	8u	8u	8u	8u
CRR		3.5	(1.0)	2.5	(1.0)	2.2	(1.0)
SMR		3.0	(1.0)	3.0	(1.0)	2.6	(1.0)
SRR		3.0	(1.0)	3.0	(1.0)	2.3	(1.0)

Figure 14.2 The base may be represented either by a census or by a sample, such as referents, the census giving full information about the size of the base units (u), whereas the sample only provides relative information. *CRR* is the crude rate ratio, *SMR* and *SRR* calculated as in *Table 14.2*. In base c, D_1 adds two cases and D_2 one case per population unit. A measure of confounding (as controlled) is obtained in terms of the confounding rate ratio, $CoRR = CRR/SMR$

as it is dependent on the model of the risk estimate; here the rate ratio as inherently based on a multiplicative model. The risk difference would take better care of an additive situation, but cannot be derived from case-referent data (*see Table 14.1*).

A difference in the overall, standardized rates between the compared populations would indicate that another determinant is also operating, or the differences could be due to chance (as subject to statistical evaluation). This other determinant is likely to be nothing but the exposure factor, when the compared populations have been defined on the basis of an exposure. Alternatively, control of confounding may be incomplete, or another factor associated with the exposure could be responsible, that is, an unknown confounding factor; this is impossible to account for as only known or suspected factors can be included in data collection and analysis. There is little justification, however, to be overly critical and speculate about the operation of various unknown confounding factors as soon as an effect is seen in a study. There should be at least some rationale for such an alternative explanation.

Technical comments on study designs

Cohort studies

Employment records or trade union registers are almost always the starting point for setting up an occupational cohort. Past cross-sectional studies of specific exposures or data for biological monitoring may also offer suitable groups for follow-up. Occupational cohorts are usually historical or retrospective in character, but a cohort may also be prospective (followed into the future). The terms 'retrospective' and 'prospective' have also been taken to indicate directionality— the prospective cohort as going from exposure to outcome, and the retrospective case-referent study as looking back into the exposure history of cases and referents.

Tracing cohort individuals

In tracing the individuals of an occupational cohort, a computerized linkage with the registers of the living population and of deaths or registered cancers would be quick and effective if possible. In many countries the use of driving licence registers, telephone directories, writing and calling those with similar family names living in the area of the factory, etc. are a cumbersome way of tracing cohort members. A reasonably successful follow-up should include some 95 per cent of the cohort, but 98–100 per cent may be traced in countries with good registers. If a follow-up with health examinations is required, the participation rate may drop to, say, 80–90 per cent or even lower causing subsequent uncertainty of the results.

Analysis of cohort material

The analysis of cohort material may be based on cumulative incidence, especially if there is a reference cohort, but more often the 'person-years method' is applied [22, 28]. Computer programs are available for obtaining the person-years distribution, for which the principle is given in Appendix C along with some statistical calculations [29]. A latency time criterion is usually applied (especially in cancer studies) by disregarding both the new cases and the cumulative person-years for a certain period of time after start of exposure. Alternatively, the cases and the person-years might be given by time periods like 0–4, 5–9, etc. years since first exposure.

Case-referent studies

Aetiological factors for rare diseases are usually best studied by case-referent studies, given a reasonably common exposure, which may be achieved by choosing the study population in an area or within a company where the particular exposure operates. However, should the exposure of interest be both scattered and rare, the case-referent approach tends to fail; this also happens if the exposure is extremely common.

The study base for a case-referent study is usually open, but may be closed if the study is nested in a cohort. An open base can be predetermined in geographical or administrative terms, but the boundaries may also be secondarily laid down by the way the cases are recruited. In the former situation, the cases may be harvested and the referents randomly drawn from a population register, but if the study base is secondary, one would have to draw the referents similarly to the cases, usually using other patients to represent the base population [30].

When such other disease entities are used as referents, a possible relation between the exposure and some of these should be considered, so that the exposure frequency of the base population should not be misrepresented by the referents and the risk ratio biased. If a mix of other disorders are used as the referents, disease entities that could relate to the exposure may have to be excluded. Should unrelated disorders also be excluded, it would not affect the ratio of exposed to non-exposed among the remaining, properly selected referents. However, exclusions may easily be misunderstood and lead to sceptical comments.

In a case-referent study, several exposures may be considered with regard to the disease under study. Since various referent disorders could be related to different exposures, the referents may have to be further refined, when multiple exposures are studied [31]. This concern is not relevant, however, if the exposure-related diagnosis was not the reason for admittance to hospital.

Three different types of case-referent studies may be distinguished: those in the open and closed populations, respectively, recruiting the cases and referents over a period of time; also the study may be based on prevalent cases with a sample of referents drawn at a particular point in time [16, 32].

In an open population study, a referent drawn early in the study period may later become a case (although rare in practice), which follows from the fact that the referents should reflect the occurrence over time of the exposure and other determinants in the base population. If the population is closed, that is, the case-referent study nested, and the referents are drawn from those remaining healthy at the end of the study period, then the disease under study has to be relatively rare if caused by the exposure, since otherwise the relation between still healthy exposed and non-exposed individuals would be distorted relative to the original situation in the base population. The estimate of the risk ratio, historically and formally taken as the exposure *odds ratio* in all types of case-referent studies, would now be an approximation of the cumulative incidence (rate) ratio. When the case-referent study is based on prevalent cases at a particular point in time and a contemporary sample of referents, the odds ratio equals the incidence rate ratio (given no influence by the exposure on duration of the disease) and not the prevalence rate ratio, as might perhaps be expected [32].

The practical implications of the fact that there are several types of case-referent studies are usually marginal. Sometimes there are also mixed designs, for example, the cases might have been collected from an open population over some time but the referents are all drawn at the end of the study period. This is acceptable, when there is only little or no change of the exposure pattern over time.

Matching has often been employed in case-referent studies for the control of confounding, but fails in this respect as discussed below (p.291). In some instances, however, matching may improve efficiency [33]. If matching has been undertaken for one reason or another, it is wise to maintain the matched pairs, triplets etc. in the analysis, unless it can be shown that the matching did not bring about any correlation in the exposure pattern between cases and referents. If there is such a correlation, this would also tend to obscure the effect of an exposure. For example, suppose that the lung cancer and mining relationship is studied in an area, with no other industrial activities but some farming and forestry by self-employed people. If the cases of lung cancer were matched on employee status, then every lung cancer case in a miner would get a miner as a referent and no effect would be seen of mining, not even if every lung cancer case had occurred among miners. This is an (exaggerated) example of so-called overmatching, which appears when matching is undertaken on exposure-related factors in a case-referent study. Furthermore, matching also makes it impossible to evaluate the effect of the matching factor.

A simple analysis of a stratified case-referent material is shown in Appendix D. Usually the Mantel–Haenszel statistic has been applied together with the calculating approximate confidence limits for the Mantel–Haenszel estimate of the rate ratio [16]. This estimate might be seen as a suitable weighing of the stratum-specific rate ratios; a programmable calculator may be used for such analyses [6].

Proportional mortality study

Somewhat related to case-referent studies is the so-called proportional mortality study. Its principle is to take the number of deaths of a particular disease out of all deaths and compare this quotient for exposed and non-exposed individuals in terms of the proportional mortality ratio (PMR). Stratifications on age etc. and standardizations may be applied. The PMR study comes close to a case-referent study, in this context called a mortality odds ratio study, where the referents are taken as deaths other than those constituting the case entity [34]. Since the case-referent study is a better concept, the design is preferable to the PMR study.

Validity in occupational health epidemiology

General remarks

Any findings in epidemiology should be subject to considerations regarding possible errors in design or analysis. If methods and data are well displayed, the reader should be able to develop a view on the quality of the design and to check the calculations. Further analysis might also be undertaken and the information merged with other studies for more reliable conclusions, especially when small-scale studies only are available. To some extent, the statistical analysis may even be thought of as a service to the reader, but the design of the study is a core issue for the investigator. So, the critique of a study focuses on design rather than on analysis, emphasizing the importance of validity in the design. A good validity means freedom from systematic errors and there are various attempts to describe this issue under a few main concepts as in the following section.

Selection bias

Systematic errors may distort the study base if they are selective in character. For example, dead individuals could have been sorted out from company records or from trade union registers, making them useless for epidemiological purposes. On the other hand, cases of a severe disease may be remembered better than those remaining healthy. This possibility should be kept in mind especially if one has to discuss fellow workmates with (former) employees, a practice not recommended. Nor is it any better to discuss past exposure for, say, particular cases with management representatives unless a blind procedure can be achieved, since the pertinence of an exposure only for cases might be underestimated.

Pre-employment health examinations and other selection of the fittest for employment, particularly in qualified jobs, tend to create a 'healthy worker effect' [35]. A selection phenomenon will also take place if the exposure somehow becomes part of the diagnostic criteria. For example, a pathologist would probably be more apt to diagnose a mesothelioma when the histology is suggestive and if the case is known to have had asbestos exposure. Similar concerns may arise as soon as there is some suspicion of a relationship, but so far there is little problem in occupational epidemiology because of the general ignorance of occupational risks.

There is no satisfactory way to cope with selection bias if present in epidemiological material. It would be wiser to abstain from a study, if some sort of selection has affected the study base, either before or in the first phase of a study.

Observational problems

The observation of disease should be as efficient among exposed as non-exposed individuals. In cohort studies with the general population as the reference, the case diagnoses should preferably be those given by the authority in order to achieve good comparability with the rates from the official statistics. However, when medical files and histopathological preparations are checked, the official diagnosis may come into question. Comparability would be best served by keeping the official diagnosis, although highly unsatisfactory when such a diagnosis is known to be wrong. There is no solution to this dilemma other than to explain the situation clearly when presenting the study.

Case-control studies are less problematic in diagnostic quality, since the researcher may set up special criteria for the case diagnosis and check both exposed and non-exposed cases, preferably blind with regard to their exposure status. Instead the crucial problem of observation in case-referent studies relates to the assessment of exposure, since both the unhealthy individuals themselves and the interested investigator may reveal pertinent exposures more efficiently among cases than among referents (recall and observer bias, respectively). This is at least the constantly repeated but poorly documented critique of case-referent studies, whereas little is usually said about the opposite phenomenon, namely that random misclassification may obscure an effect. Furthermore, questionnaire information on exposure has been found to agree well with 'objective' information from company records [36, 37] often consulted in case-referent studies of occupational risks.

Confounding

Concern is often expressed about uncontrolled confounding, especially when information on smoking is lacking. However, even a comparison population will

Table 14.3 Estimated effect of confounding from uncontrolled smoking in terms of rate ratios with regard to the fraction of smokers in various hypothetical populations (after Axelson [27])

Non-smokers (risk of 1)	*Percentage and type of smokers in the population*		Rate ratio*
	Moderate smokers (risk of 10)	Heavy smokers (risk of 20)	
100	–	–	0.15
80	20	–	0.43
70	30	–	0.57
60	35	5	0.78
50*	40*	10*	1.00*
40	45	15	1.22
30	50	20	1.43
20	55	25	1.65
10	60	30	1.86
–	–	100	3.08

*Compared to *reference population with 50 per cent non-smokers, 40 per cent moderate smokers, and 10 per cent heavy smokers. The model behind the table may also be used for correction of expected numbers, that is $I = RI_0 P_{CF} + I_0 (1 - P_{CF})$, where I is the overall rate, R the rate ratio of the confounding factor (smoking in this case), I_0 the rate among those without the confounding factor and P_{CF} the proportion of the population with the confounding factor, which can be split up into heavy and moderate smokers etc. with their respective rate ratios. Solving I_0 and adjusting P_{CF} according to smoking habits in a population provides for the figures in the table and for adjustment of I in a study.

include a substantial fraction of smokers, either as a specially selected reference-group or as part of the general population. Even at worst, the contribution of uncontrolled smoking to the risk ratio would hardly be much more than about 1.5, not even for smoking and lung cancer [27]. *Table 14.3* shows the potential confounding effect for lung cancer with regard to differing smoking habits in the index population. The formula in the footnote below this table allows adjustments of the expected rate or numbers with regard to any confounding factor for which the risk ratio and the occurrence in the population can be estimated.

Even if the effect from confounding is usually smaller than believed, control is desirable as far as possible either by restricting the study to individuals without the confounding factor or through stratification by (categories of) the confounding factor. Other possibilities would be to apply multivariate analysis or stratification by a multivariate confounder score [7, 9, 10].

For a cohort, matching may be an attractive way of creating an 'unconfounded' comparison group. Even if the number of matching factors have to be limited for practical (economic) reasons, relatively little uncontrolled confounding is likely to remain from other factors after matching on a few determinants of the disease. For example, matching on smoking would presumably also take care of much of the potential confounding from alcohol drinking and other associated 'lifestyle' factors.

As already mentioned, matching does not solve the problem of controlling confounding in case-referent studies as it does not create an homogeneous study base. The situation is even more problematic, however, since the selection of referents is influenced by random variation, so that confounding may appear in the data although absent in the base and vice versa, and positive confounding in the base may come up negative in the data or the reverse. So in principle there is no way of fully controlling for confounding in case-referent studies, but a reasonable number of controls, say about 200, would reduce random influence and reasonably well transfer confounding from the study base into the data, allowing some control by the methods mentioned [34].

Although concerns about uncontrolled confounding in occupational epidemiology often relates to smoking, drinking and other generally operating determinants, there is a more relevant aspect of confounding, namely from other industrial exposures, when the effect of a particular agent rather than an overall, job-related risk is supposed to be evaluated.

For example, haematite mining has been associated with lung cancer and haematite itself suspected to be the cause. However, exposure to radon daughters has also occurred in haematite mines and is more likely than haematite to be the main cause of lung cancer. Similarly, silica exposure may be carcinogenic to the lung, but again concomitant carcinogenic exposures are difficult to rule out—for example, soot exposure for foundry workers, radon daughter exposure for miners, etc. Only new studies from less complex industrial environments may finally elucidate the role of the various exposures.

Comparability of populations

Another issue of validity concerns the character of the reference population. For example, an industrial population other than that under study could have some totally different exposure causing the same disorder(s) as the exposure at issue. Hence, the choice of a group of copper smelter workers (similar in smoking traits, etc.) as a reference population for miners would fail to reveal fully the excess risk of lung cancer due to radon daughter exposure in the mines, since the copper smelter workers also suffer from lung cancer, although essentially due to arsenic exposure. Nor should the reference group come from an urbanized area, if the index population is rural and if, say, an increased risk of lung cancer is part of the hypothesis under study. Similar considerations are always necessary, but the circumstances may be less obvious than in these examples.

These remarks on comparability of populations may be equally relevant to both cohort and case-referent studies. Although commonly seen, it is questionable to evaluate just one exposure at a time, disregarding other determinants of the disorder operating in the unexposed sector of the study base (that is, a situation that may also be seen as one of negative confounding). Instead it is important to try and identify a population sector of the base, which is free from any *a priori* known determinants of the disease and to use this sector as the reference. For example, when miners and copper smelter workers lived in the same area, it was necessary to identify and separate these categories and to use others as the reference category [33].

Because of the problem in identifying a proper reference group, some justification can be seen for using national rates in the context of cohort studies [22]. For case-referent studies the corresponding justification would be to have just all non-exposed as the reference, but this disagrees with the principle argued above. It would be more reasonable, therefore, with at least some refinement, to exclude, for example, white collar professions from the study population when an industrial exposure is studied. By contrast, national rates inevitably include all kinds of occupations as well as unemployed with the resulting incomparability phenomenon—the 'healthy worker effect'.

Costs, power and size

Even relatively small studies are expensive, and an estimate of the costs and effectiveness of an intended study is therefore desirable. In general, a case-referent approach tends to be less expensive than a cohort, given that the exposure frequency is reasonably high, say, at least 5 per cent among the referents. For scattered and rare exposures the cohort approach is the only possibility, picking up a few individuals with exposure from a large number of companies. Since occupational health epidemiology usually deals with relatively high risk estimates, often about 2.0 or above, even rather small populations may be studied.

As a rule of thumb, it is reasonable to have at the very least 50–60 cases in a case-referent study, given an exposure frequency of some 15–20 per cent among the referents. In principle, an equal number of cases and referents would be the most efficient, but with a lack of cases, the number of referents might be increased up to about five referents per case, though little gain in power is obtained thereafter. To obtain reasonable control of confounding in the base, about 200 referents would always be required. Cohorts within the general population as the reference should preferably have a size and age structure so that the expected number of cases would be at least two or three for the particular disorder, but even so the study would yield little information, especially if it did not show a clear excess of cases.

Table 14.4 Required numbers in a study for 80 per cent probability (power) of detecting a given rate ratio at 5 per cent significance level (one-tailed)

Cohort studies; required number of expected cases		Case-referent studies; number of cases required with a case-referent ratio of 1:2 and different exposure frequencies†		
Rate ratio	Expected number*	Rate ratio	Exposure frequency† 5 per cent	Exposure frequency 20 per cent
1.3	78.6	1.3	2602	808
1.5	30.6	1.5	1041	332
2.0	9.0	2.0	327	110
3.0	2.9	3.0	116	43
4.0	1.0	4.0	67	27
5.0	1.0	5.0	47	20

*The required number of person-years is obtained by dividing this number with the (national) rate for a particular disease (or aggregated rates for certain groups of disorders like cancer, cardiovascular diseases etc.).
†The exposure frequency in the population is estimated through the referents as representing the population or, in the planning stage, from other information.

Formal calculations can be made regarding the power of a study; these may be used in the planning stage, and also for the reporting of a negative or non-positive result [39]. The size of a study and its power do not relate to the number of individuals *per se* but rather to the expected number among the exposed. Hence, even a large cohort of young individuals would not give much information about a cancer risk, simply because few if any cases would occur.

In deciding whether or not a study should be undertaken, or if a presented non-positive study has power enough to be informative, some guidance may be obtained from power calculations as reflected in *Table 14.4* [5, 11, 40]. Various priorities and circumstances in addition to power may determine why a study is finally undertaken.

Last but not least, even if the methods used in occupational epidemiology are those of epidemiology at large, a specific knowedge of the exposure conditions at workplaces is necessary for the occupational epidemiologist to design studies that provide proper information for subsequent preventive measures.

References

1. Rothman, K.J. The rise and fall of epidemiology. *New England Journal of Medicine*, **304**, 600–602 (1981)
2. Lilienfeld, A.M. and Lilienfeld, D.E. A century of case-control studies: progress? *Journal of Chronic Diseases*, **32**, 5–13 (1979)
3. Monsson, R.R. *Occupational Epidemiology*. CRC Press, Boca Raton, Florida (1980)
4. Breslow, N.E. and Day, N.E. *Statistical Methods in Cancer Research, Vol. 1: The Analysis of Case-control Studies* and *Vol. II: The Design and Analysis of Cohort Studies*. International Agency for Research on cancer, Lyon (1980 and 1988)
5. Schlesselman, J.J. *Case-control Studies: Design, Conduct, Analysis*. Oxford University Press, New York (1982)
6. Rothman, K.J. and Boice, Jr, J.D. *Epidemiologic Analysis with a Programmable Calculator* (2nd edn). Epidemiology Resources Inc., Brookline, Maine (1982)
7. Andersson, S., Auquier, A., Hauck, W.W., Oakes, D., Vandaele, W. and Weisberg, H.I. *Statistical Methods for Comparative Studies*. John Wiley and Sons, New York (1980)
8. Kleinbaum, D.G., Kupper, L.L. and Morgenstern, H. *Epidemiologic Research. Principles and Quantitative Methods*. Lifetime Learning Publications, Belmont, California (1982)
9. Miettinen, O.S. *Theoretical Epidemiology, Principles of Occurrence Research in Medicine*. John Wiley and Sons, New York (1985)
10. Rothman, K.J. *Modern Epidemiology*. Little, Brown and Company, Boston (1986)
11. Karvonen, M. and Mikheev, M.I. (eds) *Epidemiology of Occupational Health*. WHO, Regional Office for Europe, Copenhagen (1986)
12. Last, J.M. (ed.) *A Dictionary of Epidemiology*. Oxford University Press, New York (1983)
13. Epstein, S.S. *The Politics of Cancer*. Anchor Press, Garden City, New York (1979)
14. Soskolne, C.L. Epidemiological research, interest groups and the review process. *Journal of Public Health Policy*, **6**, 173–184 (1985)
15. Miettinen, O.S. Design options in epidemiologic research. An update. *Scandinavian Journal of Work, Environment and Health*, **8** (Suppl. 1), 7–14 (1982)
16. Miettinen, O.S. Estimability and estimation in case-referent studies. *American Journal of Epidemiology*, **103**, 226–235 (1976)
17. Axelson, O. Dealing with the exposure variable in occupational and environmental epidemiology. *Scandinavian Journal of Social Medicine*, **13**, 147–152 (1985)
18. Checkoway, H. Methods of treatment of exposure data in occupational epidemiology. *La Medicina del Lavoro*, **77**, 48–73 (1986)
19. Hoar, S.K., Morrison, A.S., Cole, P. and Silverman, D.T. An occupation and exposure linkage system for the study of occupational carcinogenesis. *Journal of Occupational Medicine*, **22**, 722–726 (1980)
20. Pannet, S., Coggon, D., Acheson, E.D. A job-exposure matrix for use in population based studies in England and Wales. *British Journal of Industrial Medicine*, **42**, 777–783 (1985)
21. Aitio, A., Riihimäki, V. and Vainio, H. Biological monitoring and surveillance of workers exposed to chemicals. Hemisphere Publishing Corporation, Washington (1984)
22. Gardner, M. Considerations in the choice of expected numbers for appropriate comparisons in occupational cohort studies.*La Medicina del Lavoro*, **77**, 23–47 (1986)
23. Miettinen, O.S. Proportion of disease caused or prevented by a given exposure, trait or intervention. *American Journal of Epidemiology*, **99**, 325–332 (1974)
24. Miettinen, O.S. Standardization of risk ratios. *American Journal of Epidemiology*, **96**, 383–388 (1972)
25. Rothman, K.J., Greenland, S. and Walker, A.M. Concepts of interaction. *American Journal of Epidemiology*, **112**, 467–470 (1980)
26. Miettinen, O.S. Confounding and effect modification. *American Journal of Epidemiology*, **100**, 350–353 (1974)
27. Axelson, O. Aspects on confounding in occupational health epidemiology. *Scandinavian Journal of Work, Environment and Health*, **4**, 85–89 (1978)
28. Breslow, N.E. Elementary methods of cohort analysis. *International Journal of Epidemiology*, **13**, 112–115 (1984)

29. Liddell, F.D.K. Simple exact analysis of the standardised mortality ratio. *Journal of Epidemiology and Community Health*, **38**, 85–88 (1984)
30. Miettinen, O.S. The 'case-control' study: valid selection of subjects (with dissents, comment and response). *Journal of Chronic Diseases*, **38**, 543–558 (1985)
31. Axelson, O., Flodin, U. and Hardell, L. A comment on the reference series with regard to multiple exposure evaluations in a case-referent study. *Scandinavian Journal of Work. Environment and Health*, **8** (Suppl. 1), 15–19 (1982)
32. Axelson, O. Elucidation of some epidemiologic principles. *Scandinavian Journal of Work, Environment and Health*, **9**, 231–240 (1983)
33. Thomas, D.C. and Greenland, S. The efficiency of matching in case-control studies of risk-factor interactions. *Journal of Chronic Diseases*, **38**, 569–574 (1985)
34. Axelson, O. The case-referent study. Some comments on its structure, merits and limitations. *Scandinavian Journal of Work, Environment and Health*, **11**, 207–213 (1985)
35. McMichael, A.J. Standardized mortality ratios and the 'healthy worker effect': scratching beneath the surface. *Journal of Occupational Medicine*, **18**, 165–168 (1976)
36. Pershagen, G. and Axelson, O. A validation of questionnaire information on occupational exposure and smoking. *Scandinavian Journal of Work, Environment and Health*, **8**, 24–28 (1982)
37. Hardell, L. and Sandström, A. Case-control study: soft-tissue sarcomas and exposure to phenoxyacetic acids or chlorophenols. *British Journal of Cancer*, **39**, 711–717 (1979)
38. Pershagen, G. Lung cancer mortality among men living near an arsenic-emitting smelter. *American Journal of Epidemiology*, **122**, 684–694 (1985)
39. Hernberg, S. 'Negative' results in cohort studies. How to recognize fallacies. *Scandinavian Journal of Work, Environment and Health*, **7** (Suppl. 4), 121–126 (1981)
40. Beaumont, J.J. and Breslow, N.E. Power considerations in epidemiologic studies of vinyl chloride workers. *American Journal of Epidemiology*, **114**, 725–734 (1981)

Appendix A: relation of rates

$CI = 1 - e^{-a\Sigma I}$ where CI is cumulative incidence; ΣI is the sum of incidence rates over categories, a is the width of age categories and e the base of the natural logarithms. Example: incidence rates (1960) from Appendix C gives a cumulative incidence for the age span 50–64 years as $CI_{50-64} = 1 - e^{-5(230+420+685)/10^6}$ or 0.007; also approximately $CI = a\Sigma I$.

To obtain the relationship of incidence and prevalence, assume a steady state with n cases in a population of N, that is, n new cases have to appear in a period d, equal to the average duration of the disease. Therefore $n = (N - n)dI$, where $(N - n)d$ are the person-years of healthy individuals, upon which the incidence rate acts. Divide by N and obtain $n/N = (1 - n/N)dI$, where n/N is the prevalence rate, P. Therefore $P = (1 - P)dI$ and also $P = dI/(1 + dI)$.

Appendix B: aetiological fraction

The aetiological fraction among the exposed, EF_1, is the fraction of the cases (or of the rate) caused by the exposure, that is $(R_1 - R_0)/R_0$ where R_1 and R_0 are the rates for exposed and non-exposed, respectively. With $R_1/R_0 = RR$, the rate ratio, $EF_1 = (1 - 1/RR)$ or $EF_1 = (RR - 1)/RR$ (also called attributable risk). The aetiological fraction for the total population EF, is obtained by multiplying EF_1 with the case fraction, CF, which is the fraction of exposed cases out of all cases, that is $EF = CF_1 \times (RR - 1)/RR$, an expression showing the relative contribution of cases in the total population as due to the exposure. When the joint effect of two or more exposures is more than additive, the sum of the aetiological fractions for each exposure may exceed unity, for example, for *a*sbestos (say risk ratio of 5) and *s*moking (say risk ratio of 10), $EF_a = (5 - 1)/5 = 0.8$ and $EF_s = (10 - 1)/10 = 0.9$ and also $0.8 + 0.9 > 1.0$.

Appendix C: principal procedure of cohort analysis using national rates

1. Obtaining person-years: Take Mr A born in 1910, exposed from 1951, died in 1967. Requiring 10 years of induction-latency time he would contribute half a person-year at observation in 1961 in the age group 50–54 and will continue with one person-year in 1962, 1963, 1964, but in 1965 he contributes to age group 55–59 and continues to do so throughout 1966, with half a person-year assigned in 1967.
2. By the same token all individuals in a cohort contributes person-years so that a table may be created as follows:

Person-years

Age group	Calendar year									
	1960	*1961*	*1962*	*1963*	*1964*	*1965*	*1966*	*1967*	*1968*	etc.
50–54	29	25	24	23	20	15	15	13	13	
55–59	16	18	20	17	19	24	24	25	25	
60–64	26	21	34	31	45	38	34	29	33	

Underlined figures — see point 4 below.

3. National incidence rates (for a particular disease) per 1 000 000 person-years.

Incidence rates

Age group	Calendar year									
	1960	*1961*	*1962*	*1963*	*1964*	*1965*	*1966*	*1967*	*1968*	etc.
50–54	230	239	249	243	294	261	.252	288	280	
55–59	420	431	418	427	416	410	408	421	415	
60–64	685	696	711	693	710	712	698	715	731	

4. Obtain the expected number of deaths or cases from this disease by multiplying cell by cell in the two tables and sum up $(1/1\,000\,000)$ $(29 \times 230 + 25 \times 239 + 24 \times 249 + \ldots + 33 \times 731) = 0.33$ and suppose three cases of the disorder were observed.
5. Calculate the SMR with three cases observed and 0.33 expected, that is SMR $= 3/0.33 = 9.1$. The lower and upper confidence limits for the SMR ($\underline{SMR}$; $\overline{SMR}$) may be obtained by means of a chi-square table as the confidence limits for the observed divided by the expected, E. $\underline{SMR} = 0.5[X^2_{\alpha/2}(2a)]/E$ and $\overline{SMR} = 0.5[X^2_{1 - \alpha/2}(2a + 2)]/E$, where $(2a)$ and $\overline{(2a + 2)}$ refer to the degrees of freedom, a being the observed cases, and $100(1 - \alpha)$ indicates the desired confidence interval, for example, with $\alpha = 0.10$ the 90 per cent interval is obtained. Hence, for 3/0.33 one obtains $\underline{SMR} = 0.5[X^2_{0.05}(6)]/0.33$ and $\overline{SMR} = 0.5[X^2_{0.95}(8)]/0.33$ and by means of the X^2 table $0.5[1.635]/0.33$ and $0.5[15.507]/0.33$ or 2.4–23.5 (the limits preferably abbreviated outwards).

Appendix D: analysis of stratified case-referent data

Stratified data from a study of the relationship between arsenic exposure and cardiovascular disease (after Axelson [11])

Stratum	Age	Case-referent	Non-exposed	Exposed	Total
1	30–54	C	$6 = b_1$	$7 = a_1$	$13 = N_{11}$
		R	$14 = d_1$	$2 = c_1$	$16 = N_{01}$
			$20 = M_{01}$	$9 = M_{11}$	$29 = T_1$
2	55–64	C	$25 = b_2$	$19 = a_2$	$44 = N_{12}$
		R	$16 = d_2$	$6 = c_2$	$22 = N_{02}$
			$41 = M_{02}$	$25 = M_{12}$	$66 = T_2$
3	65–74	C	$45 = b_3$	$27 = a_3$	$44 = N_{13}$
		R	$26 = d_3$	$10 = c_3$	$36 = N_{03}$
			$71 = M_{03}$	$37 = M_{13}$	$108 = T_3$
1–3	Total	C	76	53	
		R	56	18	
Crude rate ratio			(1)	2.2	
SMR			(1)	1.9	
X^2 (1) (Mantel–Haenszel)				5.52	
Rate ratio (Mantel–Haenszel)					
point estimate				2.1	
90 per cent confidence interval				1.2–3.5	

The crude rate ratio, $CRR = (53/18)/(76/56)$ or taken as an odds ratio $(53 \times 56)/(76 \times 18) = 2.2$ (and as in *Table 14.2*) $SMR = (7 + 19 + 27)/(6 \times 2/14 + 25 \times 6/16 + 45 \times 10/26)$ or $SMR = 53/(0.86 + 9.38 + 17.31) = 1.9$. Some confounding is present, because the confounding rate ratio, $CoRR = CRR/SMR$, is $2.2/1.9 = 1.2$. This value represents the magnitude of the confounding that was controlled through the stratification. The Mantel–Haenszel chi-square statistic (with one degree of freedom) has the structure $X^2(1) = (\Sigma_j a_j - \Sigma_j N_{1j} M_{1j}/T_j)^2/[\Sigma_j N_{1j} N_{0j} M_{0j} M_{1j}/T_j^2 (T_j - 1)]$ with N_{1j} etc. denoting the exposed and N_{0j} etc. the non-exposed, as shown in the table above (Σ_j means summation over the strata, j; for example $\Sigma_j a_j$ means $a_1 + a_2 + a_3 = a$, or in the example $7 + 19 + 27 = 53$.)

The Mantel–Haenszel estimator of the rate ratio (or odds ratio) is $RR_{M-H} = \Sigma_j(a_j d_j/T_j)/\Sigma_j(b_j c_j/T_j)$

The calculations can now be made as shown in the following scheme:

Stratum	a_j	$N_{1j} M_{1j}/T_j$	$N_{1j} N_{0j} M_{1j} M_{0j}/T_j^2(T_j-1)$	$a_j d_j/T_j$	$b_j c_j/T_j$
1	7	4.034	1.590	3.379	0.414
2	19	16.667	3.504	4.606	2.273
3	27	24.667	5.456	6.500	4.167
Σ_j	53	45.368	10.550	14.485	6.854

Hence $X^2(1) = (53 - 45.368)^2/10.550 = 5.52; RR_{M-H} = 14.485/6.854 = 2.11$. The approximate Miettinen confidence limits (90 per cent) are $\underline{RR}, \overline{RR} = (2.1)^{1 \pm 1.645/\sqrt{5.52}}$ or $\underline{RR}, \overline{RR} = (2.1)^{1 \pm 0.70}$ or $\underline{RR}, \overline{RR} = 1.2$ to 3.5.

Chapter 15

Survey design

K.M. Venables

Introduction: the scope of occupational epidemiology

Epidemiological surveys have an important place in occupational health and much of our present knowledge of the effects of occupational exposures on man has come from epidemiological research. Despite the relevance of epidemiology to their work, occupational physicians and hygienists are often reluctant to carry out their own surveys. This chapter aims to encourage the diffident by describing the necessary steps in planning and undertaking surveys. It concentrates on studies of morbidity, rather than mortality. The general concepts underlying epidemiological surveys are discussed in Chapter 14.

A survey may be *descriptive*, performed perhaps to aid planning about resource allocation in an occupational health service. It may be *analytical*, testing hypotheses about the relationship between occupational exposure and its effects. It may form part of the *evaluation* of a control measure which reduces exposure or aims to limit its effects. These three ways of using epidemiology are complementary and a single study may have descriptive, analytical and evaluative components. For example, describing the accident rates in different areas in a factory would give information which is useful in itself, may confirm theories about the causes of certain types of accident and also could form the first phase of an evaluation of accident prevention measures. Although even the simplest survey goes through several stages (*Figure 15.1*) they can be accomplished quickly if necessary. Few surveys are so urgent that speed takes priority over preparation and a poorly planned and executed survey could produce actively misleading results. Preparations for a study, and the analysis of results, always take longer than collecting the data, sometimes considerably longer.

Questions

The first step is the recognition of a question or questions which should be answered by an epidemiological survey. This may seem an obvious point, but occasionally surveys are proposed from a wish to 'do something' about an occupational health problem. Further thought may suggest that action, rather than research, is needed or that clinical or toxicological research would be more appropriate than epidemiology. The availability of an exposed population, set of records or series of patients may prompt the collection of data before questions

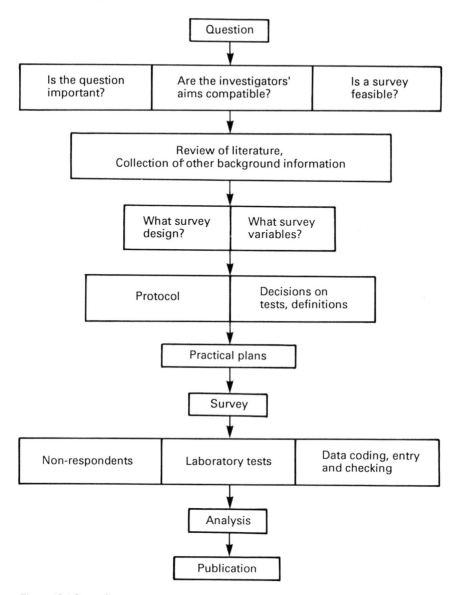

Figure 15.1 Stages in a survey

have been formulated adequately. Although such opportunities should not be neglected, careful consideration of the questions to be answered can only improve a study's quality.

There is no shortage of questions to be answered by occupational epidemiology. They are constantly generated by the results of toxicological research, and by drawing analogies with the results of research in fields other than occupational health. Of greater concern is that many ideas for surveys are raised, but, like the elephant's question in *Figure 15.2*, remain only ideas and are never translated into

Figure 15.2 A survey in search of a question. Included by kind permission of the artist Abu Abraham (first printed in *The Guardian*, 1981)

practical planning and execution. The important steps in translating ideas into action are consideration of the question's importance, discussing the idea with potential collaborators and assessing the survey's feasibility.

How important is the question?

Medawar writes lucidly and wittily on scientists' thought-processes and activities and has commented that anyone *'who wants to make important discoveries must study important problems.* It is not enough that a problem should be interesting — almost any problem is interesting if it is studied in sufficient depth' and again, 'the problem must be such that it *matters* what the answer is' [1]. Occupational medicine and hygiene are essentially problem-solving disciplines which identify and quantify problems and recommend, execute and evaluate measures to control them. It is thus often easier to follow Medawar's advice than in the fields of basic research about which he wrote.

What is the likely result of the study? Will it lead directly to action to improve workers' health, perhaps by providing information useful in setting environmental standards or in deciding an occupational health service's priorities? There may be no immediate practical result, but the survey will help in understanding the causes or natural history of disease, leading ultimately to prevention. If the survey's prospects for disease prevention and improved working conditions seem remote, one must question the value of doing it. Burr [2] has made similar points about medical research in general and also commented that little is gained by confirming what everybody believes, or by refuting something which nobody believes. Research investigates issues between these two extremes, in the area of genuine doubt.

Plans for studies are sometimes made prematurely, when reviewing the literature or talking to other with similar interests would show that the questions have already been answered. Conversely, the potential investigator may find that his ideas are fresh and will make a worthwhile contribution. Sometimes 'facts', accepted as common knowledge, are revealed as untested assumptions.

Collaboration

Few surveys can be done single-handed and at least one formal meeting of all potential collaborators should be held. It is very difficult to plan surveys by telephone, letter or in casual conversation. The purposes of this meeting are clarification of the survey's aims and a decision about its feasibility. Talking to other interested persons brings more factual information and a range of perspectives to bear on the problem and always helps to refine questions. Quite disparate aims may be compatible, as long as they are positive and clearly defined. For example, a company physician who wishes to know the prevalence of a particular disease in a factory can work easily with a research department whose primary objective is, say, the validation of a new survey technique. A vague wish to 'do research' will accomplish little.

Feasibility

Discussion may reveal that, at present, the survey is not feasible. It will require access to a population of workers or to a set or records. A population cannot be assumed accessible if the views of management and unions are not known. The adequacy and completeness of records must be assessed and confirmation sought that they can be used for research purposes. A study will not be immediately feasible if it appears to require technical expertise not available to the current collaborators. The plan can be discussed further after talks with management and unions, custodians of records or with further collaborators.

To ensure the study gets done, one person must be appointed to write a protocol and be responsible for planning and, probably, analysis. He need not have initiated the survey, but must be able to devote time to the project and be sufficiently knowledgeable to take day-to-day decisions about the study. His first practical step in a survey of an occupational population is identifying one person in the company (assuming a survey at a single company) of sufficient seniority to take decisions without lengthy consultation. Such a person is invaluable even when an occupational health service performs an in-house investigation. He should liaise between the investigators and the company, ensure the survey does not disrupt production, make practical arrangements on-site, provide information about the worksite and workforce and keep the company and unions informed about progress. Frequently the personnel officer or safety officer assumes this role, but whoever does must be committed to the success of the survey and prepared to spend time on it.

Protocol

Much information is needed to perform a survey, which is collated in a protocol, or series of increasingly detailed protocols, under headings such as in *Table 15.1*. A protocol is simply a 'road map' for the survey, a written document containing the survey's aims and methods and discussing other important issues. Even simple surveys require a protocol. Its preparation clarifies ideas about the study: it acts as an *aide-mémoire* during analysis, or if planning must be halted for a time; it is circulated to collaborators and other interested parties, thus avoiding misunderstandings; and it forms the basis for a submission for funding if additional personnel or equipment are needed.

The literature pertaining to the problem must be reviewed. If the problem is well known, with a large literature, it is wise to consult experts. Reviewing the

Table 15.1 Headings for a protocol

General objectives
Review of the literature
Other background information
Specific objectives
Design: including sampling, controls, non-respondents
Methods: including equipment, techniques
Observations: including record forms, definitions
Analysis
Timetable
Ethical and 'political' issues
Cost

literature may lead to changes in the way questions are framed. The protocol should contain a succinct account of publication in the field, focusing on the particular questions the survey asks. The review should include the methodology proposed for measuring the variables in the study, and the current legislation may also be relevant.

Information about the process, materials handled, ventilation, other protective measures and past environmental measurements is collected. The investigators must arrange to see the process for themselves. For a population survey, a personnel list must be obtained of all workers to be studied. Groups such as cleaners, security, managerial, clerical and occupational health staff, and contract, part-time and night workers, are easily forgotten in preparing these lists, which must be carefully scrutinized for completeness. Meetings must be arranged with management and unions, so that queries are answered and concerns voiced. Helpful practical suggestions and information arise from these meetings.

Information about the proposed methodology should also be collected. Manufacturers of equipment may have extensive technical literature and some may keep bibliographies of publications about their equipment or supply names of other people using it.

Survey objectives

The aims, or objectives, of the survey evolve from general, broad ideas into specific aims which are achievable. For example (and hypothetically), let us assume that clinical observation suggests that chemical industry workers may be at risk of skin rashes. The question 'do chemicals cause skin rashes?' is too general. A list of likely chemicals must be drawn up and the type of skin condition defined. Let us say it is decided to focus on acetic acid and urticaria. The question 'does acetic acid cause urticaria?' is specific, but not framed in epidemiological terms. It could, for example, be answered by studying experimental animals. Epidemiology notes associations and seeks, by the design of the study and by comparison with other evidence, to establish if an association is causal. Bradford Hill [3] has discussed the features of an association which suggest causality. 'Is urticaria associated with exposure to acetic acid?' is the right phrasing for an epidemiological study but still does not suggest a specific design. 'Is the incidence rate of new cases of urticaria higher in workers exposed to acetic acid than in unexposed workers?' would be asked in a longitudinal study. 'Do persons with urticaria report exposure to acetic acid more frequently than persons without urticaria?' would be asked in a case-referent study.

Tests of statistical significance will ultimately be applied to the data. These assume a 'null hypothesis' of no association, a concept discussed in standard textbooks of statistics [3, 4]. Some find it helpful to phrase questions as null hypotheses. In the example above, assuming a cross-sectional survey, this would be 'workers exposed to acetic acid have a similar prevalence rate of urticaria to unexposed workers'.

Design

Type of study

Discussion of survey methodology starts with design and the reasons for choosing a particular study type should be stated in the protocol. Epidemiological surveys are *cross-sectional*, *longitudinal* or *case-referent* in design. Cross-sectional surveys measure disease prevalence, longitudinal surveys measure incidence or follow changes over time and case-referent studies compare the proportion exposed in cases and referents. The *prevalence rate* expresses the number of cases with disease as a proportion of the population studied at a defined point in time (or sometimes, over a period of time). Prevalence is an appropriate measurement for chronic diseases with no clearly defined starting point. Acute events, such as accidents or acute illnesses, are best described using the *incidence rate*, the number of new cases arising in the population over a period, with both population and time in the denominator.

One of several advantages of longitudinal studies is that they follow events from exposure prospectively. Cross-sectional and most case-referent studies have no true time dimension, although they can make retrospective enquiries of their subjects. The cross-sectional and case-referent designs have a practical advantage in that studies can be undertaken quickly, whereas longitudinal surveys, particularly morbidity surveys, may take some time to set up and carry out. Case-referent studies are carried out increasingly frequently, often to study potential associations between occupational exposures and rare diseases. Schlesselman [5] discusses this design in detail. Cross-sectional surveys may produce biassed results if the disease affects 'survival' in a job. However, cross-sectional surveys are common in occupational medicine, and almost all occupational hygiene surveys are cross-sectional. Because of this, this chapter's emphasis is on cross-sectional studies, but the issues discussed are relevant to any type of survey.

Sampling and choice of referents

Other design issues are sampling and the choice of referents (controls). If the workforce is large, it may be more economical of effort to sample, usually by stratified random sampling in order to focus on groups most likely to show the effect under investigation. Statistical advice may be needed on the size of the sample. Usually, surveys look at the effects of a specific exposure and unexposed referents are available from within the workforce, but it may be desirable to include a referent group from another workforce, or perhaps from the general population.

Referent groups are often difficult to find, rarely entirely satisfactory, and the advantages and disadvantages of the group chosen should be discussed.

Surveys aim to see all of the population, or of the sample. Response rates of 85 per cent or more are acceptable, but anything less will throw considerable doubt on the results. The survey planning must encourage response, with explanatory letters and meetings for workers, physical siting of investigations close to the place of work and repeated assurances that no pay will be lost and that results will be confidential. Although all surveys are voluntary, it is reasonable to stress that the results will be important, perhaps for workers worldwide on similar processes, and will be meaningless if participation is low. Plans should be made to see as many non-respondents as possible, perhaps with only some of the main survey's tests, and for comparing them to respondents using available information from personnel and occupational health service records.

Variables

There are three types of variable: *exposure*, *response* and *potential modifiers*. Occupational exposure has two components, duration and intensity. Duration is obtained from company records or from an occupational history. There are various ways of estimating intensity. None is accurate, in the sense of measuring the concentration of a toxic agent at its target organ. The best estimates are environmental measurements made specially for the survey by, or with the advice of, a hygienist familiar with both technical considerations and the biological mechanism through which the agent exerts toxicity. The biological model may be that accumulated exposure is important, expressed as the product of duration and some measure of average intensity summed over jobs with different exposure intensities. This model is often assumed when there is little information on an agent's toxicology in man. There may be evidence for other modes of action. For example, short peaks of high exposure may be more toxic than the equivalent 'dose' as a long-term, low-intensity stimulus. The survey may make use of past environmental measurements made by the company. Sometimes, no measured values for exposure are available and intensity is estimated qualitatively, for example by ranking job titles into 'high', 'medium' and 'low' exposure categories.

That survey variables are only indirect estimators of the 'true' quality we wish to measure is emphasized when we consider response variables. For example, 'asthma' is notoriously difficult to define [6]. Because it cannot be 'measured' directly, surveys of asthma usually include several variables known to be associated with the condition, such as various respiratory symptoms and measures of lung function.

The other variables included are those thought to modify exposure, for example, respiratory protection, or modify response, for example sex, age, prescribed medication or alcohol consumption. It is particularly important to note effect-modifiers which are unevenly distributed through exposure categories and confound the relationship between the two. Smoking is often a confounding variable in studies of respiratory and cardiac disease and of cancer because smoking may, itself, cause the disease under investigation and also be a more common habit in people with high exposure than people with little exposure. This is often the case if exposure relates to socioeconomic class, and *Table 15.2* gives a hypothetical example of confounding by smoking.

Table 15.2 An apparent effect of exposure because of confounding by smoking

	Smokers	Total Non-smokers	Smokers	Cases Non-smokers	Prevalence of disease (per cent)
Exposed (manual) workers (n = 100)	80	20	40	5	45
Unexposed (clerical) workers (n = 100)	20	80	10	20	30

This hypothetical example assumes no effect of exposure, but half of smokers have the disease, compared with a quarter of non-smokers. More exposed workers smoke than in the unexposed group chosen as their referents. This leads to a 1.5-fold excess of the disease in the exposed group.

Tests, records and definitions

The choice of tests may be aided by consultation with experts. For example, there are several components of lung function which are measurable. For each there are several methods and for each method, equipment from different manufacturers. The protocol should justify the choice of test and new and untried procedures should be described in detail. In protocols which will be discussed with lay managers and union officials, an outline of routine procedures should also be provided. This readership will, rightly, wish to know exactly how each test is carried out and if any might be uncomfortable or unpleasant for individual subjects.

Method criteria

Survey methods must be safe (for example, electrically, if a factory environment contains combustible material), acceptable to essentially healthy people (a requirement which disqualifies many hospital tests used on patients) and also simple and robust enough for repetitive use during the investigation. Methods must be valid, that is provide a meaningful estimate of the quality or quantity they are measuring. It is sometimes assumed that the newest, most complex equipment must give the most valid results, but this is not so. Technological sophistication may have considerable disadvantages. Measurements should also be reproducible, and the protocol should refer to calibration, standardization and quality control procedures and discuss other unwanted variation, such as caused by the timing of biological tests where there is a chronological rhythm, or observer bias if subjective judgements are to be made. Where possible, standard methods should be used, with known validity and reproducibility characteristics, which allows comparison of the results with those from other surveys using the same methods. If this is not possible, it is desirable to incorporate formal measurement of these qualities, during the main survey or in a pilot study. The principles of these assessments are discussed by Barker and Rose [7].

Record forms

The proposed record forms should be included in the protocol, such as questionnaires or forms for recording the results of biological measurements. To keep data collection and analysis simple, the number of questions, and of tests, must be rigorously pruned until only the essentials remain. But once the tests (or questions) are decided upon, as much information as they can reasonably provide should be recorded. Some variables are continuous, for example height or blood pressure. It makes sense to record continuous variables with as much precision as the instrumentation allows, for example height to the nearest ½ cm, or blood pressure to the nearest millimetre of mercury (mmHg). The record forms should state the precision required and include enough space, or recording boxes, for the number of digits expected. Other variables are categorical: dichotomous such as 'Yes' and 'No' answers in a questionnaire, ordinal scales, such as of the severity of a symptom, or implying no order, such as categories of marital status. For categorical scales, decisions must be made on whether multiple answers are permissible and whether 'unsure' or 'unknown' answers are allowed, or if the worker (or investigator) must make a choice between the options suggested. The forms must indicate this clearly. All record forms should be tested by the investigators before the survey for clarity and ease of completion. Design is particularly important for questionnaires, both interviewer-administered and self-completed types. The reader is referred to books which discuss questionnaire design [8, 9].

Records storage

If the survey is a large one, the storage of records may be a major practical issue. Even small surveys require some thought about record storage and retrieval. A numbering system which gives each subject a unique number and which is carried across all record forms is essential. Names, addresses and dates of birth are not unique identifiers and should not be used. Every page of a multipage form should contain the identity number, as staples and paperclips have a limited life, and personal environmental samples and biological samples should share the same numbering system as other information.

Specifying definitions

Many definitions can be specified before the survey. For example, one may specify how long a subject should have stopped smoking to be classified as an ex-smoker. Others can only be stated precisely when the results are known, but the approach to definition can be specified. For example, it may be known from the expected survey numbers that grouping by, say, quartiles of measured environmental exposure will give reasonable numbers for comparison between groups.

Planning the analysis

If the survey's questions have been well formulated, the basic lines of analysis will be clear and should be presented in the protocol. 'Fishing expeditions' through large numbers of variables may give statistically significant associations by chance. Unexpected findings may, of course, be important, but the temptation to 'fish', which is greatly facilitated by the power of available computer packages, should be curbed by specifying the major analyses in advance. In a survey which examines only a limited number of variables, it may be possible to draw up a complete set of 'dummy' tables in advance.

Thinking about the analysis at an early stage has two other benefits. First, it will highlight questions or definitions which are not yet sufficiently concrete. Second, it will suggest ways of improving record forms. Answers to questionnaires, for example, can be precoded, which increases accuracy in transferring data to computer. *Figure 15.3* shows different ways of recording information from a questionnaire used in surveys of occupational asthma. The person who will be responsible for data entry, either by first transferring it to a coding form or entering directly from the survey forms, should review the survey forms. Attention to their layout will increase the accuracy of transcription.

(1)   What happens to your   Better ☐   Same ☐   Worse ☐
      wheezing on holiday?

(2)   What happens to your   (For 'better' enter 1,   ☐
      wheezing on holiday?   'same' 2, 'worse' 3)

(3)   What happens to your   Better ☐ 1   Same ☐ 2   Worse ☐ 3
      wheezing on holiday?

Figure 15.3 Recording answers in a questionnaire. The answers (better, same, worse) do not follow any obvious numbering and mistakes could be made in data entry to a computer file without precoding. The second option is not uncommon in interviewer-administered questionnaires and requires the interviewer to enter the correct code in a box. The third option is preferable as the interviewer (or subject) has to think only about the reply to the question and not about coding it

Timetables

There are two types of timetable which should be prepared and included in the protocol. One is an overall timetable, including health and environmental measurements, any subsequent laboratory tests and data analysis. Protocols for surveys involving interviews and tests on individual workers also should include a timetable for the survey participant. *Table 15.3* illustrates the considerations which may apply when deciding on the individual's timetable. Tests whose results may be influenced by other tests must always be separated. For example, an interviewer recording symptoms may be influenced by the subject's occupational history, so different investigators should administer symptoms and occupational questionnaires. As a general rule, it is always wise to take a blood sample as the last test, and to arrange some privacy. Very occasionally people faint when donating blood and fear of needles is not uncommon.

Ethical and 'political' issues

Strictly ethical issues, such as the use of invasive medical procedures, rarely arise in epidemiological surveys, but it is a sensible practice to submit protocols to the appropriate Ethical Committee. More commonly, surveys present problems with an 'ethico-political' component concerning the confidentiality of results, provision of medical advice to individual workers, provision of advice to the firm or publication of results. In general, all the environmental measurements and the

Table 15.3 Two hypothetical timetables for survey tests

Accumulated time (minutes)	Timetable 1 Subjects				Timetable 2 Subjects			
	A	B	C	D	A	B	C	D
5	1				1			
10					2	1		
15	2	1			3a	2	1	
20					3a	3b	2	1
25	3	2	1			3b	3a	2
30	3						3a	3b
35		3	2	1				3b
40		3						
45			3	2				
50			3					
55				3				
60				3				

Tests 1 and 2 take 5 minutes, but test 3 takes 10 minutes and is the 'rate-limiting' test. Either the investigators carrying out tests 1 and 2 spend an equal amount of time doing nothing (timetable 1) or there are two investigators (3a and 3b) for test 3 (timetable 3). Timetable 2 sees subjects at a faster rate than timetable 1 (at intervals of 5 compared to 10 minutes) and individuals are away from work for a shorter time (20 compared to 30 minutes) but has the potential disadvantage that four (instead of three) workers are away from their posts at any one time.

group results of biological measurements should be freely available to management and unions but the individual's results are confidential. A few subjects may need further investigation because of clinically abnormal results, such as raised blood pressure or low lung function. They must be offered advice and if the company employs a doctor, he could readily provide it. However, transmission of results to the company doctor may not be acceptable to unions, or to individual workers. The general practitioner is an alternative to whom abnormal findings may be referred. Each worker with abnormal findings must be asked individually if his results may be passed on and the reason for this explained to him. The potential conflict between confidentiality and the provision of medical advice needs sensitive handling and an agreed procedure. Although not strictly part of the survey, it may take almost as much time to perform, and to the individual worker is probably more important.

The provision of advice to the firm, on environmental control or medical surveillance, may be the proper concern of the investigators, if they have expertise in these areas. Otherwise advice may be given by the occupational health service or perhaps by the Health and Safety Executive (HSE) if the company does not have its own occupational health professionals.

Most surveys produce results of general interest which should be published. This may be an alarming prospect for management, who will be concerned that commercial secrets may be revealed or the company may receive adverse publicity. The protocol should state that the company's name need not be used in publications or presentations, and the company may see drafts so any commercial secret may be deleted or obscured by appropriate rephrasing. However the protocol must state that the company has no right of veto over publication and will not hold the copyright of published material.

Costs

Protocols submitted to funding organizations must contain detailed costings to justify the total sum requested. Even if no additional funding is sought, the protocol should contain an approximate estimate of the survey costs, including that for staff time, the use of equipment, stationery, laboratory and computer time and the cost of any travelling and accommodation. In collaborative surveys, costings provide a crude method of estimating the input of each participant department, if this seems important. Costings also focuses the mind on whether the questions, and the likely answers, justify the effort and time involved in doing the survey.

Practical plans

Detailed planning proceeds when the protocol has been drafted. Administrative and technical details must be arranged, such as rooms booked at the site, forms duplicated, equipment serviced and calibrated, supplies purchased and arrangements made for the transport of equipment and supplies. It is rare for industrial sites to have facilities for disposal of contaminated waste such as used needles, so appropriate containers must be supplied which must be taken away for incineration.

The investigators need not be the originators of the study, although greater continuity is achieved if they take part. The ideal field-worker is neutral, consistent and meticulous in following his instructions. Procedures will require training, but often do not require special technical qualifications or experience. Indeed, the well-qualified field-worker, such as a doctor, may be a liability, as he may have his own theories about the cause of the disease studied which may unconsciously bias his behaviour.

It is helpful to list every activity, however trivial, which each investigator must perform and to do a 'dry run' for several hypothetical subjects on paper or with colleagues acting as subjects. This is often revealing, suggesting that more rooms or more investigators are needed than was originally envisaged. A procedure initially assigned to one investigator may, on examination, break down into a series of tasks which contains clear switches from one mode of thought to another. It is better that he be bored doing repetitive tasks than hurried and liable to omit important items so the procedure might be better split into two or three and shared out among additional investigators. One field-worker should have enough time for liaison with the company about any problems arising at the survey.

Dates must be confirmed with all investigators and it is wise to have reserves in case of illness. Dates are then confirmed with the responsible person at the firm. Special arrangements may be necessary to cover night and weekend shifts. Some advance preparation may be necessary, such as the distribution of a questionnaire. Some firms prefer a fixed appointment system, which has advantages if many tests are to be performed or if the study group is scattered about a large site. A looser timetable, with groups arriving for survey at predefined intervals, will often suffice and is more flexible than individual appointments. Timetabling should ensure that environmental and biological measurements are made as closely together in time as practicable and all workers seen over the shortest possible time. Surveys involving several sites should also be close in time. This minimizes the effects of fluctuations in variables over time, such as changes in exposure across the week, or longer periods, or the effects of epidemic viral infections on biological measurements. This means a concentrated effort by the investigators.

The survey

The site team should arrive the night before or at least 1 h before the survey starts. Supplies should be checked and equipment, including any in reserve in case of breakdowns, checked and recalibrated if necessary. People handling blood (or other samples) should wear laboratory coats and any other protective clothing which they usually wear to prevent infection from blood-borne pathogens. No investigator wishes to be discourteous to subjects but pressures of time may mean he gives an unhelpful impression, which will reduce participation. All members of the team must be prepared to answer queries about the survey's aims and methods, though without revealing specific hypotheses, which may influence the subject's answers to questionnaires. Name-badges for investigators, notices on doors and chairs where subjects will wait all make the survey easier for the subject. If the planning has been thorough, the survey should go smoothly. The investigators' performance should be checked regularly to ensure each understands his instructions and is carrying them out, and equipment needs regular calibration. Problems should be referred to one person and close liaison maintained with whoever at the firm is coordinating the timetable.

Towards the end of the survey an approximate response rate should be calculated and the coordinator from the firm asked to establish which workers not yet seen are at work, so they can be encouraged to participate while the team is present. If a second survey is necessary to see non-respondents, a provisional date can be arranged. A poor response rate would suggest that planning was inadequate and the survey may need to be repeated.

Sufficient crude data-processing should be performed immediately after the survey for a preliminary report including at least an assessment of the response rate. Workers with clinically abnormal findings may be referred for advice, if this was not done at the survey. Any second survey of non-respondents and any laboratory tests on environmental and biological samples will be performed at this stage.

Analysis

For computer data-processing, the results will be coded into a computer-readable form and entered into a computer file. The accuracy of coding and entering must be checked by comparison with the raw data or by repeat coding of all or a sample of the results. Special programs may have to be written and checked for some analyses and some variables may need special handling, such as transformation. The analysis proceeds by asking specific questions and meetings of collaborators are helpful as it progresses. Florey and Leeder [10] discuss the documentation required, which is important, for personnel may change during a lengthy analysis. Finally, the results are sufficiently processed for publication. The experience of the survey, discussion of results and comments from other people almost always stimulate further questions, and possibly further surveys.

Acknowledgements

I am indebted to Professor Corbett McDonald's stimulating teaching and also to my students.

References

1. Medawar, P.B. *Advice to a Young Scientist*, p. 13. Harper and Row, New York (1979)
2. Burr, M.L. Medical memorandum: criteria for planning a research project. *Community Medicine*, **1**, 157–159 (1979)
3. Bradford Hill, A. *A Short Textbook on Medical Statistics*, 11th edition. Hodder and Stoughton, London (1984)
4. Colton, T. *Statistics in Medicine*. Little, Brown, Boston (1974)
5. Schlesselman, J.J. *Case-control Studies: Design, Conduct, Analysis*. Oxford University Press, New York (1982)
6. Porter, R. and Birch, J. (eds) *Ciba Foundation Study Group, no. 38.Identification of Asthma*. Churchill Livingstone, Edinburgh (1971)
7. Barker, D.J.P. and Rose, G. *Epidemiology in Medical Practice*, 3rd edition. Churchill Livingstone, Edinburgh (1984)
8. Bennett, A.E. and Ritchie, K. *Questionnaires in Medicine: A Guide to Their Design and Use*. Nuffield Provincial Hospitals Trust, Oxford University Press, London (1975)
9. Moser, C.A. and Kalton, G. *Survey Methods in Social Investigation*, 2nd edition. Gower, Aldershot (1971)
10. Florey, C. du V. and Leeder, S.R. *Methods for Cohort Studies of Chronic Airflow Limitation. WHO Regional Publications, European Series No. 12*. WHO Regional Office for Europe, Copenhagen (1982)

Further reading

Abramson, J.H. *Survey Methods in Community Medicine: An Introduction to Epidemiological and Evaluative Studies*, 3rd edition. Churchill Livingstone, Edinburgh (1984)

Chapter 16

Atmospheric monitoring

C.J. Purnell

Introduction

Occupational exposures in the workplace are rarely constant and will be subject to considerable variability during any one day depending on the type of equipment used, the nature of the process, control measures applied and the time spent in the areas. The situation is made even more complex by the action or behaviour of individual employees. This emphasizes the need to study the work practices involved from the receipt of the raw materials through to the final product including any maintenance and cleaning procedures which can, and often do, result in transient high exposures. Unlike dusts, gases and vapours do not separate into their components under gravitational forces so once vapour phase equilibrium has been achieved, gas mixtures do not separate out. Pollution sources are, however, usually localized and may be intermittent. In extreme cases, layering can occur instead of mixing. Consideration should be given to potential pollution sources, air speed and direction when setting up a monitoring programme.

In these circumstances the details of each operation and each individual's work practices can have an important influence on the overall exposure pattern. These observations permit the development of an appropriate sampling strategy which should aim to determine: whom to sample, when to sample, where to sample. How and with what to sample are often interrelated and dependent on the type of contaminant present and available monitoring resources. The cost of the measuring techniques can vary enormously from a few tens of pounds for the simple gas detector tube system to tens of thousands of pounds for sophisticated equipment such as portable mass spectrometers.

The criteria for effective evaluation of workplace exposure to hazardous substances normally involve three stages:

(1) Development of a sampling strategy for assessing workplace exposures.
(2) Selection of the most appropriate measurement equipment or sampling methods.
(3) Interpretation of the results and deciding on the basis of the results and observations whether a health hazard exists.

A sampling programme will therefore be required:

(1) To estimate personal exposure to hazardous substances.

(2) To assist in the assessment of the efficiency of engineering and process control measures.
(3) To demonstrate compliance with legal requirements.

A well-designed sampling strategy should provide information on all these aspects, especially when associated with informed observations on the nature and causes of exposure.

Inhalation is usually the most important route of entry of toxic substances into the body which places the emphasis of assessment on air sampling methods.

Air sampling methods

Air sampling in the workplace is required to assess individual exposures (personal sampling), to provide information on the effectiveness of control measures and to demonstrate compliance with relevant legal requirements. Sampling devices employed for such assessments must be able to collect sufficient material to permitt accurate analyis and, in addition, the sample taken must be representative of the conditions to which workers are exposed.

Particulate sampling methods

Inhalation is particularly significant with airborne particulate material. The size of the particles in common aerosols in a factory may vary from submicron to well over 200 μm in diameter. Of particular interest to the occupational hygienist are those particles which remain suspended in air long enough for them to be inhaled.

Respirable dust

Only particles below 10 μm aerodynamic diameter are capable of penetrating to the deep lung after being inhaled; this is known as the respirable fraction. The collection of this fraction is possible with a number of commercially available samplers such as the cyclone and simpeds which have been extensively used, for example when measuring concentrations of silica dust (respirable quartz) which tends to affect the alveoli in the lung and is capable of initiating fibrogenic disease.

Total dust

The dust fraction of particular interest here is the inspirable fraction of the total dust, that is, the fraction of the 'total' dust in the air that will be inspired by the nose and mouth and may have biological effect (*Figure 16.1*). Ideally the sampling devices used for 'total' dust measurement should be capable of reproducing the sampling characteristics of the human respiratory system. A considerable amount of research has gone into defining an inspirability curve but as yet there are no samplers on the market which give a good match to the developed curve. Performance data on the acailable dust sampling heads have shown that the seven-holed UKAEA and the open Gelman head give reasonable agreement but the single holed UKAEA head used for lead monitoring gives the poorest fit to the curve.

In practice particulate sampling is performed by drawing a known volume of air (by means of a pump) through the sampling device which incorporates a filter to

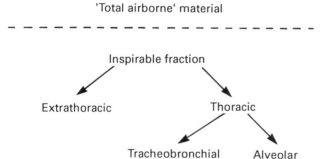

Figure 16.1 Deposition regions of dust in the lung

collect the material that enters the device. The filters used for particulate sampling are varied and include cellulose ester membrane, glass fibre or silver membranes depending on the type of analysis required. After sampling the filter is removed and subjected to analysis in the laboratory and the time-weighted average (TWA) concentration calculated.

Dust monitoring instruments

Direct reading instruments can be used to provide on-site dust measurements. Their size collection efficiencies are, however, different from those of the filter methods so a certain amount of judgement is required when using the results to assess health risks. They are particularly valuable for control purposes since they have a rapid response to changing conditions and hence can be used to identify peak emissions. They can be used as screening tools prior to deployment of filter methods.

There are three classes of instruments available. One collects dust particles electrostatically onto a piezoelectric quartz crystal vibrating at its natural frequency. As the crystal increases in mass due to the deposited dust its resonant frequency will change and this change in frequency is used to indicate mass concentration. In another instrument, particles are impacted onto a grease-coated mylar film. The mass of particles is proportional to the attenuation of beta-radiation from a carbon-14 source. The third type of instrument operates on the light scattering principle. Particle laden air is drawn into the instrument through a beam of infrared light. The particles in the air stream scatter the light, this being detected by a sensor. The amount of scatter is proportional to the particle concentration.

Light scattering instruments such as SIMSLIM and the MINIRAM are particularly good (and most expensive) and have outputs for connection to data storage systems which permit a continuous record of concentration to be obtained.

Gas and vapour sampling methods

Analytical techniques exploit some physical or chemical property of the substance being measured and since gases and vapours are more reactive than dusts they invariably lend themselves to a wide range of detection methods and hence many

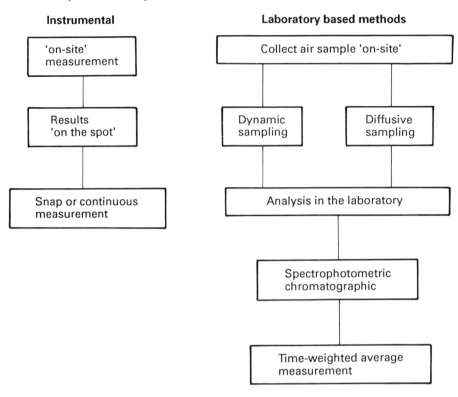

Figure 16.2 Measurement techniques for gases and vapours

different methods are available for measuring the same compounds. In these circumstances the choice of measurement technique depends on the contaminant of interest, whether the equipment needs to be potable, the concentration range, and the sensitivity and specificity of measurement required.

Some instruments are capable of providing a direct readout of concentration 'on-site' and a continuous output of concentration information with time whereas laboratory-based methods provide a retrospective time-weighted average (TWA) result: these can also be obtained from averaging a series of short term measurements or from continuous measurements. These normally involve some kind of sampling method and analysis of the collected material in a laboratory. Techniques for the measurement of toxic gas and vapour concentrations in air can therefore be classified as either instrumental or laboratory based sampling methods (*Figure 16.2*). The final choice of method depends to a large extent on available resources and degree of expertise available for carrying out the analysis.

Gas detector tubes

This method involves drawing a fixed volume of air through a glass tube containing reagent impregnated silica gel which changes colour when the gas or vapour of interest passes down the tube. The concentration of the substance to which the indicating layer is sensitive is then determined by either length or colour change or by matching the colour generated to a standard colour in another part of the tube. The magnitude of the colour change is proportional to the concentration. Colorimetric gas detector tubes are used for short term (snap) measurements. They are available for a wide range of gases and vapours with some manufacturers listing over 200 different tubes for various contaminants, mixtures and concentrations. The system is cheap and simple to use and able to provide a rapid and convenient means of determining the concentration of gases and vapours in the work environment and it may be used by semiskilled operators. Detector tubes are not suitable for measuring personal exposure and are used mostly for area or background measurement since they are 'hand held'.

Although simple, however, care needs to be taken in their use if misinterpretation of the results is to be avoided. The accuracy of the measurement is controlled by a number of factors many of which are standardized at the time of manufacture, including tube size, packing density and adsorbent/reagent system. Each batch of tubes is subjected to a rigorous quality control programme by the manufacturers prior to being released for sale. This testing includes calibration of a selection of tubes from each batch against a standard atmosphere of the particular gas or vapour. The tubes once calibrated are boxed and date-stamped at the time of production. Calibration is made for a particular pump and tube system since pumps from different manufacturers operate in different ways. The pump-tube systems are therefore not interchangeable, and the pump of one manufacturer cannot be used with the tubes of another since this will invalidate the calibration and erroneous readings may result.

Because of the nature of the adsorbent/reagent systems the tubes have a limited shelf life after which they should be discarded. This shelf life is normally in excess of 2 years at an ambient temperature of 20°C. The shelf life of the tubes is increased if they are stored at lower temperatures and it is normally recommended that they are stored in a refrigerator; they should be allowed to warm to room temperature before being used. It follows that shelf life will be reduced at higher temperatures so the tubes should not be left on top of heated radiators, or cars boots and window ledges on hot summer days.

Most of the chemical reactions used to generate colour in the tubes are not specific and will be subject to interference from other similarly reacting gases or vapours which may also be present in the atmosphere being sampled. These interferences may have a positive or negative effect on the detecting system. Most of the potential interferences are known to the manufacturers of the tubes and attempts are made in the design of the tubes to reduce their effects (by adding a prelayer to trap the interfering compounds, for example). Interference information is detailed in the data sheets supplied with each box of tubes. It is important to read these before using the tubes for taking a measurement particularly where co-contaminants are likely to be present in the atmosphere.

Finally, the tubes are calibrated to a fixed volume of air drawn through the tube. Any change in this volume will affect the reading given by the tube. It is essential, therefore, that the pump performance is checked and the volume of air it draws through a tube is checked *each* time it is used. Gas detector tubes are useful for obtaining a picture of the order of magnitude of gas or vapour concentrations at a particular time but they are not always specific and are prone to errors if not used in accordance with the manufacturer's instructions (which are different for each tube). Provided these simple precautions are followed the detector tube system may be used with confidence to measure a wide range of gases and vapours in the workplace.

Indicator tubes as described above are suitable for short-term sampling where the measurement of concentration is normally completed in a few minutes. Recent developments of the system have resulted in the production of long-term detector tubes for measuring time-weighted average concentrations up to 8 h. The pump, in this case, is battery operated and capable of drawing air through the tubes at a constant flow rate—normally 20 ml/min. The pump and tube is worn by the worker throughout a working shift, The strain length produced is directly related to the concentration of the gas or vapour and the volume of air sampled through the tube from which the TWA concentration can be calculated.

Gas detecting meters

PAPER TAPE DETECTOR SYSTEMS

These detecting meters are based on colour changes resulting from the reaction between a particular atmospheric pollutant and a chemical reagent-impregnated paper tape through which the atmosphere is passed. A fibreoptic detection system is employed to measure the absorbance difference of the reacted and unreacted paper which in turn is converted electronically to a direct reading of concentration of the gas or vapour concentration being sampled. The paper tape monitors are suitable for area monitoring, and recent advances in design have enabled the system to be miniaturized for use as personal monitors. Readings can be obtained on-site and the instruments have alarms to provide warning of hazardous concentrations. Some devices have microprocessors which allow storage of time and concentration data for eventual printout.

The tape systems either move continuously or are stepped such that a reading can be taken every few minutes. In the latter case the tape is held in the air stream for a fixed period and then moved on to the detector where a reading of concentration is taken and either displayed or stored for future electronic retrieval. Continuous personal tape monitors are also available which are capable of operating over an entire working shift. The tape can provide a time-weighted average (TWA) exposure and a complete history of the concentration to which the wearer has been exposed.

Paper tape monitors are available for a range of reactive contaminants such as vinyl chloride, phosgene, isocyanates, aromatic amines, ammonia, hydrogen sulphide and sulphur dioxide.

ORGANIC VAPOUR ANALYSERS

Flame ionization detection (FID) This type od detector is commonly employed in gas chromatography. A detector based on FID is capablen of responding to all substances which produce ions on combustion and is therefore ideally suited to the detection and quantification of organic gases and vapours. Detection meters for field use are normally portable continuous monitors responding to the organic contaminant mixture as a whole and offering little specificity. Some devices incorporate a small chromatographic system as part of the unit to improve specificity for some applications but the system can no longer be regarded as a continuous monitor when operated in this mode. The detection capability of this type of instrument is excellent with a linear response being obtained over a wide range of concentrations. In addition the units can be made intrinsically safe for use in areas which require the use of flameproof equipment.

The rapid response and sensitivity of the FID is of particular value for the assessment of on-site exposure problems and of establishing the effectiveness of the ventilation control measures. Hand-held devices are available which operate from self-contained batteries and lightweight hydrogen cylinders.

Photo-ionization detectors (PID)   These operate in a similar way to FID detectors except that ionization is achieved through the use of ultraviolet radiation. Ions produced are measured in a similar way to the FID detector. Only those compounds which have ionization potentials below the energy of the irradiating source will be detected and consequently the devices offer some enhanced specificity over their FID counterparts.

Photo-ionization has been made the basis of a portable detector. Both FID and PID devices can be used to measure very low concentrations of organic vapours. particularly solvents.

Mass spectrometry   Mass spectrometers are very sensitive and highly specific. Some attempts have been made to develop units for on-site use but these units are best described as transportable rather than portable and hence tend to be used as fixed point monitors for specific applications.

Heat of combustion   This technique measures only combustible gases. The sampled atmosphere is passed over a heated catalytic sensor which forms part of an electrical Wheatstone bridge. The heat of combustion of the compound by the catalyst causes an imbalance in the electronic circuit due to a change of resistance of the sensor filament which is proportional to the concentration of the organic vapour passing over the sensor. The technique is non-specific and will respond to a wide range of organic vapours and is particularly useful for monitoring explosion hazards or as a general 'sniffer' for leak detection purposes.

Thermal conductivity detector   This detector is based on the principle that loss of heat from a heated wire filament is dependent on the specific heat of the gas mixture passing over it. This forms the basis of a simple, non-specific gas and organic vapour detector.

CHEMICAL AND ELECTROCHEMICAL REACTION DETECTORS

A very wide range of detectors is available for gases and vapours which undergo chemical or electrochemical reactions to produce electrical signals proportional to their concentration in the atmosphere.

Chemiluminescence detectors have been developed for compounds which undergo chemical reactions that give out light the intensity of which is dependent on their concentrations. Highly specific detectors have been developed for ozone and nitrogen oxides.

In electrolytic reaction systems, the molecules of the gas to be measured are absorbed into an electrolyte where they dissociate and take part in an electrochemical reaction to generate a small electrical current. The magnitude of the current is proportional to the number of molecules that undergo reaction. A wide range of detectors is available and they lend themselves to miniaturization and hence use as personal dosimeters. Examples of gases which can be measured include oxygen, carbon monoxide, sulphur dioxide and nitrogen dioxide.

OPTICAL MEASUREMENT DETECTION SYSTEMS

Both infrared (IR) and ultraviolet (UV) detection systems are available whereby the air containing the gas or vapour of interest is passed into a gas cell. The amount of radiation absorbed by the gas or vapour is proportional to the concentration.

A portable infrared gas analyser utilizing a multi-pathlength cell with an absorption pathlength of up to 20 m offers high sensitivity for certain applications in gas and vapour monitoring. A filter monochromator provides the system with selectivity of operation and a microprocessor-controlled instrument is available which permits up to 20 different wavelength settings to be sequentially monitored. An ultraviolet meter is available for the measurement of mercury vapour.

SOLID STATE SENSORS

These devices are based on non-stoichiometric metallic oxides whose electrical properties depend on the atmosphere which surrounds them. They are relatively non-specific although selectivity can be improved by chemically doping the semi-conductors. Semiconductor sensors are incorporated in a number of instruments. These devices can be miniaturized and this makes them ideally suited for use as personal dosimeters.

As can be seen from the above discussion many different types of reading instruments are available. *Table 16.1* describes the 'ideal' monitor for occupational hygiene purposes but unfortunately none of these instruments meets all the requirements. Most are bulky and hence not suitable for personal monitoring although some attempt to miniaturize has been made in recent years. All require periodic calibration and all are expensive.

They are useful for acute acting substances since they are able to give an immediate warning of hazardous concentrations and they can be used to obtain continuous measurement of fluctuating concentration. Some examples of different types of metres available are given in *Table 16.2*.

Laboratory based methods

These methods involve taking a known volume of air into a sampling device by means of a personal sampling pump. The samplers are returned to the laboratory for analysis of the collected material and a retrospective time-weighted average (TWA) concentration calculated.

GAS PHASE SAMPLING

This method involves direct sampling into containers such as glass syringes, plastic bags and evacuated metal cans. The technique is usually used for short-term (snap) sampling but longer term samples can be taken by using a pump to fill the container or by the use of critical orifice or needle values for vacuum-filled containers. Inert

Table 16.1 The 'ideal' occupational hygiene monitor

Sensitive
Wide linear range
Specific response
Continuous measurement with facility to calculate TWAs for up to 8 h
Audible and visible alarm when dose or hazardous level exceeded
Stable calibration
Small, portable and lightweight
Cheap

Table 16.2 Instrumental monitoring methods for gases and vapours

Type of monitor	Applications
Catalytic combustion	Flammable gases
Chemiluminescence	NO_x, ozone
Colorimetry	NO_x, SO_2, H_2S, NH_3, aldehydes
Electrochemical	NO_2, SO_2, O_2, H_2S, NH_3, CO, hydrides, halogens, HCN, $COCl_2$, AsH_3, PH_3
Flame ionization	Total hydrocarbons
Infrared analyser	Organic vapours
Paper tapes	H_2S, phosgene, NO_2, SO_2, NH_3, aliphatic amines, hydrazine, Cl_2, HCN, aromatic isocyanates
Thermal conductivity	Organic vapours
Ultraviolet	Mercury vapour

containers are required if chemisorption and adsorption losses are to be avoided. The technique is only suitable for non-reactive gases and some organic vapours with a boiling point below 80 °C. Permeation losses can occur with some plastic containers especially with lengthy storage times. Examples of gases sampled in this way include methane, carbon monoxide, carbon dioxide (mine gases). Aliquots of the collected gases and vapours are then removed for subsequent gas chromatographic analysis.

LIQUID ABSORPTION SAMPLING

This sampling method involves the use of bubblers or impingers containing an absorbing reagent solution to collect the gas and vapour of interest. The reagent reacts with the contaminant to form a stable compound. Subsequent analysis is carried out using an appropriate analytical method, normally spectrophotometry or chromatography.

Most bubblers and impingers are made of glass and are prone to breakage or spillage although non-spill versions are available. Sintered glass bubblers theoretically offer the highest efficiency due to smaller bubble size and hence potentially greater transfer of gas to liquid occurring at the gas–liquid interface of the bubble. Gases and vapours have to diffuse across the bubble to reach the gas–liquid interface before they can be absorbed so the smaller the bubble size the less distance the molecules have to diffuse to reach this surface. Dissolution into the absorbent medium is then very rapid. Combination of bubble size, depth of liquid and hence time taken to reach the surface of the liquid in the impinger or bubbler are also important considerations in overall efficiency of sampling which is often less than 90 per cent. Fine frits may become blocked by particulates and so impingers with open tips tend to be used for this application. If the reagent is dissolved in an organic solvent then losses of the solvent will occur through

Table 16.3 Colorimetric liquid absorption methods

Analyte	Sorption/reagent solution
Acrolein	α-resorsinol
Ammonia	Sulphuric acid/Nessler reagent
Arsine	Silver diethyldithiocarbamate in pyridine
Formaldehyde	Sodium acetate/chromotropic acid
Hydrogen sulphide	Cadmium sulphate
Nitrogen dioxide	Saltzman reagent
Ozone	Potassium iodide in KOH solution/starch
Phosphine	Silver diethyldithiocarbamate in pyridine
Sulphur dioxide	Sodium tetrachloromercurate/formaldehyde/p-rosaniline
Aromatic isocyanates	Acid hydrolysis/N-2-amino-ethyl-1-naphthylamine (NED)

evaporation and frequent topping up with the solvent will be required for sampling times longer than 30 min; in practice, solvent evaporation effects often become the predominant factor.

This type of sampling method is most used for sampling inorganic gases such as HF, HCl, SO_2, NH_3, O_3 and H_2S or reactive organic vapours such as formaldehyde, acrolein or isocyanates. Reagent systems are normally specific. A list of appropriate systems is given in *Table 16.3*.

SOLID SORBENT SAMPLING

This is the most common method for measuring time-weighted average (TWA) personal exposures to organic vapours. The method involves drawing a known volume of air through a tube containing an adsorbent (usually charcoal or a porous polymer) by means of a personal sampler pump. Flow rates between 20 and 200 ml/min are normally used. Charcoal is the most widely used adsorbent and is applicable for sampling a wide range of organic vapours. After sampling, the tubes are returned to the laboratory where the charcoal is desorbed with carbon disulphide (CS_2), analysed by gas chromatography and the TWA concentration calculated. The maximum volume of air that can be sampled through the tube varies depending on the nature of the organic vapour and its concentration in the air. If this volume is exceeded, saturation of the charcoal can occur leading to erroneously low results. Manufacturers of the tubes normally supply information on the maximum volumes that can be sampled for particular vapours (breakthrough volumes).

The breakthrough volume on such samplers is defined as the sample air volume above which sufficient quantities, usually 10 per cent, of the sampled vapour begin to appear in the exit stream. This is dependent on the bed-depth over and above a critical bed-depth, below which there is instantaneous breakthrough. Charcoal adsorption tubes normally have two sections both of which are analysed after sampling has been completed. If more than 25 per cent of the total material collected on the tubes is found in the back-up section of the tube then the tube is deemed to have been overloaded.

Adsorbents may be classified as strong or weak. Strong adsorbents such as active charcoal have a high surface area and many adsorbent pores. Strong adsorbents might be thought of as the first choice in sampling, especially as weaker adsorbents (such as porous polymers) have lower surface areas, fewer pores and show essentially linear isotherms. This results in an equilibrium vapour pressure above the adsorbent which increases with the quantity adsorbed leading to a progressive reduction in uptake rate with dose (exposure concentration × time) in some diffusive samplers. This is especially noticeable with the more volatile analytes with a boiling point below 90 °C. However, strong adsorbents require solvent desorption using toxic and flammable organic solvents such as CS_2 whereas weak adsorbents such as porous polymers may be thermally desorbed which has advantages in both manipulation and desorption efficiency.

Active charcoal generally contains surface oxygen groups which allow the chemisorption of water molecules due to hydrogen bonding. The surface becomes hydrophilic and saturation rapidly occurs around 80 per cent relative humidity. This effect is competitive with organic vapour adsorption and leads to a reduction in capacity at high humidities. Charcoals may be chemically tailored to remove oxygen groups and they are then hydrophobic up to 90% relative humidity.

Alternatives to charcoal are porous polymer adsorbents (such as Tenax GC, PoraPak) which utilize thermal rather than chemical desorption of the collected material. This procedure eliminates the use of CS_2 which is toxic and flammable, but the equipment for desorption of the tubes is more expensive than the conventional equipment used for the analysis of charcoal tubes.

Thermal desorption offers the principal advantage that one may analyse the whole sample and not merely a small fraction of it. Thermal displacement as opposed to solvent elution will lower the limit of detection of the pollutant by a factor equal to the volume fraction taken for gas chromatography (GC) analysis.

Direct displacement onto the gas chromatography column does have a disadvantage in that it offers the analyst only one opportunity to analyse each thermally displaced sample unless an intermediate reservoir is employed. The use of a reservoir, however, reduces the intrinsic sensitivity of the method and moreover may introduce residual adsorption problems.

Diffusive samplers

The methods described in the previous section all require the use of a pump to draw a known volume of air through liquid reagent impingers or adsorbent packed tubes. In recent years an alternative to this sampling system has been developed in the form of diffusive samplers. These devices sample by gaseous diffusion onto an adsorbent thereby avoiding the need for a pump and thus offer a simpler means of measuring time-weighted average (TWA) concentrations of gases and vapours in air; their greater 'wearer' acceptability and the fact that they are safe to use in areas requiring the use of flameproof equipment are added attractions.

Fick's law governs the uptake rate (J) in terms of the diffusion coefficient gradient (dc/dx) such that:

$$J = -D \, (dc/dx)$$

where J = nM/cm^2/s
       D = cm^2/s
       (dc/dx) = nM/cm^4

In a diffusive sampler x is assumed constant and equal to the diffusion air gap (L) between the diffusive surface of the sampler and the adsorbent. It is also assumed that the adsorbent acts as a zero sink and that the concentration of the analyte at the face of the sampler is equal to the ambient concentration (c). It follows that the amount collected (Q) on the adsorbent of cross-sectional area (A) in time (t) becomes:

$$Q = D\,(A/L)ct, \text{ so that } Q = nM$$

The uptake rate of the sampler, $D(A/L)$, is dimensionally equivalent to a volume flow rate, although there is no net movement of air. Diffusion coefficients can be determined experimentally or derived from published values.

In general, provided a suitable matching of sampler design and adsorbent is chosen, diffusive samplers can be used to monitor a wide range of contaminants, both organic and inorganic and many different types are now available. The most commonly used sorbent for organic vapours is charcoal with quantification of the collected material by gas chromatography. However, although this analytical procedure is convenient and applicable to a wide range of vapours, it suffers from a number of disadvantages, principally the use of a desorbing solvent (CS_2) and dilution of the sample prior to analysis. The experience gained with the conventional pump and tube method has shown that a porous polymer/thermal desorption method can offer distinct advantages over charcoal/solvent desorption, since the method not only eliminates the solvent but vastly increases sensitivity. This in turn lead to the development of 'heat' desorbable diffusive samplers, including a 'pen' type version based on the conventional porous polymer adsorbent tube.

Liquid colorimetric reagents have been used in various designs of diffusive sampler for inorganic gases. This concept has been extended to include reagent impregnated papers as adsorbents which are easier to handle than liquid reagents and give a direct, visible colour stain. This type of diffusive sampler is particularly attractive as the stain produced may be quantified immediately in the factory by reference to a stain chart and confirmed subsequently in the laboratory by an appropriate analytical technique.

Diffusive samplers have engendered some debate as to their effectiveness and suitability in occupational hygiene monitoring. It has been shown that, provided a suitable combination of diffusive sampler type and sorbent medium is selected, diffusive samplers can be used with confidence; in many cases they have accuracies comparable with the more established dynamic sampling methods over the concentration range normally encountered in occupational hygiene monitoring and for use with a wide range of possible contaminants.

Diffusive sampling systems have enormous potential in occupational hygiene monitoring since they provide a convenient and cheap means of assessing occupational exposures in a large number of industries. It is possible to sample a much larger number of the workforce than is currently available with dynamic sampling methods because of the cost implications involved in the use of a large number of pumps. Diffusive samplers can also help provide an effective occupational hygiene monitoring database for future epidemiological studies.

Choice of monitoring system

Several alternative methods of sampling are usually possible. For instance, styrene could be sampled in a bubbler with colorimetric analysis, onto a charcoal or Tenax tube, onto a diffusive sampler, or be measured by an instrumental method such as infrared, ultraviolet or FID continuous monitoring. The choice of method will depend to a large extent on the chemical properties of the analyte of interest, the acceptability of the sampling equipment, the length of sampling period and the sensitivity of the analytical or detection system.

Very often the choice of sampling method is governed by its cost. For personal monitoring, the sampling strategy is simpler in that a continuous integrating device may be fitted in the individual's breathing zone to give a time-weighted average (TWA) of the exposure over a given period. The determination of short-term exposure variations ideally requires continuous monitoring capable of infinite resolution if instantaneous exposure maxima or minima are required. Personal samplers give only an approximation of the individual's actual exposure since the sampling device can never be in the actual breathing zone without also being in the expired air zone and hence a sensible compromise is required.

The sampling strategy for a given location will be subject to certain physical constraints. Potentially flammable atmospheres demand the use of intrinsically safe sampling or monitoring equipment. Heavy or bulky sampling equipment cannot be used for personal sampling or in restricted spaces or specialized working areas such as operating theatres.

Chapter 17

Analytical methods

R.J. Brown

Introduction

The techniques used to analyse samples taken in the occupational hygiene field cover nearly all the technique available to the analytical chemist. Obviously this chapter will not be able to cover such a broad subject in detail but will look only at the main techniques which are used in most analytical laboratories for routine analysis, with a brief outline of some of the more powerful analytical instrumentation available in major research centres.

Colorimetry

Colorimetry is one of the older techniques in analytical chemistry, having been in use for over a century. It involves the comparison of the colour density of an unknown quantity (your sample) against that of known standards. This used to be done by direct comparison by eye but now sensitive instrumentation has taken the place of the analyst's eye.

For use in making accurate measurements a coloured system must obey *Beer's law*.

$$A = \epsilon c l$$

where A is the absorbance of the solution
          ϵ is the molar absorptivity at a particular wavelength
          c the concentration of the solution being measured
          l the pathlength of the solution.

ϵ is a constant for a particular compound and l is usually constant for a particular instrument. This then gives a straight line relationship between the absorbance and the concentration of the solution; by making up a range of solutions of known concentrations and measuring their absorbance a graph of concentration versus absorbance can be obtained. This can then be used to calculate the concentration of an unknown solution once its absorbance has been measured. To maximize the sensitivity of the technique the absorbance of the solution is measured at λ_{max}, the wavelength at which maximum absorbance occurs; λ_{max} is different for different chromophores.

A chromophore is the section of the molecule that actually absorbs the incident light. Absorption takes place due to electronic transitions within the molecule. The types of transitions that are easily detectable in the visible and near ultraviolet spectra come from molecules containing any of the following chromophores:

>C=C<   −C≡C
−NO$_2$ −NH$_2$
−N=O   −N=N−
>C=O   −Br
−COOR   −COOH
−CHO

If a molecule has two or more chromophores the absorption of the molecule will be given by the sum of the absorptions of the individual chromophores if they are separated by two or more single bonds. On the other hand, conjugated chromophores show both enhanced absorption and a shift in λ_{max}. This is because the electrons in these bonds are able to move freely between all the conjugated bonds.

The following is an example of unconjugated chromophores:

>C=CH$_2$−CH$_2$−CH$_2$−CH=C<

The following is an example of conjugated chromophores:

>C=CH−CH=CH−CH=C<

A typical spectrophotometer will consist of a light source, a sample holder, a monochromator and a detector. The light source for the visible part of the spectrum is usually a tungsten lamp while a deuterium lamp is used for the near ultraviolet. The sample holders consist of glass or quartz (for measurements in the ultraviolet) cuvettes made to very high tolerances to give well-defined pathlengths. The monochromator will be either a prism in older instruments, or a diffraction grating in more modern ones. The detector is usually a photomultiplier tube (*Figure 17.1*).

It is also possible to use the infrared region of the spectra to analyse samples. Molecules absorb radiation in this region by molecular vibrations and rotations

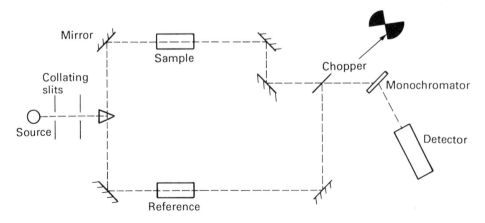

Figure 17.1 Schematic diagram of ultraviolet/vis spectrophotometer

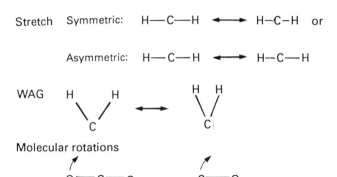

Stretch   Symmetric: H—C—H ⟷ H–C–H   or

Asymmetric: H—C—H ⟷ H--C—H

WAG   (figure)

Molecular rotations

Figure 17.2 Molecular vibrations and rotations

m₁

m₂   Figure 17.3 Spring–mass system

(*Figure 17.2*). To absorb radiation in this region the molecule must undergo a change in dipole moment during the vibration. Specific chromophores will absorb specific wavelengths. However, unlike absorption in the visible and ultraviolet region, infrared chromophores do not become conjugated; that is two adjacent chromophores do not join together to form a new system with a different wavelength of absorption. The frequency of vibration of a molecule is analogous to a spring–mass system, the masses being the atoms and the spring being the bond that joins them (*Figure 17.3*).

The frequency of vibration of a system is given by the following equation:

$$v = \frac{\pi}{2} \, c\sqrt{(k/u)}$$

where v = the frequency of bond vibration
          c = the velocity of light
          k = the force constant of the bond
          u = the reduced mass of the atoms involved.

The reduced mass is defined as

$$u = \frac{m_1 \, m_2}{m_1 + m_2}$$

where m_1 = the mass of the first atom
           m_2 = the mass of the second atom.

In practice the frequency of absorption is expressed in wavenumbers, with units /cm. This is directly proportional to frequency and the force constant.

$$\bar{v} = \frac{v}{c} = \frac{\pi}{2} \sqrt{(k/u)}$$

where $\bar{v}$ = wave number.

So, for example, a simple molecule such as methane would only give C–H vibrations, whereas a molecule such as trichloromethane would give C–H and C–Cl vibrations. The information given by infrared spectra can be used in two ways. First, as a qualitative device to give an idea of the possible identity of a compound by finding out what functional groups are present from their characteristic absorption frequencies. Second, to quantify compounds containing similar features, such as all the ketones present by monitoring the carbonyl stretch at 1600/cm or chlorinated compounds by monitoring the C–Cl stretch at 800/cm.

The basic design of an infrared spectrometer is the same as an ultraviolet/vis spectrophotometer, a source of radiation, a sample cell, a monochromator and a detector. The radiation source is usually a heated metal or metal oxide. The cells are made from salt crystals, usually sodium chloride. For special applications where large amounts of water vapour are present, silver bromide crystals are used because of their insolubility in water; however these are much more expensive than normal salt cells. The monochromators are gratings. The detectors work on a variety of principles which involve detecting changes in a physical property of the detector caused by temperature changes when infrared radiation is absorbed.

Atomic spectroscopy

Atoms can only absorb or emit light when transitions in its electron energy levels occur (*Figure 17.4*). These transitions are discrete or quantized which means that only certain wavelengths of light will be absorbed or emitted by each element. Generally, two different elements will not absorb or emit radiation of the same wavelength. At room temperature the elements are found in their ground state. The resonance wavelength of an element will be the wavelength of absorption or emission between the ground state and the next accessible energy level.

The resonance wavelength gives the strongest absorption or emission for a particular element, so will give the lowest detection limits for that element. Other wavelengths can be used if the highest sensitivity is not required. To be able to use atomic absorption or emission as an analytical technique, the element of interest has to be in the atomic state, otherwise the electrons in other atoms in the same molecule would disturb the electron energy levels of the element of interest causing it to absorb or emit at wavelengths different from those expected when in the atomic state. Whether the technique used is *atomic absorption* or *atomic emission spectroscopy* depends upon the proportions of the electrons in the ground and excited states. If the higher proportion of electrons are in the ground state absorption will predominate, if they are in the excited state then emission will predominate. This can be calculated from the Maxwell–Boltzmann equation.

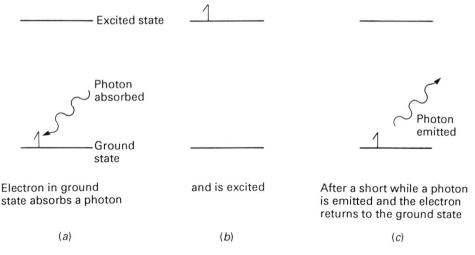

Electron in ground   and is excited   After a short while a photon
state absorbs a photon   is emitted and the electron
                                                    returns to the ground state

(a)   (b)   (c)

Figure 17.4 Electron energy levels with absorption and emission of photons

$$\frac{N_u}{N_o} = \frac{g_u}{g_o}\ e(E_u - E_o)/kT$$

where N_u is the number of atoms in the excited state
         N_o is the number of atoms in the ground state
         E_u is the energy of the excited state
         E_o is the energy of the ground state
         g_u and g_o are the probabilities that an electron will be in a particular
         energy level
         T is the absolute temperature in kelvins
         k is the Boltzmann constant.

For a given element the only controllable variable in the above equation is the temperature. So the choice between atomic absorption or emission can be made by altering the temperature of the analysis. The most common method for producing atoms is by using a flame to dissociate the molecules. Various gases are used to produce flames of differing temperatures; these are listed in *Table 17.1*.

The sample (as a liquid) is introduced into the flame by a nebulizer (*Figure 17.5*). This device breaks the liquid up into a very fine aerosol which is carried into the flame by the fuel and oxidant gases. This aerosol can be dissociated to atomic vapour in the short time it is present in the flame.

Table 17.1 Gases used to produce flames at differing temperatures

Fuel gas	Oxidant gas	Flame temperature (°C)
Propane	Air	1725
Hydrogen	Air	2045
Acetylene	Air	2250
Hydrogen	Oxygen	2677
Acetylene	Nitrous oxide	2955
Acetylene	Oxygen	3060

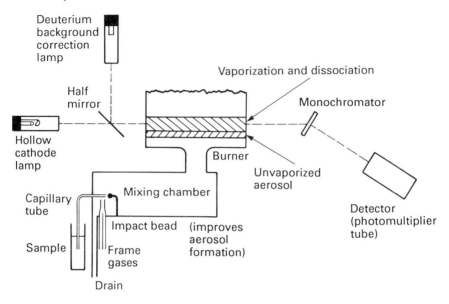

Figure 17.5 Nebulizer and burner

Graphite furnace

Another technique for producing atomic vapour is the graphite furnace otherwise known as *carbon furnace* or *electrothermal atomization*. This technique involves placing a small volume of the solution of interest into a graphite tube. This tube is heated electrically to the required temperatures. However, unlike a flame this can be set to precisely the required temperature. The heating process usually consists of three stages, drying, ashing and atomizing. After the solution has been placed in the tube it is initially heated to around 100 °C to dry the solution, then the temperature is raised to between 250 °C and 1500 °C to ash any extraneous material; finally the temperature is raised to between 1800 °C and 2700 °C to atomize the remaining material, the temperature chosen depending on the nature of the matrix and the element of interest. The amount of the element of interest present is quantified by atomic absorption spectroscopy. This technique is much slower than using flame atomization but it does have several distinct advantages. It is much more sensitive than flame (up to × 1000) and can analyse fluids such as blood and urine with very little pretreatment, but which would normally block the nebulizer and burner of an instrument using the flame technique.

Cold vapour atomic absorption spectroscopy

Mercury is the only metal that is liquid at room temperature. It also has a high vapour pressure at room temperature which enables it to be analysed without need of any form of heating. This technique is called cold vapour atomic absorption spectroscopy (*Figure 17.6*). Any mercury compounds present in solution are reduced to mercury metal using a reagent such as stannous chloride. Air is passed through the solution and carries any mercury present to a flow cell in the spectrometer where its concentration is measured.

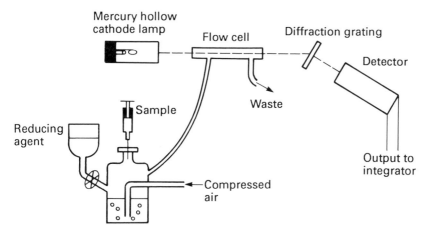

Figure 17.6 Cold vapour atomic absorption spectroscopy

This technique can be used for a wide variety of samples, including air, blood, urine and water. The major instrumental difference between atomic absorption and emission is that atomic absorption requires a light source to be absorbed by the atomic vapour. Since the bandwidths of atomic spectra are very small (of the order of 0.002 nm) it is not practical to use a continuum radiation source such as a tungsten or deuterium lamp as the energy supplied at each atomic line would be very small giving poor sensitivity. Instead a line radiation source is used. This employs the element of interest as part of the lamp, most commonly in the cathode.

These lamps only emit radiation specific to the particular element in the lamp. This gives rise to high intensity line sources which will give good sensitivities and low detection limits. The disadvantage with this technique is that a different lamp is required for each element, and only one element can be analysed at a time.

Background correction

When the sample is introduced to the flame or atomized in the graphite furnace not only atomic absorption takes place but also molecular absorption and light scattering. These processes are not predictable but would be added to the atomic absorption signal if not accounted for by a procedure known as background correction. This is achieved by measuring the background absorbance either side of the atomic wavelength. The mean of these two measurements is then subtracted from the measurement made at the atomic wavelength to give the true atomic absorption. To be able to undertake these measurements for all the elements a continuum source is required, usually a deuterium lamp. This is passed through the atomic vapour at the same time as the light from the elemental lamp. There are several other methods of background correction but the deuterium lamp method is the one in major use today.

Chromatography

Chromatography involves the separation of a mixture of substances followed by their detection and quantification. In both chromatographic techniques described here the components eluted are identified by their retention time, that is, the length of time it takes for the components to travel down the column from the injection port to the detector. The retention times of the eluted components are compared with those of standard materials prepared from pure materials; if they are the same the unknown component will have been identified. The hygienist can provide useful information to the chemist on the possible identity of vapours and gases present by looking at the various substances used in the production process. It is impractical for the analytical chemist to run several hundred standards to try to identify the vapours that are present in the sample.

Gas chromatography

Gas chromatography is used to analyse volatile solvents and gases. The apparatus (*Figure 17.7*) consists of:

(1)   A heated injection port for injecting the sample into the gas stream (*mobile phase*).
(2)   A column containing the *stationary phase* is contained inside a thermostated oven.
(3)   A detector for monitoring the effluent from the column; this will normally be connected to an integrator and chart recorder.

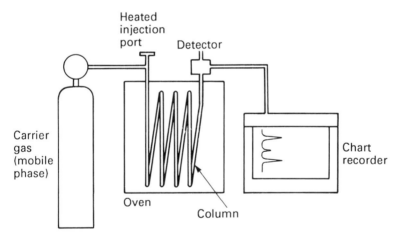

Figure 17.7 A gas chromatograph

The separation of the various components in a mixture of solvents takes place because of the different affinities each component has for the mobile and stationary phases. Heating the column would decrease the affinity of all the components for the stationary phase and increase the speed of analysis; however, this reduces the capacity of the column to resolve very similar compounds. Conversely if the column temperature is low the resolution is much improved but the time of analysis is longer (*Figure 17.8*).

The mobile phase is a gas, usually nitrogen but it may also be helium or argon. All these gases should be very pure and contain no trace of oxygen or water since

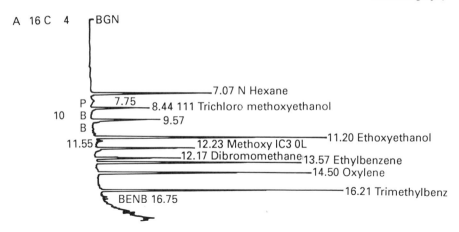

A 16 C   4   BGN

10

P   7.75
B
B

11.55

————————7.07 N Hexane
—8.44 111 Trichloro methoxyethanol
—9.57

—————————11.20 Ethoxyethanol
————————12.23 Methoxy IC3 0L
———12.17 Dibromomethane 13.57 Ethylbenzene
———————————14.50 Oxylene
————————————16.21 Trimethylbenz

BENB 16.75

Figure 17.8 Typical chromatogram from modern chromatograph

these degrade the materials used to make the column, shortening its useful life. The stationary phase consists of either a solid, such as a porous polymer, or a very viscous liquid, which is usually coated onto an inert support. There is a wide variety of these stationary phases, each designed for a particular use.

Packed and capillary columns

There are two basic designs of column, the packed column and the capillary column. Commonly a packed column has an internal diameter of 2–4 mm and a length of 1–4 m depending upon the application. It is packed with a fine granular material which is the stationary phase. The capillary column has an internal diameter of 0.2–0.5 mm and a length of between 15 and 100 m. The stationary phase is coated onto the inside of the column in a layer between 0.25 and 1.0 μm thick.

Capillary columns have a much greater resolving power than the packed columns and are being used to a much greater extent in spite of their considerably greater cost. Because of their small volume, capillary columns have a much smaller capacity than packed columns so that they would become overloaded and give poor results with normal injection volumes of 1–5 μl. To overcome this problem a splitter is incorporated into the chromatograph between the injection port and the beginning of the column. This reduces the amount getting onto the column, the remainder going to waste. The splitter can produce split ratios down to 500:1; so that only 0.2 per cent of the injected material will get onto the column (*Figure 17.9*).

Apart from a heated injection port for syringe injection, other methods are available for placing the sample onto the column. With a gas sampling valve, gas is sampled into a reservoir of known volume, which is then switched into line with the column and the gas flushed onto the column by the carrier gas.

A head-space analyser can be used to sample volatile components in a non-volatile medium, such as solvents in water, blood, or urine. The sample is heated in a vial, the more volatile components are concentrated in the head-space and a sample from this head-space is then injected onto the column.

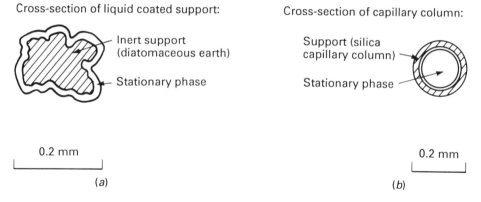

Cross-section of liquid coated support:

Inert support
(diatomaceous earth)

Stationary phase

Cross-section of capillary column:

Support (silica
capillary column)

Stationary phase

0.2 mm

(a)

0.2 mm

(b)

Figure 17.9 Liquid coated support and cross-section of capillary column

A thermal desorber can be used to desorb solvent vapours that have been sampled onto porous polymers by heating them while flushing them with carrier gas which carries the solvents onto the column. After being desorbed the tubes can be reused. This technique cannot be used for charcoal tubes as they cannot be thermally desorbed with reasonable efficiency.

Detectors

FLAME IONIZATION DETECTOR (FID)

As the separated components leave the column they enter the detector. Various detectors are in use on gas chromatographs, the most common now in use being the flame ionization detector (FID) (*see also Chapter 16*). This consists of a small hydrogen/air flame with two electrodes on either side of the flame. When components enter the flame they burn to form ions which are attracted to one of the electrodes to cause a small current to flow between the two; this current is amplified prior to output to the integrator and chart recorder. The FID will detect any combustible compound so it is a good general purpose detector for occupational health purposes.

ELECTRON CAPTURE DETECTOR

Another detector in common use is the electron capture detector (ECD) which has a small beta emitter, usually ^{63}Ni, as the negative electrode. Electrons emitted from this electrode are accelerated across the column effluent to the positive electrode where they produce a standing current. When electronegative compounds are eluted from the column they 'capture' some of the electrons thus reducing the standing current at the positive electrode. This change in the standing current is amplified for output to the integrator and chart recorder. This detector is more selective than the FID in that it will detect only electronegative compounds—those containing a halogen atom, a nitrogen atom, or an oxygen atom. ECD has the greatest sensitivity for halogenated compounds which makes it useful for detecting trace levels of pesticides which often contain chlorine atoms.

Many other detectors are in use, but most of the others are for specialized applications and unlikely to be encountered in general use.

Liquid chromatography

Liquid chromatography is used for the analysis of non-volatile or thermally labile compounds that cannot be analsed by gas chromatography.

In liquid chromatography the mobile phase is, as the name suggests, a liquid; the stationary phase is a fine powder. This can be alumina or silica, known as *normal phase*, or alumina or silica capped with octadecylsilane (ODS), known as *reverse phase*. Normal phase columns are polar and have a non-polar mobile phase, usually made from organic solvents such as hexane, tetrahydrofuran, or iso-octane. These columns will retain polar compounds longer than non-polar compounds. Reverse phase columns are non-polar and have a polar mobile phase, usually made from water and water-soluble solvents such as methanol and acetonitrile. These columns retain non-polar compounds longer than polar compounds. Columns of intermediate polarity can be produced by not capping all the hydroxyl groups on the silica and alumina with octadecylsilane. These intermediate columns are treated as reverse phase columns.

The most commonly used detector in liquid chromatography is the ultraviolet detector. This will detect any eluted component that absorbs ultraviolet light at the frequency set on the instrument. Modern ultraviolet detectors have adjustable frequency control so can be set to the optimum wavelength for a particular application. The most versatile detectors will take the entire ultraviolet spectra of each component as it is eluted.

Mass spectrometry

Mass spectrometers are powerful instruments that can give the absolute identity of an unknown compound. The compound is bombarded by high energy electrons which break the compound into fragments. These fragments are then passed through crossed electric and magnetic fields and separated according to their mass to charge ratio. The pattern of fragments produced is specific to each compound. Mass spectrometers can now be attached to gas and liquid chromatographs as detectors, resulting in some powerful analytical tools that can be used to help solve some of the more complex problems found in occupational health, such as the identification of polyaromatic hydrocarbons in coal tar pitch volatiles, or the identification of unknown odours. This technique used to be very expensive and would be used only in cases where the more routine techniques could not cope; however, cheaper benchtop mass spectrometers have become available over the past few years. These new instruments do not have the capabilities of the larger machines, having a smaller mass range of up to only 1000 atomic mass units. However, they are very useful for solving problems in the occupational health field, particularly where gases and vapours are concerned. Because they are cheaper than their larger counterparts to purchase and to run mass spectrometers will certainly be used much more in the future to solve problems that may have been out of reach because of the cost involved.

Mass spectrometers are ideal detectors for use with gas chromatographs since the sensitivity of an electron capture detector is available with the ability to detect and positively identify any compound within its mass range from its computer library.

Quality assurance

The results that an analytical chemist gives to an occupational physician or hygienist will determine any remedial actions: these might, for example, consist of the installation of costly ventilation equipment to reduce air concentrations, or the suspension of a worker with high blood metal concentrations. Therefore, it is important that these results are as accurate and precise as possible. To ensure this a good laboratory will undertake a quality assurance programme consisting of three parts.

(1)   The management structure and running of the laboratory.
(2)   External (interlaboratory) quality assurance.
(3)   Internal (intralaboratory) quality assurance.

The management of a laboratory plays an important role in quality assurance, from ensuring that new staff are properly trained in the laboratory technique to making sure there are no typing errors in reports (a decimal point in the wrong place, for example). The National Physical Laboratory has a National Testing Laboratory Accreditation Scheme which many laboratories in the occupational health field are joining.

External quality assurance compares the results of many laboratories having analysed identical samples. Each laboratory is given a proficiency score after each round of the quality assurance. If the laboratory score is outside a critical range then the laboratory has to investigate the reasons for failing to meet the required criteria.

Some current external quality assurance schemes available in Britain are shown in *Table 17.2*.

Internal quality assurance samples are run by the laboratory themselves, usually on a more frequent basis than the external samples. These samples are made up in large batches with a number being analysed immediately to obtain a baseline and standard deviation; the remaining samples are then stored in a way that prevents sample loss or contamination and analysed over the following months. The results obtained with each sample are compared with the baseline figures to see if they meet the criteria of acceptance.

Table 17.2 External quality assurance schemes available in Britain*

UKEQAS	Blood lead
AQUA	Toluene on charcoal
	Lead on filter
	Isocyanates
RICE	Asbestos on filter
HERMES	Urinary mercury
WASP	Benzene, toluene and xylene on charcoal
	Benzene, toluene and xylene on Tenax
	Lead, cadmium and chromium on filter

*Further details on these schemes are available from the Health and Safety Executive.

Further reading

Curry, A.S. (ed.) *Analytical Methods in Human Toxicology*, Parts 1 and 2, MacMillan Press, London (1986)

Manual of Analytical Methods, National Institution for Occupational Safety and Health (NIOSH), United States

Methods for the Determination of Hazardous Substances (MDHS), Health and Safety Executive (HSE) United Kingdom

Chapter 18

Sickness absence

S.J. Searle

Introduction: responsibilities, costs and definition

The need for control of absence attributed to sickness is crucial to the economic viability of any business. The monitoring and control of sickness absence is primarily a function of line management with the occupational health service and the personnel department providing advice and support. As a result of a 'hard-nosed' style of management and accountability related to the bottom line of the balance sheet, involvement with the medical aspects of sickness absence forms a large part of the work of the occupational physician in the present industrial climate. In order to provide effective advice he needs a good understanding of the many factors affecting sickness absence in both individuals and groups of employees.

This chapter considers the recording and measurement of sickness absence together with the wide range of factors which influence it, and indicates how an occupational health service can help management carry out its responsibility for control. The level of sickness absence in an organization is only one of the factors which reflect the overall 'health' of the organization, together with accident rates, labour turnover and the industrial relations climate. Good cooperation between the occupational physician and management is crucial if proper control of sickness absence is to be attained.

As proved by feedback from many organizations regular discussion between members of the occupational health service and managers at all levels enables the most effective use to be made of the occupational health service in helping managers to control absence. For many years now the Post Office has published an annual report on sickness absence and medical wastage (deaths in service and medical retirements) which in 1986 [1] included an appendix giving guidance to managers on the 'medical aspects of sickness absence control' to enable them to make the best use of the information and services available.

Responsibilities

It is the responsibility of management to identify individuals with adverse patterns of absence and to initiate appropriate action to remedy and control the problem. This involves the consideration of the individual, his working environment, and also groups of employees who may be similarly at risk. To do this effectively it is necessary to have adequate sickness absence records with details of both frequency and duration of sickness absence for each individual employee, rather than more generalized information such as changes in the percentage of working time lost.

It is the responsibility of the occupational health service to give clear advice to management about individual cases referred to it, based on factual information obtained from a combination of recent individual assessment by a doctor or nurse of the occupational health service and where necessary reference to the patient's own general practitioner or specialist. This enables them to advise management on the physical capabilities and limitations of the employee and the prospects of effective work performance and attendance in the future, the occupational health service interpreting the medical information into functional terms that management can understand. It is essential that employees realize that medical information will be kept strictly confidential and that any advice which is given to management will have been discussed with them first. When an employee's work prospects are in doubt final advice to a manager must be from the occupational physician. There has been much debate about medical ethics in occupational medicine recently [2] and guidance has been issued by the Faculty of Occupational Medicine of the Royal College of Physicians [3].

If there are particular absence problems affecting groups of employees or types of workplaces then doctors and nurses within the occupational health service are in a good position to liaise with line and personnel managers, occupational hygienists, safety officers and workplace safety representatives, engineers and ergonomists in trying to diagnose and control the problem.

Costs

The direct costs of sickness absence in payment of sickness benefits are large. In 1982–83 in the United Kingdom benefits amounted to 780 million pounds for the 530 000 individuals who were sick on any one working day [4]. In both 1981–82 and 1982–83 more than 370 million working days were lost in the United Kingdom due to certified incapacity for work because of sickness or injury [5]. In one large nationalized industry the direct costs of sickness absence in 1984 and 1985 were equivalent to £1 per head of all employees for each working day.

The indirect costs of sickness absence more than equal the direct costs and cannot be ignored. Factors contributing to this are the cost of substitution for absent employees, increased levels of overtime and reduction in the quality of products or services provided by the company. The administrative costs in terms of involvement of personnel departments and advisory services such as the occupational health service are also substantial. Considering both direct and indirect costs it is likely that the drain on the United Kingdom economy from sickness absence annually exceeds £2000 million.

Definitions

Sickness absence is defined as absence from work which the employee attributes to sickness or injury and the employer accepts as such [6]. This emphasizes the fact that the phenomenon primarily concerns absence due to incapacity for work, as declared by the employee, the presence of a medical condition in the individual being only one of many factors behind the inception of a spell of absence. The absence may or may not be supported by a doctor's certificate. It is usual to distinguish between certificated and non-certificated absence and this is discussed below. Absence due to normal pregnancy or confinement is not normally included as sickness absence.

Absence is a repetitive event and as such must be measured in terms of frequency (spells) and duration (severity). A spell of absence is an uninterrupted period of absence from its commencement, irrespective of duration. Duration of absence in any period should preferably be counted in calendar days although some analyses have used working days or shifts. The recording and measurement of sickness absence is discussed below.

Absence records: personnel and occupational health records

Personnel absence records

There is a wide variation from one company to the next in the records which are maintained and which may be suitable for recording absence. There are basically two approaches to this, one being the recording of attendance, such as by clocking in, and the other the recording of absence reported by employees and supervisors. The former usually applies to hourly paid workers and the latter to salaried employees. Recording of attendance is by far the more reliable approach and so the reliability of absence recording reduces as individual status rises in an organization.

The quality of records maintained in industry varies enormously and makes the collection of information about absenteeism across a broad spectrum of industries on a national basis almost impossible. A recent survey by the Industrial Society [7] was compiled from questionnaires sent to 1100 personnel managers. Only 26 per cent were returned completed, the remainder being returned uncompleted due to either a complete lack of absence records in the organization, or records that did not allow the collection of the relatively simple information required.

The only practical means of recording absence is to have a record for each employee on which basic details can be recorded for each absence. This may either be kept by the supervisor or within the personnel department, although in both cases the input of the first-line supervisor is the key to the accuracy of the record. The record may be a card maintained manually, or a computer file. The use of computerized recording of absence is increasing but it is vital that computerized personnel systems are updated on a daily basis and that accurate information about an individual's absence record is readily retrievable by line and personnel managers and the occupational health department. The great advantage of computerized recording of absence is the facility it provides to examine easily the effects of the many different factors involved in absence in both individuals and groups. For instance, it allows an easy printout of all the individuals taking absence on particular days of the week, perhaps in relation to local sporting events! It also enables groups of workers repeatedly exposed to particular hazards to be monitored more easily in terms of their absence records.

The personnel record card should contain at least the following personal information: name, company, identity (for example, payroll number), national insurance number, date of birth, sex and marital status, occupation, department, hours of work (full-time, part-time or shift) and date of entry to the company. For each spell of absence the date of onset and duration, the final diagnosis (if medically certified or the employee's reason for absence from the self-certificate

P 3606
(Formerly M 549)

ABSENCE RECORD

Grade ....(Long term/Temporary/Part time)	Surname (In capitals)	Mr/Mrs/Miss
Where employed	Forename(s)	
Date of birth		
Date employment commenced	Other Information (e.g. Address, if required, etc)	
Date long term employment commenced		
Period of trial		
Whether re-employed pensioner	Whether regist'd under Disab'd Persons' Act	

National Insurance				Summary of Sick Leave during previous years							
National Insurance Number	Date of entry to Nat. Ins. and type of contribution paid	Whether signed E68 (PO) option (Yes or No)		19..................		19..................		19..................		19..................	
				Number of		Number of		Number of		Number of	
				Absences	Days	Absences	Days	Absences	Days	Absences	Days·

Period of Absence		Number of Days			REMARKS
From	To (inclusive)	Sick Leave §		Special Leave ‡	eg Nature of illness (as shown on medical certificate) and certification†: Accident on or off duty: special leave (show type): half pay, pay at pension rate or without pay etc.
		Cert	Self Cert		

§ Intervening non-working days included:
‡ Working days only unless unpaid special leave

† Abbreviations to be used:- NAT = National Insurance,
P = Private, H = Hospital, NC = No certificate,
NCE = No certificate - due for early duty.
SC = Self Certificate

Figure 18.1 Record card

of explanation to the supervisor), whether or not the cause of absence was related to work (for example, accident on duty), and whether the absence was medically certificated should be recorded.

An alternative to the type of record which lists spells of absence is a series of annual calendar records for each employee in which there is a box for each working day which remains empty unless an absence occurs. A letter code is used to indicate the type of absence (for example, C for certificated absence, U for uncertificated absence). Holidays and lateness can also be recorded on such a form and it can be a very useful tool for the first-line supervisor, giving an immediate visual impression of the amount and pattern of individual absence in a year. This system is described in detail by Behrend [8]. Although it may be useful to line management it is less useful in an occupational health department than the 'list of spells' type of record card, which gives more detail of medical diagnoses and shows long-term patterns of absence, especially if numbers of spells and days of absence are summarized at the end of each calendar year (*Figure 18.1*).

Recent changes in employer's obligations for the payment of statutory sick pay in the United Kingdom may improve the quality of absence records now that employers are responsible for the payment of statutory sick pay for the first 28 weeks of any absence. The financial implications of this to an employer may also stimulate interest in obtaining suitable medical advice about problem cases of absence as discussed by Kearns [9].

The occupational health record

The occupational health service will require information additional to that in the personnel record, such as the extent of smoking and alcohol habits, exposure to any toxic substances (including biological and physical hazards) at work, immunization status, known chronic disabilities, family responsibilities, place of residence, name and address of general practitioner and NHS number. The latter is particularly useful for purposes of record linkage with NHS records when undertaking an epidemiological study.

It is useful to the occupational health service to collect such information on all employees who are seen, but any study on a particular absence problem requires the collection of such information for either the whole population at risk, or in a large organization an adequately selected sample. There is little point in knowing that half the spells of absence after injury at work occurred in smokers, drinking more than 30 units of alcohol a week and with no dependents, without knowing how many with these characteristics did not have absences from injury at work.

Recent proposals in the United Kingdom for the health surveillance of workers exposed to hazardous substances [10] are relevant to the structure of the occupational health record, which should have the facility for regularly updating changes in employment status and exposure to toxic substances, biological or physical hazards in the workplace. Such changes in employment status should also be regularly updated in the personnel record. There should also be the facility for cross-reference of any relevant occupational hygiene data obtained on individuals exposed to particular hazards. Potential legislation regarding the maintenance of such records may pose a problem in terms of storage space and costs for both manual and computer records as it has been suggested that such records should be kept for a minimum of 50 years.

Measuring absence: basic statistics and misconceptions

Basic statistics

Sickness absence can only be measured in terms of both duration (in days) and frequency (number of spells in a period). The use of this type of measurement in three typical patterns of absence over a calendar year can be seen in *Figure 18.2*. The absence in A is a single spell of 90 days certificated absence following a myocardial infarction with complete recovery and no further absences after return to work. The pattern in B is of repeated spells of certificated absence due to chronic bronchitis, worse in the winter months, and C shows a pattern of repeated short self-certificated spells of absence due to a variety of self-limiting causes with no significant underlying medical diagnosis.

It can be seen that summarizing these patterns in terms of numbers of days and spells in each year gives an impression of the type of absence involved. Generally absences with a larger medical basis have longer but fewer spells and absence patterns with a greater behavioural basis show shorter, more frequent spells. It is clear that in any absence study both the severity rate and frequency rate are fundamental.

$$\text{Severity rate (mean days per person)} = \frac{\text{Total days of absence in period}}{\text{Average population at risk in period}}$$

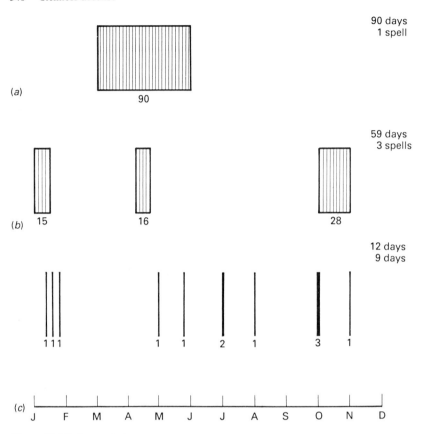

Figure 18.2 Patterns of absence

$$\text{Frequency rate (mean spells per person)} = \frac{\text{Total number of new spells of absence in period}}{\text{Average population at risk in period}}$$

The period normally used in calculating these rates is one year (either calendar year or business year) but longer or shorter periods can be used depending on the problem under investigation. Rates can be calculated separately for medically certificated and uncertificated sickness absence and then combined to give total absence rates.

Another way of expressing sickness absence which is commonly used in industry is as 'percentage working time lost'. This is a point prevalence rate expressing the number of people absent on a single day as a percentage of the total population who should have attended work on that day. The daily percentages can be averaged over weekly, monthly or annual periods to give an overall measure of time lost (severity of absence) in a workforce. The mean days per person in the period is equal to the percentage of working time lost multiplied by the number of working days in the period under study. The use of 'percentage working time lost' figures is limited to providing rough comparisons between different working units and to budgetary purposes. They give no indication of whether the problem is of long or short-term absence and are not helpful in defining the causes, either medical or managerial, behind the problem.

One factor which produces a marked variation in absence rates quoted by different organizations is whether absence is recorded as calendar days lost or working days (or shifts) lost. Calendar days are mainly used in medical literature on sickness absence and by some large organizations [1]. This system does, however, tend to overstate the amount of absence (including weekends when the individual may or may not be fit to attend work) and reflects 'incapacity' rather than the true amount of absence from working duties. Sickness absence is very different from true morbidity and where a new system of absence recording is being set up there is considerable justification for using working days lost, as this more truly reflects the direct financial cost to the business as well as the potential operational problems resulting from absence.

The major problem in expressing absence in terms of mean severity and frequency rates is that these give no indication of the tremendous variation in absence between individuals within a group. The use of frequency distributions gives a much better picture of the absence variation in a group, expressing the distribution either graphically (*Figures 18.3 a and b*) or as a table (*Table 18.1*). It is relatively easy, without requiring extensive statistical knowledge or a computer, to plot on graph paper the number of employees taking 0, 1, 2, 3, etc. spells or days of absence in a period. This then enables easy identification of those few employees with the highest severity and frequency of absence. It is clear that both severity and frequency of absence is highly skewed, with the majority of individuals taking little or no absence and the majority of absence resulting from very few individuals.

The highly skewed distribution of sickness absence does cause some difficulty in statistical analysis. This is because it is not possible, when comparing patterns of sickness absence between different groups, to use statistical methods based on the 'normal distribution', such as t tests. There can also be problems in using more sophisticated techniques such as analysis of variance (ANOVA) and multivariate

Table 18.1 Distribution of self-certificated absence: postal staff 1984–85

Number of days or spells	Percentage of days of absence	Cumulative percentage of days of absence	Percentage of spells of absence	Cumulative percentage of spells of absence
0	29.5	29.5	29.5	29.5
1	11.2	40.7	25.4	54.9
2	11.0	51.6	19.6	74.5
3	10.0	61.7	13.1	87.5
4	7.1	68.7	6.7	94.2
5	6.8	75.6	3.0	97.3
6	5.5	81.1	1.4	98.7
7	4.5	85.6	0.7	99.4
8	3.6	89.2	0.3	99.7
9	2.8	92.1	0.1	99.8
10	2.0	94.1	0.1	99.9
11	1.7	95.7	0.0	99.9
12	1.1	96.8	0.0	100.0
13	0.9	97.7	0.0	100.0
14	0.6	98.3	0.0	100.0
15–21	1.5	99.8	0.0	100.0
Over 21	0.2	100.0	0.0	100.0

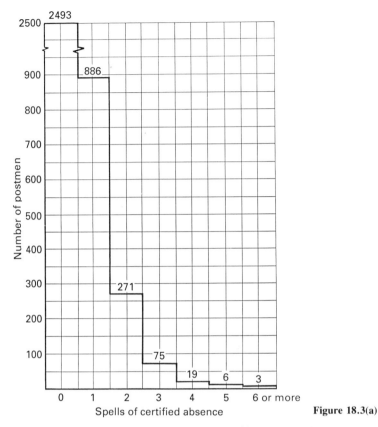

Figure 18.3(a)

Figure 18.3 Number of postmen taking x spells (a) and x days (b) of certificated absence in 1982–83 (total: 3753 postmen)

analysis, as these also depend on a 'normal distribution' of the data being analysed. There are, however, a number of published studies which have used such techniques, especially in the field of behavioural and psychological aspects of sickness absence [11, 12]. If such methods are envisaged in the design of a sickness absence study then professional statistical advice should be sought at the outset.

While it is difficult to directly compare mean rates of severity and frequency it is possible, by using frequency distributions, to use statistical techniques for comparing qualitative data, such as chi-square tests. In this way the proportion of individuals with particular absence characteristics, such as high or low absence severity or frequency can be compared. It is a relatively simple technique and allows a valid comparison between groups, as used in studies by Pocock [13] and Searle [14]. Using frequency distributions also allows curve-fitting analysis in testing hypotheses of causation such as 'proneness' and 'pure chance' as described by Froggatt [15].

Misconceptions

There is a common misconception within most organizations that the vast majority of staff take some sickness absence during a year. It is clear however from *Table*

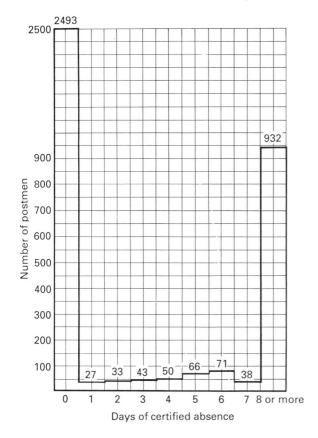

Figure 18.3(b)

18.1 and *Figure 18.3* that it is in fact a minority of staff who take any absence during a period. These examples show that in a large group of postmen, a group undertaking mainly manual tasks, one-third took no self-certificated absence and two-thirds took no certificated absence during the year under study. This misconception probably relates to the use of mean rates to describe absence in groups as these overstate the problem, being so heavily weighted by those taking large amounts of absence. A better measure of the 'average' absence is given by the median rate which is the number of days or spells exceeded by half the group under study. In a recent Post Office publication such rates are given for different grades of staff (*Table 18.2*), together with the number of days and spells exceeded by the worst 5% of the population. This gives managers some guidance as to what amount of sickness absence can be realistically expected from the majority of staff and also helps them to define the characteristics of those whose attendance may be considered unsatisfactory. Using either the worst 5% or the worst 10% (upper decile) of the distribution can help define those where action may be needed either managerially or medically, enabling those with an adverse pattern to be offered help by the occupational health service to see whether or not there are any medical factors which can be modified to try and improve the absence pattern, as discussed later in the chapter.

A further misconception is that long-term absence is primarily due to medical

Table 18.2 Absence criteria for grades median and upper 5% values 1984–85

Sex and grade	Median days	5% Days	Median spells	5% Spells
Men				
Senior managers	0	16	0	2
Executives	1.2	36	0.4	5
Clerks	2.3	29	0.7	4
Sorters	4.0	48	1.3	5
Postmen	4.4	53	1.4	6
Other grades	1.9	35	0.7	4
All postal men	3.4	47	1.1	5
Women				
Executives	2.5	33	0.8	4
Clerks	4.0	43	1.3	5
Postwomen	6.3	70	1.4	6
Other grades	5.3	76	1.2	5
All postal women	4.9	60	1.2	5

causes and short-term absence is usually due to behavioural factors, sometimes known as 'attitudinal' absence. However, every occupational health practitioner will have come across situations where long-term absence is clearly not related to a medical condition and where short-term absence may be the earliest indicator of a relatively serious medical problem. An example of the former was an employee who had been absent from his duties as an engineer for over 6 months, certified by his general practitioner as suffering from 'capsulitis of the shoulder'. When assessed by the occupational physician it was clear that there was no serious medical problem in the shoulder from a functional point of view. An example of the latter was a manual worker with frequent short-term absences which were the first indication that he was receiving regular chemotherapy for a lymphoma. It is by no means the case that all long-term absences have a medical basis nor that short-term absences are necessarily due to behavioural factors. Any organization which operates absence control procedures and has an occupational health service available is wise to refer all cases of unsatisfactory absence, of whatever pattern, to the occupational health service so that important medical factors are not overlooked.

The population at risk

In any absence study it is clear that the numerator for calculating rates is the amount or frequency of absence and the denominator is the 'average population at risk' in the period under consideration. It is often more difficult to ensure the collection of adequate information about the whole population at risk than to collect information on those who take absence as most occupational health departments have records confined to those attending for advice but not on all employees. For adequate analysis in a study on sickness absence certain minimum information must be available for the whole population at risk: person-years, sex distribution, age distribution and job grade distribution. Many other variables which are discussed later need to be included, but the three variables of sex, age and job grade account for by far the largest variation in absence rates between groups and must always be determined.

Person-years

This is the determination of the 'average population at risk' during the period, being the denominator for calculating the frequency and severity rates. In a static population a mid-year census may be adequate but, as most occupational groups include new recruits and others who leave during the period of study, this can be unreliable. It is more accurate to calculate person-months or days for each individual employed in a period and to add these together to give a mean annual population. In constructing frequency distributions only the population at risk for the whole period can be considered and those joining or leaving during a period have to be excluded in such analyses. It is generally not possible to derive any useful conclusions from an absence study with a population of less than 100 person-years.

Sex distribution

It is commonly found that sickness absence rates are higher for women than for men, despite increasing equality in other respects. Women with family commitments, particularly those with children under 5, may have higher rates than women without these responsibilities, and it may be necessary to distinguish between these two groups.

Age distribution

The frequency of absence, particularly for short spells, tends to fall with increasing age whilst the severity of absence increases with advancing age. When studying large populations 5 yearly groups should be used (for example, number of employees less than 20, 20–24, 25–29, 30–34, etc.); 10 yearly groups can be used for smaller populations, and even in a small study the groups should at least be split between those up to age 39 and those aged 40 and over.

Job grade

Job grade has a marked effect on absence behaviour with more manually demanding tasks usually resulting in a higher absence rate than relatively sedentary tasks. Taylor [6] states that job status in an organization may account for up to threefold differences in sickness absence rates, even allowing for age and sex differences. Separate rates should at least be calculated for 'white collar' or staff employees and for 'blue collar' or manual workers. In a large organization the job structures may determine the occupational groups, but it should at least be possible to differentiate between managers, skilled manual, skilled non-manual, semiskilled and unskilled workers.

Standardization or stratification techniques can be used to minimize the influence of these three major variables when comparing groups to study other factors influencing sickness absence.

Factors known to influence sickness absence

The many factors known to influence sickness absence have been summarized by Taylor [16] and are shown in *Table 18.3*. It is interesting to note that of over 30

Table 18.3 Some factors known to influence sick absence

Geographical	Organizational	Personal
Climate	Nature	Age
Region	Size	Sex
Ethnic origin	Industrial relations	Occupation
Social insurance	Personnel policy	Job satisfaction
Health services	Sick pay	Personality
Epidemics	Supervisory quality	Life crises
Unemployment	Working conditions	Medical conditions
Social attitudes	Environmental hazards	Alcohol
Pension age	Occupational health service	Family
Taxation	Labour turnover	responsibility
		Journey to work
		Social activities
		Length of service

factors only three are strictly medical: mainly the occurrence of epidemics, the effect of environmental health hazards, and the presence of a specific medical condition in the individual. The occupational health service can help to modify the effect of these factors. Advice on public health measures to control epidemics may be a large component of the work of an occupational health practitioner in developing countries and may contribute greatly to the success of an undertaking. In developed countries influenza epidemics are the main problem in this respect and influenza vaccination programmes can be cost-effective, but only if the uptake is sufficient. Achieving sufficient uptake of influenza vaccination in a workforce becomes more difficult as time increases since the last influenza epidemic.

A large part of the work of the occupational health practitioner today involves close liaison with occupational hygienists in monitoring the effectiveness of measures to control environmental hazards in the workplace. Biological monitoring may in certain circumstances provide the first clue that control measures are not being effective, enabling action to be taken to prevent further risk to employees.

There are several ways in which the occupational health practitioner can influence the effect of medical conditions in an individual on sickness absence. Health education and promotion in the workplace has been seen as a major initiative by some occupational health services, such as the Post Office project on alcohol education and management training in the workplace [17]. These initiatives seek to minimize the incidence and severity of preventable medical conditions not specifically related to work, and encourage employees to assume a greater responsibility for their own health. By being easily available for employees to consult if they have worries regarding their health in relation to their work, the occupational health service can help to resolve problems at an earlier stage before absence occurs. The individual can also be helped to make the best use of available health services. Good links between management absence control procedures and referral to the occupational health practitioner also help the employee to obtain advice on the control of medical problems at an early stage, possibly preventing the inception of further spells of absence.

Sickness absence involves the interaction of the individual with his total environment and as such many factors have an effect. In an epidemiological study it is not necessary to control for all these factors when comparing groups for absence rates, especially within the same country and the same organization or workplace. The geographical factors relate to political, social and economic differences and these can result in major differences in absence rates, such as reported by Prins and De Graaf [18] between the Netherlands, Germany and Belgium. The organizational factors relate to both the whole company and to individual workplaces and are affected by the function (product or service) of the company and its personnel policies. The size of the working unit is more relevant than the size of the whole company, smaller working units having better absence rates associated with greater peer support and better quality of supervision. The strong influence of age, sex and job grade have already been mentioned and these factors must always be considered. The other personal factors are many and varied but job satisfaction and personality are major factors behind the decision to return to work after a spell of absence and indeed in the inception of absence.

The certification of sickness: the doctor or the patient?

One important variable in the operation of sickness benefit schemes is the degree to which medical certification of sickness is required by the employer and the state. The inception of a spell of sickness absence depends upon so many factors other than the individual's medical condition that the decision not to attend work is primarily one made by the patient rather than the doctor. The fact that the vast majority of certificates of medical incapacity are issued at the start of the working week supports the view that the patient rather than the doctor decides when the absence will start. In the United Kingdom more medical certificates to refrain from work are issued on a Monday than other days of the week and the likely day for returning to work on a final medical certificate is also a Monday. It is interesting that in Israel, where Sunday is the first working day, more certificates are issued on that day [19]. Medical certification for short spells of absence is somewhat irrelevant as in minor self-limiting conditions there are often no physical signs on which a physician can base an opinion and he is merely countersigning or 'approving' the absence. Self-certification for short spells of absence has been used in various forms in different organizations and countries for some time and is not generally abused if adequate control measures are taken by management [20].

In the United Kingdom in 1982 a scheme of self-certification was introduced for the first 7 days in a spell of absence. Despite fears amongst some employers that this would result in a marked rise in sickness absence this scheme has operated successfully without such a rise becoming apparent. There is some evidence in the Post Office that employees are in fact taking shorter spells as a result of self-certification, returning to work when they themselves consider they are fit to do so rather than waiting for the expiry of a medical certificate. Lim [21] studied a group of postmen for 2 years before and 3 years after self-certification and found no change in the pattern of severity or frequency of absence, using a frequency distribution technique. The Industrial Society [7] reports that many businesses have found the introduction of self-certification helpful in stimulating effective absence control procedures.

Medical certification of absence does not provide a true picture of morbidity in a population as alternative diagnoses or euphemisms may be used, particularly in cases of psychiatric disorder. Jenkins [12] found a 33% prevalence of minor psychiatric morbidity in a young clerical workforce, and on reviewing their medical certificates there was no indication of any overt psychiatric diagnosis. Despite this there was a positive correlation between the presence of minor psychiatric morbidity and a high severity of absence. To build up a true picture of morbidity in a population in relation to sickness absence studies it is essential to include a clinical survey of the whole, or a valid sample, of the population. This may also reveal interesting information about those who do not take absence such as found by Taylor [22] who discovered that many individuals who take little or no absence have significant medical conditions.

In a large population, however, the coding of spells of absence by diagnosis from medical certificates can give an overall impression of relative causes of morbidity. The accuracy of the diagnosis can be improved if employees are seen on return from longer absences by the occupational health practitioner. As well as providing a check on diagnosis this allows advice to be given both to the individual and management on suitable work arrangements following return from sickness. For the effective analysis of spells of absence by diagnosis it is essential that each spell of absence is coded to the rubric of the International Classification of Diseases (ICD) [23] rather than into broad diagnostic groups. This facilitates the

Table 18.4 Postal business 1984–85: main diagnostic causes of sickness absence

| Diagnostic group | Days (thousands) | | | | Days per employee | | | |
| | Men | | Women | | Men | | Women | |
	1984–85	1983–84	1984–85	1983–84	1984–85	1983–84	1984–85	1983–84
Respiratory	427	390	83	78	2.77	2.54	3.08	2.85
influenza	(170)	(137)	(30)	(22)	(1.11)	(0.89)	(1.12)	(0.81)
bronchitis	(74)	(63)	(11)	(10)	(0.48)	(0.41)	(0.41)	(0.36)
Musculoskeletal	366	359	71	87	2.38	2.33	2.63	3.16
Injuries	294	308	65	63	1.91	2.01	2.41	2.31
(at work)	(77)	(80)	(25)	(27)	(0.51)	(0.52)	(0.93)	(0.97)
Cardiovascular	192	182	11	20	1.25	1.19	0.40	0.74
(coronary heart disease)	(51)	(86)	(4)	(6)	(0.33)	(0.56)	(0.13)	(0.20)
Digestive tract	243	217	33	37	1.58	1.41	1.22	1.37
Psychiatric	82	70	31	40	0.53	0.46	1.16	1.46
Other causes	441	380	154	137	2.87	2.48	5.71	5.01
All causes	2045	1907	448	463	13.27	12.41	16.61	16.88

retrospective analysis of records to examine trends in specific diagnoses over a period which would be impossible if diagnoses were only grouped into broad categories. The individual codes can then be combined into broad groups, usually based on the main chapters of the International Classification of Diseases, for presentation in a report or paper. Such a summary is shown in *Table 18.4* giving figures for the Post Office in 1983–84 and 1984–85. Notable points are that absence from injury is predominantly due to injury not sustained at work, and the high rates for musculoskeletal disorders relate to the difficulties in undertaking a demanding manual task (the majority of staff are postmen) while suffering from such disorders.

The occupational physician will often be asked by management to advise on the prospects for return to work of an employee who has been absent and may appear to be taking a longer sickness absence than expected from the diagnosis. Obviously some general practitioners are more amenable than others to issuing medical certificates on request once the patient is approaching the stage of fitness to return to work. In general they accept the word of their patients and are aware of the difficulty in disproving symptoms such as back or head pain when the patient maintains that these are persistent. It is also a fact that patients may be unaware of the possibilities of modification of work and rehabilitation in a workplace and may overstate or understate the physical and emotional demands of their job to their family or hospital doctor. In this situation careful clinical assessment by the occupational physician together with good communication with the general practitioner or specialist providing treatment is essential for returning the patient to effective work as soon as possible. Close liaison between the occupational physician and the treating physician in this way is of considerable benefit to the patient as well as to the company.

Positive action: diagnosis, treatment and prevention

Diagnosis

Diagnosis of sickness absence relates to the identification of the real causes of the inception of sickness absence in an individual or a group. As previously discussed these may relate to factors in the individual or his environment but especially to the interaction between the two. Accurate diagnosis always requires an appreciation of the patient in his total environment and in occupational health terms this requires the occupational health practitioner to have a comprehensive working knowledge of the culture of his organization and the nature of the work carried out. A particular problem which has to be increasingly considered is that of problem drinking. It is estimated that about one in ten of adults in employment may have some personal, social or employment problems associated with their level of alcohol consumption. It is such a common problem that it behoves companies to develop a specific policy on how to handle employees with alcohol problems to ensure that they are identified early. This involves a commitment to education of the workforce in general and training of management and supervisors in particular.

Personal and psychological problems can impact on absence and performance and these can be explored initially by the occupational health practitioner as a first-line counsellor. Counselling has been shown to be effective in reducing the frequency of sickness absence in groups of employees, and the provision of counsellors in the workplace is increasing, initially following the success of employee assistance programmes in the United States. A major initiative is being undertaken by the Post Office in providing specialist counsellors in the workplace with the appointment of two specialist counsellors in separate locations, with careful monitoring of the effects over a 3 year trial period [17]. Feedback on problems affecting groups within the organization from those undertaking counselling can be just as useful in psychological terms as the provision of group occupational hygiene data in relation to environmental hazards is in preventing organic occupational disease.

Treatment

In terms of sickness absence this does not relate to medical treatment in the accepted sense but to the role of the occupational health service in preventing the inception of spells of absence and reducing the duration of those that occur. It also involves the effective return to work of employees after illness and helping those whose medical problems make them unfit to return to their previous occupation in terms of resettlement or retirement on medical grounds.

Examples of the prevention of inception of absence include the provision of influenza vaccination to groups of staff, providing there is sufficient uptake, and also health education initiatives not related specifically to occupation but using the workforce as a captive audience in order to stimulate awareness of more healthy lifestyles. The increasing awareness of the adverse effects of smoking in the workplace is an example. Mention has already been made of the role of the occupational physician in liaising with general practitioners and specialists to ensure the most rapid and effective return to work of employees after illness by arranging any necessary modifications to duty and providing information to the training physicians about working conditions.

Ensuring that those who return from sickness absence are effective at work is a major task of the occupational health service and occupational health nurses in particular are in an excellent position to observe the individual once he has returned to work and ensure that he is not having continuing problems. If work problems related to recovery are noted then suitable advice can be given to both the individual and management. This may help to prevent the inception of a further spell of absence.

Prevention

The best predictor of future absence is past absence and it may be necessary to emphasize to personnel departments the importance of adequately assessing at the recruitment stage an employee's past attendance in previous employment or at school. Pre-employment assessment by the occupational health service of those who are known to have particular health problems may help to ensure that the work allocated will not compromise the individual's health status or that the health problem will not affect the performance at work. It is important for the recruitment department to realize that this approach is much more effective than blanket rejection of all applicants declaring a particular medical problem. If this happens the word soon goes round the locality that this is current business practice, then applicants declare that they have had no past medical problems and the business may end up with the very problem that it has tried to prevent.

Conclusions: control of sickness absence

The phenomenon of sickness absence is clearly multifactorial at whatever level it is studied, geographical, organizational or personal. In assessing any absence problem in an individual or a group some of these factors have to be taken into consideration. It is rare that medical diagnosis and prognosis alone is all that is required. The primary responsibility for the control of all forms of absence from work, including sickness absence, lies with line management supported by the personnel department. The occupational health practitioner should be resistant to any suggestion that he should take a disciplinary role in sickness absence. His responsibility is clearly advisory, both to the individual and the business, and he can only undertake this role effectively if his advice is impartial, confidential, and based on an accurate assessment of both medical and occupational factors.

It is impossible for either managers or occupational health practitioners to make a useful contribution to control of absence unless reliable and comprehensive individual absence records are kept. These must allow the measurement of both frequency and severity rates in groups of employees which can then be used to establish acceptable standards of attendance. The first-line supervisor is the key individual in the control of sickness absence and it is necessary for organizations to develop and implement appropriate personnel and industrial relations policies and procedures to meet local and individual needs and to provide support at this level. Such procedures should deal humanely but firmly with problem employees and allow the occupational health service to play its advisory and supportive role professionally and effectively. When dealing with individual problems of sickness absence it cannot be overemphasized that careful individual assessment of all the factors are required by both the occupational physician and the manager concerned and that a mechanistic approach is not only counterproductive but is not understood by the employee or his union representative.

It is a common and quite erroneous belief that an employee cannot be disciplined or dismissed because of undue sickness absence. Certainly in the United Kingdom it is the case that employees can be dismissed for unsatisfactory attendance at work, but it is imperative that they are given an opportunity to state their case and try to improve the situation before a final decision is made. It is also encumbent upon the management to seek appropriate medical advice as to when or whether the employee may be able to return to work and whether or not he will be able to regularly and effectively undertake his duties.

There are two keys to the effective control of sickness absence by management and the support given by the occupational health practitioner. These are adequate records and good communications.

References

1. The Post Office. *Annual Report on Sickness Absence and Medical Wastage 1984-85*, Post Office and National Girobank, London (1986)
2. Watterson, A. Occupational medicine and medical ethics. *Journal of the Society of Occupational Medicine*, **34**, 41–45 (1984)
3. Faculty of Occupational Medicine, Royal College of Physicians. *Guidance on Ethics for Occupational Physicians*, Royal College of Physicians, London (1986)
4. Office of Population Censuses and Surveys (OPCS) *Social Trends No. 13*. HMSO, London (1983)
5. Central Statistical Office. *Regional Trends 20*, HMSO, London p. 88 (1985)
6. Taylor, P.J. Absenteeism, definitions and statistics of. *Encyclopaedia of Occupational Health and Safety*, pp. 8–16. ILO, London (1983)
7. The Industrial Society. *Survey of Absence Rates and Attendance Bonuses*. The Industrial Society, London (1985)
8. Behrend, H. *How to Monitor Absence from Work, from Head Count to Computer*. Institute of Personnel Management, London (1978)
9. Kearns, J.L. Statutory sick pay, a briefing for the physician and manager. *Occupational Health Review*, **June/July**, 2–5 (1986)
10. *Control of Substances Hazardous to Health. Consultative Document*. HMSO, London (1984)
11. Nicholson, N., Brown, C.A. and Chadwick Jones, J.K. Absence from work and personal characteristics. *Journal of Applied Psychology*, **62**, 319–327 (1977)
12. Jenkins, R. Minor psychiatric morbidity in employed young men and women and its contribution to sickness absence. *British Journal of Industrial Medicine*, **42**, 147–154 (1985)
13. Pocock, S.J. Relationship between sickness absence and length of service. *British Journal of Industrial Medicine*, **30**, 63–70 (1973)

14. Searle, S.J. Sickness absence and duration of service in the Post Office 1982–83. *British Journal of Industrial Medicine,* **43** 458–464 (1986)
15. Froggatt, P. Research in industrial morbidity, principles and opportunities. *Transactions of the Society of Occupational Medicine,* **18,** 89–95 (1968)
16. Taylor, P.J. Absenteeism, causes and control of. *Encyclopaedia of Occupational Health and Safety,* pp. 5–8. ILO, London (1983)
17. The Post Office. *Annual Report on the Post Office Occupational Health Service by the Chief Medical Officer, 1985/86.* The Post Office, London (1986)
18. Prins, R. and De Graaf, A. Comparison of sickness absence in Belgian, German and Dutch firms. *British Journal of Industrial Medicine,* **43,** 529–536 (1986)
19. Weingarten, M.A. and Hart, J. Sick leave certification in general practice. *Australian Family Physician,* **13,** 702–711 (1984)
20. Taylor, P.J. Self-certification for brief spells of sickness absence. *British Medical Journal,* **1,** 144–147 (1969)
21. Lim, L. *Comparison of patterns of sickness absence before and after self-certification amongst postmen in a head office.* MSc in Public Sector Management, University of Aston in Birmingham (1985)
22. Taylor, P.J. Personal factors associated with sicknes absence. *British Journal of Industrial Medicine,* **25,** 106–117 (1968)
23. *International Classification of Disease (ICD)* World Health Organization, Geneva (1975)

Chapter 19

Toxicity testing of industrial chemicals

A.G.Salmon

Introduction

Toxicology has always been a political subject, although the use made of this branch of knowledge in the ancient world seems to have had to do with the promotion of private gain rather than public good. However, knowledge of toxic phenomena with potentially important public health impacts long preceded either the will or the social mechanisms to do anything about them. The effects of herbal and other medicines, and the adverse effects of various metals and their ores on miners and metalworkers were reported by alchemists and herbalists centuries ago. The toxic effects of occupational lead exposure were well known in the eighteenth century, and Pott's classic observation of the carcinogenic effect of soot on chimney sweeps also dates from this period. In spite of this the development of scientific toxicology, forming the basis of legislative controls on the handling of chemicals in industry, is a twentieth-century phenomenon. Even now, although there are international agreements and guidelines promoted by international organizations such as the WHO and OECD, the existence and enforcement of toxic substance control legislation is largely confined to the developed nations.

The following discussion is based on the current format of legislation in the United Kingdom. In most respects this legislation closely resembles that applicable in other countries of the European Community, being the local implementation of community-wide directives on control of occupational and environmental toxicity problems. The legal systems in force in Scandinavia, North America, Japan and other OECD countries involve more divergent systems for regulation of workplace health. That part of the legislation concerned with toxicity testing has tended to be harmonized in response to OECD guidelines designed to ensure that repetitive testing in different countries will not be necessary. This objective has been mainly but not entirely achieved. The principles of toxicity testing have, in practice, also tended to follow the lead provided by the United States, which was the first country to impose comprehensive testing requirements on new chemicals. These principles as applied to industrial chemicals have also borrowed heavily from the practices required of the pharmaceutical industry and of the agrochemicals (particularly insecticide) sector, which were subject to voluntary or statutory requirements before these were made applicable to general industrial chemicals.

Control of toxic substances in the United Kingdom was initially by means of restrictions on specific substances found to cause health problems, such as lead, asbestos or aromatic amines. Detailed regulations were made in response to identified problems under the old Factories Acts and similar legislation. The attempt to identify and eliminate such problems before they cause disease, suffering and economic loss is a more recent initiative, and only when this was undertaken was systematic toxicity testing of chemicals required. The first United Kingdom legislation to embody this preventive approach to chemical and other hazards in the workplace was the Health and Safety at Work Act of 1974. It was considered necessary to apply the systematic screening earlier encouraged or required for drugs and pesticides to all industrial chemicals. This has involved not only a major increase in the amount of work performed, but also the development of new and quicker methods, particularly for identifying carcinogenicity, which has become a major and sensitive issue as a result of the identification of aromatic amines, and later vinyl chloride as occupational causes of human cancer. Development of systematic and preventive approaches continues both within the science of toxicology, where new methods based on biochemistry and molecular biology are replacing the old 'corpse counting' approach, and within the legislation controlling workplace chemicals hazards.

Legislative provisions for toxicity testing

Testing of chemicals for toxicity is both a scientific and a legislative activity. The types of test used and the circumstances in which they are required are obviously matters of scientific judgement, but these judgements have been formalized and standardized so as to provide a common standard for assessment and control of chemical hazards. The UK requirements for toxicological testing of industrial chemicals are specified in the Notification of New Substances Regulations, a statutory instrument providing for the exercise of powers granted under the Health and Safety at Work Act 1974 (see Appendix). These regulations correspond to an EEC directive (67/548/EEC) and a more recent amending directive popularly known as the 'Sixth Amendment' (79/831/EEC).

As may be deduced from their title the regulations apply to newly introduced substances. An inventory of substances already in use when these regulations were introduced was also developed, and their prior use was assumed to indicate that such substances were safe for their intended use. They were therefore automatically registered without further testing. Where problems are subsequently identified these will be subject to further research and review, leading to possible restriction. However, in practice, it is inevitable that there are gaps in the data on 'existing' substances, and these are generally less well studied than 'new' substances. The list of 'existing' substances has now closed, and all unregistered substances must be tested according to the New Substances Regulations even if a prior use has been identified.

The requirements for registration of industrial chemicals are dependent on the scale and type of intended use. No requirements are imposed for substances produced or imported for toxicological testing, research or analysis in quantities less than 1 tonne per annum, and a statement of identity, quantity of use and recommended labelling only are required for other uses in quantities up to 1 tonne per annum. Testing may, however, be required if it is known or suspected that the material is 'toxic' or 'very toxic', terms to be defined later. (Note: this and subsequent references to quantities in use refer to all uses throughout the European Community.) For all other substances (other than those covered by the separate

Table 19.1 Toxicology testing base set

Effect	Test	Route or level
Acute toxicity (rat)	Approx LD_{50} or limit test	(1) Oral
		(2) Dermal or inhalation
Skin irritation (rabbit)	Patch test, scored at 24, 48 and 72 h	0.5 g Solid or 0.5 ml liquid
Eye irritation (rabbit)	Topical application, scored at 24, 48 and 72 h	0.1 g Solid or 0.1 g liquid
Skin sensitization (guinea-pig)	Various, for instance 'maximization' test	(determined by test protocol used)
Subacute toxicity (rat)	28 Day exposure: find a 'no-effect level'	Oral or dermal or inhalation
Mutagenicity	(1) Ames test	*in vitro*
	(2) Chromosome aberrations	*in vitro* or *in vivo*

legislation for medicines and pesticides) there is a base set of toxicological data required. This is described in *Table 19.1*.

The quantity of material expected to be used must also be declared (in bands with break points at 10, 100 and 1000 tonnes per annum, or 5, 50 and 5000 tonnes in total). Further testing may be necessary if use is increased so as to enter one of the higher bands, or if potential hazards are identified by the base set experiments. In particular, studies of reproductive toxicity and carcinogenesis *in vivo* may be required. The special regulations for drugs, food additives and pesticides require all these tests and others besides. The New Substances regulations require basic ecotoxicological data in addition to the studies of mammalian toxicity: these consist of acute LC_{50} measurements for a fish species (the rainbow trout, *Salmo gairdneri*, often being used) and *Daphnia*, plus degradation studies (percentage ready biodegradability, biological and chemical oxygen demand and hydrolysis as a function of pH).

Performance of test procedures

Specific procedures are recommended for each of the required tests, and it is necessary that these be understood not only by those responsible for actually performing these tests, but also by those who may need to interpret the results and determine the implications for particular circumstances and risks. In most legislative systems for toxic chemical control the methods are specified fairly closely, and variations from the specified protocol are not acceptable. However, the UK Health and Safety Executive has taken a different, and perhaps a more scientific view. Recommended methods are described in the Approved Code of Practice *Methods for the Determination of Toxicity* (*see* Appendix) and it is assumed that they will be followed unless there is a good reason to do otherwise. However, where there is scientific justification for a protocol variation or even an alternative test method, this is accepted.

Since the results of toxicological testing may form the basis of legal proceedings, safeguards are required by all OECD legislative systems to ensure that the performance of the tests is competent, and that reporting of procedures and results is complete. Standards for the organization of testing laboratories, the qualifications of responsible staff, and the conduct and reporting of studies have been (more or less!) agreed internationally. The UK version of these standards is described in the Approved Code of Practice *Principles of Good Laboratory Practice* (*see* Appendix). Adherence to these principles should ensure an acceptable quality of data for regulatory purposes. Many of the provisions are basically commonsense measures which a competent scientific organization would implement anyway. However, special requirements which go beyond the intuitive notion of good practice are the provision of an audit procedure ('quality assurance') by independent observers during the conduct of a study, and the retention of extremely comprehensive archives so that the exact day-to-day progress of a study may be reviewed several years after it was conducted. Handling of the test substance is also required to be tightly controlled to establish its identity, and preserve its integrity during extended test procedures.

Details and interpretation of test procedures

Acute toxicity

The acute oral LD_{50} test has been used for many years, and is the simplest possible test procedure. In its basic form, the test animals (normally rats) are each given a single dose of compound. A range of dose levels is selected to include a dose at which no deaths occur, and also higher doses at which some or all animals die. A statistical analysis (formerly graphical but now usually employing one of the many computer programs available for the purpose) provides an estimate of the median lethal dose (LD_{50}), that is, the dose at which in a large population of animals half would die and half survive. For an accurate estimate of this value several dose levels are required including both a dose at which there are no deaths, and a dose at which all or nearly all the animals die. The number of animals per group is also required to be large if the estimate is to be accurate. The test in this form has been extensively criticized on several grounds. The observation of lethality is a very extreme and relatively uninformative measure of toxicity; it is to be hoped that it represents a situation which would never be relevant to an industrial exposure. The attempt to derive statistically accurate estimates of the LD_{50} involves the use, and possibly suffering and death, of large numbers of experimental animals, which is seen as ethically undesirable both by experimental scientists and by those concerned for animal welfare. The statistical accuracy sometimes sought is also spurious in that the phenomena producing death are intrinsically variable, and LD_{50} values are seldom repeatable within a factor of about 2 or 3. Also, for less toxic materials where the LD_{50} may be several grams per kilogram (g/kg) body weight death may not be due to any relevant toxicological process, but rather to asphyxia or intestinal obstruction; clearly such data reveal nothing of relevance to occupational exposures.

In response to these criticisms various changes in the test procedures for acute toxicity have been proposed, and the test specified for UK regulatory purposes incorporates these modifications. First, the requirement for a high level of statistical accuracy has been abandoned in favour of an approximate value, for which ten rats per dose group (five of each sex) are sufficient. Second, if there is no mortality in a group receiving 5000 mg/kg then no higher dose is used: this is the so-called 'limit test' for compounds of low toxicity. Observations of mortality are made within a specified period (14 days being the minimum, or longer if delayed toxicity is evident). The most important difference from earlier mortality study protocols is that cage-side observations of sublethal toxic effects, and necropsy of both dead and surviving animals, are demanded. Such observations are most important in providing an indication of the nature or site of action of toxic effects, and are likely to be far more relevant to the identification of potential occupational toxicity problems. Although acute lethality or severe toxicity is implausible as a result of routine industrial exposures except for a minority of highly toxic materials, the importance of these data for assessment of the hazards to be expected from accidents and major environmental releases due to engineering failures must not be underestimated.

Acute toxicity measurements are required for oral dosing with liquid or solid materials, or inhalation for gases, vapours and aerosols where inhalation exposure is of occupational significance. In the latter case the limit test specifies a dose level of 20 mg/l of a gas, or 5 mg/l of a liquid aerosol. Acute toxicity of liquid or solid compounds by skin absorption is also required to be studied, where a dose is applied to and held in contact with the skin for 24 hours using a porous gauze dressing. As with the other exposure routes, observations and necropsy indications are required. The relationship between the oral and skin absorption LD_{50}s is an important indicator of the potential for skin penetration by a toxic chemical; if the skin value is low it may indicate extensive absorption by this route. If this is found, further testing including pharmacokinetic measurements of skin absorption may be indicated.

Skin irritation

A determination of the potential for causing skin irritation is required. This is usually performed on rabbits, whose skin structure and response to irritation are considered to provide a better model for human skin than the rat. Whether this assumption is entirely true is a matter for debate, but at least this species has the advantage of long precedent in the grading of responses, which is done according to a standard table in which erythema or eschar formation (tissue injury in depth), and oedema are separately scored on a scale 0 (no effect) to 4 (severe effect). As for the assessment of dermal LD_{50}, a gauze patch is used to hold the material in contact with the skin, but normally only for a period of 4 h. Assessment of the irritant response, if any, is normally made immediately and at 24, 48 and 72 h after a single exposure, although variations to this protocol might be indicated to model more accurately specific features of the expected human exposure in a proposed use. A few compounds will be found to have much more severe effects on the skin including tissue destruction, and will be judged to be corrosive. If a corrosive rating may be anticipated on physicochemical grounds (for example strong alkalis or acids) this is assumed and no actual testing is required.

Eye irritation

Determination of eye irritation is also required, and this uses a similar scoring system after application of a dose of the material to the rabbit eye. Opacity of the cornea is scored 0 to 4, structural change or paralysis of the iris is scored 0 to 2, and for the conjunctival reactions redness and chemosis (swelling) are separately scored 0 to 4. Observations are made 1, 24, 48 and 72 h after exposure, with further observation if irritant responses are still apparent at 72 h.

This procedure (based on the Draize assessment scale) has been criticized on several grounds. It has been claimed that excessive pain and suffering may be caused to the experimental animals, and that this cannot be justified for some applications such as cosmetic testing for which it is extensively used. Whatever the correctness of that argument, the occupational physician will be aware that the frequency and severity of industrial eye injuries by chemicals is such as to require thorough testing of these materials. Research is in progress on alternative methods using isolated tissues and other techniques *in vitro*, and it is to be hoped that these can be developed successfully, perhaps as a preliminary screen to be used in a similar way to the short-term mutation tests for carcinogenicity. However, no such tests have been sufficiently developed to be acceptable for regulatory purposes at present.

A further difficulty with the eye irritation test is that there is considerable interspecies variation in both the structure and the response to irritants of mammalian eyes. The rabbit eye is used partly because it is anatomically a better model of the human eye than the eye of rodents, and also, being larger, easier to dose and examine. However, the extremely severe reactions scored in the Draize type of assessment do not have a direct counterpart in responses of the human eye. It has been suggested that the interspecies differences are exaggerated by the use of high exposure levels in these direct instillation tests. The use of lower doses coupled with more sensitive assessment procedures such as fluorescein staining of corneal erosions or functional tests could provide better models of human response, and permit the use of cheaper experimental animals such as rats. This might also reduce the ethical objections perceived by some to certain uses of the test procedure.

Skin sensitization

Skin sensitization is a common occupational problem where even small dermal exposures occur repeatedly. The control of this problem has not been helped by the absence of adequate testing for this phenomenon until relatively recently. Sensitization is a function of the immune system and as such shows considerable interspecies variation. Various test methods and test species have been proposed, and it would be unwise to claim that any were entirely satisfactory. However, the 'guinea-pig maximization test' has been established as the most widely applicable and reliable method, and is chosen as the reference method for UK regulatory purposes. This uses a complex multiexposure sequence in which the immune response is enhanced by Freund's complete adjuvant (FCA), a well-known activator of such responses. The protocol is shown in *Table 19.2*.

It is obvious from the experimental design, and the use of the guinea-pig, a species noted for its sensitivity to immunological challenges, that this test is intended to provide an indication of any possibility that sensitization might occur, rather than any estimate of the frequency of such occurrences among exposed humans. In practice, it seems to provide a reasonable indication of significant risks for most types of sensitizing compounds, but it must be borne in mind that human immunological responses tend to vary greatly between individuals. It is probably true to say that an occasional person will respond to repetitive exposure to virtually any organic compound, and also to many inorganics. Toxicological testing is unable to predict these idiosyncratic responses; indeed there is no procedure which will

Table 19.2 Guinea-pig maximization test

Group size:		Treated, minimum 20 animals Control, minimum 10 animals (guinea-pig, either or both sexes)
Day	*Group*	*Treatment*
Induction		
0	Treated	Three pairs of intradermal injections in the shoulder region, one of each pair on either side of the midline: (1) 0.1 ml FCA (2) 0.1 ml Test substance in vehicle (3) 0.1 ml Test substance in FCA
	Control	Three pairs of intradermal injections as above (1) 0.1 ml FCA (2) 0.1 ml vehicle alone (3) 0.1 ml vehicle in FCA
7	Treated	The test area is shaved, test substance is spread on a filter paper in a vehicle (or alone if liquid) and held in contact with the test area by a suitable dressing for 48 h
	Control	Vehicle only applied similarly
Challenge		
21	Both	The flanks are shaved. A patch or chamber containing the test substance is applied to the left flank only and held in place for 24 h by a dressing. A similar patch or chamber containing the vehicle, if any, is applied to the right flank
Observation		
23	Both	The challenge areas are cleaned 21 h after removing the patch, and examined 3 h later (48 h after the start of the challenge application). Any skin reactions or unusual findings are noted.
24	Both	A second observation is made (72 h after the start of the challenge application)

do this. A positive result in the maximization test will, however, indicate the probability of a more frequent and consistent induction of sensitization.

Subacute toxicity

Subacute toxicity investigations are designed to identify the occurrence and nature of toxic effects when repeated exposure occurs, and particularly to identify tissue damage or other changes at sublethal levels. If correctly designed, they are likely to provide an indication of the problem which might be encountered as a result of routine industrial exposures. The design of these studies will include careful observation of exposed animals during an extended period of repeated dosing, and also detailed study of clinical chemistry. At autopsy, both gross pathology and histopathology are investigated. If organ-specific toxic effects are identified, functional tests may be indicated.

The study design required for UK registration of industrial compounds is aimed at determining sublethal toxic effects. Species is not specified, although the rat would be the normal choice in the absence of specific indications that another would be more useful. At least ten animals (five of each sex) are required for each dose group. Doses used should include a 'no-effect' level, a clearly toxic but sublethal level and an intermediate level. The study is required to continue for 28 days. When there is no significant bioaccumulation it is relatively straightforward to estimate the required doses from the acute toxicity data. However, the lethal doses on repeated dosing may be much lower than expected where there is extensive bioaccumulation, as is found with some inert and fat-soluble materials. Much longer repeat-dose studies, and also pharmacokinetic investigations, may be indicated where bioaccumulation is identified. Such studies are required in every case for medicines and pesticides, where bioaccumulation in man or animals is a more frequent and serious problem than with most industrial chemicals.

The dose route is chosen to most closely represent the probable route of human exposure resulting from the intended use of the compound. Thus either oral, inhalation or skin permeation routes may be appropriate. In practice skin permeation studies of this type are difficult to perform satisfactorily, and another route (such as inhalation, which avoids any first-pass effects in the hepatic portal system) is generally used.

Where no toxic effects are seen at high dose levels (generally 1000 mg/kg/day for oral studies, or other levels chosen on the basis of likely human exposure), the 'limit test' principle is applied in the same way as for acute toxicity, and no further testing is required.

In the interpretation of data from these studies clues to organ-specific sites of toxic action are obviously important. These may take the form of histopathological findings, or obvious functional damage such as neuropathies or renal failure. However, important indications may also be obtained at lower dose levels from the clinical chemistry data. Serum transaminases may indicate liver damage, and changes in blood cations or urea may indicate kidney problems earlier than histopathology. This latter point may be important in that interpretation of rat kidney histopathology is not always easy, since even the controls may show substantial degenerative changes, depending on age and strain. Clinical chemistry data on experimental animals are generally similar in value and significance to those familiar in human medicine, but there are some specific differences in both normal values and organ-related changes.

Mutagenicity

The requirement for mutagenicity testing is based on the belief that such observations provide an indication of possible carcinogenic effects, as well as concern for mutagenicity as a deleterious process in its own right. In fact the latter aspect is rather a theoretical concern. Deleterious heritable mutations have never been shown to have been induced in a human population by industrial chemical exposures, although this may be as much due to the difficulty of demonstrating such a process as to its rarity. Chemical carcinogenesis on the other hand is a demonstrated human health problem, with several well-known instances in the occupational field.

The use of a short-term bacterial mutation assay as a quick and inexpensive indicator of carcinogenic potential was first developed by Ames, and the test protocol bearing his name continues to be the most widely used. The Ames test is a reverse mutation assay. Histidine-dependent mutant strains of *Salmonella typhimurium* are placed on agar plates which are almost devoid of histidine (a very small amount is in fact included so the bacteria do not actually die, but are unable to grow significantly). The test compound is added, usually as a solution in a suitable low-toxicity solvent, although suspensions and gas or vapour exposures have also been used successfully. If the compound is mutagenic, revertants will appear which are able to grow on the agar and after a suitable period can be identified as macroscopically visible colonies. The control rate of reversion, and thus the background rates of colony formation is low even when substantial numbers of test bacteria are placed on the plate; thus the test is sensitive because hundreds of thousands of test organisms can be exposed in a single experiment.

Most carcinogens and mutagens are active only when subject to metabolic activation, and this is provided by including in some test plates a liver homogenate subfraction ('S9'—the supernatant after centrifugation at $9000 \times g$ for 10 min). This contains microsomal and supernatant enzymes responsible for foreign compound activation in the liver; since these require nicotinamide adenine dinucleotide phosphate (reduced form) to work the fraction is supplemented with $NADP^+$ and glucose-6-phosphate (substrates for the glucose-6-phosphate dehydrogenase (G6PD) present in the fraction).

The bacteria used have been genetically modified to include specific types of lesion causing histidine dependence. Thus frameshift mutations are reversed only by mutagens causing frameshift type DNA damage (insertion of deletion of base-pairs), while base change mutations are reversed only by mutagens causing single base-pair substitutions. Also some other departures from the wild type have been included; these include DNA repair deficiencies and a weak and permeable outer cell wall, both of which changes increase the sensitivity to externally generated chemical mutagens. Some strains also have the sensitive gene located on a plasmid, which increases sensitivity, at the expense of a higher background reversion rate. A further strain specific character is the possession of the bacterial enzyme nitroreductase, which is important in producing positive results with some aromatic nitrocompounds. Tests normally include several different strains of test bacteria to cover these various possibilities. The strains known as TA1535, TA1537, TA98 and TA100 are specified in the UK regulatory recommendations, but others may also be used. A range of dose levels (minimum five) is used for the test compound, and also an appropriate range of positive and negative controls. Some recommended direct-acting positive controls are given in the UK guidelines, but it is actually better to choose positive controls which resemble the test compound chemically, including routes of metabolic activation, for a final evaluation, where this is possible.

Any positive result, usually defined as a reversion rate five times the background rate (with or without activation) is considered indicative of carcinogenic potential, and this conclusion is strengthened when consistent results in several strains provide information about dependence or otherwise on metabolic activation, type of mutations generated and so on. A positive result normally shows a clear positive dose–response relationship up to a maximally effective dose, beyond which toxicity of the compound to the bacteria causes a fall in the apparent reversion rate. Any failure to demonstrate such a dose–response relationship for the test compound or the positive controls suggests that the experiment has not worked correctly and cannot be interpreted.

Variations on the test are numerous, the most important being the use of alternative bacterial test organisms (particularly *Escherichia coli*, using a tryptophan-dependent mutant test strain), and single-celled eukaryotic test organisms (usually yeast). These variations are useful for certain classes of compound for which the standard conditions are unreliable. However most registration experiments will employ a standard protocol.

The other type of mutation test which is considered important is that which measures chromosomal changes rather than point (single base-pair) mutations. It is necessary to test separately for chromosome changes since such processes have no counterpart in the prokaryotic systems normally used for detection of point mutations. Cytogenetic (chromosome number and structure) changes have been demonstrated in response to chemical exposures *in vivo* in experimental animals and *in vitro*. A few cases have been identified where occupational exposure to mutagenic chemicals have produced cytogenetic changes in circulating lymphocytes in workers.

Tests for cytogenetic changes *in vitro* may use established cell lines, or lymphocytes from human blood activated by phytohaemagglutinin. Exposure to the test chemical is made to cover the whole cell division cycle either by extended exposure or timed observations to study effects separately in different phases of the cycle. In order to observe chromosomes the dividing cells are arrested in the division phase by addition of colchicine, then fixed, spread and stained so as to display the chromosomes of those cells arrested in metaphase. Incidences of chromatid aberrations (gaps, breaks or interchanges of one component of the paired chromosomes) and chromosome aberrations (gaps and breaks affecting both chromatids of a chromosome, and abnormalities such as fragments, rings and dicentric or polycentric chromosomes) are noted and analysed.

The testing of cytogenetic changes *in vivo* uses similar methodology, but exposure of living rats, mice or Chinese hamsters to the test chemical is followed by analysis of dividing cells in the bone marrow extracted from the femur. Exposure may be a single dose, or multiple dose if it has been shown that the compound is not toxic to the bone marrow. A single dose level may be used if this can be chosen on the basis of data available from other studies as a maximum tolerated dose, or a dose producing partial inhibition of mitosis. Alternatively a multiple dose level design may be employed. Each dose level group should consist of at least five animals of each sex.

The full cytogenetic analysis of a series of bone marrows is a time-consuming operation, and produces a lot of data some of which are difficult to interpret (particularly when only chromatid type lesions are observed, or the dose response is unprogressive). One attempt to overcome this problem is the 'micronucleus' test. Procedurally similar to the *in vivo* cytogenetic experiment, the observation made is of the number of micronuclei (isolated chromosomal fragments lagging at mitosis) left behind in polychromatic erythrocytes in the bone marrow. These fragments often persist for some time after the expulsion of the main nucleus from the developing erythrocyte.

Interpretation of individual mutagenicity test results may be somewhat difficult. A positive Ames test result has been found to correlate well with carcinogenicity for many types of compound, and false negatives are significantly rarer than false positives. On the other hand the cytogenetic data in isolation do not correlate so well with carcinogenicity. However, where a test compound produces positive results in a bacterial mutation assay and in a point mutation or cytogenetic assay in a eukaryotic system the correlation is high. A compound having these results would probably be assumed to be carcinogenic and abandoned or used only with great care, unless there was some other source of data (such as a history of extensive use with no epidemiological evidence for carcinogenicity) to offset the mutation test findings. Where only one positive result is obtained or there are difficulties with the data it may only be possible to resolve these problems by performing a full carcinogenicity test *in vivo*. This process is lengthy and extremely expensive, and so is only undertaken when justified by the uncertainty and the importance of the compound.

Other tests

Where problems of interpretation of data exist, or where large-scale use of a material is envisaged under circumstances where human exposure may occur, further testing is likely to be required. Three main types of additional test are commonly requested: reproductive toxicity (fertility and teratogenicity); extended repeated dosing (subchronic or chronic toxicity) studies; and carcinogenicity testing. Also repeats of previous tests in other species, and pharmacokinetic and metabolic studies may be required to assist the interpretation of data obtained by the standard methods.

Reproductive toxicity studies

These take various forms for different purposes, and in fact there is more variation between different regulatory authorities with regard to this type of effect than most others. The simplest type of test is that for teratogenicity, or production of malformations (including nervous system and behavioural abnormalities) in the young. The pregnant female is exposed to the test compound during the early stages of fetal development which are sensitive to teratogenic effects. The offspring are examined when the pregnancy is terminated just before birth is due; this is necessary since most laboratory animals will eat dead or deformed offspring at birth. The test species is variable: in the immediate aftermath of the thalidomide tragedy the rabbit was considered the most suitable. However, it is in fact coincidental that this compound is effective in the rabbit and in the human but not in the rat. The rat is no worse as a model species for chemicals in general, although in fact there is no experimentally useful species whose known sensitivity matches that of men exactly. Primates are sometimes used in reproductive studies for medicines, but this cannot usually be justified either ethically or economically for industrial chemicals. The choice of species is therefore based on a combination of analogy and intuition, and most often one of various rodent species (rat, mouse, hamster) or the rabbit is used.

Fertility studies may take various forms and be combined with teratogenicity (multigeneration studies) and mutagenicity (dominant lethal) tests. Depending on the protocol used either or both parents may be exposed either continuously or for discrete periods within the cycles of gametogenesis, fertilization and fetal development. In some studies the females are allowed to give birth and the effect of chemical exposures on the immediate postnatal development of the young is observed. The assessment of fertility involves determination of the number of live young produced in each pregnancy (and for the dominant lethal 'mutation' test the number of late fetal deaths as well); for animals such as the rodents where many young are normally present in a single litter this may vary independently of the number of pregnancies established.

The number of industrial chemicals which have been effectively screened for reproductive hazards is not as great as those subjected to the 'base set' screen, but reproductive toxicity is increasingly seen as an important occupational and environmental problem. Already a number of studies of teratogenicity have been undertaken, and emphasis is now being given to fertility studies especially since the problems with dibromochloropropane (DBCP) were identified. This material, a pesticide, was found to have caused male sterility among exposed workers. Recent experimental work on glycol ethers has also identified the ability of 2-methoxyethanol to cause testicular atrophy and male sterility, although fortunately in this case laboratory data became available in time to avoid any demonstrated examples of human injury of this type. This latter case is also an example of the fact that detailed studies of toxic mechanisms, including the metabolism and distribution of toxicants, can allow more confident prediction of which compounds present a hazard to man, and help the search for safer alternatives.

Subchronic and chronic toxicity studies

The normal 28 day (subacute) study is adequate to detect many of the problems that would appear on chronic exposure, but there are cases where this period would be inadequate. This might be the case for toxic actions which take a long time to appear, such as hexane or methyl *n*-butyl ketone-induced peripheral neuropathy, or ocular toxicity by nitrophenols or dimethyl sulphoxide. Where compounds with chemical analogues with these materials, or causing similar toxic effects, are being tested, a longer evaluation period may be advisable. If previous toxicity studies show substantial bioaccumulation, or toxicokinetic measurements show a long excretion half-life for the compound or its metabolites, longer repeat-dose studies may be necessary. There may be special features of the human exposure associated with the intended use which suggest that very prolonged exposure is likely, and therefore the effects of long-term exposure must be considered. It may also be desirable to prolong the normal subacute study, with or without continued dosing, when toxic effects are noted after 28 days but they have not shown saturation or completion of the relevant process. Such investigations may need to be specially designed, but for simple cases protocols are defined for 90 day repeat dose studies. The route of exposure may be oral, inhalation or dermal as most appropriate. Larger dose groups (for rats, ten of each sex) are recommended, with further increases in group size if interim sacrifice for histology or other testing is required. In other respects the methodology is similar to that for the 28 day studies, although special functional tests or anatomical examinations (such as ophthalmoscopy or nerve conduction velocity measurements, depending on the expected effect) may also be required. Again the usual species used are rodents, although others may be chosen if previous research on related compounds suggests that they would be more appropriate.

Carcinogenicity studies

Although the short-term tests using mutagenicity and cytogenetic techniques are a useful indicator, the only final test of a compound's carcinogenicity or otherwise is a full long-term study in appropriate test species. Many different designs and dose routes have been used for research purposes, but most regulatory require-ments indicate a study similar to that used by the US National Toxicology Programme (NTP). This programme has included a series of carcinogenicity bioassays of industrial chemicals and other materials which are regarded as suspect or possible carcinogens. A common basic design of bioassay has been used, in which dosing is for 2 years at a 'maximum tolerated dose' (MTD) and at one-half of this dose, with a large group size (around 50 of each sex at least) and appropriate controls. The dose level is defined by means of a preliminary subchronic study; the MTD is defined as a dose producing no lethality and minimal or no specific organ toxicity, but resulting in slight inhibition of weight gain (target range 10–20 per cent).

Dosing may be oral or by inhalation, but if the oral route is used the dose is given by gavage (direct instillation by stomach tube) rather than by incorporation in the diet or drinking water. The latter methods have often been employed in long-term studies, and are considered technically adequate, but suffer from uncertainties about the exact dose given, and the possible inhibition of dietary intake (a variable known to affect background tumour incidences). These problems can only be overcome by elaborate paired-feeding protocols, which are at least as labour-intensive as direct dosing. At the end of the dosing period, observation of the animals is continued until 80 per cent mortality in each group occurs, when the remaining animals are sacrificed. All clinical, dietary and weight-gain observations are recorded, and extremely thorough autopsy procedures with gross pathology and histopathology reports on all major organs are required both for animals dying during the study and at termination.

The NTP bioassay protocol specifies two specific strains of test animals: the Fischer rat and a special hybrid mouse (C57 Black/C3H Swiss F1). This consistency of strain has been one of the strengths of the NTP bioassay programme, since no other comparable volume of carcinogenicity data exists where comparisons between compounds can be made without the important confounding factor of different strains, experimental protocol and so on. On the other hand the particular choice of strain has been criticized, particularly in the case of the mice, since both strains were selected as 'high-sensitivity' strains rather than for any evidence of comparability between these animals and man. This is certainly consistent with the underlying purpose of the programme, but problems in interpretation of the results have arisen in some cases. This is particularly true where liver tumours have been observed in the mice (and sometimes but not always the rats), since rodents in general and the Swiss mouse in particular appear to be much more sensitive to liver carcinogenesis than man. There is not only a greater incidence of tumours at a given dose level for agents known to be effective in both species, but also many compounds such as halogenated hydrocarbons produce hepatocellular tumours (especially nodules and adenomas) in the mouse when there is no evidence of a liver carcinogenic effect in man even where there has been extensive and well-studied occupational exposure.

Carcinogenicity studies for regulatory purposes are routinely undertaken for pesticides and medicines, in rodents and, in the latter case in non-rodent species including primates (most frequently small ones, such as marmosets). Use of non-rodent species is useful in resolving problems of interpretation where positive results in the standard study are suspected of being rodent-specific. They will also be required for some special cases such as aromatic amines; since rodents do not retain urine they are very insensitive to bladder carcinogens such as 2-naphthylamine compared with man or animals such as the dog which do retain their urine in the bladder. However, non-rodent carcinogenesis bioassays are seldom undertaken except where absolutely required because of their cost, duration and (usually) poor statistical power due to small group size and greater interindividual variation. These problems become greater with increasing size of test animal: non-rodent species used have included dogs and large primates, but small primates such as marmosets, or other non-rodents such as the ferret or guinea-pig, would be preferred if suitable.

Interpretation of carcinogenicity studies is a complex and contentious business, not least because such lengthy and expensive studies are only undertaken for industrial chemicals where previous evidence is inconclusive, or where major financial and political issues are at stake. Detailed consideration of individual studies requires examination of the original experimental data, to determine whether the technical execution of the experiment was correct (no mean feat in a study of this size, duration and complexity), and whether the statistical analysis was able to provide valid conclusions from the data. One persistent difficulty even with well-conducted studies of weak or doubtful carcinogens is that chronic toxicity or infectious diseases may cause extensive non-tumour related deaths in treated groups, so the size of the group surviving to the end of the dosing period is inadequate. The available statistical power of the study becomes critically important when the point at issue is a possible negative, rather than a clear positive finding.

Many of the materials currently in use in industry which have been suspected of carcinogenesis and for which experimental or epidemiological evidence exists, have been reviewed by the US National Toxicology Programme and/or the International Agency for Research on Cancer (IARC). The monographs produced by the latter organization provide thorough expert analyses of these problems, and sometimes also offer firm conclusions. There are, however, a very significant number of cases where data are inadequate or interpretations too conflicting to allow a definitive conclusion to be reached.

Metabolic and toxicokinetic studies

Knowledge of the route of metabolism, and the rate of uptake distribution and excretion of a toxic chemical, are sometimes required for registration of chemicals, particularly where the standard subacute studies have indicated the possibility of bioaccumulation. They are also advisable when the material under test is analogous to other materials known to produce chronic toxicity or to accumulate either in man, in animals or in the environment. Even when not actually required by regulatory authorities they can be very useful in assisting the interpretation of toxicity testing data, and would usually be undertaken as part of a testing programme for any material for which subchronic or carcinogenicity studies were likely to be necessary. They would also be useful in assessing the significance of any result indicating organ-specific toxicity, since they are an important step towards the understanding of underlying mechanisms of toxicity.

Specific guidelines for such studies are not usually given (although standard bioaccumulation measurement protocols exist for pesticide registration studies), since the experimental programme needs to be designed specifically to suit the circumstances. Metabolism studies may begin with simple measurements *in vitro*, but usually both metabolism and kinetics are measured *in vivo*. Radioactive tracers and sophisticated chromatographic techniques are generally employed to determine levels of compound and metabolites, and metabolite identification frequently requires resort to mass spectrometry. Interpretation depends partly on an understanding of the biochemical processes involved in toxic chemical metabolism, and may also need detailed knowledge of the chemical structure and properties of the test compound.

Hazard classification

Analysis of all the occupational health implications of data obtained from toxicological testing obviously involves consideration of the detailed results obtained, and also of the circumstances of use of the material. However, as a convenient first step, simple schemes are used for the classification of chemicals according to the general nature of the hazards presented. These general classifications are used to indicate the appropriate form of warning labelling required for transport by land, sea or air and for supply within the jurisdiction of various regulatory authorities. Toxicological data are generally considered along with physical and flammability data in these schemes. Within the European Community a standardized scheme of hazard classification of any pure substance is used, which in the United Kingdom is specified by the Classification, Packaging and Labelling (CPL) Regulations. Classifications of mixtures or formulations is a more complex issue which has not so far been fully standardized within this system, although the UK regulations have used a formula for dealing with mixtures for some time. A recent Community directive has now established the required format for use throughout Europe, and a draft of the modified UK regulations has been published. Labelling classifications and standard warning labels are specified for supply (defined as sale *or* movement from one site to another), and also for road transport; standard codes are used to identify materials to emergency services for the latter purpose. Use of the standardized classifications is also made in the preliminary identification of major accident hazards (Control of Industrial Major Accident Hazard Regulations 1984).

Table 19.3 Toxicity classification criteria*

Category	LD_{50} (mg/kg) Oral rat	Skin absorbed rat or rabbit	LC_{50} (mg/kg) 4 h Inhalation rat
Very toxic	≤ 25	≤ 50	≤ 0.5
Toxic	> 25–200	> 50–400	> 0.5–2
Harmful	> 200–2000	> 400–2000	> 2–20
Not hazardous	> 2000	> 2000 (or no fatalities in 'limit' test)	> 20

*If significant toxic effects other than lethality are observed these have to be taken into account in assigning a classification.

The classification system derives a rating of non-toxic, harmful, toxic or very toxic, according to the LD_{50} values (*Table 19.3*), and requires warning symbols and risk phrases describing these properties. Any material harmful or worse is considered dangerous for supply, and must be appropriately labelled, and the related Transport Regulations are also applicable. Specific hazards, including flammability, irritant or corrosive properties (as determined by the results of the skin and eye irritation tests) are also explicitly covered. Although account is taken of carcinogenicity, mutagenicity, sensitization or other chronic or irreversible toxic effects, these are not so easy to incorporate satisfactorily into the scheme.The reliance on acute LD_{50} values is defensible for transport and major accident hazard purposes, but the relevance to problems likely to be encountered in the workplace is not so clear. Unfortunately the sublethal acute and chronic toxic effects are much more difficult to quantitate on a general basis. Where there are important hazards other than acute lethality the classification becomes much more a matter for expert judgement. Classifications by the UK authorities of a number of widely used substances are published for reference (the authorized list). It cannot be supposed, however, that the basic classifications used for labelling, transport and so on provide adequate hazard identification for analysis of occupational health risks where extensive use or substantial hazards are involved.

Reference

Pott, P. *Chirurgical Observations related to the Cataract, Polypus of the Nose, the Cancer of the Scrotum, the Different Kinds of Ruptures and the Mortification of the Toes and Feet*, London (1975)

Appendix: UK Documentation and Legislation

Regulations under Health and Safety at Work etc. Act 1974

Notification of New Substances Regulations (1982)
Control of Industrial Major Accident Hazard Regulations (1984)
Classification, Packaging and Labelling of Dangerous Substances Regulations (1984; amended 1986).

Guidance notes on these regulations are available.

Approved Codes of Practice and Approved List

Classification and Labelling of Substances Dangerous for Supply and/or Conveyance by Road
Principles of Good Laboratory Practice (COP7)
Methods for the Determination of Toxicity (COP10)

Revision no. 1 to the approved list (information approved for the classification, packaging and labelling of dangerous substances) incorporating information approved for the conveyance by road of dangerous substances in packages etc. *Note:* Revision no. 1 does not include the parts relevant to classification for supply, which remain as in the previous list, now out of print. However, a new revision of the complete list is in preparation.

Further reading

Ballantyne, B. (ed.) *Perspectives in Basic and Applied Toxicology*, Wright, London (1988)
Bridges, J.W. and Hubbard, S.A. Principles, practice, problems and priorities in toxicology. In *Current Approaches to Occupational Health* (ed. W. Gardner), Wright, Bristol (1982)
Brown, V.H.K. Acute toxicity testing–a critique. In *Testing for Toxicity* (ed. J.W. Garrod), Taylor and Francis, London (1981)

Chapter 20

Ergonomics

E.N. Corlett

Introduction

Ergonomics is an area of scientific and professional interest which arose as a result of an observed lack of success of many new and complex weapon systems during the Second World War. Much of their failure was traced to the inability of the user to cope with the increased complexity, speed of operation or other demands of the system. Work by combined teams of engineers and people drawn from the human sciences revealed that the requirements specified by the designers to operate these systems did not match the abilities or capacities of people. As a result much of the user's effort was taken up with trying to overcome the inadequacies of the system rather than in achieving the system's objectives.

Over the last 50 years ergonomics has become a worldwide discipline, supported by an international association and the emergence of professional or registered ergonomists in many countries. In spite of this, much industrial equipment in both the factory and the office displays clear evidence of its inadequacies for the user, and the results of this frequently are also evident in the occupational health clinic. Consequently it is desirable that a text on occupational health practice should incorporate a section on ergonomics, to alert occupational health practitioners at all levels to the contribution that ergonomics can make to the prevention of occupationally related disorders.

An outline of the broad field of ergonomics may be gained from *Figure 20.1*. It will be seen that the field can be crudely divided into the physiological links between a person and his or her working environment, the psychological factors involved and the organizational aspects. These are of course not clear-cut and separated, each aspect can affect and be affected by the other two, but this framework will be used to shape the rest of this chapter. A feature of ergonomics which is not evident from *Figure 20.1* is that its central focus of interest is the individual. The primary concern is arranging the physical relationships, the environment and the information loads to be within the capacities of the individual himself/herself. This is not to say that the objective is to minimize the loads, this would clearly be an inadequate approach since too low a level of activity can be as harmful in its own way as too high a level. Ergonomists recognize that there is a range within which safe and healthy activity can be carried out, in all the fields expressed in *Figure 20.1*, and endeavour to so design the work situation that the loads remain within the acceptable range.

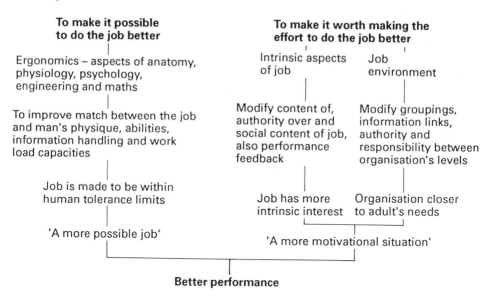

Figure 20.1 Two directions for the improvement of the experiences of work, and of its performance; although improvements will result from effort put into the left hand column of the figure, long-term benefit requires both columns to be given equal attention

It will be seen that this approach is diametrically opposite to the so-called 'Tayloristic' viewpoint, still widely accepted within industry. F.W. Taylor, an engineer active in the United States at the turn of the century, investigated the design of work situations so that the maximum amount of work could be obtained from the individual in any given case. He did this experimentally and christened his developments 'Scientific Management', a term that is now widely used but which represents a procedure that is far from scientific. The focus of Taylor's attention was on the work output: the behaviour of the individual to achieve the output was optimized from a short-term and mechanistic point of view, that is the individual was seen as a sort of biological machine which manipulated the components in the work situation. In order to achieve maximum output work was structured to require minimal learning, attention or physical or intellectual activity.

The results of the scientific management approach were not wholly bad as far as industry was concerned, although the human effects were, by today's standards, unacceptable. Perhaps a major benefit was to draw management's attention to the fact that it was possible by thought and investigation to improve the opportunities which people had for achieving their work objectives. The fact that initially these were done to a great extent at the expense of the worker concerned was the obverse side of the coin. The practice of industrial management began to grow up and the beginnings of an intellectual approach were evident.

Today the position in many countries in the world is different. The part on the right-hand side of *Figure 20.1*, concerning work organization, is a major concern of ergonomists in advanced countries. People are now viewed holistically rather than as just a pair of hands or a source of energy, and a greater understanding of what is desirable and healthy for an individual at work is being achieved. In consequence new approaches to working conditions, working relationships and management structures are providing work situations in which it is possible for an educated and sophisticated workforce to find satisfaction and interest in their normal working lives.

Interaction between ergonomists and occuptional health personnel at all levels can be very fruitful. The ergonomist is in a position to identify the workloads, activities and objectives of people in their jobs and also identify when those relationships are adverse. On the other hand, the occupational health professional can provide ergonomists with valuable information on the trends in diseases appearing in the surgery, and can alert the ergonomist to the need for redesign or other change as a result of the occurrence of particular incidents.

Physical effort

As mentioned earlier, the objective in ergonomics is not to reduce the amount of effort an individual is required to produce to an absolute minimum. However, what is the minimum is much more open to question than what is an acceptable maximum. Work physiologists, particularly in Britain, Germany and Sweden, have contributed to extensive research which has identified how to measure physical effort in the workplace, using both heart rate measures and psychophysical techniques, for example [1]. They have also made great strides in identifying what is an acceptable rest pause and how overload may be identified. Although modern electronics has provided highly portable heart rate monitors, simple techniques can cope with the great majority of problems. That by Brouha [2] utilizes the recovery pulse in a 3 min period after the cessation of activity to estimate the workload as a result of that activity. An evaluation of the trend in a series of such measures can indicate whether recovery is sufficient during the non-work periods, as well as assessing whether the workload is within the worker's capacity.

Concern has been expressed on numerous occasions concerning the levels of fitness of such groups as mine rescue workers or firefighters. They may have to engage in high levels of physical activity with intervening intervals, with no extreme effort, of weeks or months. Studies (for example, Kilbom [3]) have illustrated the inadequacy of the physical condition of such specialist groups and the increased risks to their health in consequence. Others have recommended both fitness training (for example,Brown [4]) or equipment redesign [5] to reduce the risks.

While it is probable that the extremes of physical effort required of people in industry have gone down over the decades, there are still many occupations which require exertion of forces, quite often in the most awkward of postures. Agricultural workers, those in stores or engaged on maintenance duties, as well as retail delivery workers, can be called upon to lift, carry or push heavy loads, often with bent back, knees or in a twisted position. One typical study in this area, undertaken by the Finnish Institute of Occupational Health, was into the loads on the backs of loaders engaged in stowing cargo into modern aircraft [6]. High levels of back disorders had been reported, but redesigned methods of working and handling were responsible for a considerable reduction in the risk.

In all such cases the ergonomist would endeavour to reduce the peak loads by design. For example, to aid in the maintenance work in a steel works, researchers from the Laboratory of Industrial Ergonomics of the Royal Institute of Technology in Stockholm developed a mobile workstation [7]. This was capable of assisting in the lifting and manipulation of the heavy roll bearings of rolling mills, taking part of the load of welding equipment, providing adequate lighting and all the facilities which a maintenance worker required. The pushing of trolleys, straining to lift bearing housings from awkward places and the manipulation of welding equipment were all reduced or eliminated, and a direct gain was seen in both efficiency of maintenance and increased health.

In the area of manufacture, low levels of effort are often mistakenly considered to indicate that the job is safe and easy. It is still customary to see advertisements for people to undertake 'light work', which means that small loads have certainly to be manipulated, but they have to be manipulated many hundreds or thousands of times a day. Increased speed of working, together with a reduced range of working activities, has focused increased attention on the so-called 'repetitive strain injury'. This can be found to occur in assembly workers, office workers, and even university computer operators. Most industrial medical departments will have seen one or the other manifestation of such a disease and will be familiar with its symptoms. Although research still continues into the causes of the various types of repetitive strain injury, the ergonomist endeavours to attack the problem on two fronts. First of all, of course, to rearrange the work to minimize the number of repetitions. This is to a certain extent a fairly late attack on the problem; an attack much earlier in the causative chain is to redesign tools to minimize the need to exert forces while joints are in extreme positions. *Figure 20.2* shows a pair of pliers designed for use by people wiring circuit boards who had to hold two wires together and twist them. The design is at least 20 years old and was developed by one of the branches of the General Electric Company in the United States.

In the case of the widespread form of the disease which arises from the operation of computer keyboards, redesign of the shape of the keyboard, its placement in relation to the operator and the position of all the other components of the work, for example the script or the screen, can reduce the probability of injury.

Figure 20.2 The effect of recognizing the need to avoid the flexed wrist, arising from the need to point the conventional pliers at the work point, enabled the designer to bend the plier handles instead; the consequence is that hand and arm are in line and such problems as carpal tunnel syndrome are avoided

Extensive studies over the last 8 or 10 years at the STK Kabelfabrik factories in Norway, in particular by Aaras and Westgaard [8] have demonstrated clear relationships between certain aspects of the posture and force exertion by workers engaged in repetitive tasks and the subsequent incidence of musculoskeletal problems. Recent work has demonstrated a correlation between EMG levels, positions of the arms in space, and shoulder and neck problems. Such studies lead to the possibility of identifying the more dangerous postures, dangerous in the

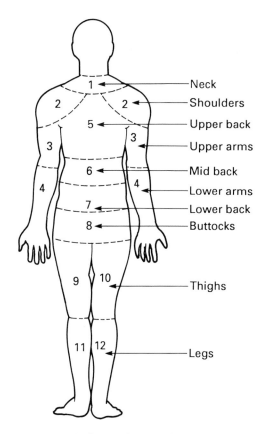

Figure 20.3 A diagram for exploring the subjective feelings of discomfort during work. It should be sketched in a sitting position if the worker is normally seated. The divisions should be chosen after discussion with workers so that appropriate areas are marked out. Each area should be given a different number. In the case where this diagram was used the arms were working together sharing the loads equally

sense that they are the ones likely to lead to early occurrence of diseases, as well as opportunities to specify what are desirable or undesirable working postures.

Although relatively sophisticated techniques, such as EMG, can be used in the workplace, they are not often convenient and are certainly expensive in expertise. In recent years a greater understanding of the use of psychophysical methods has developed, such as that mentioned earlier and developed by Borg [1]. While his work dealt with the perceptions of physical effort, the discomfort recording procedures proposed by Corlett and Bishop [9] were particularly oriented to the effects of postural loading. The technique requires questioning of operators at intervals about the position of discomfort, using a 'body map' (*Figure 20.3*). The relative increase of discomfort in various parts of the body indicates to the ergonomist which areas are most heavily loaded, and thus of major concern for redesign, while the level of discomfort reached gives indications of the severity of the problem. An example of the use of the procedure is given in *Figure 20.4*. After changes a repeat of the procedure will demonstrate the effectiveness of the changes as far as the worker's comfort is concerned.

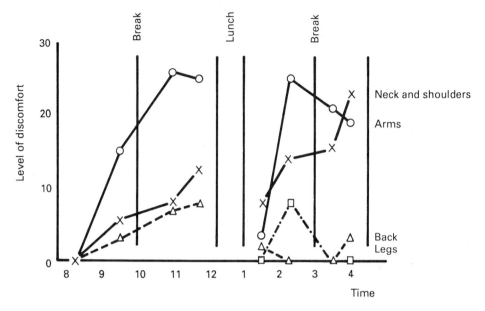

Figure 20.4 The effects of discomfort, recorded from four young women working engraving machines. The subjects were asked to rank their discomforts using a 7 point scale, a value of 28 representing maximum discomfort. It will be seen that the lunch break had little effect on recovery, the morning and afternoon graphs for neck and shoulders, as well as arms, are virtually continuous

Static effort

It is perhaps a bit anomalous to put lifting and handling problems under the heading of 'static effort'. However, much of the work, particularly that conducted for the National Institute of Occupational Safety and Health [10], has utilized biomechanical analysis as a major technique in the investigation of safe or unsafe loads. Biomechanical analysis is usually pursued by the use of static biomechanic formulae, that is, not including inertial loads, and it is customary to refer loads to the level of L3 or L5/S1. Calculations of the moments around these lumbar points in relation to the moments exerted to maintain the posture of the body and to overcome the lifted load can guide the ergonomist as to the adequacy or otherwise of the work design. The NIOSH publication quoted [10] gives recommendations for lifting and handling which are in extensive use in the United States. In part the dynamic aspect of the problem has been considered by once again calling on psychophysical methods [11]. The studies required a large number of subjects to estimate the levels of load which they could lift, under certain conditions and from various positions, without feeling stressed. *Figure 20.5* indicates some of the results, taken from Snook [11] and modified for a graphical format. This psychophysical method has been used in Europe and provides a complement to other techniques in the study of lifting.

The introduction of increased levels of automation, of robots which move material about and the presence of computers in the office have all reduced the amount of physical movement which people need to make at the workplace. In the office, for example, the secretary can produce the typing needed, refer to files and

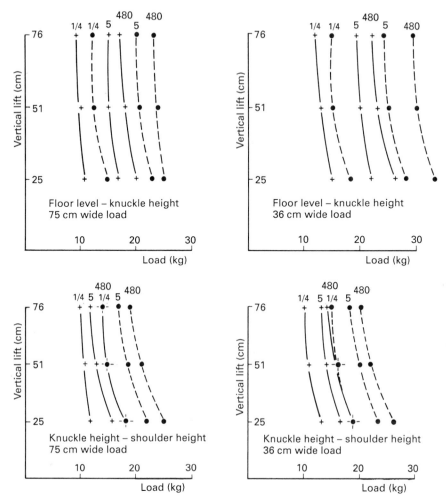

Figure 20.5 Maximum acceptable weights of lift for 90 per cent of population showing effect of height of lift, frequency of lift, width of object, and male and female differences. ● = Male 90 per centile; + = female 90 per centile; interval between lift in min. (From Snook [11])

submit messages to other offices without ever leaving the keyboard. As a result it is not just the repetitive injury arising from increased use of the keyboard which is important, and can be a problem, but the long periods of sitting put an increased load on the back and the legs. Winkel's study [12] of leg swelling in secretarial workers demonstrated the importance of periods of exercise after relatively short intervals of sitting, as illustrated in *Figure 20.6*. Once again the need is for careful design so that periods of exercise occur in the natural course of the work. The introduction of artificial pauses or the attempt to regulate people's behaviour are both unsatisfactory substitutes for this.

The problems arising from extended periods of sitting have focused attention once more on the design of seating, both for the office and factory floor. Recent work [13] has demonstrated the importance of taking into account the work

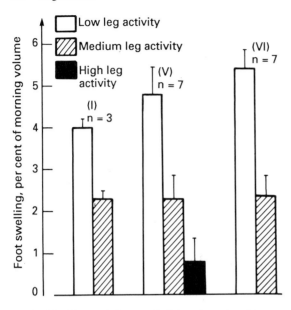

Figure 20.6 The mean increase in foot volume in subjects engaged in sedentary tasks lasting 8 h. Blank columns were the effects of constrained or inactive sitting, the hatched columns had the conditions of the blank columns interrupted by (I) four 2 min periods of walking per hour, or (V) 'semi active sitting', or (VI) a period of walking at least once every 30 min. The black column was for active sitting, where the subjects had a leg activity device provided under the desk. Vertical bars are standard errors of the mean. (From Winkel [12])

activities and other workspace dimensions in the specification of the seat. *Figure 20.7* outlines some of the important factors in seat specification, some of them barely taken into account in seat selection. Indeed, attention has been focused for so long on the question of lumbar support and seat height to avoid under thigh pressure that manufacturers sell so-called 'ergonomically' designed seats to customers who are unaware of the overriding importance of the work tasks themselves in seat selection.

It will be seen that one of the points mentioned in *Figure 20.7* is that a good seat assists in unloading the back. It is evident that forces exerted by the hands, together with the inertial and gravity loads imposed by the movements and weight of the upper trunk, head and arms, must all be transmitted to the seat or to the floor via the buttocks or legs. Virtually all must pass through the lumbar spine since the contribution of intra-abdominal pressure to support of the trunk is relatively small. Much industrial work which is done seated can be eased if the design of the seat back is such that some of these forces can be transmitted through it rather than to the seat surface itself. Studies by Eklund [14], illustrated in *Figure 20.8*, demonstrate the effect of pursuing light assembly activities on a stool or on a properly designed seat where the back can be partially unloaded. The technique is that of using a very precise stadiometer, shown in *Figure 20.9*, which measures changes in total stature to within 0.5 mm. The effect of different designs of workspace or work activities can be compared in relation to the amount of spinal shrinkage incurred. These experiments demonstrated the close relationship between loads transmitted to the trunk and spinal shrinkage, thus it was possible

Seating model

Sources of demand	Responses		Measures
	Initial	*Subsequent*	
Task			
Size—i.e. positions of hands in space to do task	Spinal shape	Discomfort	Shrinkage
Forces ⎱ level	Spinal load	Disease	Biomechanical load
⎰ direction of	Neck/shoulder muscle load	Performance change	EMG
			Anthropometric assessments
Repetitiveness	Effects of surface pressures		
Length of time on task ⎱ Time and space	Displacement of internal organs		Rating scales
Constant posture requirements ⎰ constraints			Epidemiology studies
Visual requirements			Seat loads
Workplace			
Height, leg clearance, visibility, job aids, body support, clothing			Foot loads
Individual characteristics			Loads on seat back
			Seat usage and work performance

Figure 20.7 The factors which influence, and affect, the design and choice of an industrial seat. It will be noted that seat and table heights are influenced by, for example, work requirements, and cannot be chosen independently

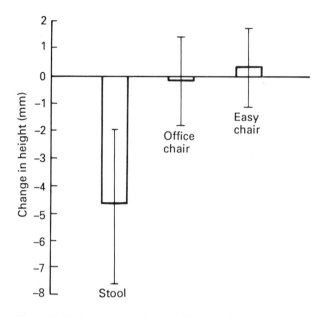

Figure 20.8 The spinal shrinkage which occurs if subjects are seated on different forms of seat. Light assembly work was done and work heights adjusted to give the equivalent work heights for each form of seat. The absence of a back rest and the effect of the resulting kyphosed back when using the stool are very evident.

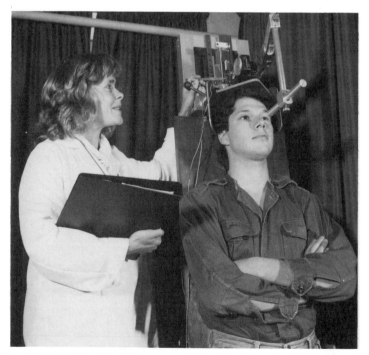

Figure 20.9 Measuring the spinal shrinkage using the Nottingham precision stadiometer. Close control of the posture when on the instrument is necessary. Readings are taken to 0.01 mm

wide range of working activities other than just the design of seats. This permits the ergonomist to identify situations which will minimize the likelihood of low back pain and disease. The technique, readily usable in the field, promises to provide new information for the investigation of conditions known to cause lumbar damage.

Information handling

The introduction of massive opportunities for electronic information handling has led to major changes in work activities at all levels of business and industry. It is common knowledge that computer control in the factory is steadily reducing the numbers engaged in manipulating machine tools and requiring fewer but more skilled people to deal with their programming and the maintenance. An even greater revolution is occurring in the office. Where once there were armies of clerks and typists there are now armies of people seated in front of VDUs. It has already been commented that this change in technology has led to new forms of disease, or rather to increases in certain diseases as a result of the increase of keyboards and the decreased activity in the office as a whole. There have also been major changes in the handling of information. Whereas documentation was once done slowly, using ledgers or proformas which were familiar, and eventually business machinery which substantially changed the activity only in the sense that the data were entered by key rather than by pen, now the forms themselves are provided on the computer screen and may not exist on paper at all. Increasing sophistication in computer hardware and software has led to increasing sophistication of office requirements, and the use of the VDU in the office now requires the operator to maintain a mental map of the system he or she is operating. This mental model of the system can be complex and requires the programmer to have some understanding of how people learn such systems and handle information in order to produce what is called 'user-friendly' software. Much software is still produced with little understanding of the needs of 'user-friendliness'. It is considered that the provision of extensive menus, together with some graphical display of the system will make it 'friendly'. The result is that learning is slow and errors more frequent than are necessary. Furthermore, the amount of attention that has to be devoted to the system, the concentration needed, is unnecessarily increased.

There is both anecdotal evidence and results from stress studies to show that working in a computerized office has its own problems. One major mail order warehouse in North America operated a very relaxed system whereby an operator could walk off the job at relatively short notice in order to counteract the steady build-up of strain caused by long periods spent in the entry of complex strings of numbers. These numbers were automatically transmitted to warehouses at various points in the United States and Canada, from which the goods were immediately despatched. An error in entry could lead to considerable problems both for the company and the customer, and the continual entry of these numbers from a myriad of order forms arriving daily presented a high working stress.

In some respects the modern office has become like the old-fashioned assembly line, and can be too easily made more like it than the health and efficiency of its workers can stand. Pinned to their seats by the requirement to operate a keyboard and fed with an unending supply of documents, the operator has little time for change in activity, for social interaction and in some cases even for relaxation.

With the international transmission of data it is possible that such work can be transferred to Third World countries, much as mass production assembly is being transferred, where workers have little protection from the attention of rapacious employers. They are a challenge to the ergonomist as well as to the production engineer or industrial engineer to achieve working conditions which are both efficient, safe and attractive for people in the western world, rather than taking the easy route of exploiting those who are unprotected. This area, while being seriously explored, still has major problems the understanding of which requires urgent research. There is much that the ergonomist can do, even so, but as yet not enough.

Accidents and errors

A dramatic incident such as the Three Mile Island or Chernobyl disasters focuses the attention of the world on the potentially tragic effects of human errors. Our industrial accidents procedures still spend too much time in seeking who is to blame, an approach which is well recognized to lead to distortion in the investigations as those involved attempt to avoid being saddled with the blame. However, the ergonomic approach to accidents is at once more human-centred, and starts from the initial premise that an accident must have had a reasonable cause. By this is meant that it is, for instance, quite possible that the individual concerned undertook the acts which led to the accident under the misapprehension that those acts, at the instant they were undertaken, were appropriate for the situation which existed. We are all familiar with the distortion in perception which can be engendered by bad light, sleeplessness, or other influencing stressors. By investigating accidents from this point of view, that is, that some stress may have distorted the perception or judgement of the persons concerned, many may yield information which will provide the ergonomist or engineer with ideas which can be used to design out the likelihood of an occurrence.

For many years both railway and aircraft accidents have been investigated with this point of view as one of their major objectives. As a result, there has been a steady increase in the safety of both rail and air transport. A similar comprehensive approach is not yet evident in the industrial accident field, although there have been some dramatic improvements in safety in many areas—mining, for example. The introduction of an ergonomic analysis as a necessary component of any accident investigation should lead to a similar contribution across a broader front.

The physical environment

Illumination

The ergonomist is particularly interested in the lighting, the noise and vibration levels and the thermal condition of the environment, since all of these have a major effect on human performance. Lighting problems have been prominent in recent years in relation to the introduction of VDUs. There are still many instances where a VDU is introduced into an office without an appropriate understanding of the difficulties of viewing a screen in the conventional brightly lit office. The contrasts, even on a modern screen, between the letters and the background are lower than would normally be found on a typed or printed page. As a result ordinary room lighting on the screen will tend to increase the luminosity of the background and make the information presented on the screen more difficult to read. The fact that the screen is vertical means that the operator is viewing it more or less horizontally, rather than the downward directed gaze when using paper. As a result the lighting in the office can be more or less in the same line of sight as the screen. These high intensity sources of light will cause the eye to adapt, reducing still further the ability to read the displayed text. Having the VDU so placed that the viewer sees the reflection of the office windows, or of light surfaces behind the operator, will also have the effect of reducing the contrast.

There are many texts now available which provide sound advice on the design of VDU workplaces (e.g. Cakir *et al.* [16]). There is still room for further work, but enough has already been done to enable most of the problems of the computerized office design and layout to be satisfactorily overcome.

Noise

As a result of extensive activity over the last two or three decades by the Health and Safety Executive, the problems of noise and its control are much better understood than heretofore. There is however still a regrettable tendency to rely on ear defenders as a first rather than a last resort.

The problems of noise, however, are not just those of industrial deafness. Communication in factory or office is still very much a matter of voice communication, and the presence of high levels of noise increases the difficulties of this. If there is a possibility of mishearing and consequent error, people will be reluctant to maintain the use of ear defenders. Also, immersed in a high level of noise the attempt to hear can in itself add to the working strains. Of course, it is not just voice communication, but many pieces of equipment communicate information about their performance and condition by the noises they make. These can be important sources of feedback for the user, and masking by irrelevant noises can increase the load on the operator.

Heat

The thermal environment, also well covered elsewhere in this book, if adverse, adds another stress to the worker. The technique by Brouha [2], described earlier in relation to physical effort, has been used successfully in several cases where the physiological cost of adverse temperature conditions has been under study. The combination of heat and physical effort can move an otherwise feasible task into the 'extremely difficult' region; the relative simplicity of the measures required to assess this usually cause these problems to be tractable.

The design of protective clothing is one thing, its effective use is another. Monitoring by management of the correct methods of wearing must be maintained, but a thorough programme of involvement for the workforce in the selection, and then on the introduction of the clothing, together with a presentation of information about its relevance and its method of functioning can be even more important. The ultimate effective monitor is the wearer. Proper use can only be achieved if the equipment protects without hindering the performance of the work itself.

Vibration

The two major areas of interest here are whole body vibration and hand–arm vibration. The problems of the former range from seasickness at very low frequencies to damage, and probably high spinal loads, at higher (4–12 Hz) frequencies. Problems of control and monitoring also exist outside these frequencies, particularly for the operators of military vehicles.

The possibility that increased spinal loading can arise from the oscillation, in or out of phase, of major body segments at around their natural frequencies is arousing research interest. The prevalence of back pain in vehicle drivers, in spite of the advances in seat design in recent years, spurs the search for other possible spinal stressors.

Hand–arm vibration, particularly from hand-held tools, has received a lot of attention, and the design of tools to reduce vibration transmission to the user is well advanced. In the case of fettlers who hold the component rather than the grinding wheel, attempts to automate the process are the most likely direction for ameliorating the major incidence of vibration white finger amongst such workers.

Work design and organization

In the last half century there has been a revolution in social attitudes, and not only in Britain. The opportunity to make the most of one's life, once the privilege of the rich, is now the expectation of everyone. Since the 1950s this has been the common experience, and present-day adults recognize that they can run their own lives. Consequently, the operation of industry and business cannot be carried on by managers taking the decisions and giving the orders and workers operating as obedient doers.

It it these experiences and expectations which have major influence on people's attitudes to simple, dull or repetitive work. The comparative isolation and the closed society of many factories is an unattractive milieu to those who see their friends working in cleaner and more social surroundings. Advanced companies in Europe, such as for instance Volvo, have seen this trend and more than 20 years ago set to work to design factories and working conditions that were acceptable to a modern educated European workforce. Moving ahead from the basic environmental improvements, a major aim was to provide some autonomy and responsibility for every individual, the capacity to have as much opportunity as possible to direct their own working activities.

Clearly, such developments must lead to changes in management organization and management behaviour. From being somebody who decides what to do and orders people to do it, management becomes a facilitator and provider of the means and requirements necessary to complete the workers' tasks. Factory management structures become less hierarchical and flatter, while those engaged in doing the work in the shop have the responsibility and opportunity to produce better work with less stress.

Such developments are not a separate procedure, to be operated by some personnel management group independently of the rest of the factory system. As illustrated on the right-hand side of *Figure 20.1*, it is a natural component of the ergonomic scene, where the matching between environment and individual is in the field of psychological needs, which are as relevant and important as matching in the field of physical effort. The Work Research Unit, part of ACAS, is dedicated to encouraging and monitoring developments in this area and like its counterpart in many other western countries is well aware of the importance of the ergonomic component.

Introducing ergonomics

The introduction of ergonomic changes is not a matter for specialists, working remotely from the plant or office, delivering new equipment and leaving it to be introduced by the foreman or office manager. People often do not take kindly to having strangers telling them how they can do their job better, but on the other hand are very happy to talk about their own work, what they do and what its difficulties and attractions are. It is essential that changes in working equipment and working practices must be decided in conjunction with those who are involved in them. Early discussions concerning the difficulties and changes will be much more rapidly focused and far more fruitful if these are conducted directly with the workers and their representatives. Any who are involved from management should be people concerned with the problem itself, not concerned to eavesdrop and exercise a monitoring role for control purposes. Developments of new ideas, as they progress, should be discussed with a working group and the final mock-up should be tested by the people who will use it, if possible, in the factory or office environment itself. Running trials to match the equipment to the particular working population concerned will enable comment to come from the future users, and cause changes to be made, before the final design is fixed, thus saving money and unnecessary errors. In a very real sense people 'own' the equipment that they use at the workplace. This feeling of ownership must be developed by involvement in the equipment's creation.

To introduce ergonomics as a function into a factory or office is seen by most management as an unnecessary luxury. Provided as they are with the necessities to perform their tasks well, and with the power to introduce changes in their working environment if they find they are hindered in their professional activities, realization of the difficulties of others who do not have these opportunities is often shrouded.

However, an ergonomics function which can provide a link between the medical services, the production people and the engineering staff can have many benefits. At a public relations level, and provided of course that the ergonomist is indeed allowed to pursue his ergonomic activities, it can demonstrate a real concern by management for the conditions under which its workpeople are operating. The exploration and improvement of working conditions, perhaps a more important aspect of the work, can very quickly pay for the ergonomist. It is comparatively easy to achieve a 5% improvement, and sometimes a very much greater improvement than this, in the performance of people simply by removing the obstructions which hinder them from doing the work which they have come to the factory to do. It does not take many such improvements, which are usually permanent, to cover the costs of the ergonomist. Indeed, in one instance, at a German factory, the development of new workplaces and work features caused the company to start a completely new business, in the manufacture and sale of ergonomically designed working equipment. Such a development is by no means unique.

Ergonomics and the occupational health professional

Many years ago the Philips company in Eindhoven developed an occupational health service. This was very specifically a health service and was separate from their occupational medicine section. Its purpose was to develop better working conditions so that the health of employees was maintained rather than dealing with their sicknesses, and the separation was maintained for many years. A major group in this occupational health service was that for ergonomics. It is probably true to say that the ergonomics group at Philips was responsible for the development of ergonomics in Holland, and certainly produced the first general textbook on ergonomics in that country.

The advantage of this 'separate but equal' link was that the two services could work together, feeding information across to each other. In a small plant this is not feasible, and of course it is not the only direction. In Denmark all industrial physiotherapists have to have a course in ergonomics and the training schools give some practice in applying ergonomics during the course [15]. This is not extensive, but it does introduce them to the subject's possibilities. Recently in Britain those members of the Chartered Society of Physiotherapists working in industry have introduced ergonomics as a requirement of their own course.

It should be clear from this chapter that the activity of ergonomists is directly complementary to the work of occupational health practitioners. In their different ways the aim is to make the workplace a safer, healthier and more satisfactory environment. The occupational health professional sees the results of not achieving this desirable end and will endeavour to influence management to introduce changes. The role of the ergonomist can be seen as perhaps more proactive in this direction in that an inadequate condition is actively sought out, its effects measured and changes proposed to reduce the adverse consequences. This is a positive role to redesign the situation before its results are seen in the surgery. If business and industry are to continue to be acceptable sources of employment for large numbers in the population, and it is difficult to see how this is not to be, then this positive role in the design of good and satisfying work must be more widely adopted. The social changes and their resulting attitudes towards life which were briefly touched on earlier will not be reversed, and will increase the demand for better working conditions. The operation of an efficient and competitive economic society must be done in a way that those who operate it do so willingly. In the creation of such a situation ergonomics has a major role; it is the only discipline which adopts this human-centred view of the application of science and technology to the total design of human work.

References

1. Borg, G.A.V. Subjective aspects of physical and mental load. *Ergonomics*, **21**, 215–220 (1978)
2. Brouha, L. *Physiology in Industry: Evaluation of Industrial Stresses by the Physiological Reactions of the Worker*. Pergamon Press, New York (1967)
3. Kilbom, Å. Physical work capacity among firefighters. *Arbete och Hälsa*, No. 12, 26pp, National Board for Occupational Safety and Health, Stockholm (1979)
4. Brown, A. An exercise training programme for firemen. *Ergonomics*, **25**, 793–800 (1982)
5. Hoag, L.L. Human factors design problems of firefighters. In: *Proceedings of the 22nd Annual Meeting of the Human Factors Society, Detroit, Michigan*, pp. 129–132 (1987)

6. Stalhammår, H.R., Leskinen, T.P.J., Kuorinka, I.A.A., Gauntreau, M.H.J. and Troup, J.D.G. Postural, epidemiological and biomechanical analysis of luggage handling in an aircraft luggage compartment. *Applied Ergonomics*, **17**, 177–183 (1986)

7. AML. *Designing Plants for People*. Laboratory of Industrial Ergonomics, Royal Institute of Technology, Stockholm (1986)

8. Westgaard, R., Waersted, M., Jansen, T. and Aaras, A. Muscle load and illness associated with constrained body postures. In *The Ergonomics of Working Postures*, edited by E.N. Corlett, J. Wilson and I. Manenica, pp. 3–18. Taylor and Francis, London (1986)

9. Corlett, E.N. and Bishop, R.P. A technique for assessing postural discomfort. *Ergonomics*, **19**, 175–181 (1976)

10. NIOSH (National Institute for Occupational Safety and Health) *Work Practices Guide for Manual Lifting*. DHSS (NIOSH) Publication No. 81–122, Government Printing Office, Cincinnati (1981)

11. Snook, S.H. The Ergonomic Society's Lecture 1978: The design of manual handling tasks. *Ergonomics*, **21**, 963–985 (1978)

12. Winkel, J. On foot swelling during prolonged sedentary work and the significance of leg activity. *Arbete och Hälsa*, Stockholm, **35**, 84pp, National Board for Occupational Safety and Health, Stockholm (1985)

13. Eklund, J.A.E. and Corlett, E.N. Experimental and biomechanical analyses of seating. In *The Ergonomics of Working Postures*, edited by E.N. Corlett, S. Wilson and I. Manenica, pp. 3–18. Taylor and Francis, London (1986)

14. Eklund, J.A.E. *Industrial seating and spinal loading*. Unpublished PhD thesis, University of Nottingham (1986)

15. Blom, L. and Hauch, A. Physiotherapists in ergonomics. *Applied Ergonomics*, **13**, 289–292 (1982)

16. Cakir, A., Hart, D.J. and Stewart, T.F.M. *Visual Display Terminals*. John Wiley, Chichester (1980)

Chapter 21

Protective clothing

J.R. Allan

Introduction

It is self-evident that all clothing is to some extent protective, even if only of the wearer's modesty or the sensitivities of others. However, with an approach as broad as that the topic would become unmanageable and it is therefore necessary to draw some boundaries. The first of these is between clothing worn in the normal course of day-to-day activities in the usual range of environmental conditions found in the United Kingdom, and clothing worn as specific protection against environmental or weather extremes or other hazards in the working environment. In modern industry, the range of such hazards is truly vast and growing year by year and an attempt to classify them is made below. The second boundary that must be drawn is one of detail. It would be impossible in a single, succinct chapter to detail the characteristics of each item of protective clothing and the various forms of protection will therefore be dealt with in broad terms rather than in detail. Specific forms of protection against noise, vibration, the pressure environment and some forms of electromagnetic and high energy particulate radiation are dealt with elsewhere in this book. They will therefore receive only a mention by way of including them in the classification below.

These limitations in the scope and detail of this chapter are, however, not considered serious. This is because it is possible to describe a methodology for assessing the need for protective clothing systems, and for the development or selection of specific items which is general rather than specific to one particular type of hazard. This approach coupled with the details given in the references and under the further reading section, should enable the occupational physician to formulate appropriate advice for each specific situation. Indeed, it is not unknown for the lack of such a generalized methodical approach to lead to wholly inappropriate protective clothing policies, it being all too easy to become buried in the vast array of details provided by the manufacturers of protective clothing and by the official bodies concerned with its use.

Classification of protective clothing

Protection against hazards in the working environment

These are as follows:

(1) Protection against the thermal environment including weather.
     (a) Heat and cold

(i) Passive systems—insulation, ventilation, radiant shielding
(ii) Active systems—personal conditioning clothing
(b) Contact protection (hot or cold surfaces)
(c) Wind, rain and spray.
(2) Protection against chemical hazards.
(3) Protection against other aspects of the physical environment.
(a) Electromagnetic and high energy particulate radiation
(b) Noise
(c) Vibration
(d) Pressure.

Protection against accidents

These are as follows:
(1) Protection against fire
(2) Impact protection—safety helmets, footwear, body armour, protection against cutting and penetrating injuries
(3) Eye protection and splash protection—goggles, visors, antisplash garments
(4) Protection against falls—safety harnesses
(5) Protection against accidental immersion—lifejackets, cold shock suits, immersion/survival suits.

Protective clothing systems: brief descriptions

In this section it is proposed to give broad descriptions of each of the categories of protective clothing given in the above classification excluding those that have been covered in other specialist chapters in this book.

Protection against the thermal environment

Heat and cold

PASSIVE SYSTEMS

The two important variables that may be adjusted by appropriate selection of protective clothing are insulation and clothing ventilation.

Insulation or thermal resistance is given in $°Cm^2/W$ under the SI system of units. Much of the literature on clothing, however, uses the 'clo' unit which was originally defined as the insulation of a comfortable man sitting in an environment of 21 °C, <50% RH and 0.1 m/s air movement. Descriptively it is the insulation of a light business suit with the usual underwear. More recently the clo has been redefined in physical terms as $0.155 °Cm^2/W$. Another unit of insulation which the reader may come across is the 'tog' which is widely used in the clothing industry, for example to describe the insulation of duvets and sleeping bags. The 'tog' is one-tenth of an SI unit, i.e. 1 tog = $0.1 °Cm^2/W$.

The thermal insulation provided by a garment is mostly determined by the thickness of the air layer trapped within the material fibres and between the garment and the skin. Very little insulation is attributable to the properties of the material fibres themselves. For this reason it is possible in practice to equate clothing insulation with clothing thickness, the relationship being approximately 0.025 °Cm²/W per mm of thickness. However, the total insulation provided by a garment consists not only of the insulation of the garment itself but includes the insulation of the boundary air layer immediately next to the garment through which heat must pass to get to the general environment. It is a matter of everyday experience that clothing insulation is significantly altered by the presence of wind or artificial ventilation in the environment. This comes about because the insulation of the boundary air layer is markedly decreased as wind speeds increase and, unless the item of clothing includes a windproof or impermeable outer layer then wind penetration of the clothing displaces the warm air trapped within the garment. In warm environmental conditions leading to sweating, this would of course be an advantage as it would promote both convective cooling through the garment and sweat evaporation. This would be unusual, however, since more often the garment would simply be removed in warm conditions. In the cold, however, any wind penetration of the insulation layer is a disadvantage since it will significantly increase convective cooling. The effect of wind and water penetration of insulating layers is dealt with in more detail below.

Similarly insulation may be effectively reduced by air penetration of the clothing system induced by bellows action during exercise. Indeed clothing designers will attempt to promote this characteristic where appropriate by careful design of openings or ventilation holes.

ACTIVE SYSTEMS

In this category are included the whole range of artificial heating or cooling garments collectively known as *personal conditioning clothing*.

Air ventilated suits   Air ventilated suits for personal cooling have been used for nearly 50 years in the Services and in industry. They work either by cooling convectively or by promoting the evaporation of sweat, although many systems do both. To cool convectively an air ventilated suit must be supplied with air at a temperature sufficiently below the skin temperature of the worker to bring about effective cooling. A mass flow of air of approximately 1 kg/min at an inlet temperature of 15–20 °C will produce substantial convective cooling. Effective sweat evaporation requires a rather different quality of air supply since it is the humidity of the air supply that is of prime importance in promoting evaporation. Even if the supply air is at a relatively high temperature, say 40 °C but dry, there will be a net balance of evaporative cooling.

Personal conditioning systems designed to cool convectively or evaporatively also differ in the way the air is distributed over the body surface. Convective cooling systems are best when a majority of the air is distributed peripherally over the limbs and only a small amount over the trunk, as in *Figure 21.1*. If too much air is distributed over the trunk, then the wearer will often complain of cold spots. By contrast, an evaporative system requires the air supply to be distributed more or less evenly, that is equal volume flows of air to equal skin areas as in *Figure 21.2*. This is because human beings sweat more or less evenly over their body surface and the most effective use of their sweating is achieved by producing maximum evaporation.

396

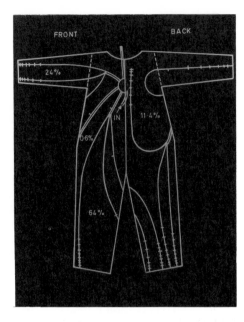

Figure 21.1 Air ventilated suit for convective cooling. Insert shows the mainly peripheral distribution of air

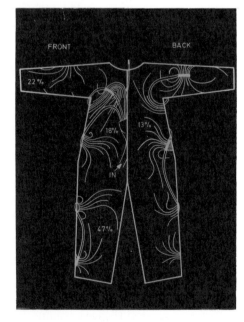

Figure 21.2 Air ventilated suit for evaporative cooling. Insert shows the distribution of equal volumes of air to equal skin areas

Supply systems for air ventilated suits usually involve connection via an umbilical to a source of air. High pressure air lines may be used in conjunction with body-mounted vortex tubes to produce cooling or, in more sophisticated systems, with air cycle refrigeration systems in which high pressure air is expanded through an air turbine. Simple evaporative supplies can sometimes be provided by an electrically driven fan or air pump.

Although it is theoretically possible to use air ventilated suits for heating purposes, the low specific heat of air requires very high mass flows to achieve effective heat transfer. In practice, it is generally easier to provide a heating system through the use of electrically heated garments or a liquid-conditioned suit.

Liquid-conditioned suits   The use of garments containing networks of small pipes through which liquid is circulated so as to achieve cooling or heating was developed at the Royal Aerospace Establishment, Farnborough in the mid-1960s [1]. These systems have been shown to be highly effective and were rapidly adopted by NASA for use in the Apollo space programme. They are, however, rather expensive and complicated for day-to-day industrial use although some recent developments are overcoming these disadvantages. It has been shown, for example, that effective cooling at a rate of about 150 W can be achieved with a liquid cooled vest (*Figure 21.3*) which is considerably cheaper to manufacture than a whole body suit (*Figure 21.4*). Modern techniques for constructing these garments include radiofrequency welding of the fabrics to provide the channels which locate the small liquid-containing pipes. A number of supply systems have been developed and these

Figure 21.3 A liquid-conditioned vest based on a network of small pipes mounted on an undergarment

Figure 21.4 A whole body liquid-conditioned suit

include vapour cycle refrigeration systems, systems based on the use of thermoelectric cells, and those that depend on latent heat sinks such as ice or carbon dioxide snow.

Electrically heated garments   A variety of electrically heated garments is available and this includes whole body suits as well as socks and gloves (*Figure 21.5*). A number of different techniques have been developed including garments in which an electrical wire is knitted into the fabric, garments with panels of woven electrical conductors and garments manufactured from a variety of conducting fibres with a high carbon content. The latter have not yet proved successful in practice and the most reliable system is probably still the knitted wire system. With this system it is possible to provide up to 200 W heating to the trunk and limbs and 25 W to each hand and foot (300 W total).

Contact protection

Contact protection is really a subdivision of protection against heat and cold. It consists principally, though not exclusively, of gloves worn to prevent burns or cold injuries caused by contact with surfaces. The prevention of contact burns involves the use of protective devices ranging from the housewife's simple oven gloves, through a variety of thick fabric or leather gloves such as are used by blacksmiths and welders, to the highly insulating and fire-resistant gloves such as might be worn by steel workers or firemen. Two factors are important in determining the need

Figure 21.5 An electrically heated suit with socks and gloves. This example is manufactured by knitting the wires into the garment

for protection against contact burns, the temperature of the surfaces and the material from which they are made. Generally speaking metal surfaces will cause contact burns if they are at temperatures of 45 °C or higher, since the skin will burn at approximately 46 °C. However non-metallic surfaces, of plastic or wood for example, are less likely to cause contact burns since they are relatively poor conductors of heat. The relationships between the type of material, its temperature and the duration of contact, in the causation of contact burns can be predicted [2].

Injuries through contact with cold surfaces are common when environmental temperatures are below about −20 °C. As with contact burns, the characteristics of the surface in terms of the material from which it is constructed and its temperature are the important factors and metal surfaces constitute the real risk. Serious injuries may be caused by an unprotected worker placing bare skin against a very cold metal surface and finding he has frozen to it by the time he realizes what is happening. Another common source of this type of cold injury is when the mechanic, who has been in the practice of mouthing nuts and bolts during his work, goes to a very cold environment for the first time and finds that the nuts and bolts freeze to his lips or tongue. Protection of the hands from cold contact injuries can be achieved simply with a relatively thin pair of gloves; silk contact gloves are commonly used for this purpose.

Wind, rain and spray

Protection against wind, rain and spray is necessary in a wide range of activities which includes land-based activities such as police work, agriculture and building

Figure 21.6 A modern suit of foul-weather clothing developed for the Royal National Life-boat Institution by Musto Ltd

construction as well as the more obvious marine activities in the fishing industry, merchant marine and yachting.

The essential feature of such foul-weather clothing, as it is known, is that it will effectively exclude wind-driven rain and spray [3] and thus maintain the effective insulation of the clothing worn beneath; *Figure 21.6* shows an assembly developed for the RNLI by Musto Ltd. It has been mentioned above that the insulation of clothing is largely attributable to the air trapped within it, and if the external protection leaks then water will displace the air from the clothing underneath and cause a serious loss of insulation. The design of foul-weather clothing poses a number of problems in that exclusion of wind-driven rain and spray has to be achieved with a design that is compatible with the activity being undertaken in the clothing, both from a ergonomic point of view and also from the point of view of thermal comfort. Thus a totally impermeable external covering may well be effective in excluding water, but will lead to the progressive accumulation of sweat in the clothing beneath unless there is some form of ventilation. The insulation of clothing is just as effectively destroyed by sweat as it is by water leakage from outside.

A specification for foul-weather protective clothing will be in three main parts. First, there must be a specification of the material from which the suit is made; second, there must be a specification of the water excluding performance required; and third, there must be a specification of the ergonomic requirements of the suit in terms of job compatibility, visibility, etc.

The material specification will lay down limits for such qualities as tear strength, resistance to wear or resistance to water penetration and tests for these functions and many others have been laid down in British Standard (BS) 3424 [4]. The most generally used materials at present consist of neoprene or polyurethane-coated nylon fabrics or polyvinyl chloride (PVC) although a range of water vapour permeable fabrics are increasingly used in non-marine environments. Depending on how effective the water exclusion is required to be, the specification will need to pay attention to such details as the way in which the seams are stitched and sealed or welded.

The next part of the specification dealing with water-excluding performance is more difficult since, at present, there are no generally agreed tests of this function even though this is the main purpose of the clothing. British Standard 3424 [4] lays down a hydrostatic head test for water penetration, but that is a test of the performance of the material itself and the seams. A suit made from wholly waterproof material may nevertheless be highly ineffective at excluding water intrusion via such openings as the zips, wrists, ankles and neck. Attempts to devise a suitable test of this overall function are beginning to appear [3]. One is based on the artificial reproduction of appropriately severe weather conditions. If such a test is not available in an individual case then it may be necessary to resort to practical wearer tests, but it is essential that these are undertaken in appropriately bad conditions.

The final part of a specification for foul-weather protective clothing details the various ergonomic requirements appropriate to the particular application. These will always include such items as the fit of the clothing, compatibility with other items of protective clothing, general comfort and compatibility with the work to be undertaken, and non-interference with hearing and visibility.

Protection against chemical hazards

Protective clothing for chemical hazards ranges in complication from a simple PVC apron to a full body coverage suit with boots, gloves, helmet and respirator together with a self-contained breathing system. What is necessary in a particular case obviously depends on the hazard involved and in particular whether the protection required is simply against the risk of splash over the body surface and the eyes or whether there is also a dangerous vapour risk requiring a gas-tight suit with a protected breathing system.

Where the protection required is confined to protection against splash, for example in the handling of a range of acids, bases, oxidizing agents or solvents, PVC-based fabrics are the most generally useful. It will, however, always be necessary to check that the fabric proposed for the protective clothing is in fact effective against the particular chemicals involved. Splash protection will include the hands and feet and the head and eyes of the wearer, and the example of a simple splash proof suit shown in *Figure 21.7* would be worn in conjunction with suitable gloves, boots and a face mask.

Where the chemical risk includes a vapour hazard it will usually be necessary to provide some form of respiratory protection. This may vary from a simple face mask with a filter to systems that include a breathing system supplied with air from an airline or a backpack gas cylinder. The detailed requirements of such breathing systems may be found in a number of British Standards which are detailed under the further reading section, (*p.414*).

Some chemical contaminants, in the form of vapours or dust or fibres, may be sufficiently corrosive or carcinogenic or have other dangerous properties as to

Figure 21.7 A simple Butyl splashproof suit for use by paint sprayers

make it necessary to exclude them totally from all possible contact with the body surface or breathing system. Protection against such hazards requires a completely gas-tight suit, and it may frequently be necessary to provide the suit with positive air pressure with outboard leaking to be absolutely certain of preventing any ingress of dangerous substances. It is self-evident that such protective suits will include a protected breathing system and very often they will also include some form of cooling system as an integral part of the assembly. An example of such a gas-tight chemical suit is shown in *Figure 21.8*.

Protection against other aspects of the physical environment

Protective clothing for use against electromagnetic and high energy particulate radiation, noise, vibration and pressure changes is described elsewhere (*Chapters 6, 7, 8 and 22*).

Protection against accidents

Protection against fire

Fire is a hazard in many working environments including steel making, welding and cutting, chemical manufacture and the oil industry. It is also a hazard following accidents in vehicles and aircraft, and a particular hazard to those whose job it is

Figure 21.8 A gas-tight chemical suit with breathing system and built-in cooling (by kind permission of Trelleborg Ltd, Rugby)

to fight emergency fires—namely the firemen and those who work in specialist teams such as mine rescue teams.

Fire protective clothing may be considered at three functional levels: simple spark or flame-lick protection; a slightly greater level of protection required where there is significant flame-lick risk coupled with very high radiant heat loads; and the fully flameproofed protection of the kind required by firemen.

SIMPLE SPARK AND FLAME-LICK PROTECTION

The main requirement here is an outer layer of clothing which will not support combustion or be seriously degraded or melted by brief contact with sparks or flame. Welders are a good example of a group of workers who require this form of protection. A number of fabrics and treatments are available to achieve a suitable level of protection. For example, both wool and cotton may be treated to reduce flammability. A good example is the Proban treatment of cotton in which a flameproofing agent becomes incorporated within the fibres of the cotton fabric.

Figure 21.9 (*a*) and (*b*) Examples of welder's suits manufactured from 'Proban' treated cotton for flame retardance (photographs kindly provided by Albright & Wilson Ltd, Warley, West Midlands)

A welder's suit made from Proban-treated cotton fabric is shown in *Figure 21.9*. Nylon which melts at 180 °C (nylon 11) or 250 °C (nylon 6.6) should be avoided in all clothing systems likely to be exposed to spark or flame. However, there are a number of manmade fibres which have been developed specifically for their flame-resistant and fire-retardant qualities. Of these the group of aromatic polyamides (Nomex, for example) are well known examples.

MODERATE HEAT AND FLAME PROTECTION

In a number of industrial conditions, such as the handling of molten metal, the risk of contact with flame and sparks is combined with severe radiant heating, sufficient to singe or ignite many common fabrics or cause burns beneath thin external layers. Protection in these circumstances requires not only an effectively flameproof outer layer, but also sufficient insulation beneath to prevent burning of the skin. The insulation must also be resistant to degradation or melting by heat. For this purpose the aromatic polyamide materials are highly suitable for the external layer since they are flameproof and do not melt at any temperature, even though degradation may commence at about 370 °C. Aromatic polyamide fabrics are generally quite thin, and a single layer will not provide protection against skin burning during significant exposure to flame or severe radiant heating. To provide this level of protection requires either multilayered garments of flameproof material or the provision of some suitably treated insulating layer beneath the flameproof outer layer. For this purpose one of the treated wool fabrics would generally be suitable; nylon should be avoided.

FULLY FIREPROOF CLOTHING ASSEMBLIES

A good example of a fully fireproof clothing assembly is the fireman's suit shown in *Figure 21.10*. This consists of an aluminized asbestos outer layer which is not only flameproof but also reflects considerable amounts of radiant heating and is sufficiently thick to provide an effective level of insulation. Insulation may be further increased by wearing appropriate assemblies of underclothes which again should be selected from fabrics that would not be degraded or melted by high temperatures. Such suits are obviously required to cover all body surfaces including the face and therefore require flameproof visor systems. They often incorporate an integral breathing system, since smoke and fumes will be additional hazards.

Figure 21.10 Fireproof clothing manufactured from aluminized asbestos. The suit includes boots, gloves and a hood with metal gauze face piece (by courtesy of the Royal Aerospace Establishment Fire Service)

TESTS FOR FLAMEPROOF MATERIALS

Methods for testing the flammability of fabrics are described in British Standard 5438 [5, 6], and it is possible to specify the required flame-resistance in terms of the duration in seconds before ignition. However, such tests are not always suitable for assessing the protective qualities of whole garment assemblies, and for this purpose highly specialized manikin systems have been developed for use in conjunction with experimental fire facilities in which the exposure to heat and flame can be closely controlled. The advantage of such thermal manikin tests is that they give an indication of the likelihood of skin burns beneath a flameproof external covering and they may be used to devise clothing assemblies which will give sufficient protection.

Impact protection

Safety helmets or hard hats are far and away the commonest form of impact protection found in industry and there are many rules and regulations controlling their use.

The two main functions of a protective helmet are to provide penetration resistance and impact protection. Penetration resistance is required to protect the wearer from injury caused by penetration of his skull by sharp objects in his working environment, either during the normal course of work or in an accident. Penetration resistance is also required to protect the wearer from sharp falling objects. The resistance of a given helmet to penetration may be tested by striking it against a conical anvil having a 0.5 mm radius tip. The impact energy used in this type of test should be related to the type of risk to which the worker is exposed and the helmet must be capable of withstanding this impact without penetration of the shell.

Impact protection is achieved by spreading the impact load over the surface of the head and by increasing the time during which the impact is attenuated. These results can be achieved either by suspending an external shell on a head harness with a gap between the harness and the shell or by providing a deformable foam liner within the external hard shell. The performance requirements for a specific helmet should be determined in relation to the risk against which it is required to provide protection. Specification and test methods for penetration and impact resistance are laid down in British and International Standards, for example in BS 6658: 1985 [7] for helmets for vehicle users, or ISO 3873 for industrial helmets [8].

Helmets, although the most common, are not the only form of impact protection. Another common requirement is for protection of the feet against dropped heavy objects and this is generally provided by boots having protective toe caps. The performance of these is laid down in British Standard 1870 [9].

An extreme form of impact protection is found in the modern flak jackets or bulletproof vests as they are more commonly known. While these may be regarded as highly specialist items they are increasingly commonly used by police forces and security services. Most flak jackets are constructed from glass fibre composites or from a range of ceramic materials which are highly resistant to penetration by bullets.

A final category of impact protection includes protection against cutting tools such as carpenters' tools or knives. A good example of this type of protection is the chainmail apron worn by butchers and abattoir workers and the leather wrist protectors worn by several groups of workers who use cutting tools in their jobs.

Eye protection

The use of goggles and visors to protect the eyes against metal or chemical splash or other hazards in the environment is commonplace in industry. Such devices are very often part and parcel of overall protective assemblies.

Protection against falls—safety harnesses

Safety harnesses and belts for protection against falls from heights are very widely used in industry, particularly in construction and for workers in such activities as power line erection and maintenance and window cleaning. Safety harness systems have four important components which should each be considered in recommend-ing a system for a particular purpose: the body-mounted harness or belt system; the link or tether system joining the harness to a strong point or anti-fall system; the strong point itself; and finally a variety of components, collectively referred to as anti-fall systems, which serve to break a worker's fall so as to limit any injuries.

SAFETY BELTS AND HARNESSES

Safety belts and harnesses may be classified into five main groups as follows:

Pole belts   These are designed for use by linesmen and others working on poles and are intended to be continuously loaded. They generally consist of simple waist belts with a lanyard for attachment to the pole and are not intended to arrest falls of more than 60 cm, indeed they would be likely to lead to back injuries if used for that purpose.

General purpose safety belts   These are similar to pole belts except that the safety lanyards incorporate attachment devices for fixing to strong points rather than around poles. As with pole belts they are not suitable for arresting drops of greater than 60 cm.

Chest harnesses   These harnesses incorporate a chest belt with shoulder straps, suitably linked, and are intended for use with lanyards permitting a certain amount of mobility of the user. They are designed to arrest a maximum fall of 2 m.

General purpose safety harnesses   These incorporate thigh straps in addition to shoulder straps and are used with safety lanyards attached to anchorage points or anti-fall devices. They are intended to allow some freedom of movement of the wearer, but without an anti-fall device falls are limited to a maximum of 2 m.

Safety rescue harnesses   These are designed for use by workers in confined spaces such as coal bunkers, slack hoppers, grain silos and the like, where there is a risk of being overcome by noxious gases or fumes or of immersion in the material in which the worker may be standing. They are used in conjunction with a rescue line designed for withdrawing the individual in the event of an accident. Rescue harnesses are designed for a maximum drop of 60 cm, and should be considered as extraction devices.

The general requirements for safety belts and harnesses, including the strength of materials used in the belts and the lanyards, are laid down in British Standard 1397 [10]. In general, webbing for safety harnesses should have a minimum break force of 9 kN and rope used for lanyards should have a minimum breaking force of 29.4 kN. It should be noted that where a safety lanyard is required to arrest a fall between 60 cm and 2 m the use of natural fibre rope or chain is unsuitable unless there are adequate shock-absorbing fittings included in the system. Polyamide, nylon or polyester filament ropes are to be preferred unless there are specific contraindications such as the risk of contact with acids or alkalis.

STRONG POINTS AND ANTI-FALL SYSTEMS

The design of a safety harness system should give close attention to the strong points or anti-fall systems to which the safety harness is attached. Clearly these should be manufactured to adequate strength to resist the maximum likely fall of a heavy individual. They should also be carefully considered from the point of view of ergonomic requirements so that they are fitted in convenient positions for the work intended.

A number of simple shock absorbers are available for incorporation into safety lanyards and British Standard 1397 [10] suggests that these should not be capable of extension by more than 65 cm. However, where there is a risk of falls through considerable distances such as when the nature of the work being undertaken requires substantial mobility and a long safety line, then an anti-fall breaking system should be considered. Three classes of anti-fall system have been described as follows.

Class 1   This is a system in which the safety lanyard is attached to a sliding and blocking device which moves vertically on a metal rope cable or rail.

Class 2   This consists of a type of inertia reel for rolling, unrolling and blocking a safety lanyard.

Class 3   These systems include a range of kinetic energy absorbers built into the harness or tether. An example is a length of webbing stitched together in such a way as the stitches progressively break during deceleration of the falling subject.

The basic aim of an anti-fall system is to limit the decelerating force acting on the worker to levels within human tolerance. Standards differ somewhat in different countries. For example, for a safety harness system the standard in the United States is a maximum force of 17 000 N or a deceleration of $25g$. In the United Kingdom the maximum deceleration suggested is limited to $10g$ and in France to $5g$. Clearly these figures would be substantially reduced if the worker wears a simple safety belt as opposed to a harness.

Accidental immersion

Specific protective clothing for those at risk of accidental immersions has been used by the Services for a considerable time, notably in the Navy and for military pilots who may eject over the sea. In recent years, however, increasing use has been made of these systems in the industrial field and a number of mandatory regulations require immersion protection. The two groups of workers particularly involved are those in the offshore oil and gas industry and the Merchant Navy and fishing fleets.

The two main risks of accidental immersion are drowning and hypothermia, for which the corresponding protective clothing items are lifejackets or flotation aids and survival or immersion suits respectively.

LIFEJACKETS AND FLOTATION AIDS

The important functional requirements of a lifejacket are as follows:

(1)   It should provide adequate buoyancy to maintain the individual's mouth at a minimum height of 120 mm above the water surface and his body at an approximate angle of 45° to the vertical. British Standard 3595 [11] requires a minimum buoyancy of 15.8 kg.
(2)   The lifejacket should be capable of turning an unconscious survivor from a face-down position to a face-up postion within a reasonable time.

Lifejackets are sometimes manufactured with some inherent built-in buoyancy although this is usually not to the full lifejacket requirement of 15.8 kg. The advantage of built-in buoyancy is that an individual falling into the water unconscious will have some flotation prior to operation of an inflation system. Many lifejackets, however, do not have inherent buoyancy and rely on CO_2 inflation systems which may be either manually operated or automatic. The manual systems are the commonest and require the survivor to pull a lanyard or toggle to operate the system. A number of automatic systems have been developed which operate when in contact with water. These either operate by using a water soluble pellet or disc of some kind which releases the firing mechanism or, in the case of more modern systems, they use a firing system in which an electrical circuit is completed by the conductivity of water. Most automatic systems are required to operate within a maximum period of 5 s following immersion.

Lifejackets are often provided with a number of integral survival aids such as an automatic light, a whistle, oral inflation tube, reflective patches and a loop to assist in retrieval of the survivor from the water. More recently it has become recognized that drowning as a result of wave-splash is a significant risk especially for weak, hypothermic survivors. To protect against this several current lifejackets are provided with splash-guards.

IMMERSION OR SURVIVAL SUITS

Suits designed for thermal protection in water may be classified into three main groups: cold shock suits, short-term survival suits; and long-term survival suits.

Cold shock suits   If an unprotected individual falls into water cooler than about 10 °C the sudden peripheral cold stimulus may lead to a phenomenon known as cold shock [12]. The main features of cold shock are an involuntary, uncontrolled, gasping respiration coupled with other uncoordinated body movements which can lead to rapid drowning even in a capable swimmer. This reaction can be prevented by relatively simple suits which may not have any significant long-term survival potential. *Figure 21.11* shows a trunk and arm wet suit designed as an anti-cold shock suit for use in the North Sea in circumstances where rescue is guaranteed to be rapid, although recent evidence suggests that this type of partial coverage suit may not be sufficiently effective to allow breath-holding under cold water as required, for example, for escape under water following helicopter ditching. A further example of an anti-cold shock suit is the Royal Navy's once-only suit shown in *Figure 21.12*.

Short-term survival suits   The International Maritime Organization [13] have laid down requirements for immersion suits which envisage two main types. The first, described as an uninsulated suit, is intended to provide protection for a maximum period of 1 h in water at 5 °C. The second, described as an insulated immersion suit, is designed to provide protection for a period of 6 h in water between 0 and 2 °C. Those in the first category may be described as short-term survival suits and the essential requirement is that they should exclude water adequately so as to preserve the insulation of normal clothing worn beneath. It has been suggested [14] that this can be achieved if the suit, after allowing for any leakage of water, will provide an external insulation in water of approximately 0.05 °Cm²/W.

Figure 21.11 An anti-cold shock suit employing wet suit material over trunk and arms only

Figure 21.12 The Royal Navy 'once-only' suit. The suit is designed to fit over a life jacket

Long-term survival suits   This category includes the suits designed for survival for longer than 1 h, such as those under the IMO regulations for insulated immersion suits providing 6 h survival in water between 0 and 2 °C. The chief differences between these suits and short-term survival suits are that they must be more reliably waterproof and that they should include built-in additional thermal insulation or be used in conjunction with highly insulating other clothing. It has been suggested [14] that a suit providing this level of protection should have an immersed insulation value of approximately $0.1 °Cm^2/W$. For an excellent general description of the requirements for immersion suits the reader is referred to a consultative document prepared by the UK Department of Energy [15].

A methodical approach to the selection of protective clothing

From the above brief account of the range of protective clothing items it will be apparent that each protective clothing problem has its own specific features.

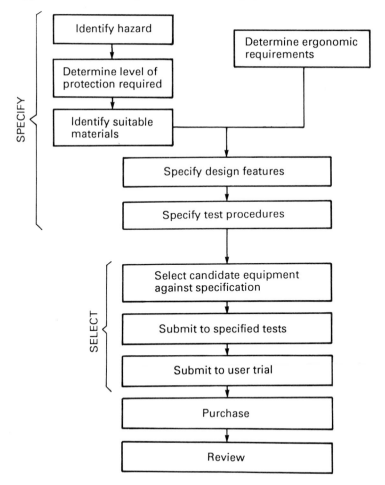

Figure 21.13 A flow chart giving the main stages in the development of a protective clothing system

Nevertheless, this does not preclude a reasonably methodical approach the broad features of which are common to all protective clothing problems. In this final section of this chapter such an approach will be described.

Figure 21.13 is a flowchart in which the various steps involved in specifying and selecting protective clothing are outlined. These stages are as follows.

Arriving at a specification

Identify hazards and the level of protection required

This is reasonably self-explanatory but it is surprising how often protective clothing is selected without a clear idea of what it is intended to protect against. In the case of a suit designed to protect against chemical contamination, one should start with an exact description of the chemical hazard involved in terms of type, concentration, vapour, liquid, etc. It is then necessary to determine the maximum level of contamination acceptable and this will usually be done separately for skin contamination, eyes, mucous membranes and the respiratory system, since the quality of protection required may differ considerably in each case.

Identify suitable materials and specify material performance tests

Having identified the hazard and the level of protection required the next stage is to survey the available materials that will provide suitable protection. At this stage it is best to construct a list of all possible candidate materials solely on the basis that they will provide the required protection. Such other considerations as cost, strength, durability or comfort will come later in the selection process.

Determine ergonomic requirements

Under this heading are considered all those aspects of the protective clothing assembly which are important in making it a suitable item for the work required. These will include such items as bulkiness, snagging hazards, imposition of thermal stress, restrictions to visibility, interference with hearing, interference with manual dexterity or other physical activity involved in the work.

Determine the design features

At this stage the information gathered together in the first three stages can be used to draw up the main design features of the clothing assembly. This will include such aspects as which parts of the body are to be protected, whether this will be achieved with separate items or with a single one-piece suit, and special design features such as hoods, neck closures, glove assemblies and footwear required to meet the performance and the ergonomic requirements. It will also be necessary at this stage to determine such matters as the size-roll of garments required in relation to the workforce for whom it is intended. This important matter is very often neglected and otherwise highly satisfactory protective clothing items prove unacceptable or inefficient in use merely because they are wrongly sized or there are insufficient sizes to fit all members of the workforce.

Specify functional performance tests for whole protective assembly

Having drawn up a list of the performance requirements and design features required it is now necessary to decide on the tests that will be used in order to select items against the specification drawn up. Many of these tests will be laid down in national standards such as the British Standards. Occasionally, however, it will be necessary to devise special tests in order to look at some specific feature important in the particular work situation being considered. Once the test procedures have been agreed upon it should be possible to prepare a specification document incorporating all the above information.

Trials

Having agreed the specification, a number of items will be obtained either from the protective clothing industry as off-the-shelf items or by specific development in the case of special requirements. These items will have been submitted to tests against the various specification clauses but these tests will generally have been done under laboratory conditions by only one or two individuals. Before committing large expenditure on protective clothing it is sensible to conduct some form of limited user trial in which the actual workforce will try the proposed items. With the best will in the world, and no matter how much care and attention is paid during the development stages of protective clothing, it is possible that certain features may have been incorporated which will prove unacceptable in practice to the labour force. These should be revealed by a user trial.

User trials are generally conducted by issuing a limited number of test items to a cross-section of the workforce who will then use them in their day-to-day activities for an adequate period of time. In some cases it may be important to try items at different seasons of the year if temperature effects could be important. The reaction of the labour force is best obtained by using written questionnaires. These should ask for comments on all the main features of the equipment, but it is most important that adequate space is left for individuals to record comments and views on aspects that may not have been considered by the design team.

Post-introduction review

After a reasonable interval following introduction of a new piece of protective clothing there should be a review of its performance. This review should include workforce comments, which may have changed in the light of longer experience with the equipment, and will also include information on such items as the durability of the equipment, wear and tear, and any maintenance or servicing problems. A number of economic aspects may then be reviewed such as the total cost of ownership of the equipment.

References

1. Burton, D.R. and Collier, L. *The Development of Water Conditioned Suits.* Royal Aerospace Establishment, Farnborough, Hants Tech. Note. No. ME 400 (1964)
2. Stoll, A.M., Chianta, M.A. and Piergallini, J.R. Thermal conduction effects on human skin. *Aviation, Space and Environmental Medicine,* **50(8)**, 778–787 (1979)
3. Allan, J.R. *A Water-exclusion Test for Foul-weather Clothing.* RAF Institute of Aviation Medicine, Farnborough, Hants, AEG Report No. 531 (1986)

4. British Standards Institution. *Testing coated fabrics*, BS 3424 (all parts) London (1982)
5. British Standards Institution *Fabric flammability burning accidents and relevance of BS 5438, PD 2777*. London (1977)
6. British Standards Institution *Methods of test for flammability of vertically oriented textile fabrics and fabric assemblies subjected to a small igniting flame*, BS 5438. London (1976)
7. British Standards Institution *Protective helmets for vehicle users*, BS 6658. London (1985)
8. International Organization for Standards *Industrial safety helmets, International Standard ISO 3873* Geneva (1977)
9. British Standards Institution. *Safety footwear*. BS 1870. London
10. British Standards Institution *Specification for industrial safety belts, harnesses and safety lanyards*, BS 1397. London (1979)
11. British Standards Institution *Specification for Lifejackets*, BS 3595. London (1981)
12. Golden, F.StC. Problems of immersion. *British Journal of Hospital Medicine*, **23**, 371–373 (1980)
13. International Maritime Organization: *International Convention for Safety of Life at Sea* Chapter III (revised) (1974)
14. Allan, J.R. and Hayes, P.A. *The Specification and Testing of the Thermal Performance of Immersion Suits*. RAF Institute of Aviation Medicine, Farnborough, Hants, AEG Report No. 512 (1984)
15. Department of Energy *Offshore Immersion Suits—Draft Guidance and Supporting Report*. Marine Technology Support Unit Report MATR/28, London (1986)

Further reading

British Standards Institution. *Breathing apparatus*, BS 4667. BSI, London
Industrial Safety Protective Equipment Manufacturers Association *Reference Book of Protective Equipment*, 5th edition. Industrial Safety Protective Equipment Manufacturers Association, London (1978)

Chapter 22

Health effects of hyperbaric environments

J. King

Introduction

Uncritical hyperbole and wide-eyed wonder leading to headlines such as 'Wonders of the deep' and 'Tunnelling into danger' characterize the popular conception of diving and tunnelling operations. Even medical colleagues submit to the romance of the underwater world, and discussion of one's chosen specialty ends with the comment 'bends and all that'. Hyperbaric environments, however, are affected by the same physical laws that govern the natural world, and an appreciation of the effects on health of work and play at high pressure begins with a thorough knowledge of those laws. The health effects, not all adverse, are many and widespread, and some recent work suggests that there are some long-term effects whose significance is yet to be evaluated [1–3]. This chapter looks at some of the adverse effects likely to be seen by the occupational health specialist, and in more orthodox medical practice, and also considers some of the ways in which more esoteric problems in the field are met.

Exposure to high ambient pressures occurs in professional and sport diving, and in tunnel work under compressed air through water-bearing ground. The professional diver and the tunnel or caisson worker are of social, commercial and military importance. It should also be remembered that changes in ambient pressure occur in aviation; the middle ear can be damaged almost as easily in climbing to 2430 m as in descending to 2.4 m in a swimming pool. Sir Robert Boyle saw bubbles of separated gas in the seventeenth century but the effects of pressure at gas–fluid interfaces and the role of gas bubbles in the production of dysbarism had to await the late nineteenth century for understanding.

Paul Bert in *La Pression Barometrique* [4] notes how Boyle observed bubbles in a viper's eye during its exposure to low pressure; he had also subjected freshly drawn blood to a partial vacuum observing it to 'boil'. Diving dress however was still primitive and experimental, and pumps capable of applying pressure to subaqueous tunnels were not within the bounds of the then technology, leaving Boyle's observations of little immediate value. Commercial applications of diving and compressed air tunnelling had, however, been considered as practice or theory in the late seventeenth century and a number of diving salvage operations had been undertaken.

Halley died experimenting in his own diving bell in 1785. Lord Cochrane patented a lock system for tunnels in 1831, with Triger first using the system in France in 1839. About then, Augustus Siebe improved diving dress by making it fully enclosed. Men could finally work under pressure for long periods, only to become victims of what was described as 'rheumatism' after exposure. In caissons (French for a big box) such pain became known as caisson disease, but the cognomen of 'bends' was awarded by the workers on the Brooklyn Bridge by virtue of the similarity the sufferers bore to a popular but affected fashionable posture called the Grecian Bend.

Not only pain, but also fatalities, paralysis, disorders of balance and vision, and serious interference with respiration occurred. Men became deaf. Some chronically ill men were reputed to become fit!

Because of poor appreciation of the cause of the illness, treatment was haphazard and ineffective; prophylaxis for caisson workers was with galvanic armour. It did not work. It had been noted that relief of symptoms often followed a return to the working chamber; Sir Ernest Moir introduced the first medical lock onto the Hudson River Tunnel contract in 1889, and formalized treatment. In 1871, Smith, later quoted by Snell in 1896 [5] had considered the role of physical fitness and obesity. By the end of the nineteenth century a wide spectrum of disease, with a narrow aetiology, truly occupational, had made itself known. These diseases are now, 100 years later, of economic importance, and of increasing complexity as diving goes deeper. Full understanding of the dysbaric diseases is essential to the successful completion of many transport and public health works, and a growing number of weekend sportsmen are being injured, some permanently, by the failure of their medical advisers and sport instructors to appreciate the risks. No one should enter compressed air works for fun; the distinction between the professional and amateur diver is often difficult.

Decompression sickness or DCS

'The bends' is used loosely to describe a variety of symptoms of common aetiology and similar pathology. DCS is now the preferred term. The symptoms can be easily confused with other symptoms of dysbaric illness, and with naturally occurring disease. A useful classification of the symptoms of DCS was made by Griffiths [6, 7], and is used extensively by the tunnelling and diving industries (*Table 22.1*).

Not all cases however can be firmly ascribed as either type 1 or type 2 disease. Some cases are neither. At its simplest this classification puts all cases with pain as the only symptom into type 1. It does not matter how many joints are affected, only musculoskeletal pain is included. All other symptoms are called type 2 disease. Some skin manifestations cannot be ascribed to either category. Pain is a clinically noisy and persistent symptom and will mask other more subtle and sinister symptoms [8]. The range of these symptoms can be found in *Table 22.1*.

Depending on the source, the incidence of the disease varies from 0.3 per cent to 15 per cent of exposures, and type 1 cases will make up 50–95 per cent in any series. Type 1 is more common in tunnel work. All cases can be expected to develop within 24 h of leaving high ambient pressure, with some cases developing during the final stages of decompression [6]. Occasional cases will develop late, especially if aggravated by ascent to altitude in civil aircraft. It is not unusual for symptoms to be unreported for as long as a week.

Current thought on the pathophysiology of DCS suggests that the following events are important. In obedience to the gas laws (Boyle, Charles, Henry and Dalton), exposure to a compressed gas environment causes inert gas to be dissolved

Table 22.1 Classification of decompression sickness (not including symptoms due to barotrauma or arterial gas embolism)

Type 1
Pain in a joint or joints

Intermediate type 1/2*
Skin rashes, mottling and bruising
Oedema of trunk and limbs
Skin itching

Type 2*
Cerebral symptoms
Acute loss of consciousness
Acute confusion or disorientation
Headache
Vertigo, nausea and vomiting
Tunnel vision, flashing lights
Altered gait ('staggers')
Spinal symptoms
Numbness
Paraesthesiae
Weakness/paralysis
Girdle pain, thoracic or abdominal
Respiratory symptoms
Cough, wheeze
Shortness of breath
Cyanosis
Retrosternal discomfort or pain
Cardiovascular
Faintness
Pallor/shock
Palpitations

*If any symptom is reported in the type 2 or intermediate type 1/2 classification, the patient must be treated as for serious decompression sickness

and distributed to all body tissues, until a new equilibrium is reached. A gas burden is created, which must leave as it entered, through the lungs. The reduction in pressure induces a state of supersaturation and in many cases bubbles will form in tissues and blood. The number, site of origin and size will partially determine whether symptoms will occur. The inert gas is usually nitrogen. Helium is used extensively in deep diving operations (*see below, p.423*) and other gases have been used experimentally. Hydrogen is currently being evaluated on account of price and availability. Helium is the safest of the inert gases which make up the difference between the partial pressure of oxygen to be supplied, and the hydrostatic head of water to be balanced. Occasionally two gases may be used together. Oxygen cannot be used in isolation, save for certain military applications since it has pulmonary and neurotoxic effects at depth.

Full equilibration or saturation takes about 8 h, but for practical and safety purposes a 4 h exposure is considered as reaching saturation. Different tissues have different saturation/desaturation times. In modelling uptakes for preparation of decompression tables, times as short as 5 min or as long as 240 min to reduce the gas load by half are taken. On the various premises alluded to here, a wide variety of decompression tables have been produced. Some are for special circumstances such as diving at high altitude. All tables however aim at minimizing the incidence of DCS; none is perfect and an acceptable 'bends' rate is between 1 per cent and

2 per cent over the whole range of pressure or depth exposures covered. Tables with a worse incidence are usually not commercially viable and are ethically unacceptable. A zero incidence of DCS is theoretically possible but in practice unattainable and would certainly be unviable commercially.

Most decompression schedules use a series of steps at reducing pressures with stops of increasing duration although some linear and logarithmic tables have been used. One technique, surface decompression, and its tunnelling equivalent decanting, is rightly criticized for producing higher rates of DCS and other complications such as dysbaric osteonecrosis.

DCS is a multifactorial illness. It occurs only after exposure to pressures around and in excess of 1 bar or 2 ATA (atmospheres absolute). It is frequent during the early shifts or after holidays, it comes after unusually hard work, after injury, in tiredness or exhaustion, in the obese and unfit. It seems to increase with falls in both ambient air pressure and temperature. It may also increase in very hot weather or during exposure in badly controlled tunnels and it occurs for no apparent reason. Poor decompression and safety practices increase the rate, and all too often the introduction of *ad hoc* safety margins to the established tables not only causes cases, but inhibits adequate analysis of possible causative factors, often because they are not admitted.

It is clear therefore that effective prevention of DCS as a cause of industrial morbidity is an amalgam of good subject selection, good training of men, supervisors, and those running the decompression schedules, vigilance at all stages, use of best available decompression schedules and the avoidance of extreme exposures. A high standard of fitness is required of both divers [9] and tunnellers [10, 11]. A major contribution to safety in divers is the use of oxygen for decompression. This acts by reducing the partial pressure of inspired inert gas, so promoting the rate of release of dissolved gas.

All decompression sickness should be treated. The patient most frequently makes the diagnosis by reporting symptoms. All symptoms occurring after a dive or an exposure to compressed air are decompression sickness until proven otherwise. The primary treatment is recompression and a recompression facility is mandatory in most circumstances. Again a wide variety of therapeutic tables exist and are used variously depending on the precipitating exposure, type and severity of symptoms and on the delay in reporting for treatment. The use of oxygen for tunnel DCS both in prevention and therapy finds little favour in the United Kingdom due to formidable safety problems and a low incidence of acute symptoms. Saturation techniques find favour for refractory cases in both disciplines and British doctors have gained a wide international experience in the management and salvage of difficult or neglected cases. Treatment tables and algorithms are to be found in most standard texts and in the various Navy diving manuals [10, 12, 13].

All diving contractors are required to have recompression facilities either on site or within a reasonable distance of the dive site [14]. Tunnelling calls for a chamber on site where pressure exceeds 18 psig (lb/in^2 or 124 kN/m^2) (1958 regulations) or 1 bar (CIRIA report no. 44). The latter is regarded as best practice. It is a peculiarity of hyperbaric work that serious illness with potential for death or serious disability is treated on site, some distance from immediate medical help, by the victim's colleagues who may have had little practical experience and often only a limited appreciation of the theoretical aspects of the disease. Recourse to medical advice is at the discretion of the supervisor; on him legal duties and obligations are laid and his overall control of a case cannot and should not be usurped by a doctor. While most tunnel contracts are not large enough to support a full-time medical adviser, the medical relationship is more orthodox.

Dysbaric osteonecrosis

This condition has been previously known as aseptic necrosis, bone rot, caisson disease and so on. The above title however is precise, and applies to a disease induced by exposure to high ambient pressure and following a discrete distribution affecting the shafts of tibiae, femora and humeri, and the head, neck and upper shafts of humerus and femur. Occasional lesions are found in other bones. Disease has been reported in the radius, lower tibia and the os calcis [15]. The pathology, diagnosis and differential diagnosis has been fully discussed by Davidson [16] and incidence and contributory factors reviewed by the MRC Decompression Sickness Registry [17] for divers. Information on tunnel workers is less amenable to analysis, but there is little doubt that tunnellers are at risk of more severe disease. Two leading opinions now believe that failure to observe established decompression procedures is the major factor in most tunnel disease.

It is a chronic form of decompression sickness and has been attributed to the formation of gas emboli in the marrow leading to death of the marrow and bone cells. Medullary bone is insensitive to pain. Efforts at natural repair lead to overgrowth of dead trabeculae by new bone and a net increase in mineralization on X-ray. This thickening of trabeculae interferes with blood flow and at the joints, hip and shoulder, the articular cartilage is compromised, leading to breakdown of the surface and eventually joint destruction and secondary arthritis. The time from insult to first radiological changes may be as short as 14 weeks, and breakdown with severe symptoms can be as short as 6 months or as long as 20 years after the believed incident. Each exposure to pressure is considered capable of causing a lesion or lesion complex.

The advent of radioisotopes has meant that damage of the nature found in dysbarism can be detected within a few hours [18, 19], and the progress of lesions can be followed by a process which involves a lower dose of radiation and is more sensitive than radiology. Interpretation of X-rays and scans is, however, a specialist activity, and should be undertaken in specialist units. The MRC DCS Panel made recommendations for the routine screening for the disease by X-ray long ago [20] and is currently engaged in making similar recommendations for the use of other methods of detection, including scanning.

Most lesions are in the shafts of the long bones and are called B lesions. They cause no disability or symptoms: they may be the site of later malignant change but such cases are very rare [21]. Lesions close to and in the femoral and humeral heads may still be B lesions, but when close to the joint surface are called juxta-articular or A lesions. These may break down and cause disability and pain. Hip disease is more common in tunnellers, rare in divers, a difference explained by the relative weightlessness of divers in water. Tunnellers, however, spend longer under pressure and divers are subject to a more rigorous discipline. Snell in 1896 [5] commented on the desirability of shorter exposures for tunnellers to limit the severity and frequency of acute DCS!

Divers currently have radiological examinations before joining the profession, and at such intervals as recommended by the approved doctor at annual examinations [9]. In tunnellers no statutory requirement is yet made although

CIRIA 44 [22] suggests that best practice is to X-ray before starting on a contract and at intervals of 12 months, then finally 12 months after the end of a contract. This periodicity will amply cover the 14 week period mentioned above for changes to occur. It is not possible on radiological examinations to predict which lesions will cause joint disease but computerized scanning may make this possible.

Attitudes to osteonecrosis and fitness to continue diving and tunnelling have changed in the last 10 years. It is now acceptable for divers with juxta-articular lesions showing discontinuity of the joint surface to continue diving, the criteria being that the lesion be identified and that no safety risk is presented. The certificate may need an endorsement. Bone damage is rare in less than 50 m sea water. The tunneller with juxta-articular disease however should be barred. It has not been shown that shaft lesions predispose to the production of joint lesions but all men with lesions should be most carefully assessed. Where only shaft lesions are found a strong reason must be adduced before a man is deprived of his job and his experience lost to the industry. This is a facet of medicine ripe for risk factor analysis.

Any employer or medical adviser who fails to examine the bones of prospective tunnellers or divers before they are employed must be considered negligent. Either traditional X-ray or scanning is acceptable. The other interested parties, insurers and courts have not yet expressed their views on the acceptability of shaft lesions in men seeking work, and it seems unlikely that they will ever do so.

Other effects due to gas absorption and decompression

Rozsahegyi found a high incidence of neurological sequelae in workers employed on the Budapest tunnels in the 1950s to 1960s. These men had not all suffered acute decompression sickness. The pre-exposure status of his subjects was not recorded and so a sceptical reception attended his comments in 1966 [1].

More recently Palmer and Calder [23] have studied the brains and spinal cords of goats and divers exposed to experimental and operational decompression sickness respectively. Diffuse microfoci of inflammation, and gliosis similar to the punch-drunk boxer was found; it was absent from the non-exposed controls. Anecdotal evidence from divers' families supports the view that minor damage with impairment of some psychological functions may occur, but forgetful and aggressive behaviour is not detectable on the routine annual examinations. This loss is supported further by formal testing of subjects on experimental deep dives in the United Kingdom, United States and Norway [2]. Recently completed work in London and Lancaster has shown impairment of cognitive function in the older and more experienced divers [3]. Data on tunnel workers are yet to be evaluated, but in them the social use of alcohol also seems to be important. It seems therefore that some occult damage is occurring but its significance lies not so much in the social damage to the family but in the less obvious fact that decompression schedules and routine hyperbaric practices are not free of risk, and that type 2 decompression sickness is but the tip of an iceberg of what must now be seen as subclinical disease. The need to identify and effectively manage these problems is yet to be appreciated, but adds further strength to arguments for the concentration of hyperbaric expertise in a moderate number of centres.

Acute decompression sickness, dysbaric osteonecrosis and these barely defined poorly understood psychological changes constitute the group of diseases that may be attributed to the solution of gas in body tissues. All in their way might compromise a career in diving or in tunnelling. The part of the workforce most at risk in both disciplines is of course the younger men who spend most time at the sharp end of the work.

Effects of pressure/volume changes

Some effects of hyperbaric environments are due entirely to pressure/volume changes and not to the nature of the breathing mixture. These adverse effects are felt entirely at gas–fluid interfaces in the lungs, middle ear cleft, paranasal sinuses and to a lesser extent the gut. The gut being in a compressible cavity escapes injury during compression, but may be uncomfortably distended during decompression as those who drank champagne in the compressed air at Blackwall in 1896 discovered.

During compression, gas-filled cavities must be supplied with gas at the new ambient pressure for as long as there is increase of pressure. Free venting must be possible during decompression. Without prior venting distortion of tissues with consequent potentially irresistible stresses will occur. Most relevant work seems to have been done on the decompression phase with particular emphasis on the lungs. Malhotra and Wright [24] showed that the lungs of fresh chilled cadavers could withstand a pressure difference at the surface of about 1.5 m sea water. This work has not been repeated under pressure, but no rational argument suggests that the tensile strengths of the alveolar wall, and their supporting basal membranes and interalveolar septa increase with depth. At full expansion there is a resistance to overexpansion which is constant for each part of the lung irrespective of depth. Local or general overexpansion of the lung at any depth by the equivalent of about 1.5 m is therefore capable of rupturing the lung. There is a popular misconception that because the volume change involved in rupture of the lung is large at the surface (in the average man some 0.8 litres from 1.5 m to zero), the same volume is needed to cause lung rupture at great depth. To do the same damage at 100 m therefore an ascent with closed glottis and fully expanded lung would need an ascent of more than 10 m. That this is not so can be deduced from several sources. Careful history-taking from victims suggests that only a few metres is needed to develop symptoms; the elasticity and tensile strength of lung tissue does not change, and the expansion of gas at all depths represents a release of stored energy, and it is the energy released which overcomes the lung stroma.

If a fully expanded lung is thus further decompressed, or stressed without the potential gas expansion implied by $PV = k$ free to escape, then it will rupture. The stress may be as slight as a hard working man raising his intra-abdominal pressure while holding his breath. The end result can be catastrophic causing mediastinal emphysema, pneumothorax with death during subsequent ascent from tension, and gas embolism with leakage of gas into the pulmonary veins and thence to the cerebral circulation. These are listed in order of ascending severity. A number of patients will have pulmonary barotrauma without any other signs than a little haemoptysis, retrosternal discomfort or a slight shortness of breath. These represent the minor adverse effects.

Gas embolism

Gas embolism is rare in the tunnelling industry but is too frequent in professional and sport diving. Much knowledge has been gained from cases occurring during submarine escape training. Recompression treatment for embolic disease is mandatory, all cases occurring within 15 min of the return to surface, and symptoms ranging from sudden death to loss of sensation or power in localized areas. Treatment of pneumothorax and mediastinal emphysema can be on more orthodox lines, but where discomfort is severe some significant relief can be obtained by recompression. When diagnosed under pressure, decompression must be most carefully controlled, in conjunction with orthodox measures.

After barotrauma, fitness to dive must be most carefully assessed. Gas embolism should be an absolute contraindication as should pneumothorax. Lesser illness should be assessed by careful and repeated spirometry, flow loop studies, assessment of transfer factors, and CT scanning in inspiration and expiration. About 3 months should be allowed between the incident and returning to diving, but experience suggests that 6 months is needed for full recovery. If the CT scan shows areas of lung failing to empty on expiration, no further diving should be allowed. CT scans are also useful in the investigation of those with doubtful fitness in whom no incident has occurred.

Sinus disease

Sinus disorders of minor significance in normal life are potentially very dangerous in hyperbaric work. Obstruction of the ostia during compression will cause swelling of the mucosae with pain and bleeding. If sinus pain occurs on compression, that compression must be aborted. Decongestant nasal drops may be needed to relieve symptoms and may enable the victim to perform the dive as planned. Such drops should not be used regularly, nor should they be used prophylactically on a regular basis. The causative disease should be treated.

During decompression similar symptoms may occur, with pain, bloody discharge and occasional orbital emphysema. This has led on to orbital cellulitis. Not only should post-dive sinus disease be treated seriously, with analgesia, decongestants and possibly antibiotics, but again, the underlying disease must be treated before the diver or tunneller returns to duty.

Ears

Failure to equalize middle ear pressure to the ambient causes a wide range of injury to the tympanum and inner ear. This injury is in part dependent on the pressure difference and the rate of change but may cause disruption of the ossicular chain, perforation of the drum or cochlea and vestibular damage with round window rupture. Mild haemorrhagic changes on the drum may be the only sign. All divers and tunnellers must be taught how to clear their ears properly by a recognized technique. They should be instructed in the symptoms to be expected if they cannot do so at any time, and of the long-term consequences. Damage occurs on both descent and ascent. Barotrauma to the basal turns of the cochlea causes high frequency hearing loss. Where the vestibular apparatus is damaged, the injury is direct and local. It does not appear to be related either by physiology or aetiology to isolated decompression sickness of the inner ear—which almost certainly is a more central lesion.

Uncompensated pressure changes cause much damage and are responsible for many men leaving the industries. Like pulmonary barotrauma, they can occur from as shallow a depth as 1.5 or 1.8 m. Hearing and balance can be lost, but both may recover to some degree. Losses should be assessed by pure tone audiometry and electronystagmography respectively; recovery should be monitored by the same methods, and maximal recovery seems to take up to 3 months. While hearing losses themselves may be insufficient to justify terminating a career, an ENG deficit of more than 25 per cent is significant and it may be unwise to carry on diving. Two patients in the author's clinic received their injuries with total loss of vestibular function from performing six or eight shallow dives, in the course of a single day.

Divers and tunnellers are also exposed to risk of explosion. Divers cutting with oxy-arc are used to frequent low energy explosions which become unnoticeable and cause no damage. Occasionally a build up of gas occurs and a violent explosion is set off. Where the diver survives, large hearing losses and vestibular disturbance are found. Although the symptoms of vestibular dysfunction may disappear and be compensated by other mechanisms, they will reappear in zero visibility conditions and sudden disorientation may lead to loss of life.

Breathing gases

Nitrogen

Nitrogen as the inert component of compressed air is the commonest gas to pad out divers' and tunnellers' oxygen. Indeed tunnellers use only compressed air. Divers also breathe helium, often with some nitrogen added, to make heliox and trimix respectively. Neon, and argon have been tried, hydrogen is being evaluated. All breathing mixtures must be free of solid particles, noxious vapours such as lubricating oils from compressors, and if it is to be used or transported in cylinders it must be as dry as possible. Various BS Standards and HSE requirements cover this aspect. Compressed air in diving, however, is often compressed on site and fed to the diver through a receiver. The intake for compressors must be sited away from exhausts of both the compressor and other engines.

Nitrogen is a narcotic gas and its effects which are measurable at 3 bar (4ATA) have reached 8–15 per cent performance decrement by 5 bar or 50 m. This can be a serious safety problem for a gas breathed on a regular basis by some 60 000 persons annually in the United Kingdom and many millions worldwide. Narcotic deaths make up a small number of the unexplained diver deaths annually, with carbon monoxide intoxication causing more morbidity and mortality.

Helium

Helium is an expensive gas but has no appreciable narcotic effect below about 1200 m of sea water. Man has so far reached 686 m on a simulated dive. Helium affects voice communication even at the surface causing the voice to sound like Donald Duck and at depth, unscramblers are needed to make speech intelligible. It also has a very high thermal conductivity which increases as depth and hence density of the gas increases. The working diver is therefore subject to a constant heat drain of some magnitude, far in excess of that experienced with nitrogen, and this must be countered by heating both the diver and his breathing gas. Hot water systems are currently most reliable for both purposes. When diving in the saturation mode, until recently reserved for work below 50 m, divers live under pressure in a system of chambers on a deck on the vessel or installation. Even where shallower saturation with nitrox is undertaken there are thermoregulatory requirements on the system to ensure that divers remain euthermic. With helium especially the band of thermal comfort narrows from the normal 10–12°C to become less than 2°C. A potentially disabling but fully recoverable affect on some nervous systems by helium is called the *high pressure nervous syndrome*. It occurs at about 200 m.

The choice of breathing gas and the manner and its use is the result of many factors. It must be available, as cheap as possible, and have few problems attached to its use. It must not be too dense to breath comfortably at high pressures and it must be possible to ensure adequate mixing of oxygen with it. It is after all only a carrier gas. The other gases mentioned have been associated with high levels of decompression sickness. Every gas has a diffusion constant through body tissues, and if gases are changed during decompression local supersaturation of tissues such as the ear may occur, causing isobaric decompression sickness [25]. These problems are confined to divers.

Oxygen

The gas for which all others is the carrier is oxygen. It is both neurotoxic and pulmonary toxic. At 50 m pressure, compressed air contains 1.2 bar partial pressure oxygen, exposure to this level for 10 h reduces vital capacity in 50 per cent of subjects. Shift patterns and cylinder size limit the time for which most divers are exposed and legislation prevents tunnellers being exposed to depths greater than 33 m or 3.4 bar. Annual spirometry must be performed on all persons exposed to hyperbaric environments in the course of work. In saturation work the upper permissible limit of oxygen at work is limited to 600–800 millibars, with about 500 mb allowed at rest; two-and-a-half to four times the normal inspired levels are therefore available. The gases must be properly mixed.

Oxygen is also used in routine decompression from air dives and in the treatment of acute decompression sickness. It is breathed by mask at 100 per cent. For routine use 40 fsw (feet of sea water) is permissible, giving a total Po_2 of just over 2 bar, but in therapy it is used at 60 fsw or 2.8 atmospheres. At this pressure some individuals will develop fits which are relieved by stopping the oxygen. Sensitivity to oxygen fits varies from man to man, and from day to day in each man. Oxygen is therefore breathed for periods of about 25 min with at least 5 min on air.

It is not easy to predict oxygen neurotoxicity although oxygen tests at 2.8 ATA are often performed. They have doubtful value, and it is better to prevent the effects by careful use of oxygen. Pulmonary damage is also largely preventable but the risk can be in part predicted by the use of tables of unit pulmonary toxicity dose (UPTDs) [26].

Diving chamber environment

Typical diving chambers are small, made of special steels and are often in exposed places on decks. Saturation diving uses a complex of chambers joined together with a diving bell allowing the transfer of divers to their work site under constant pressure. The divers live in the system for several days or weeks, and the grossly uneconomic costs and time spent on decompression from daily dives are eliminated and diver availability is increased. Up to 8 h in water availability is obtained for each diver and as much as 22 h at the work site can be achieved in each 24 h.

There is a cost, however, in that the system needs a life support team to see that the divers are well fed and watered, that they observe proper personal hygiene, that the chambers are kept clean (the actual work being done by the divers) and that gas composition, mixing and humidity remain within the limits set for the particular contract. The need to maintain temperature at a constant level has been mentioned earlier. The original systems were deck mounted, and these are still found in overseas locations, where legislation is poor or non-existent. Most European systems are now between decks or covered in to protect them from the weather. Such protection is needed as much in hot climates as in cold. Hypothermia is a well-recognized hazard of North Sea operations; hyperthermia is not so well recognized but is a serious hazard for tropical, for example, Middle East operations [27]. Several deaths have been reported through hasty and ill-advised recompressions in chambers exposed to the full heat of the sun. In temperate climates heat regulation is by a combination of insulation, air conditioning and underfloor heating. In the tropics, enclosing the chamber in an airconditioned room is the best course, but most chambers are still exposed and kept cool by pumping sea water onto sacking.

Chamber gas composition is important at all depths and for all types of diving, the level of sophistication needed to achieve satisfactory levels of control increasing as depth and complexity of the operation increases. In saturation chambers the Po_2 is only 500 mb, there is a small amount of residual nitrogen, much helium and a number of men producing carbon dioxide, carbon monoxide, methane and water. They have also a personal flora of bacteria which is soon modified by the conditions and normal commensals are suppressed with overgrowth of pathogens such as *Pseudomonas pyocyanea* being a constant risk. The selected divers must be free of overt infection and it is now usual to screen them bacteriologically before they enter the system. *Pseudomonas* infections cause economic havoc to the system, since divers with otitis externa are in severe pain and incapable of diving; the close proximity in which they live contaminates the others. A high standard of personal hygiene and rigorous toilet discipline is essential. Clothing and towels must be changed at least daily.

The level of oxygen in chambers and systems constitutes an ever-present risk of fire. It is therefore essential that only flammable substances essential to the job or to life should be present in the chambers and then only for as long as required. Only non-flammable lubricants are allowable for such things as valves and door hinges. Drench systems of fire control are now required for most chambers.

The compressed air tunnel environment

By contrast, tunnel working chambers are wet, dirty, full of men and materials, noisy, misty with odours due to cement, burning of metal fabrications, and regrettably still in many places, cigarette smoke. The machinery is usually hydraulic but much electrical equipment is being introduced. The number of men involved at any time, being compressed and decompressed, or worse being decanted, is great when compared with diving. On large contracts as many as 50 men may need decompression at any one time and the statutory requirements for space for each man have to be met. Educational attainments in the industry, save for the engineers and professional men involved are uniformly low, the work being unattractive to many people, and sadly lacking the machismo image of diving. Discipline and control may be difficult for this reason.

Atmospheric control is usually easy, since losses of pumped air through the ground are high and meet the requirements for air changing. On occasions however, with a closed tunnel face, during maintenance periods, levels of carbon dioxide and carbon monoxide may rise.

Personal hygiene is not always at a desirable level and toilet facilities are not popular with the men due to problems of privacy.

Fire is a constant risk partly as a result of increased oxygen, partly due to much unavoidable flammable material in the tunnel, and much due to failure to observe simple rules. Clothing for instance should all be of the least flammable material, but the most serious burns in recent years have been to a man who wore nylon clothing against advice, it being ignited by a stray spark from his burning torch.

In temperate countries tunnel temperatures are fairly steady between 15°C and 20°C. In tropical countries maintaining temperatures compatible with comfortable work requires refrigeration plant, to cool the pumped air. Even so, temperatures at the tunnel face will occasionally rise to the mid-30s or higher. CIRIA 44 recommends no work over 27°C. In Singapore the upper limit is 29°C. Often the surrounding ground is at a temperature higher than permitted in the local regulations.

Other aspects of hyperbaric environments

Noise and divers

The professional diver has been required to have a pure tone audiogram annually ever since the introduction of the Offshore Installations (Diving Operations) Regulations in 1975. It has been suggested that they suffer noise-induced hearing loss. The sources of noise to the diver are many. Chamber noise during compression and decompression may reach 110 dB. In many helmets and masks, the noise levels go up to 115 dB [28]. Hydrojets, some burners and other underwater tools are capable of inducing significant temporary threshold shifts, sometimes with tinnitus. However the total annualized equivalent dose of sound energy may not reach levels causing damage. Total days worked may not pass 120, and the actual exposure may be shortlived and not often repeated.

Curiously, being under pressure seems to protect hearing by inducing a shift of threshold for gas conduction [29, 30] with bone conduction not affected, and the changes are not sensorineural or a true temporary threshold shift. Typically, maximizing at 100 fsw, the levels of protection are some 30–40 dB across the typically tested sound spectrum. A survey in the author's clinic in 1981 showed that the auditory sensitivity of approximately 1000 divers had no identifiable losses attributable to the noise of diving. There was, however, a positive correlation with previous head injury, previous noise exposure such as gunfire or explosion and most important, with diving accidents involving barotrauma to the ears [31].

Noise and tunnellers

Noise levels in tunnels have been observed as high as 117 dB(A). Men with long histories of tunnel work usually show appalling losses and may be socially quite deaf. The losses become detectable at an early stage, sometimes as soon as 6 months after starting. The use of ear defenders is fortunately becoming fashionable with younger men and pit bosses leading the way. Compression noise can be as high as 110 dB, and lasts for about 3 min or for 10 per cent of the LEq for the day! The men work an average five shifts per week for 46 weeks a year, and when not in tunnels with or without compressed air will be found on road gangs with comparable noise levels. The protective effect of pressure seems to have no part to play in these men.

Such serious hearing losses present difficulties in assessing fitness for compressed air work. The men often know no other type of work and may not have the education or ability to change. It seems unfair to deprive them of work for what to them seems an acceptable part of their life, and to the occupational physician and hygienist is a preventable disorder.

Tunnellers do not exhibit much vestibular damage, by contrast with the numbers seen in diving practice. One subject, however, has copied diving colleagues in obtaining a total left vestibular paralysis from a series of exposures to 6 psig undertaken to ensure the safety of the works.

Conclusions

This review of hazards and health effects of hyperbaric environments is necessarily brief and reflects the author's experience over 14 years and 35 000 assessments of fitness for exposure. The amount of morbidity and mortality is still more than is desirable despite increasing legislation.

In the early 1970s many divers died in the North Sea. This number is down to an annual one or two, but some 12–20 sport divers die every year from gas embolism, panic, poor training or poor fitness, many more suffer symptoms with little or no treatment, often through lack of knowledge on their part or their medical and sport advisers. Men in tunnels still get 'bent'. In the commercial world, these figures could be much worse, only vigilance and medical services which are models for the rest of the world helping to keep them down.

In the various sections hazardous areas have been highlighted, but for all types of work a basically fit subject is required. Where disease is present, it may constitute an absolute bar. Lungs must not have a risk of air trapping, eustachian function must be normal, balance especially for divers must be acceptably normal, since visual acuity and perception are altered underwater. Some waters have so much suspended matter that they are zero visibility media.

Criteria of fitness are set out in a wide range of books and pamphlets. The complexity of making the assessments will be recognized by perusal of the Guidance Note MA1 *The Medical Examination of Divers* [9]. This document is contentious in places, going too far for some and not far enough for others. Guidance documents are not tablets of stone, and must give room for areas of indecision or debate to grow and mature. This sub-specialty of occupational health is a live science.

The current British regulations do not include Navy divers or sport divers. Some special groups who have good cause to be free of regulation have voluntarily accepted the spirit of the regulations. Compressed air workers, subject to statute since 1958 [32] may soon have new legislation but meantime a further revision of CIRIA 44 is under way and is likely to make the scope of the examinations approximately to the diving standard as already happens in some clinics. Until new law applies to them, however, contractors will continue to apply to the factory inspectorate for certificates of exemption to allow the use of modern decompression tables, and clients will specify compliance with the better standards voluntarily embraced by most of the industry.

It seems pertinent to ask at the end of this catalogue of hazard, risk and potential disaster if there are any positive benefits of hyperbaric work. The early writers described cures and improvements in the health of some workers but these could have been the result of regular wages and good food. At present compressed air workers appear not to benefit, but they are more likely now to seek and accept medical advice, and to use ear defenders. Socially alcohol still plays a large part, but its use seems more reasonable, and the number of legendary characters has decreased.

In the diving industry, regular medical surveillance reduced the number of unfit men within 2 years of the introduction of the 1975 regulations [33]. The later regulations and recommendations [9, 14] have tended to maintain the steady and effective interest in personal health and fitness, with the diver through better basic training being more aware of the hazards he faces. Each man now has a clearly defined interest in ensuring that he is as fit as he can be, both physically and mentally, since his own safety and that of his colleagues may depend on it.

This review is necessarily brief and leaves many important topics untouched or mentioned only in passing. Hyperbaric medicine demands understanding of physical laws, of anatomy and physiology, lateral thinking in pathology, and a wide appreciation of several branches of medicine. Often the effective practitioner does not know whether he is physician or engineer. He must certainly be astute in perception of legal and industrial relations problems. Such is the rate of change and of innovation that the standard texts are soon out of date and no one person can compass the steady flow of literature. The references contain all the standard works and the main legal references for the United Kingdom. Further information is available from the European Undersea Biomedical Society in Europe and the Undersea and Hyperbaric Medical Society in the United States. Both Societies however are as international and catholic as the diving and tunnelling which they serve.

References

1. Rozsahegyi, I. Neurological damage following decompression. In *Decompression of Compressed Air Workers in Civil Engineering*, edited by R.I. McCallum, pp. 127–137. Oriel Press, London (1966)
2. Bennett, P.B. Possibility of residual effects from saturation dives deeper than 300 meters. In *Long Term Neurological Consequences of Deep Diving. EUBS Workshop*, edited by T.G. Shields, pp. 69–86. EUBS (1983)
3. Morris, P. and Leach, J. *Psychological and Neurological Impairment in Professional Divers*. Final Report to the Department of Energy (August 1987)
4. Bert, P. *Barometric Pressure; Researches in Experimental Physiology* (1877), English translation (1943)
5. Snell, E. *Compressed Air Illness*, p. 194. H.K. Lewis, London (1896)
6. Golding, F.C., Griffiths, P. *et al.* Decompression sickness during construction of the Dartford Tunnel. *British Journal of Industrial Medicine*, **17**, 167–180 (1960)
7. Elliott, D.H. and Kindwall, E.P. Manifestations of the decompression disorders. In *The Physiology*

and Medicine of Diving, edited by P.B. Bennett and D.H. Elliott, 3rd edition. Baillière Tindall, London (1982)

8. Kidd, D.J. and Elliott, D.H. Decompression disorders in divers. In *The Physiology and Medicine of Divers,* edited by P.B. Bennett and D.H. Elliott, 2nd edition, p. 472. Baillière Tindall, London (1975)

9. MA1 *The Medical Examination of Divers; Information and Advice from the Health and Safety Executive's Medical Division* (revised) (August 1987)

10. *Medical Code of Practice for Work in Compressed Air. Report no. 44,* 3rd edition. Construction Industry Research and Information Association, London (1982)

11. Form F755 *Memorandum on the Medical Aspects of Work Under Increased Atmospheric Pressure.* H.M. Factory Inspectorate (1960)

12. BR2806 *Diving Manual.* Ministry of Defence, London (1972)

13. US Navy *Diving Manual* Navy Department, Washington (1979)

14. *Diving Operations at Work Regulations* (1981)

15. Williams, E.S., Khreisat, S. Ell, P.J. and King, J.D. Bone imaging and skeletal radiology in dysbaric osteonecrosis. *Clinical Radiology, 38,* 588–592 (1987)

16. Davidson, J.K. *Dysbaric Osteonecrosis in Aseptic Necrosis of Bone,* pp. 147–212, Elsevier, New York (1976)

17. Decompression Sickness Registry. A report from the decompression sickness central registry and radiological panel 1981: aseptic bone necrosis in commercial divers. *Lancet,* **ii,** 384–388 (1981)

18. MacLeod, M.A., McEwan, A.J.B., Pearson, R.R. and Houston, A.S. Functional imaging in the early diagnosis of dysbaric osteonecrosis. *British Journal of Radiology, 55,* 497–500 (1982)

19. Pearson, R.R., MacLeod, M.A., McEwan, A.J.B. and Houston, A.S. Bone scintigraphy as an investigative aid for dysbaric osteonecrosis in divers. *Journal of the Royal Navy Medical Service,* **68,** 61–68 (1982)

20. MRC Decompression Sickness Panel. Radiological skeletal survey for aseptic necrosis of bone in divers and compressed air workers. *Radiography,* **47,** 141–143 (1981)

21. Michael, R.H. and Dorfman, H.D. Malignant fibrous histiocytoma associated with bone infarcts. Report of a case. *Clinical Orthopaedics and Related Research,* **118,** 180–183 (1976)

22. *Medical Code of Practice for Work in Compressed Air; CIRIA Report No. 44,* 3rd edition, p. 15 (1982)

23. Calder, I.M. Pathological findings in the central nervous system after decompression sickness. In *Long-term Neurological Consequences of Deep Diving,* edited by T.G. Shields, pp. 133–150, EUBS (1983)

24. Malhotra, M.C. and Wright, C.A. Arterial air embolism during decompression and its prevention. *Proceedings of the Royal Society of London,* **154,** 418–427 (1960)

25. D'Aoust, B.G. and Lambertsen, C.J. Isobaric gas exchange and supersaturation by counterdiffusion. In *Physiology and Medicine of Diving,* edited by P.B. Bennett and D.H. Elliott. Baillière Tindall, London (1982)

26. *The Underwater Handbook,* edited by C.W. Shilling, M.F. Werts and N.R. Schandelmeier. Plenum Press, New York (1976)

27. Cox, R.A.F., McIver, N.K.I., King, J.D. and Calder, I.M. Hyperthermia in hyperbaric activities. *Lancet,* **8207,** 1303–1304 (1980)

28. Summitt, J.K. and Reimers, D.D. Noise: a hazard to divers and hyperbaric chamber personnel. *Navy Diving Experimental Unit Report,* pp. 5–71 (1971)

29. Thomas, W.G., Summitt, J. and Farmer, J.C. Human auditory thresholds during deep saturation helium oxygen dives. *Journal of the Acoustic Society of America,* **55,** 810–813 (1974)

30. Farmer, J.C. and Thomas, W.G. Auditory and vestibular function in diving. In *The Physiology and Medicine of Diving,* edited P.B. Bennett and D.H. Elliott. Baillière Tindall, London (1975)

31. Rodriguez, M.A. *Can commercial diving in the United Kingdom be considered a hazard to normal hearing levels?* Dissertation for MSc in Occupational Medicine, London (May 1982)

32. *The Work in Compressed Air Special Regulations 1958; Statutory Instrument 1958, no. 61.* HMSO, London (1958)

33. *Offshore Installations (Diving Operations) Regulations.* HMSO, London (1974)

Chapter 23

Treatment and first aid services

P.J. Taylor* and A. Ward Gardner

Introduction

Whatever occupational physicians consider their main function should be, it is in the therapeutic role that they are most often judged by both employers and employees. It is a common dilemma of doctors working in industry that their own perception of duties may differ substantially from that of those among whom they work. There are many who believe that, except in emergencies, 'treatment' should not be included among the functions of an occupational health service. On the other hand, society usually expects a doctor to be a clinician, and it is as a clinician that he or she can most readily establish a reputation. Few actions of a doctor win the confidence and support of workers and management more than prompt and expert treatment of serious injury or illness occurring in the factory. The first medical director of the Slough Industrial Health Service maintained that the successful growth of the service was due to the considerable emphasis placed upon providing efficient first aid and follow-up treatment [1]. Nevertheless, it can be argued that this is an incorrect use of scarce medical resources which could be used to better advantage for larger populations by restricting the activities of occupational health services to preventive medicine. The whole issue of treatment provided as part of an occupational health service, including the question of cost-benefit studies of treatment services, has been discussed in the booklet *Occupational Health Services: The Way Ahead* [2].

The extent to which treatment sources form part of an occupational health service's daily work varies widely between organizations and also between countries. Treatment is expressly forbidden in France, although unofficially a number of French services actually do provide some treatment. It was actively discouraged in the ILO recommendation no. 112 (1959). However the recent revision (1985) which also produced Convention no. 161, Recommendation no. 171 (para 16 and 17) states:

> 16. Taking into account national law and practice and after consultation with the representative organizations of employers and workers, the competent authority should, where necessary, authorize occupational health services, in agreement with all concerned, including the worker himself and his own

*P.J. Taylor died on 6 January 1987.

doctor, to undertake or to participate in one or more of the following functions:

(a) ambulatory treatment of workers who have not stopped work or who have resumed work after an absence;
(b) treatment of the victims of occupational accidents;
(c) treatment of occupational diseases and of health impairment aggravated by work;
(d) medical aspects of vocational re-education and rehabilitation.

17. Taking into account national law and practice concerning the organization of health care, and distance from clinics, occupational health services may engage in other health activities, including curative medical care for workers and their families, as authorized by the competent authority in consultation with the representative organizations of employers and workers.

Thus the ILO has now recognized reality and the wide differences that inevitably exist in the needs for and the abilities to provide treatment services in industry.

In this chapter, the wide range of treatment facilities which can be found in occupational health services today is discussed and some guidelines for the services of the future suggested.

Factors influencing the establishment of treatment services

There are at least six important factors which influence the establishment of treatment services within an occupational health setting.

The level of medical care in the community

There are wide divergences both within and between countries in the availability of individual medical care for the sick and injured, and there are important differences in political philosophy that affect the administrative framework for the provision of medical care. Some countries have a comprehensive national health service, while in others, particularly developing countries, there may be in certain areas a virtual absence of any diagnostic or medical care. In developing countries it has become common practice for an organization wishing to establish a factory to be required to provide for the full medical care of all the employees, their dependants and sometimes for others in the local community. Some international companies have considerable experience in this field and the agreement to set up such a service often includes a provision for the ultimate transfer of medical responsibilities to the local government. At a symposium on the health problems of developing countries [3] it became clear that the introduction of general medical services linked to occupational health services was an effective and economic method of raising the health status of a community. This has been recognized by WHO in its global plan called *Health for All by the Year 2000* which has recognized the essential part that occupational health services must play in developing countries.

Isolation from other medical services

Even in countries where medical care is generally available, circumstances arise where the place or work is relatively isolated and it is impracticable to rely upon the general medical services to provide treatment without a lengthy journey to a general practitioner or hospital. Even when isolation is not a serious problem, the provision of prompt medical care, including first aid and medical aid, is an essential function of any occupational health service. Breakdown in emergency care can lead to immediate emotional problems in a workforce, while a good performance in this field can earn a high reputation, confidence and gratitude.

Cost-benefits of treatment services

Very little has been done to measure the cost-benefit of occupational health services, but many managements believe that, apart from any of the less tangible effects on employee morale, a good case can be made out for the provision of a wide range of medical and ancillary services to save employees' time away from work. In our experience the provision of treatment facilities for minor conditions that can be treated at work eliminates the need for visits to the family doctor and enables people to stay at work, thus saving large amounts of time.

Minor sepsis, sprains and strains, minor trauma of all kinds and many other conditions can be treated successfully at work. For example, courses of injections, say for allergic desensitization, which have been prescribed by the family doctor, can more conveniently be given at work. The cumulative effect of such time saving in absences from work can be considerable. It also helps to 'sell' the medical services in general to the 'customer', and to make people aware of the facilities that are present for their use.

Specific hazards of the factory

Some factories have particular physical or chemical hazards, the effects of which can be mitigated by an efficient treatment service on the spot. For example, a large glass factory may have special arrangments for the definitive treatment of severe lacerations and tendon injuries, and chemical manufacturers will need to pay special regard to the immediate treatment of chemical burns both of the eye(s) and of the skin.

Medical ethics and rules

Another factor to consider is the effect that the attitude of the medical profession as a whole can exert upon the activities of an occupational health service. This can range from an absolute prohibition of any treatment other than first aid (as in France) to the permissive view of other countries. Most national medical associations have laid down sets of rules or advice to doctors working in occupational health (see, for example, [4] and [5]) which try to ensure that the medical responsibilities of general practitioners are not usurped by doctors working in industry.

Attitudes of occupational physicians

The attitudes and interests of occupational physicians can and do differ widely. These are influenced, above all, by their training and experience and how they perceive their role in the organization. Not infrequently, those who transferred from general practice without a period of formal training tend to concentrate on the care of individuals rather than of the group. It is one of the attractions of occupational medicine that it allows the doctor to combine the roles of looking after people in groups, practising preventive medicine and administering a service with the more traditional diagnostic and therapeutic roles in relation to the individual. It is vital, however, that a correct balance is achieved.

Matching the service to the need

It would be unrealistic to recommend one set pattern for all occupational health services. Each organization must be considered in relation to its needs taking the six general points already described into consideration. If treatment services are to be provided, their scale should be commensurate with the requirements of the organization objectively assessed in relation to the facilities within the community as a whole. To duplicate or compete with the therapeutic services already available in the area is not only wasteful of scarce medical and nursing manpower but can seldom be sustained by cost-benefit analysis. Undue concentration on treatment can, and often does, reduce the effectiveness with which the occupational health service can perform its unique task of protecting the health of all workers in the enterprise. A description of the treatment services provided in British industry gives a general indication of their scope and their limitations and thus offers some guidance to industries both within and outside the United Kingdom.

Treatment services in the United Kingdom

In a country where family doctor, hospital and pharmaceutical services are available to all, one might imagine that there would be little need for elaborate treatment services in industry. Although no domiciliary services are provided—and a few organizations which did undertake these before 1948 have ceased to do so— there are a number of occupational health services that do provide what amounts to a comprehensive out-patient service. Even so, such organizations represent a very small minority of all factories. In the United Kingdom there is a wide divergence of treatment services from the statutory minimum of first aid to the large and prestigious medical centres providing care by nurses and doctors, drugs, laboratory and radiographic facilities, visiting specialists, dentistry, physiotherapy, and chiropody. Since the cost is wholly borne by the employer, it is inevitable that the most services are found only among larger companies. Others equally large often provide very much less.

Why do some firms spend so much? Which of these activities are essential and which are merely an extravagance? How much do they cost, and have any cost-benefit calculations been done? Clear answers are difficult or impossible to obtain, but the more important activities will be considered with these points in mind.

Treatment by first-aiders

Although there have long been legal requirements for first aid boxes to be available in most places where people are employed, and in many of them there must also be someone trained in first aid, the quality and effectiveness of first aid has varied very widely. British law in relation to occupational health and safety has always been designed on the principle of setting minimum standards to which all employers must adhere and thus many of the more progressive companies have provided first aid or other treatment to a much higher standard than the law required. The law assumes, for example, that a holder of a valid certificate in first aid (that is, a certificate which is less than 3 years old) is capable of providing effective first aid to a casualty. Sometimes this is certainly the case, but it is also true that the ability to answer questions about first aid theory and to apply a triangular bandage in an examination does not always indicate competence in a crisis to treat an injured colleague when the blood and pain are real. A good first-aider must have a sense of vocation for this work and this cannot be obtained by attendance at a course.

The objectives of first aid at work have traditionally been: first, to preserve life and to minimize the consequences of serious injury or major illness until professional medical or nursing help can be obtained; and second, to treat minor injuries which might otherwise go untreated. There are two further objectives at least as far as occupational first aid is concerned: the provision of redressing or other simple continuation treatment; and the treatment of minor illnesses arising at work. Although these extra responsibilities have been undertaken by some experienced first-aiders for many years, this policy has its critics since the last proposal, treatment of minor illnesses, in particular implies that the occupational first-aider must be trained to make some sort of diagnosis, even if it is merely the exclusion of conditions for which the services of a doctor would be mandatory.

There is certainly a case to be made for education of the whole workforce in

(1) *Priorities*—doing the correction actions in the correct order.
(2) *Lifesaving first aid*—what to do for a casualty who is bleeding, unconscious and/or not breathing.
(3) *First aid to prevent worsening* which requires to be carried out immediately— for example, washing a chemical out of an eye or cooling a heat burn.

If this education is given to the *whole* workforce—and it probably takes no more than a 1½ h session to do this and a refresher of about an hour a year—then the remaining first aid can be carried out when the ambulance arrives. Lives can be lost, however, if the correct actions are not carried out in the correct order (priorities) and if the lifesaving first aid is not done by the first available person. Similarly if the *immediate* first aid to prevent worsening is not done at once, the condition of the casualty can deteriorate rapidly and needlessly. So this policy of training everyone in the workforce to do a little may be much more sensible and efficient than training a few people to a higher standard and may yield much better dividends in terms of lives saved and needless deterioration of injuries [6].

The Health and Safety (First Aid) Regulations (1981) require an employer to provide equipment and facilities which are adequate and appropriate in the circumstances for enabling first aid to be rendered to employees for the purposes of life and minimizing the consequences of injury and illness until help is obtained. First aid is also defined as meaning the treatment of minor injuries (a misnomer really, because first aid implies *second* aid, and the definite treatment of minor injuries would be better defined as only-aid).

If first-aiders are appointed, they must be suitably trained and hold qualification(s) approved by the Health and Safety Executive (HSE). Should special or unusual hazards be present at the workplace, additional training may be required, appropriate to the circumstances. These regulations are very broadly framed and avoid, deliberately, setting down details of training, what is appropriate, how many people be trained—and so on. The acid test, for which responsibility is firmly fixed on the employer, is 'do the arrangements work well and efficiently to save lives and to prevent worsening in cases of illness and/or injury?' If they do, then the arrangements are satisfactory—and vice versa.

The employer must also inform his employees of the arrangements made including the location of equipment facilities and personnel. A number of other minor matters are dealt with in these regulations which need not concern us here, but which should be examined by anyone setting up first aid provisions and arrangements for the first time.

Treatment by nurses

The provision of emergency treatment to employees taken ill or injured at work is only one of the accepted functions of the occupational nurse or doctor, although for some nurses employed in industry in the past this has been their only function. In many smaller factories the nurse may actually have no formal qualification at all, but in others the nurse may be state enrolled or state registered. The employment of a highly trained registered nurse with an occupational health nursing certificate (OHNC) solely to provide treatment is a gross misuse of scarce resources, and in practice such a nurse would probably not wish to remain for long in this restricted type of job. On the other hand, an enrolled or registered nurse without occupational health training can provide professional and competent treatment for a wide range of injuries and ailments, thus saving employees taking time off work to see their doctor or to attend the local hospital. They can also provide continuation treatment prescribed by a doctor including such things as courses of injections.

There are, however, some medicolegal problems if a nurse has to work entirely alone without a doctor available, on a part-time or 'on call' basis. The Medicines (Prescription Only) Regulations 1978 have greatly restricted the range of drugs, and in particular injections, that may be obtained and administered to patients without a doctor's prescription and written instructions. The Royal College of Nursing, Society of Occupational Health Nursing [7] has issued a useful guidance note for occupational health nurses which should be consulted for further information.

For the large factory where there is justification for the provision of a nurse-staffed treatment service because of the hazardous nature of the work, the best solution might be to have a senior nurse with an OHNC to provide the full range of occupational health care, and a treatment room staffed by enrolled nurses who are professionally responsible to the senior nurse. In such a situation there would probably be an occupational physician, either full-time or part-time, who would also be able to provide professional support and guidance in the day-to-day problems that can arise in the treatment room. Such elaborate arrangements and staffing, however, will be found only in a minority of factories and could never be justified in terms of need or cost-benefit in the majority of workplaces. This is particularly true in the United Kingdom where the freely available facilities of the national health service exist.

The cost of employing a nurse is largely determined by the salary scales laid down for guidance by the Royal College of Nursing. To this must be added the usual overheads for all employees, and a moderate revenue expenditure for drugs and dressings. Some capital outlay will also be required to provide a surgery or treatment room with basic furnishings, equipment and adequate washing facilities.

Treatment by a doctor

While it is unnecessary to specify what treatment can be provided by an occupational physician, it may be of value to indicate some of the limitations. There are important differences in practice between services where the doctor is full-time or part-time, visiting the workplace only once or twice a week.

An occupational health physician must not come between an employee and his or her own family doctor. Any treatment provided by the factory doctor should only be in the nature of first aid or, where it is not, the patient's own doctor should be consulted or informed. Since occupational physicians do not provide domiciliary services, any condition that may require medical attention at home should not be definitively treated by them. While a close and friendly liaison with local family doctors is necessary for most activities, in none is this more important than in the field of treatment. Patients should only be referred direct to hospitals or specialists in an emergency, and then the family doctor must be informed; at any other time such referrals should be only after consultation.

Occupational physicians not infrequently find themselves being used as 'second opinions' by employees. This, although understandable on the part of the patient, can be a pitfall for the unwary. Some patients succeed in playing one doctor off against another, with unfortunate consequences which can affect all three. In most cases, however, the employee only wants an explanation of what may be a standard medical procedure.

The provision of an 'open door' consultation service by the factory doctor undoubtedly limits the time available for activities more strictly related to the practice of occupational health. It is difficult to lay down guidelines but it seems that in areas where general practitioners' lists are large this may on average amount to between ½ and 1 h a day for each thousand employees. Where a nurse and a doctor work together, it is usually possible to reduce this by suggesting that patients first see the nurse, who will then refer a selected number to the doctor.

The cost of employing a doctor can be considerable; with the salary range laid down by the British Medical Association, a full-time experienced doctor in a large factory may be paid as much as the senior members of management. As with the nurse, an appreciable capital outlay is also required. A part-time doctor may share the nurse's accommodation, but a full-time doctor will need his own examination room. One must not presume, however, that these expenses are all incurred under the heading of 'treatment' since most of the activities requiring extra expenditure, including clerical assistance, will be incurred from medical examinations, environmental control, and advice to management on many problems unconnected with treatment.

The ability to provide an open consultation service without jeopardizing the doctor's ability to undertake other occupational health responsibilities must clearly depend on the size of the population. For a full-time doctor in an average factory this may be somewhere between 3000 and 5000 employees, depending on the routine examination load and other factors. Where possible, some degree of personal consultation service is of the greatest value, since it enables the doctor to get to know some employees quite well and to obtain the confidence of the workforce in the easiest and most traditional way. It also provides the doctor with the possibility of recognizing previously unknown hazards of the physical or psychosocial environment.

This activity is one of the more stimulating aspects of the routine work in an occupational health service. In some cases, problems may first be raised by the employees themselves, but in others astute observation by the nurse or doctor may suggest that a problem exists. Two new cases of dermatitis of the hands from one workshop may be an obvious clue that there is a hazard. The recognition that several patients with assorted minor complaints all come from the same working group may suggest a problem in morale of employees or management. If one is to verify a hunch of this sort, it is essential that adequate records of treatment be kept. These will be discussed below (*p.441*). Both consultations and treatment records provide invaluable data for epidemiological studies.

Medical supplies

A glance at the drug cupboard and medical stores can be an effective way of assessing the scale of treatment services provided by the medical and nursing staff. These can range from basic supplies of analgesics and antacids, and mixtures for coughs and diarrhoea, through to a range of drugs that would not disgrace a pharmacist's shop. Except for underground mines and ships at sea, for which special provisions are made by law, dangerous drugs such as morphine are only held in departments with medically qualified staff.

Although expenditure on drugs is a relatively small item in the budget compared with salaries and overheads, this is one of the areas in which savings can often be made. As in all other medical establishments, the contents of the drug cupboard bear witness both to fashions in prescribing and to the preferences of previous doctors or nurses. A stock check for valuation might prove a salutary exercise in those departments where it is not normally practised. The use of standard preparations instead of proprietary equivalents can usually result in appreciable savings, while realistic forecasts of consumption can allow bulk purchasing, particularly for those organizations having several medical centres, and this can usually be arranged at an appreciable discount.

Similar arguments may also be applied to dressings, but the benefits of prepacked sterile dressings in terms of a lower rate of secondary sepsis have probably saved more money by reducing time off work than the increased cost of materials involved. In some larger medical centres or groups of centres, it can be more economic to use an autoclave and pack department dressings during slack periods, such as on the night shift, rather than to buy from commercial sources.

Radiography

Many large occupational health centres have their own radiographic apparatus. Although there can be little justification for elaborate equipment of hospital standard which is rarely, if ever, used to its full potential, most organizations with radiographic apparatus are convinced of its value. The capital outlay is considerable, not only for the apparatus but also for its screening, the space it requires, and the accompanying darkroom. Few departments employ a full-time radiographer, but they may have one who also undertakes other duties, or more frequently a nurse or technician has been taught to take the relatively simple films usually required.

There are two reasons for having radiography in a medical centre: to take films of injuries which might be fractures; and to take chest and other films for routine pre-employment or periodic medical examinations. The former is most relevant in the context of this chapter, but the latter can provide a substantial reduction in expenditure at the local chest clinic if, as in many companies, a normal chest radiograph is required before accepting a new employee, and labour turnover is high. For most populations in the United Kingdom today, however, routine chest X-rays are not necessary.

In a country such as Britain with a national health service, there is no reason why any employee with a suspected fracture should not attend the local hospital for a radiograph to be taken, and this is routine procedure in most workplaces throughout the country. But where hospital facilities are distant much time and therefore money can be saved by taking films on the spot.

Were such films requested only when there was clinical evidence of bone injury, the argument for radiography in a factory medical department could not be sustained. Unfortunately, and largely for medicolegal reasons, it is now the practice to obtain a radiograph of all injuries which might conceivably show an abnormality. This applies particularly to occupational injuries, but also to a lesser extent to other types of injury. If between 20% and 40% of all occupational injuries reported to the medical department (many of them minor) come under the diagnostic group of 'fractures, sprains or bruises', the number requiring a radiograph can be appreciable.

The loss of working time involved in sending an employee to the local hospital, and the cost of transportation there and back, can be substantial. When the majority of such patients will prove to have no bone injury and should return to work, the unnecessary loss of working time will also be high, particularly if, as is often the case, a number of them then take a day or more off work. With a population of several thousand manual workers in a high risk industry, there would be little difficulty in demonstrating the cost-benefit of simple radiographic facilities, even taking into account the depreciation on the capital investment.

Specialist treatments

Physiotherapy

The main justification for providing physiotherapy is also one of saving in time lost from work. In contrast to the provision of radiography, however, the capital outlay is not so high and the amount of time provided can more easily be adjusted to meet the needs of the population at risk.

As with all treatment services other than first aid, it can be argued that physiotherapy is freely available at the local hospital. Here, however, the demand is considerably greater than the supply, and treatment twice or three times a week is the rule, not because it is ideal but because of the pressure of work. It is also true that for the majority of patients regular attendance at a hospital physiotherapy department involves continued sick absence until the course of treatment is complete.

Cost-benefit studies in a number of organizations have indicated a real saving by the provision of physiotherapy. The numbers of hours per week required will depend on the size of the workforce and the degree of manual labour involved. It is most important, however, that close liaison be maintained with the orthopaedic and physical medicine consultants in the area so that treatment prescribed by them can be undertaken by the occupational health service. One solution which works well is a joint appointment for the physiotherapist, shared between the service and the local hospital.

Dentistry

Relatively few dentists are employed in occupational health centres because of the shortage of dental surgeons and the heavy capital outlay involved in equipping a dental surgery.

The economic justification depends once again upon a saving in lost working time. It has become almost impossible in the United Kingdom to obtain a dental appointment outside normal working hours and for many people each visit involves the loss of half a day from work. Even though only a small proportion of the working population attends regularly twice a year, the numbers doing so are rising. The independent contractor status of dental surgeons and their payment on a fee per item of service by the national health service allows a dentist who works in industry to claim the usual fees from health service funds. In some cases the employer pays the dental surgeon a salary and recoups fees from the state, and in others the dentist is paid a retainer and collects the fees himself.

The size of the capital investment, however, usually restricts dental services in industry to organizations employing several thousands, or to group health services covering similar numbers.

Ophthalmic services

Some of the larger occupational health services have arranged regular visits by an ophthalmologist or ophthalmic optician to undertake the testing of vision, refraction and the prescription of spectacles. This is relatively inexpensive since the only requirement is for a suitable room to be made available for the sessions. The system of payment under the national health service is, like dentistry, based on a fee per item of service and the consultant obtains his fees direct. It is usual for a dispensing optician to attend so that employees can be fitted for the spectacle frames. The standard health service charges are made and so the employer is involved in no regular financial responsibility. The provision and fitting of safety glasses, both with plain and prescription lenses is often made easier if ophthalmic services are available.

The rationale for such services is twofold: not only is it important that all employees should be able to see clearly for their work, but, as in many of the other treatment services, this arrangement can also save a substantial loss of working time since without them such appointments usually involve half a day away from work.

The use of visiting specialists

A few occupational health services have arrangements with local consultants to do occasional outpatient clinics at the workplace. Although this is not widely done, the specialties usually provided are orthopaedics, physical medicine, dermatology and psychiatry. The main reason for this development in some parts of the United Kingdom has been the excessive waiting time for first appointments with such specialists within the national health service. Many of the treatment services provided by industry could not be justified if the national health service facilities were close to the factory and prompt treatment and rehabilitation were readily available. Unfortunately the waiting lists of some specialists can amount to 2 months or more for a case that is not medically urgent. Sometimes, however, the patient may remain certified unfit for work (or unfit for his usual work) during this waiting period. The real necessity for this may be questionable, but general practitioners and occupational physicians may find it difficult to insist that the patient is fit for work while requiring him to see a specialist for an opinion, for example, on the advisability of a meniscectomy, or treatment for recurrent episodes of back pain.

Chiropody

In the past decade several occupational health services have introduced chiropody services for employees. Although they started in organizations such as department stores and factories with a predominantly female population who had to stand for most of their work, they are now to be found in other industries. Many people have minor foot ailments and there is little doubt that efficient chiropody can alleviate and sometimes cure the condition. Even so, some chiropody services in industry have closed because, after initial enthusiasm, the number of employees wishing to attend have dropped below the level to justify them.

Capital outlay is small since a part-time chiropodist can use a treatment room chair. In some cases the chiropodist is paid by the employer and the service is offered free, but in others a charge (often subsidized by the employer) may be made. This is because chiropody is only provided free by the national health service to a limited number of patients referred by doctors, such as those attending diabetic clinics.

Concentration of treatment services

It is the experience of all who have been involved in the provision of treatment services in industry and elsewhere that the most efficient and the least expensive are those which concentrate trained medical, nursing and other resources as far as possible. Much better work will be done by a few highly skilled people in purpose-designed premises than can be done by more numerous but less efficient first-aiders in workshops or offices. The rates of secondary sepsis, return treatments, and the cost of dressings will all be lower, with a much greater chance of consumer satisfaction.

This will require both efficient and rapid means of transport for the injured to the treatment centre, whether by trolley or wheelchair in the workshop or by a motor vehicle in more dispersed sites—and also the training of as many employees as possible in emergency life-saving first aid, but not necessarily in the whole rigmarole of an officially approved first aid course.

Medical records

Treatment records

The need for the maintenance of adequate personal treatment records is accepted in all types of medical, nursing and ancillary medical services, but the particular needs and functions of an occupational health service may require more than is customary elsewhere. The records must be available, not only to ensure efficient care of the individual, but also for epidemiological investigations for the identification of health hazards and the evaluation of services. In the United Kingdom only one record is required by law in every place of employment: the *accident book*. This must be kept in a prominent place and be available for the recording of all occupational injuries.

The accident book is a requirement of the Social Security Act 1975 and is mandatory in all places where there are ten or more employees. It must be accessible to injured persons *at all times* (in other words, not merely on day shift) and must be filled in by the injured person or someone such as a first-aider or nurse acting on his/her behalf. Books must be preserved for at least 3 years after the last entry. The accident book is quite distinct from the obligation on the employer to report to the Health and Safety Executive and record on the *factory general register* details of death, serious injuries, absences due to injuries lasting more than 3 days, dangerous occurrences and, most recently, the 28 reportable occupational diseases spelt out in RIDDOR (Reporting of Injuries, Diseases and Dangerous Occurrences Regulations, 1985).

Records kept by the medical department for internal use can be considered in two categories: individual record cards (*Figure 23.1*) or files, and those of a chronological nature, listing all attendances for treatment, usually called *day sheets* (*Figure 23.2*).

Personal records

A centralized system of filing in which all medical records about an individual employee are kept in the same folder or envelope is the obvious and most efficient arrangement, but it is surprising how few occupational health services actually do this. Since a central registry administered by a clerk-receptionist is the usual arrangement in most hospitals and health centres of the national health service, it is of interest to consider why this is not so widely found in industry. Individual treatment record cards (*see Figure 23.1*) are often held in a filing box or in drawers or, sometimes, in rotating drums in or close to the treatment room itself. The other personal records of routine examinations, doctors' letters, reports of investigations and so on, are often kept in another place. While the disadvantages are obvious, the reason stated is usually that waiting time is of the greatest importance for visits to the treatment room and further, that the records of these visits are usually maintained by nurses, medical orderlies or first-aiders.

It is important to insist that all attendances and all treatments are noted on the records. Some departments note only the first attendance on the individual's record and put reattendances down on the day sheet. Since there are strong medicolegal reasons why all attendances for occupational injuries should be recorded, however trivial they may appear on first attendance, it is advisable to maintain scrupulous records for all attendances.

Daily records

All treatment rooms should keep a record in summarized form of every attendance (*see Figure 23.2*). Although this may be done in a ledger, it is more usual for them to be entered on large 'day sheets' with printed columns, which greatly facilitate

INDIVIDUAL TREATMENT RECORD

NAME	PERSONAL NO.	DEPARTMENT
ADDRESS	SUPERVISOR/FOREMAN	OCCUPATION
	DATE OF BIRTH	PATIENT'S OWN DOCTOR

DATE	TREATMENT	Occup.	Non occup.	DISPOSAL

Figure 23.1 An example of an individual treatment record card

DAILY ATTENDANCE RECORD

TIME	NAME & INITIALS	PERSONAL NUMBER	DAY OR SHIFT		SUPERVISOR/ FOREMAN	AREA OF INJURY	CAUSE (IF AN ACCIDENT)	DEPARTMENT

Figure 23.2 An example of a 'day sheet'

the subsequent preparation of statistical returns (*Figure 23.3*). This system should be accompanied by a separate entry on the employee's personal file—a duplication of effort that is frequently omitted by busy treatment room staff. Alternatively, an individual record may be made in duplicate at the treatment centre (*Figure 23.4*). One copy is kept at the centre as an individual record. The duplicate copies are available for statistical analysis at the end of each month or a longer period (Grant Macmillan, personal communication).

It is the responsibility of the person in charge of the treatment room to ensure that an entry is made for every attendance. The basic information required consists of the date and time, the name, identification number and department of the

SUMMARY OF ATTENDANCES

Sheet no..............

Day..................   Date...................

	OCCUPATIONAL								NON OCCUPATIONAL											OTHER				REFER		DISPOSAL					
CUTS BRUISES ABRASIONS OF SKIN	BURNS AND SCALDS	SPRAINS, STRAINS, SUSPECTED FRACTURES	OCCUPATIONAL DERMATITIS	EYE CASES — FOREIGN BODIES	EYE CASES — OTHERS	CHEMICAL INJURIES	OTHER INDUSTRIAL ACCIDENTS CONDITIONS	RESPIRATORY DISORDERS	DIGESTIVE DISORDERS	RHEUMATISM GROUP	FUNCTIONAL NERVOUS DISORDERS	NON WORKS ACCIDENTS	BOILS AND CARBUNCLES	NON OCC SEPSIS	NON OCC SKIN CONDITIONS	CONDITIONS OF EYE AND VISUAL DEFECTS	CONDITIONS OF EAR	OTHER NON INDUSTRIAL CONDITIONS	RETURN FROM SICKNESS	MEDICAL EXAMINATION	INOCULATION	OTHER REASON	RE ATTENDANCE	DOCTOR	PHYSIOTHERAPY	WORK	HOME	HOSPITAL	REFER TO OWN DOCTOR	OVERTIME	INITIALS
A	B	C	D	E	F	G	H	I	J	K	L	M	N	O	P	Q	R	S	T	U	V	W	X	1		2	3				

Signature ....

No. attendances [112]

Figure 23.3 An example of a form summarizing attendances. This summary is completed during, or at the end of, the day or shift by placing a tick in the appropriate box. Totals for the day (or shift) or longer periods may be summed without difficulty for statistical returns

DOCKYARD MEDICAL CENTRE Serial No. 1[030717]6   Name

| 7 Date | Yd | Yard Number 20 | Centre | Age | Sex | Employer | Follow-up No. 37 |

| 38 | Occupation | 57 | Date Injured/Illness | Illness | Time of Injury 68 |

| 69 | Part(s) of body injured | 84 | 85 | Cause of Injury | 104 |

For foot, head and eye injuries ONLY, was:

a. Relevant protection issued? ☐ 105

b. Relevant protection worn? ☐ 106

Medical notes

109
Treated by   ☐☐☐

Direct return to work ☐

112
No. attendances   ☐☐

Signature ....

Figure 23.4 An example of an individual record of treatment made in duplicate. One copy is for the individual's file and the other for statistical (computer) analysis

patient and, for first attendances, a brief description of the condition, its treatment, its cause if an injury, and a note of disposal. Where more than one person works in the treatment room, the identity of the therapist should also be included. This information can subsequently be used for statistical analysis of treatments given, the number and nature of occupational injuries, and the departments in which these occurred. Not infrequently they prove to be important if, at a much later date, any enquiry is raised about the attendance of an employee with a specific injury. These records should therefore be kept for not less than five years, and longer if space is available.

Records of treatment or attendance at other parts of the medical centre such as radiography, physiotherapy and so on are also required. It is important to ensure that the entire set of one person's medical documents are kept in one place and can be rapidly collected together. Clearly, the more separate records there are, the more advisable it is to have a central filing system.

Rehabilitation and resettlement at work

Rehabilitation is the process of restoring as much function as is possible in the person or in the disabled limb or other part of the body. Resettlement means placing the person in suitable temporary or permanent work.

The processes or rehabilitation and resettlement following injury or illness may be the responsibility of an occupational health service but they will require close and extensive cooperation with hospital specialists, family doctors and social workers.

The doctor in industry should be aware of the schemes in the community—both in the health and social services and those organized by employment ministries such as the Manpower Services Commission (MSC)—which can help the disabled, and of the assistance which can be given by other groups concerned first with special disabilities such as blindness, deafness and limblessness, and secondly with such diseases as diabetes, alcoholism and multiple sclerosis.

Within industry there are many different kinds of schemes for rehabilitation ranging from the elaborate formalized to *ad hoc* arrangements. In general any scheme is only as good as the interest of those who supervise it. There is no substitute for personal concern on the part of doctors, nurses, and supervisors for the restoration to the affected person of full physiological function and a satisfying job—meaningless labour is not good enough. The medical and work aspects must be given equal attention if success is to be assured.

Doctors working in industry are in a privileged position with regard to rehabilitation and job placement: the doctor has—or should have—access to all relevant facts about the medical condition and the job requirements. He can also discuss the problems with supervisors, management and union and thus stimulate interest and catalyse activity in rehabilitation and resettlement.

The aim should always be to get the person back to his normal job as soon as possible. It is usually best to modify normal work to suit the worker's medical condition rather than to uproot him from his regular work and social environment. Although the work task may be easier in specially created jobs, the social content is often much less good and usually less helpful in achieving forward-looking attitudes or attitude-change on the part of ill or injured people. The man who is among his colleagues will generally be motivated to pull his weight as soon as he can and will more often have a sympathetic group around him. Transferring a man to modified work in another department can have disastrous results since his new colleagues may resent an impaired 'outsider' whereas his own friends would have helped him.

Should it be necessary to carry out rehabilitation away from the workplace, the occupational physician can still show interest and help to guide his medical colleagues at the place of treatment by giving the sort of information which will help them to apply their efforts in the best direction and to enable them to feel happy about sending patients back at an early date to suitable work.

Job placement

In communicating with managers and supervisors about job placement, the doctor should try to avoid all such vague phrases as 'light work'. Many different interpretations of this phrase are possible, and can lead to people being less well placed than is desirable or to supervisors creating 'jobs' which consist of sitting down and doing nothing. It is best to try to spell out exactly what a person can and cannot do. An example of this approach is shown in *Figure 23.5*.

Another useful approach to supervisors is to remind them not only of what is not working—for example, 'He has a bad leg'—but of what is working—for example, 'Although he has some trouble with his left leg and cannot climb stairs, he has no other disability.' The focus on *ability* should be at least as sharp as that on *dis*ability. In this way, useful productive and creative work can be arranged

CONFIDENTIAL

NOTIFICATION OF RESTRICTED DUTY OR OF CHANGE OF MEDICAL CATEGORY

1. OHS records
2. Supervisor
3. Personnel records

From: OHS

To: ....

No: ....

Date: ....

Restricted duty for

1. No climbing of vertical ladders	} See remarks
2. No work at unguarded heights	
3. To work at ground level only	
4. No heavy lifting (40 lb. +)	
5. No stooping or bending	
6. No heavy work	
7. No field work	
8. Sedentary work only	
9. No work near moving machinery	
10. No driving of cranes/automotive equipment	
11. No arc welding	
12. No work involving the use of right/left hand/arm/leg	
13. To avoid contact with	
14. No overtime	
15. Should not work more than .... hours overtime per week	
16. Should work on days only	
17. Should do no work involving rapid action or decision taking	

The doctor would be grateful if this employee could be given work with the restriction(s) indicated for the next .... days. If this should prove difficult or impossible, please refer the employee back to the Medical Centre

Signed ....

OHS

This employee's medical category has been changed to ....

Remarks:

Figure 23.5 An example of a certificate of notification of restricted duty or change of medical category

which is within the capacities of the disabled person. Should there be a large number of problems, job analysis as well as disability and ability analysis may be formalized to aid administration. This sort of approach will only be required in very large organizations. Special rehabilitation workshops are occasionally found in very large organizations. They either use modified machinery to manufacture items for part of the factory's production schedules, or are set up to make things such as safety gloves. Only when the total employed population on one site is in the order of 20 000 or more can such special workshops be justified on economic grounds. For the smaller factory a great deal can be achieved by enthusiastic medical staff, an understanding management and a sympathetic work force.

References

1. Eagger, A.A. *Venture in Industry. The Slough Industrial Health Service, 1947–1963*. Lloyd-Luke, London (1965)
2. Health and Safety Commission. *Occupational Health Services. The Way Ahead*. HMSO, London (1977)
3. Ross Institute of Tropical Hygiene and TUC Centenary Institute of Occupational Health. *Proceedings of the Symposium on the Health Problems of Industrial Progress in Developing Countries*. London School of Hygiene and Tropical Medicine, London (1970)
4. British Medical Association. *The Occupational Physician*. British Medical Association, London (1980)
5. Faculty of Occupational Medicine. *Guidance on Ethics for Occupational Physicians* 2nd edition, Faculty of Occupational Medine, London (1982)
6. Ward Gardner, A. and Roylance, P.J. *New Essential First Aid*. Pan, London (1987)
7. Royal College of Nursing. Society of Occupational Health Nursing. *Information Leaflet No. 11* (third issue). Royal College of Nursing, London (1979)

Further reading

Health and Safety Executive. *Health Surveillance by Routine Procedures, MS 18*. HMSO, London (n.d.)
Health and Safety Commission. *Occupational Health Services: The Way Ahead*. HMSO, London (1977)

Chapter 24

Emergency medical treatment

H.A. Waldron

Introduction

Three main categories of medical emergency may occur at work: heart attacks or strokes which are incidental to work; accidental occurrences which are the direct result of the work carried on in the establishment, and catastrophes resulting from an accident on a grand scale.

The ability of an occupational health department to deal with any of these emergencies wil:, of course, depend on the facilities available, on the type of staff employed and on their skill and training. It is obviously impossible to cover all eventualities in a short chapter, and so I will deal only with general principles.

Emergencies incidental to work

Common medical emergencies are almost as likely to occur at work as elsewhere and some contingencies must be on hand to deal with them. The nature of the emergency will depend upon the demography of the workforce at any particular location; in an ageing workforce, cardiovascular emergencies are more likely than in an establishment where the majority of the employees are under 40 years of age. Some emergencies are more or less independent of age, epileptic attacks, for example.

For some conditions—and epilepsy is one of these—the likelihood of an emergency can be more or less predicted and it is useful for the occupational health department to have an 'at-risk' register. The register should certainly include all those who are known to have epilepsy, diabetes, severe asthma, psychotic illnesses and, as a counsel of perfection, any condition for which regular medication is required. Clearly the occupational health department will only know of those employees who have already come to their notice either during a pre-employment examination, or because their illness has caused them to loose time from work and they have required an assessment before their return; others may have presented themselves to the department for advice about some aspect of their illness.

The purpose of keeping an at-risk register is so that first-aiders can be given appropriate training in the recognition of signs or symptoms which the occupational health physician or nurse knows may present at some time in the future, and also so that some appropriate medication can be kept in the department for use if the need arises. For example, if there are any employees with poorly controlled epilepsy on the staff and it is considered that there is a risk of one of them entering status epilepticus, then a supply of diazepam for intravenous use is a sensible precaution to adopt. Similarly, with brittle diabetics amongst the workforce, some consideration should be given to keeping intravenous glucose for use.

The most important aspect of the treatment of medical emergencies is to have a policy for dealing with them. In a small organization where there is no medical or nursing cover some first-aiders should be trained in the recognition of serious medical symptoms and given the authority to summon an ambulance or a company car to take an ill employee to the nearest hospital with a casualty department. Where there is a nurse but no doctor, then he or she must have this responsibility. It is always preferable that occupational health nurses have access to some medical support to which they can refer for any further information or training which they may consider they need. Where the company has its own physicians, they would deal with emergencies in the normal course of events but would not necessarily be expected to give much in the way of treatment unless they are working in a location in which access to a hospital is difficult.

The policy for dealing with emergencies should be written and agreed within the company and the occupational health department must determine the extent to which it is willing and able to cope on its own with matters such as the resuscitation of those who have had a myocardial infarct. In most occupational health departments in the United Kingdom there are only limited treatment facilities and only a few drugs will be held. There must be emergency provision for dealing with anaphylactic shock if vaccinations or immunizations are given, however, and all those who give them must be trained to deal with anaphylaxis if it occurs. Some departments will hold limited supplies of drugs which employees on their at-risk registers may be taking—it is helpful to be able to give a day's supply of anticonvulsants rather than risk someone with epilepsy going without treatment or having to return home to get the tablets he forgot in his hurry to get to work. The decision to hold limited supplies of drugs, however, must be taken by individual occupational physicians and, if nurses are to be permitted to dispense them, the conditions under which they may do so and the dosage must be covered in proper standing orders.

Emergencies directly related to work

Examples of emergencies directly related to work may arise from accidents with machinery, to falls, or to overexposure to chemical or physical hazards.

Accidents involving machinery or falls

Accidents involving machinery and falls are still all too common in industry; particularly dangerous occupations are farming and construction. It is most important that attention is given to safe working practices and that workplaces are regularly inspected to ensure that these are being complied with, that safety equipment is worn and that guards on machinery are working properly. It is not at all uncommon to find that workers will jam machine guards if they find that these make their work more difficult, and this is particularly the case if they are paid piece work then compliance with safety procedures may slow down their rate of working. Workers on construction sites can often be found without hard hats on, wearing soft shoes and with no safety goggles.

Part of the duties and responsibilities of the occupational health department are to ensure that the working conditions are as safe as they can be and that those working in hazardous areas understand the risks they run if they do not comply with working practices or if they do not wear safety equipment provided. It goes without saying that the occupational health department should ensure that any protective equipment provided is adequate for the task. This is nowhere more important than in the provision of respiratory equipment where the wrong type of face mask is frequently given.

In any machine room there must be first-aiders who can tend immediately to anyone who does have an accident and arrange for further treatment or referral to hospital; everyone must know where the switches are to isolate individual machines. The resources available in the occupational health department will obviously vary from one company to another but the policy for dealing with serious accidents must be known to all and acted upon if the damage to the individual is to be minimized.

Accidents on construction sites and on farms are more difficult to deal with than those which occur in the confines of a factory. All building sites *must* have first-aiders present at all times, and particular care must be given to their training since their actions may determine whether or not the victim of an accident lives or dies. On the farm, many accidents involve children who play on or around tractors or other machines. Obviously no farm is going to have any first aid or other facilities except those normally found in the home, and government agencies and farmers' organizations must take the responsibility to ensure that farmers do understand the dangerous nature of their work, that they prevent children or others from coming too near machines, that they have training courses in using dangerous equipment (particularly chainsaws) and that all their machines are equipped with appropriate safety devices.

Chemical accidents

Chemical accidents arise in a number of differing ways and may involve an enormous range of substances. Amongst the most commonly encountered, however, are solvents, metal fumes and gases.

Solvents

Almost any worker exposed to solvents may experience some acute effects, including headache, vertigo and nausea. When working with solvents which are themselves irritant or contain irritants in a mixture, some upper respiratory tract symptoms may be noted and sometimes chemical conjunctivitis. These symptoms invariably clear up rapidly when exposure ceases. Some solvent workers may also be overcome by the fumes if they are working in confined spaces, or if they enter closed systems to repair machinery without wearing proper breathing apparatus. The most important thing to do under these circumstances is quickly to remove the affected individual to the fresh air. If he is deeply unconscious then he should be placed in the recovery position and it is important that this is made clear to anyone who may have to deal with such an emergency. No further action may be necessary since recovery is likely to be prompt and complete. If the individual has been splashed by the solvent and his clothing is soaked, this should be removed since some solvents are absorbed through the skin. The victim should not be covered if the solvent-soaked clothes are *not* removed as the heat will encourage the solvent to evaporate and this may lead to more solvent vapour being inhaled.

ADDICTION

In almost any factory which uses solvents there will be some individuals who become addicted; this is particularly likely to happen with some of the chlorinated and aromatic solvents. The sniffers may not necessarily work directly with the solvents but will have access to the department in which they are used. It is not an easy matter to track these people down and the first the occupational health department knows of them may be when they are found unconscious in a lavatory or some other secluded spot where they have been sniffing, or when they have an accident on the way home from work. Short of maintaining a high index of suspicion, there is often no preventive measure which can be taken, although once sniffers do come to light they must not be permitted to continue to work where they are.

SUDDEN DEATH SYNDROME

One other rare complication of solvent exposure which may affect those legitimately and surreptitiously exposed is the so-called sudden death syndrome (SDS). In a typical case, a moderate to high exposure is followed by some relatively strenuous exercise which causes the sudden collapse and death of the individual concerned. It is thought that the solvents sensitize the myocardium to circulating catecholamines and that their release brings on an attack of ventricular fibrillation. In experimental animals, this phenomenon can be reversed by calcium.

There is very little that can be done for the victim; prompt expert cardio-pulmonary resuscitation offers the best prospect but this may not be available when the accident happens. The best way to avoid such an occurrence is to ensure that all solvent workers and all those responsible for their supervision know that it may happen and therefore do all they can to ensure that exposure is kept to a minimum.

Metals and metal fumes

Metal poisoning does still occur at work, although much more rarely than in the past. Those most likely to be implicated are lead, cadmium and mercury. In almost every case symptoms do not come on suddenly but after a period of prolonged and heavy exposure. Lead colic may manifest itself at work and the pain can be relieved by the intravenous administration of calcium gluconate. Further investigation is mandatory in such a case and chelation therapy may be instigated if the blood lead is exceptionally high. Chelation therapy with calcium ethylene diamine tetra-acetic acid (EDTA) or with penicillamine is usually effective in lead poisoning, but in some refractory cases one or other of the newer chelation agents such as 3-dimercaptopropane sulphone (DMPS) or dimercaptosuccinic acid (DMSA) may be required. These latter agents have sometimes been tried in cases of mercury and cadmium poisoning but with much less effect. Chelation therapy requires careful monitoring and must be undertaken in hospital; *under no circumstances* should it be attempted with the patient still at work.

THALLIUM

Thallium is a metal used in the electronics industry and it may still be found as a component of some rodenticides. Poisoning at work is unlikely but deliberate administration is not an uncommon means of attempting to commit suicide; in some countries it is especially favoured for this purpose. Treatment of thallium poisoning is by the oral administration of Prussian blue (potassium ferrihexacyano-ferrate) with intravenous potassium sufficient to keep the plasma concentration between 4.0 and 4.5 mmol/l. It should always be undertaken under skilled clinical management in hospital.

CLINICAL PNEUMONITIS

A variety of metal fumes and those of some metal compounds may give rise to chemical pneumonitis which produces pulmonary oedema either immediately or after a delay of some minutes or hours. Those principally involved are shown in *Table 24.1*. Where the risk exists within a factory with a comprehensive medical department then resuscitation equipment should be available and first-line drugs should be held to treat accident victims prior to their removal to hospital. In all other circumstances, prompt removal to hospital is called for.

Table 24.1 Metals or their compounds which may give rise to pulmonary oedema

Compound	Time factor
Antimony compounds	
Bismuth pentachloride	
Boron compounds	
Cadmium vapour	Lag of up to 10 h
Cobalt vapour	
Lithium hydride	
Mercury fumes	May be a lag of several days
Nickel carbonyl	Lag of 8–36 h
Osmium tetroxide	
Selenium dioxide	
Organotin compounds	
Titanium anhydride	
Vanadium compounds	
Zinc chloride	
Zirconium tetrachloride	

Gases

The most serious chemical accidents at work probably involve gases of one kind or another. The risk may be from simple asphyxiation or from interference with cellular respiration or from profound haemolysis (*Table 24.2*). There are very few specific antidotes except in the case of cyanide poisoning where prompt treatment with dicobalt EDTA may save the life of someone who has been gassed. No-one should be permitted to enter a building or any other structure alone, or without being provided with adequate breathing apparatus where there may be a risk of exposure to a toxic gas. Moreover, where the risk exists, all those working in the area should know what first aid measures are to be taken if one of their colleagues is overcome. Where there is a risk of exposure to cyanide, the occupational health

Table 24.2 Acute sequelae of exposure to some gases and vapours

Gas or vapour	Result	Time factor and concentration
Acetaldehyde	Pulmonary oedema	High concentrations
Acid fumes	Pulmonary oedema	
Acrylic anhydride	Pulmonary oedema	
Ammonia	Laryngeal spasm, pulmonary oedema	
Arsine	Intravascular haemolysis	
Bromomethane	Pulmonary oedema	
Carbon dioxide	Simple asphyxia	At concentrations equal or greater than 10% in air
Carbon monoxide	Forms carboxyhaemoglobin (COHb); unconsciousness and respiratory depression	At concentrations >50% COHb
Chlorine and its compounds Chlorine dioxide Sulphur chloride Thionyl chloride Chlorsulphonic acid Phosgene	Pulmonary oedema	Lag period of up to 48 h
Cyanogen	Inhibits cytochrome oxidase and intracellular respiration	
Dimethyl sulphate	Pulmonary oedema	
Dioxane	Pulmonary oedema	
Methyl isocyanate	Pulmonary oedema	
Nitrogen dioxide	Pulmonary oedema	Lag period of up to 30 h
Ozone	Pulmonary oedema	
Polytetrafluoroethylene (PTFE)	Pulmonary oedema	
Stibene	Intravascular haemolysis	
Sulphur dioxide	Pulmonary oedema	
Trimellitic anhydride	Pulmonary oedema	

department must ensure that there are enough first-aiders trained to give appropriate treatment; it may be too late to wait for the nurse or doctor to arrive.

Other chemicals

Overexposure to organophosphorus pesticides gives rise to a variety of effects including muscle weakness, some central nervous system symptoms, visual disturbances, sweating, salivation, bradycardia and, eventually, respiratory arrest. The underlying pathophysiological event is the inhibition of cholinesterase and this can be overcome with the administration of atropine or with a cholinesterase reactivator such as 2-pyridinealdoximemethochloride (2-PAM).

PARAQUAT

Paraquat has achieved a certain notoriety over the years. It may be taken deliberately or accidentally. It is rapidly absorbed from the gut and the outcome depends upon the plasma concentration. Where this is sufficiently high, the patient develops hepatic and renal tubular necrosis, toxic myocarditis and pulmonary fibrosis, all within a few days of the original episode. If treatment is to have any hope of success it must be started as soon as possible in order to impede uptake from the gut. A variety of methods has been tried including the administration of Fuller's earth, bentonite, gastric lavage, forced diuresis, haemodialysis and charcoal haemoperfusion. None is hugely successful.

Catastrophes

Modern industrial plants are by no means immune from catastrophe as the recent events at Chernobyl and Bhopal have shown. It is safe to assume that no man-made system is foolproof, and in industries where a catastrophic failure in a safety system would have devastating effects on both the workforce and those living around it, plans must be made to deal with the worst-event case.

The occupational health department must have an input into any disaster planning and may be called upon to cooperate any action should the need arise. Even where the risk of an accident is considered to be remote in the extreme, contingency plans must be made. In these days of terrorist activity the risk of deliberate sabotage must be considered.

Since a catastrophe at a chemical plant or a nuclear installation, for example, will inevitably result in a potential threat to health over a wide area many agencies outside the industry concerned will have to be involved. These may include the district health authority, the ambulance service, the fire service, the police, some government ministries, various parts of the government health and safety organization, the meteorological service and perhaps road and rail authorities. The actual composition of the planning committee will vary from industry to industry and from country to country.

Conclusions

I have tried to suggest in this brief consideration that the most important part of dealing with emergencies at the workplace is through anticipation and by careful planning. It is reasonable to assume that if something *can* go wrong then it *will* go wrong and use this as the basis from which to act.

Any workforce represents a particular cross-section of the population at large, and the ills to which the larger whole is subject will also affect those at work and are as likely to manifest themselves there as elsewhere. The measures which are taken to cope with common medical emergencies will depend upon the sophistication of the occupational health department and the availability of adequate medical services elsewhere.

A knowledge of the processes within the factory and the chemical and physical hazards to which the workers are exposed will determine the likely emergencies which may result from these exposures and the measures needed to deal with them. Similarly, catastrophic events will be predicted on the basis of these kinds of consideration.

The occupational health department has the most important role of anticipating emergencies, of preparing policies for how to deal with them at the local level, and of having an input into disaster planning.

It follows that this important task can be undertaken only by occupational health professionals who are extremely conversant with what goes on within the plants for which they have responsibility and part of their duty must be to ensure that they keep up to date with any changes which take place. Those who do not have this knowledge are failing in their professional responsibilities.

Further reading

Grant, H.D. and Murray, R.H. *Emergency Care*, 4th edition. Prentice Hall, Englewood Cliffs, New Jersey (1986)

Robinson, R. and Scott, R. *Medical Emergencies. Diagnosis and Management*, 2nd edition. Heinemann Medical Books, London (1987)

Stine, R.J. and Marens, R.H. *A Practical Application of Emergency Medicine*, 2nd edition. Little, Brown and Co. (1987)

Chapter 25

Vocational rehabilitation and resettlement

M. Floyd

What do we mean by vocational rehabilitation?

As with most seemingly innocuous questions, this one also has a fairly straightforward answer, which begs a number of important issues and a more complicated one, which attempts to look at some of these issues. Let us begin, though, with the first kind of answer. Later on in this chapter we will come back to some of the questions that it begs and try to deal with them. By the end of this chapter, we shall at least have come up with a better and fuller understanding of what is meant by 'vocational rehabilitation', as well as some impressions regarding the way it is currently practised in this country and overseas.

The straightforward answer first: we can begin by defining 'vocational rehabiita- tion' as the process by which people with disabilities are helped to get back to work following some kind of injury or illness. We can make our answer a little more complicated, or less straightforward, by distinguishing between rehabilitation and resettlement. If we do this, we can regard as *rehabilitation* the help given to people which renders them employable again by reducing or eliminating their disabilities, and *resettlement* as the help which enables them to return to their previous job, or obtain another job, and to remain in it.

An alternative way of defining what is meant by vocational rehabilitation is to point to the services and people who provide it. The best known of these are probably the Training Agency's Employment Rehabilitation Centres (ERCs) and the Department of Employment's Disablement Resettlement Officers (DROs). But many local authority day centres and adult training centres offer rehabilitative work programmes to their clients. Then there are the industrial therapy units in psychiatric hospitals, many of whose patients are referred, not just for therapeutic reasons, but also as part of a rehabilitation programme.

Some unanswered questions

Helpful though both these answers are to the question 'what is rehabilitation?' it should be evident that they do beg a lot of questions, for instance:

(1) Does rehabilitation include vocational training?
(2) What is the difference between medical and vocational rehabilitation?
(3) Is vocational rehabilitation only concerned with getting people into employ- ment?
(4) What is meant by employment handicap?

Key concepts

Disability and handicap

Let us begin by examining the concept of 'employment handicap'. All too often the terms 'disability' and 'handicap' are used interchangeably. An understanding of the crucial distinction between them is, however, essential if a proper understanding of the process of vocational rehabilitation is to be obtained.

The distinction is most easily grasped in the context of an example, such as that of someone who has lost a leg in an accident. As a result of this they are unable to walk. Their inability to walk—a very common physical disability— means in turn that there are many jobs that they cannot do; they are, in other words, handicapped with regard to employment. There are several important points that are worth noting in this relatively straightforward example. First, the disability is not necessarily a fixed quantity, unlike the actual absence of a limb—the impairment. Many people learn to walk using artificial legs so that the extent of the disability is reduced. Second, even if they are unable to walk—that is, they remain disabled— the extent to which this restricts their employment prospects will depend upon many other factors. A fairly obvious one is their educational level. If this is such as to make it possible for them to carry out clerical tasks, not involving a great deal of walking about, a wide range of jobs may still be open to them.

Many such jobs, though, will probably require some kind of vocational training. This training can be a very significant means of opening up a much wider range of employment opportunity to someone whose choice of jobs has been greatly restricted by a disability, such as the loss of a limb. Thus, just as the physical disability caused by an impairment is not a fixed, permanent quantity, so too can the employment handicap vary a great deal and be subject to external and environmental influences.

Primary and secondary disabilities

It is important to recognize that disabilities can be of two kinds. So far we have only considered the primary disability, the inability to walk, resulting from the loss of a leg. The task faced by rehabilitation though can be made much more difficult because this initial disability gives rise to other disabilities. These other, or secondary, disabilities furthermore are often less straightforward and less easily overcome. Thus, for example, the loss of a leg will almost certainly have resulted in a quite lengthy period off work. The ability of the individual concerned to get up in the morning, apply himself to his job and so on, may deteriorate during this time. Or the pain he experiences while getting used to wearing an artificial leg may be such as to make him depressed and resentful of the way the world has treated him. Secondary disabilities of this kind, far from being the exception, are very common. For example, a recent study found that less than half of the physically disabled clients attending Employment Rehabilitation Centres were free from psychiatric symptoms. In other words, more than half of them could be said to have a secondary psychiatric disability.

A reverse sequence of causation, with psychiatric disabilities resulting in behaviour which leads to someone becoming physically disabled, is also quite common. Suicide attempts are perhaps the most dramatic illustration of this, but poor concentration—sometimes experienced by psychiatric patients as a consequence of their medication—can result in accidents that may leave them with a permanent physical disability. In the same way, people with sensory disabilities and with epilepsy are more likely to have accidents of this kind.

Rehabilitation and resettlement

Having made this important distinction between impairment, disability and handicap and between primary and secondary disabilities, we are now in a position to delineate more clearly the boundaries between various stages in the process of getting a disabled person back to work.

Implicit in what has been said already is the idea that the first stage in this process involves reducing the extent of the disability resulting from an impairment. This, I would suggest, is essentially what medical rehabilitation services attempt to achieve. In the case of the person who has lost his leg, they provide him with an artificial leg and they help him to learn to walk with it. They will probably also provide physiotherapy to help him develop muscles which have become weakened through lack of use in the interval between his accident and the fitting of the prosthesis.

If medical rehabilitation were always completely effective and able to eliminate the disabling effects of impairments, and secondary disabilities were not a problem, that would be the end of the story. In most cases, however, a degree of disability remains—whether it be primary or secondary—and people need additional help if they are to return to work. Two forms of help are possible and, again, these have been implicit in what has already been said. On the one hand, the employment handicap resulting from the disability can be reduced. This is basically the aim of vocational rehabilitation services. Just what may be involved here is the subject of later sections. However, on the other hand, it is frequently not enough just to get someone to the point at which, in theory, they are able to do the job. In many cases they will also need help in finding it and in obtaining it. And, occasionally, it may be necessary actually to modify the workplace so that their disability does not prevent them from getting to where the job is. All these aspects of help tend to get lumped together under the general heading of 'resettlement'.

Vocational assessment and guidance

Two things are missing from this—admittedly oversimplified—account of the rehabilitation process: assessment and guidance. Assessment is widely regarded as a key element in vocational rehabilitation services, but assessment of what? There are two possible answers to this question, each reflecting a rather different perspective—or even philosophy—of rehabilitation. Perhaps the most obvious, and the most prevalent, would be to say assessment of the individual's disabilities. Increasingly, though, those representing disabled people are emphasizing the need to look at their abilities, not their disabilities, and this orientation is beginning to influence professional rehabilitation practice as well. So a proper vocational assessment will be as much concerned with what a disabled person can do, as with what he cannot do.

Such an orientation also implies a greater emphasis on vocational guidance. This is needed to help people think through just what it is they are looking for in a job. Do they want security and a pension? Or is a higher wage the critical factor? Do they want to work out of doors? Do they have particular aptitudes?

The provision of such guidance has been neglected to a considerable extent hitherto, but is likely to become an increasingly prominent feature of rehabilitation services in the future.

Current services in Britain

The Tomlinson Committee and the 1944 Act

Vocational rehabilitation services in Britain today have been developed within a framework of government legislation and policy that was laid down over 40 years ago in an Act of Parliament which was, in turn, based very closely on the recommendations of a report prepared by the Tomlinson Committee. Since then several reviews of various aspects of the legislation have been carried out and a number of other committees have deliberated—most notably the Piercy and Tunbridge Committees—but resulting changes in policy and in the vocational rehabilitation services provided by government have been relatively insignificant. This is in marked contrast to what has happened in other areas, such as education where much more radical shifts in both policy and services have taken place.

This resilience of the system could of course simply reflect the soundness of the Tomlinson Committee's thinking and their ability to foresee the needs of disabled people well into the future. Unfortunately, however, there is a growing awareness that this is not the case, and a number of well-founded criticisms have been levelled at the assumptions underlying their recommendations and the legislation and policy emanating from them. Perhaps the most fundamental of these is that the Committee's thinking was unduly influenced by the needs of ex-servicemen returning from the war. These needs were mainly those of people with physical disabilities. The needs of people with other kinds of disability, such as those arising from sensory and mental impairments, were not given the consideration they merited.

Closely linked to this failure to examine the needs of the broad spectrum of disabilities found in the population as a whole was the assumption that rehabilitation would be concerned with helping someone who had recently become disabled and who, after rehabilitation, would be able to compete on equal terms with the 'able-bodied' for employment. Such a procedure failed to take into account the particular needs of people disabled by congenital conditions which, in many instances, would have interfered with their education and meant that their experience of employment was very different from that of someone who acquired a disability later on in life. It also ignored the even more complex problems presented by disabilities that result from conditions such as mental illness and epilepsy. Such disabilities are not fixed and unchanging but fluctuating and episodic in nature and clearly require a different kind of rehabilitation programme.

Employment Rehabilitation Centres

Employment Rehabilitation Centres (ERCs)—formerly industrial rehabilitation units—represent the main form of vocational rehabilitation provision in Britain. There are 27 and they are to be found in most major towns and cities. In London, for example, there is one in Perivale (North-west London) and parts of southern London are also served by an ERC at Waddon (south of Croydon). There is also an ERC at Egham in Surrey, but this is residential, catering for disabled people living in areas that are not within reasonable commuting distance of an ERC. Many disabled people, especially those in rural areas, fall into this category and a second residential ERC if located near Preston.

Employment Rehabilitation Centres, which are run by the Training Agency, tend to be relatively large—several having over 100 places—and offer a rehabilitation programme that varies only in relatively small ways between each centre. Thus, clients stay at an ERC for a period of between 6 and 8 weeks. The first week or two is spent in an 'intake' group, during which they work on a range of activities and an initial assessment of their abilities and problems is made. They are then moved on to one of the workshops or the clerical section, where they will probably remain until their course of rehabilitation is terminated.

At the end of their stay a report is prepared, summarizing the conclusions reached by the ERC staff regarding their vocational potential. This report, which usually concludes with a fairly specific recommendation as to the kind of work that should be found for them, is sent to the Disablement Resettlement Officer (DRO), who will have previously referred them to the ERC.

A very wide-ranging and systematic evaluation of the rehabilitation programme offered by the ERC and the effect on their disabled clients, was carried out by the Employment Rehabilitation Research Centre, which was set up in 1976. Its report was extremely critical of many aspects of ERC practice and suggested that their clients were not benefiting to any marked extent from attendance. This has led to a number of new approaches being tried out in different ERCs, some of which— such as the use of work samples (*see below, p.468*)—are now being introduced in them all. The basic approach to rehabilitation though remains unchanged and the Agency appears to be reluctant to abandon fundamental aspects of the programme, such as its length, the quota on ex-psychiatric patients (25% maximum), and the emphasis on assessment as opposed to actual rehabilitation, or work adjustment (*see below, p. 469*).

Disablement advisory services

Just as the ERCs constitute the main rehabilitation resource in Britain, the disablement advisory services represent the main resource available to resettle disabled people in open employment. Within this service it is the Disablement Resettlement Officers (DROs) who provide the most direct and personal assistance to them. There are over 450 DROs, each being attached to a job centre—or in some cases to several. They have have a large number of clients, not all of whom will choose to register as disabled. Out of 100 or more clients though there will probably be only a dozen or so that the DRO is at any one time actively helping to find work. The majority will either already be in employment or not currently seeking it for a variety of reasons.

In the past, DROs have been criticized for concentrating their efforts on placing clients who were least handicapped and therefore easiest to place. This was, to some extent, inevitable given the regular monitoring by management of the number of placements achieved by DROs and their colleagues, the employment advisers dealing with 'able-bodied' clients. More recently the service has been substantially changed so that less severely handicapped clients are dealt with by the Employment Advisers, leaving the DRO to devote more time and effort to the more severely handicapped. This has been accompanied, however, by reductions in the numbers of DROs and there have been complaints that the service offered by DROs has deteriorated.

An important part of the DRO's job used to be liaising with employers and monitoring their compliance with the quota scheme. This aspect of their work has been transferred to the Disability Advisory Service (DAS) teams, who serve an area, rather than a single job centre. The DAS teams are also responsible for dealing with applications from employers for financial assistance in adapting premises—for example, to make possible wheelchair access—and for special aids, such as Braille typewriters.

Sheltered work

The Tomlinson Committee assumed that most disabled people would be able to obtain work in open employment, but recognized that there would be some people so severely handicapped by their disability that this would not be possible. For them it would be necessary to provide some sort of sheltered employment.

In the case of sheltered employment the Department of Employment does not itself provide any; however, it does finance it, and is thus largely responsible for its current level and the form it takes. Most sheltered employment is provided by the workshops run by Remploy, an agency that was set up by the government and receives a grant from the Department of Employment to cover its substantial losses. Remploy employs over 8000 disabled people, far more than any other organization in Britain. Their workshops used to manufacture a wide variety of products but during the last few years they have become increasingly dependent on subcontract work. An employee of Remploy receives a wage that, although not high in comparison with, say, the average wage in Britain, is nonetheless substantially above that of many unskilled workers in open employment. Their jobs are also a good deal more secure.

Some local authorities, and a few voluntary organizations, also run sheltered workshops, and together they provide an additional 4000 places. As with Remploy a worker in these workshops has to be judged to be so severely handicapped as to be incapable of working in outside employment. At the same time they must be able to achieve a level of productivity at least 30% of that of an 'able-bodied' person.

Although criticisms can, and have, been levelled at the sheltered workshops—especially their emphasis on manufacturing as opposed to the provision of services—they do nonetheless meet a real need. A much more significant criticism of government policy and provision in this area is simply the inadequacy of it. Had the Tomlinson Committee been more aware of the needs of people with congenital and mental disabilities they would have recognized that the proportion of disabled people needing some form of sheltered employment was far from small. It is probable that the need exceeds the current level by a factor of at least ten.

In the absence of adequate provision in this area there has grown up a large number of facilities, which, though often charged with some kind of rehabilitation function, are in fact primarily meeting the need for some kind of work activity in a sheltered setting. Here we can include the industrial therapy units in the psychiatric hospitals, the industrial therapy organizations—which provide similar kinds of work outside the hospitals, the day centres and adult training centres run by the social service departments of local authorities, and a wide variety of work-oriented centres run by voluntary organizations. The incomes of disabled people working in such centres are mainly derived from social security or employment benefits, though these will usually be augmented by small bonus payments which are sometimes, but by no means always, constrained by the limits imposed by the rules governing social security benefits.

Vocational rehabilitation overseas

West Germany

Quota scheme

Vocational rehabilitation services in Britain have been developed without much heed being paid to practice in other countries. In this respect vocational rehabilitation is not alone. In most areas of social policy there are those who emphasize the uniqueness of the British situation and the irrelevance of the ideas and approaches adopted by other countries. Learning from the experiences of other countries is of course not always easy. In the case of vocational rehabilitation and resettlement, differences in the labour market, in the way government services are organized and in the educational system will all conspire to make a straightforward application of overseas practice inappropriate.

Nevertheless I believe that we can learn much from what has been done in other countries. To begin with, it can help us to question some of the assumptions on which British practice is based, and a good example of this is to be found in the way vocational resettlement services and policy have been developed in West Germany, and in particular, its quota scheme.

This is based on an assumption totally at odds with that made by the Tomlinson Committee. German policy acknowledges that a significant number of disabled people are likely to be substantially handicapped with regard to employment, in spite of the best efforts of the rehabilitation services. This means that organizations employing large numbers of disabled people might be at a commercial disadvantage with respect to other organizations with a smaller proportion of disabled employees in the workforce. In West Germany, therefore, organizations (including those in the public sector), which do not employ their quota (6%) of disabled people, have to pay an 'equalization levy'.

In Britain, on the other hand, where the rationale for a quota scheme is much less clear, its implementation has become increasingly ineffective, so that very few organizations actively employ their quota (3%) of disabled people. A recent review of the scheme proposed that it should be abolished, but the government refused to do this. Instead they proposed looking into ways of making the scheme more effective. Surprisingly though, in the course of a lengthy and often heated debate, no attempt was made to examine carefully the West German system and the lessons to be learned from it.

The West German system is also interesting with regard to the additional protection against dismissal that is afforded to disabled people. If an employer wishes to dismiss a disabled employee they must first obtain the permission of the German equivalent of the job centre. This in turn ensures that they must demonstrate that they have explored the possibility of alternative work situations.

At the same time the employer must involve the disabled persons' representative. All organizations must appoint someone to occupy this position and they are able to play a key role in ensuring that disabled employees receive fair treatment, as well as the additional benefits that are accorded to them—such as longer holidays—in Germany.

Rehabilitation and training

Of even greater relevance to the development of vocational rehabilitation and resettlement services in this country is the emphasis that is placed, in West Germany, on vocational training as a means of helping disabled people back to the workforce. This emphasis is a natural consequence of the recognition that a disability will often continue to be a substantial handicap in finding employment and in carrying it out. The West Germans therefore argue that this negative factor must be compensated for by providing the disabled person with a positive advantage, a vocational skill.

Thus rehabilitation centres are very different from the British ERCs. Instead of the 6-week ERC course, the German centres offer courses that are at least a year in duration, and sometimes as long as 2 years. The reason for this is simple. The rehabilitees are receiving a very thorough and intensive course of training, which will enable them to acquire a vocational qualification identical with that obtained by the 800 000 'able-bodied' youngsters, who leave school each year and progress on to apprenticeships in industry and commerce.

To a visitor from Britain these centres appear more like universities than rehabilitation centres. They are residential and the quality of the physical environment and of the equipment in the workshops would be a cause of envy to the staff and students of several of the less well-endowed British universities.

It should be acknowledged though that, impressive though such a system is, it has been bought at a price that many in Britain would not wish to pay. The German rehabilitation centres cater primarily for the less severely disabled and for those, moreover, with physical disabilities. Very few people with mental handicap or psychiatric problems will find their way there.

Nonetheless one lesson is very clear, namely that however worthy an aim integration may be, training in a sheltered environment may well be more effective and appropriate for many disabled people. In Britain, only a relatively small proportion of disabled people get as far as starting a training course; and many of those who do start one never complete it. At a time when, due to technological change, vocational training is becoming even more important, such a state of affairs is simply intolerable. Some kind of shift towards the kind of system found in Germany has to take place.

North America: vocational rehabilitation

The German system is interesting primarily because it leads us to question some of the key assumptions on which our own system—and especially that pertaining to resettlement—is based. When it comes to the 'nitty-gritty' of rehabilitation practice—the techniques and tools for carrying it out—we must look elsewhere, to North America, where the development of a more systematic and sophisticated approach to vocational rehabilitation is many years—if not decades—in advance of practice in Britain.

The approach to vocational rehabilitation in North America is interesting for a number of reasons. First and foremost is the fact that it is established as a profession, or rather, a series of professions. In every state in the United States of America those wishing to pursue a career in vocational rehabilitation will be able to avail themselves of undergraduate and postgraduate courses in such areas as rehabilitation counselling, vocational evaluation, rehabilitation engineering and work adjustment. In the State of Georgia—hardly noted for its sophistication or the advanced state of its educational system—there are at least four such courses.

North American practice is also noteworthy for the way in which it divides rehabilitation into two quite distinct, but closely interrelated processes: evaluation and work adjustment. Evaluation can be equated, to some extent, with what in this country is called assessment. But it encompasses a much wider range of activities than is usually associated with assessment, and furthermore its orientation is very different. The evaluator's role is to help clients to explore a wider range of employment possibilities and, at the same time, enable them to determine whether they would be able to perform satisfactorily in the jobs that appeal to them; or could do given additional help, such as vocational training or work adjustment. Work adjustment services follow on naturally from those of evaluation, and may have identified gaps between how an individual needs to behave in a particular work situation and the way they currently behave.

What is worth emphasizing here is the difference in orientation between evaluation, as practised in North America, and assessment as it is usually done in Britain. The latter is essentially something that is done *to* the disabled individual, while evaluation is very much an activity that is carried out with them. In evaluation they are not being put through a series of tasks, or 'hoops', but instead being helped to discover their vocational aspirations and the feasibility of achieving them.

Vocational evaluation and work adjustment

Planning an evaluation programme

In this section is described some of the main characteristics of vocational evaluation and work adjustment. Inevitably this will represent an inadequate and sketchy outline of what is involved for, as mentioned earlier, a comprehensive and thorough account can form the basis of a year postgraduate programme. Nevertheless it is hoped that it will at least serve to convey some of the key aspects of what is believed to be a much more coherent, systematic and appropriate approach to vocational rehabilitation.

One of these key aspects is the emphasis placed on the careful planning of a rehabilitation programme which is geared very closely to the particular needs and problems of the individual. Two rather different kinds of programme can be distinguished. The first kind concerns individuals who have recently, as a result of an injury or an illness, become disabled and are waiting to return to their previous job or one similar to it. The question evaluation must answer is then simply whether their disability, or disabilities, are likely to prevent this or, at least, make it difficult.

Information on the job they did before can be obtained in two ways. If the previous place of work is not too distant the evaluator may actually be able to visit it and carry out an analysis of the mental and physical demands of the job. If this is not feasible the evaluator can refer to the *Dictionary of Occupational Titles (DOT)*, which is published by the US Department of Labor. This contains detailed information on over 20 000 different kinds of job. Armed with this information the evaluator can begin to formulate a series of very specific questions regarding the mental and physical abilities of their client. Answers to these questions may be obtained in a number of ways, outlined below.

This may show that, in spite of their disabilities, the disabled person should be able to carry out his/her previous job. It may reveal though that in a number of areas there is a significant gap between what the job requires and the individual's current performance in that area. If the evaluator believes that these gaps can be closed by a work adjustment programme their client will then be referred to this next stage in the rehabilitation process.

If, however, it is felt that there is little likelihood of the client ever being able to carry out his/her previous job, the evaluator will try to help the client to identify alternative types of work that are within his/her current capabilities or, at least, are not so far beyond them that work adjustment, or vocational training, would not eventually enable them to carry it out.

Exploring the wide range of vocational possibilities and finding those that the disabled individual can do is the most challenging aspect of the evaluator's work. Many clients will require this, especially those with congenital disabilities, who have had great difficulty in finding any settled occupation.

Assessment of clients

At the end of this first stage in the evaluation process the evaluator will be seeking answers to a number of specific questions regarding the mental and physical abilities of the disabled individual, who has either selected a job—or group of jobs he would like to be employed in—or is wishing to return to the job they were doing previously. The next stage is to obtain answers to these questions.

This can be done in a number of ways. Perhaps the most obvious is simply to place the individual in the desired job, or one that is very similar. This approach, job site evaluation, however, is not always practical. It requires, first of all, that such a job is in the neighbouring geographical area and that it is currently vacant. Furthermore the employer must be willing to cooperate and be prepared to tolerate the possible inconvenience that may result if the disabled individual is not, in fact, able to perform the job satisfactorily.

In many cases, though, it may be possible to find a situation, perhaps within the hospital or authority providing the evaluation service, that gives the opportunity to assess an individual's ability to carry out certain aspects of the job in question.

A different approach to assessment involves the use of psychometric tests, the most familiar of which is the intelligence test. Many others though have been developed and thoroughly validated, making possible a comparison of the individual's performance with that of a wider reference group. There is a growing recognition, however, that psychometric tests, while undoubtedly a valuable tool for the evaluator, do have serious limitations and, used in isolation, are of little value.

One reason for this lies in the difficulty of relating performance in the psychometric test to performance in the actual job. An intelligence quotient may tell us something about an individual's ability to operate a lathe, say, but clearly on its own it is not enough. A job such as this calls for other abilities that are not readily measured by psychometric tests. To meet this need, work samples have been developed that are designed to test an individual's ability to carry out a task, or range of tasks, that may be involved in a number of different jobs. Using work samples the evaluator can assess performance in work-like activities, while at the same time being able to use a standardized procedure for administering and scoring performance. In addition, as with psychometric tests, comparisons of performance can be made with appropriate reference groups and the validity and reliability of the measures obtained can be examined in a statistically rigorous way.

Work adjustment services

So far we have discussed just two of the main stages in the vocational rehabilitation process, as it is practised in North America: a first stage in which a job goal is identified, its physical and mental requirements defined and a corresponding set of questions, regarding the abilities of the client, formulated; and a second stage during which the answers to these questions are obtained. It remains for me to briefly outline the third stage in the process, work adjustment.

As I have already indicated it will not be necessary for every individual to go through this stage in the process. In many instances the individual will be found to be capable of performing the job they wish to do. Or it may be that all they require in order to do the job is vocational training. But in a significant proportion of cases a gap between the job requirements and individual abilities will be revealed that cannot be bridged by vocational training alone.

Instead it may turn out that the individual is deficient in regard to one, or more, of what are called 'critical vocational behaviours'. These critical behaviours include several that will be familiar to rehabilitation practitioners in Britain: for example good timekeeping, and regular attendance. Others will be less well recognized. These include behaviour towards fellow-workers, behaviour in relation to supervision and the amount and quality of the work done by the individual.

The point to emphasize here is that work adjustment is concerned primarily with aspects of the individual's work performance that can be described objectively and, if possible, quantified. For example, evaluation may have revealed that the client only arrives on time 20% of the time and will specify a target for work adjustment or raising this figure to, say, 90%.

New directions

The Vocational Rehabilitation Department at Queens Park Hospital, Blackburn

Impressive though I believe the North American approach to vocational rehabilitation to be, it might be all too easy for professionals here to dismiss it on the grounds that it has been developed in a rehabilitation and economic climate that is very different from that in Britain. How relevant, therefore, is the approach to the particular context of British rehabilitation and the current economic climate, with over three million people unemployed?

A clear answer to this question is provided by the pioneering work being done by the Vocational Rehabilitation Department at Queens Park Hospital in Blackburn.

Some 8 years ago the manager of the department, once an Industrial Therapy Unit, was becoming increasingly dissatisfied with the service his unit was offering to clients. Since then, he and his staff have gradually introduced the principles and the techniques outlined above and created what is probably the finest vocational rehabilitation facility in Britain today. Although there has obviously been the need to adapt the North American approach to the particular circumstances of a hospital-based facility, and the absence of other agencies providing services to which a comparable unit in North America would have access, no major problems have been encountered in adapting the approach and pioneering its use in this country.

Apart from the belief of the staff that they are now offering a far superior service to what was being offered before, evidence of this is provided also by statistics on resettlement rates. Not only do these show a much higher proportion of clients going back to open employment than was previously the case, they also reveal that the department is now able to offer its services to a much larger number of disabled people than had previously been possible.

The ASSET centres

One of the greatest virtues of the health services in Britain is the degree of autonomy they enjoy, with central government dictating the level of resources available but not able to dictate practice 'on the ground' to any significant extent. Hence the ability of the Blackburn unit to experiment with new approaches.

In the case of the main providers of vocational rehabilitation services, the ERCs, the situation is very different. Their rehabilitation programmes have to follow very strict guidelines which are laid down by headquarters staff. The manager of a local ERC interested, say, in using work samples would not be able to just go ahead and use them even if he could find the financial resources to do this. As a result of this strong central control, changes in ERC programmes, since they were set up following the 1944 Act, have been relatively few and then of only a minor kind.

More recently though a number of more radical and innovative developments have been introduced. Foremost amongst these has been the setting up of three new rehabilitation centres, the ASSET centres, at East Ham in London, Wrexham in Wales and Gillingham in Kent. These centres make extensive use of work samples in assessing the abilities of their clients. Perhaps the most significant difference in their approach however is that, whereas a large part of the costs of the ERCs arise out of their workshops, the ASSET centres have no workshops at all. Instead they have approached local employers and experienced considerable success in gaining their cooperation in providing what was earlier termed 'situational assessments' for the centres' clients. This now forms an important part in their rehabilitation programme and represents a radical break with the past and its heavy reliance on the simulated environments of the workshops, which more often than not failed to reproduce the demands of a real work situation in open employment.

Encouraging though these developments are—and the introduction of work samples into ERCs—it is necessary at the same time, to express some disappointment that there is little evidence so far of the ASSET centres or the ERCs adopting the more systematic and individualized planning of clients' programmes that characterizes the American approach and that of the Queens Park Hospital unit.

The impact of new technology

A great deal has been written regarding the impact of new technology on the employment of disabled people. Much of it probably overestimates the benefits, in terms of new employment opportunities, and underestimates the negative impact, resulting from the disappearance of many unskilled jobs and higher levels of unemployment. Excessive attention has perhaps been paid to the well-educated person in a wheelchair who can become a computer programmer or work from home using a remote terminal. Nonetheless the possibility does exist for many disabled people, who have not been able to achieve their full potential, getting a larger share of the new jobs being created as a result of technological change, especially those involving the use of computers.

In order for this to occur, though, their access to training must become greater and comparable to that found in West Germany. Just how this might be achieved has been shown by the Information Technology Centres (ITeCs) set up over the last few years to train school-leavers. But although these cater for disabled school-leavers they are not open to disabled adults, except, that is, for the Reading ITeC where, thanks to assistance from an EEC project, ten places are reserved for adults with a wide range of disabilities. Their success in getting disabled adults back into open employment—with over 70% of trainees finding jobs during or after training—demonstrates the key role that training can play in 'cancelling out' the disadvantage arising from a disability.

New information technology may also have an equally significant impact for a quite different reason. This stems from its application to vocational rehabilitation itself. Already computer software is being developed for use in assessment and for planning of assessment programmes. Increasingly information on job vacancies, on job characteristics and on careers is being made available on computerized databases. And within the next 10 years, hopefully we shall see the development of computer-based vocational guidance systems that will provide guidance of a level and quality that not even university graduates, let alone disabled people, have access to at present.

The Sheltered Placement Scheme (SPS)

A change of a very different kind, but probably of equally great significance for the resettlement of disabled people, is the new ground being broken by the Sheltered Placement Scheme (SPS). Until recently, sheltered employment has been synonymous with employment in sheltered workshops. It is true that some jobs were created for 'enclaves' of 'severely disabled' people, working in sheltered employment in an ordinary factory or warehouse. But these 'enclaves', or 'sheltered industrial groups' (SIGs) as they came to be called, never employed more than a handful of people. More recently though the Department of Employment, who subsidize the wages of employees in both the sheltered workshops and the SIGs, made a small, but extremely important change in the rules for the funding of SIGs. No longer was it necessary for disabled people to be part of an 'enclave'; instead a SIG could consist of just one individual. And then, as many such individual SIGs came to be set up in non-industrial situations the term 'industrial group' was seen to be increasingly inappropriate and the term 'sheltered placement' was adopted.

A disabled employee in a sheltered placement receives the same wage as anyone else doing that same job. The employer however only 'pays for what he gets'. So, if a disabled employee is said to be capable of only producing 50% of the amount of work of an average 'able-bodied' employee in that job, the employer contributes just 50% of the wage. The difference is made up by the Department of Employment.

This more flexible approach to providing sheltered employment has resulted in a tremendous increase in the number of such sheltered placements, and there are now over a thousand in existence, with nearly half being operated by the Shaw

Trust. This organization, which began as a small charity in Wiltshire, but is now becoming a national organization, owes its success to the extremely business-like way it has organized itself and its relationship with the Department of Employment. Perhaps the most interesting aspect of this is the way in which the salaries of its development officers, who negotiate with the employing organization to establish the placement, and its support officers, who provide whatever emotional or other form of support is needed by the disabled person, are funded out of the financial contribution to each placement. Thus, unlike a DRO, a Shaw Trust Officer has a strong personal interest in maintaining his clients in employment, and not just finding them a placement.

Conclusions: the need for professional education and development

These developments, and many others which may be found in the further reading section below, are undoubtedly encouraging. Before concluding though, a final word must be said regarding the continuing lack of training and development for professionals working in vocational rehabilitation. As long as this persists, together with the absence of a professional body to advocate more resources and disseminate good practice, progress will be limited.

Further reading

Bolderson, H. The origins of the disabled persons' employment quota. *Journal of Social Policy*, **9**, 169–186 (1981)

Cornes, P. *Employment Rehabilitation.* Manpower Services Commission, Sheffield (1982)

Floyd, M. and North, K. *Disability and Employment in Britain and Germany.* Anglo-German Foundation, London (1985)

Grover, R. and Gladstone, F. *Disabled People, A Right to Work.* National Council for Voluntary Organizations, London (1981)

Chapter 26

Migrant workers

N.M. Cherry and J.C. McDonald

Introduction

Article 10, International Labour Organization Convention 143 [1] states:

> Each Member for which the Convention is in force undertakes to declare
> and pursue a national policy to promote and to guarantee, by methods
> appropriate to national conditions and practice, equality of opportunity and
> treatment in respect of employment and occupation, of social security, of
> trade union and cultural rights and of individual and collective freedoms for
> persons who as migrant workers or as members of their families are lawfully
> within its territory.

Beginning in 1949, member nations of the International Labour Organization
agreed on a series of Conventions and Recommendations to secure and protect the
human, civil and political rights of migrants. Although there can be few countries
where the letter, let alone the spirit, of these obligations has been fulfilled, the
existence of international moral pressure is a step forward. Migration to find work
and a better way of life is as old as mankind; poverty is a great evil and there will
always be men and women who seek other lands to avoid it. Nevertheless, labour
is not a commodity to be bought and sold without regard for emotional and cultural
ties or basic requirements for health and safety.

Migration is not confined to those who move from one country to another for
employment; there are other more complex social, political and psychological
reasons. Internal migration may have effects which are just as serious, although of
less international concern. The major problems of migration, however, affect those
men and women, predominantly poor and unskilled, who seek work in another
country, usually more industrialized and economically advanced than their own.
The phenomenon is not readily defined; decisions to stay temporarily or
permanently, to move to other jobs or countries or to return home, are all subject
to change. A broad and useful definition is that migrant workers are persons who
do not possess citizenship in the country of employment. This covers a wide range
of workers of which four groups in particular are important for occupational health
practice:

(1) Those who travel on business for relatively short periods to countries with
    different health standards and habits from those to which they are

accustomed. The reason for such travel is frequently corporate and adequate health provision must be made.

(2)   Seasonal workers, particularly in agriculture and plantation work, who are employed for short periods in rural, even primitive conditions. They are seldom accompanied by their families, and often face special occupational hazards.

(3)   Project migration, where agencies or companies undertake to supply the knowledge, equipment and labour for entire operations, often in wealthy but underdeveloped countries. Major projects carried out by European and Asian companies in the Middle East are examples of this. Workers imported simply for the purpose usually have no rights within the host country and little effective recourse to those responsible for their working conditions. They may be housed in camps in primitive areas far from urban centres and their health needs fall almost entirely on the occupational health services, if any, provided by the contracting company.

(4)   Migrants recruited individually or in groups to take jobs with employers who are unable to find sufficient workers in their own country. These workers have need for social services and employment rights within the new country; their health problems are a responsibility not only for the recruiting employer but also for both sending and receiving countries.

These four groups, for which the legal position under international conventions differs, face difficulties distinct from those entering a country for other reasons. Refugees, frontier workers and illegal migrants, with no legal status in the host country, have similar or worse problems. Their needs will not be considered here because for them the services and procedures required for legal and voluntary migrants are neither sufficient nor wholly appropriate.

The serious questions posed by migrant labour have been recognized for many years and have been of concern to the International Labour Organization (ILO) since its foundation in 1919. In recent years international standards have been considered more fully and conventions agreed by such bodies as the ILO and Council of Europe. These agreements now offer some protection to those in group (4) above, but not in groups (1), (2) and (3). Specifically, the ILO Convention gives workers who enter a country for work regarded as permanent if filled by indigenous employees the right, after a provisional period, to change jobs, to claim unemployment benefit or worker's compensation, to equality in employment and social security and to be joined by their families.

Problems with which the migrant worker must cope can be viewed in two phases. First, there is the period after arrival of an inexperienced unskilled worker in an unfamiliar country, without support of family or friends, who is faced with jobs which local workers are unwilling to do. The second period begins when, established and partly adapted to the host country perhaps with wife and children, the immigrant attempts to better his position. Occupational health services need to understand these differences for worker and employer, in each of these two phases.

Magnitude of the problem

During the Second World War, vast sections of industry were destroyed in Europe and Asia and many millions of persons fled, were displaced or were transported.

In subsequent decades of recovery and industrial expansion many returned home, but others did not and never will. Meanwhile other more localized hostilities and persecutions have continued with similar effects. Other determining factors have been the changing geographical patterns of industrialization, and exploitation of energy and raw materials, in part associated with the dissolution of former colonial systems and the appearance of other forms of economic domination. Fear, want and hope in varying proportions have always forced men to emigrate on a truly enormous scale.

In a recent ILO publication [2], it was estimated that some 20 million persons were working outside their own country. Of these six to seven million were in Europe (together with a somewhat larger number of dependents), four million in South America, four to five million in North America, three million in the Middle East, more than one million in West Africa, and smaller number in South Africa, Australia and New Zealand. Refugees were excluded from these figures but in 1981 it was estimated that there were some eight million political refugees and a further 4.6 million displaced persons; half of these found asylum in Africa. Such refugees, if they are to be resettled in employment, will face all the problems, and more, of the migrant worker recruited for employment in another country, except that they cannot return home.

These statistics are given simply to indicate the extent of migration. In any one country, whether receiving or sending workers, figures are frequently inaccurate to a degree that makes the epidemiological evaluation of health problems extremely difficult. The quoted figures refer only to first-generation migrants; their children when old enough to enter the labour market are of even greater concern as they may have no right to employment and yet are without cultural or other link to their country of parental origin.

In reviewing the demography of migration, it should be appreciated that the position is not static. The flow of workers into Europe, high in the 1960s and early 1970s, has now almost stopped, apart from those entitled under agreements of the European Economic Community to move freely to seek work in other member countries. The 'guest worker' (*gastarbeiter*) problem of the earlier era has not disappeared completely. A high proportion of such workers have settled in European countries, and their children wish to remain. A new influx of workers from less wealthy parts of the expanded community in southern Europe can also be expected to produce many of the same social problems, if not legal difficulties, as the earlier flood of unskilled workers from North Africa and elsewhere.

The potential risks

The risks faced by migrant workers have much in common with other occupational hazards which men and women accept. On the face of it, they make their choice voluntarily and for their own benefit, or at least for that of their families. It can be argued, as it has with most other occupational risks, that regulation and control will force the 'industry' out of the market and, directly or indirectly, deprive workers of the only opportunities open to them. This is not the place to explore these arguments which contain some short-term truths, but rather to recognize that migrant workers tend to share certain basic characteristics.

The migrant is keen to work, even if forced to break the law to do so. He or she is typically a young person in the prime of life, physically fit, enterprising and highly motivated. Often from a predominantly rural and economically backward environment characterized by high levels of infant and child mortality, the migrant is triply selected. First, he is a survivor, which is no mean achievement in some of the countries which provide most migrant labour. Second, he is one of a minority of his fellows who chose to emigrate, to leave the security of family and friends, and to persevere sufficiently to succeed in this intention. Third, he has usually met the entry requirements of the host country and has the skills or other requirements of the foreign employer. Why then should such generally healthy and enterprising young men and women be at special risk? What in particular should occupational health services be looking for in pre-employment and other examinations of these workers?

Most of the reasons for their vulnerability are obvious enough and are essentially either personal or environmental. The migrant may look healthy, but his nutritional status is often of marginal adequacy with little reserve. Although a survivor of many infections some, such as tuberculosis, remain latent and liable to activation by adverse conditions or undue physical stress. He may well have immunity to many infections but these reflect the ecology of his own land and not that of a more industrialized country with quite different climate and population density. In an analogous way, other patterns of behaviour and defence mechanisms acquired from past experience constitute a threat to the migrant. He has learned to deal with one set of problems at home or at work in ways which do not necessarily apply in the new country; he meets unfamiliar hazards which he does not recognize or know how to avoid. At home, when sick or in trouble he knew what to do; in the new land he neither knows nor can find out easily. To this must be added the personality and psychology of the migrant. It is a big step, requiring emotional strength, to try one's luck in an unknown country, about which there may have been serious misinformation. Many who take it are driven either by bad even desperate economic circumstances, or by more than average ambition and adventurousness. For either reason, anxiety, emotional instability and unrealistic or foolhardy optimism are common in migrants. There is also loneliness due to the fact that they are often not allowed to bring their families with them during the difficult early months. Thus, they are specially vulnerable when, as may happen, they meet disappointment, frustration, discrimination and hostility.

The new environment in which the migrant finds himself is potentially hostile in several ways. Jobs taken by migrants are generally the least attractive and poorest paid. Many work in mines and quarries, agriculture, forestry or construction, and in a variety of industries where local labour is lacking. Most of these jobs carry specific hazards of occupational disease and accident against which particular preventive measures may not be properly taken. Ignorance, unfamiliarity, language difficulties and sometimes climatic and nutritional factors or latent infections put the migrant at risk, especially during the first few months of employment. The migrant's domestic arrangements may be equally unsatisfactory since accommodation provided by the employer or obtained by the worker himself are often substandard. Adequate opportunities for sleep, recreation and cultural activities are frequently lacking. Because of low income and the necessity to save and send money home, there may be insufficient left to maintain a reasonable diet and living standard. Finally, and perhaps most seriously, the migrant may find the local population and native fellow workers unfriendly and without sympathy for his difficulties. The social isolation, loneliness and stress which result from this complex of factors may lead to depression, mental breakdown and psychosomatic disorders, and indirectly to alcoholism and even suicide.

Work accidents

It seems probable that migrant workers suffer more accidents at work than their fellows, although this is not well documented. It is less certain that the excess remains after proper allowance is made for age, duration of employment and type of work. These corrections are necessary because it is generally true that certain jobs are more dangerous than others, that young workers have more though perhaps less serious accidents than old workers, and that the first few months in any job constitute a period of high risk. All three factors operate against the migrant.

Data presented by various authors [3–6] suggest that crude accident rates among migrants in several European countries are two to three times greater than for the native-born workers. This applies particularly to minor accidents, less to the more serious and disabling accidents and not, apparently, to fatalities. There is no convincing evidence of any important difference between the various racial groups nor that differences between accident rates of migrant and native workers present a consistent pattern in various types of industry.

The surveys reported by Opferman [6] of the German Federal Ministry of Labour and Social Affairs were more searching than most and correspondingly informative. The points which emerge are that the more closely the national and migrant workers were matched, job for job, and in other ways, the more similar were their accident rates; such differences as there were reflect mainly the experience of newcomers to the job. For a variety of reasons, migrants are probably more mobile than local workers and thus more frequently find themselves in new and unfamiliar work; after several years the differences in accident frequency disappear.

Occupational disease

Migrant workers who stay in one country for many years (or even more briefly) and later develop occupationally related illness cause particular difficulty, both for compensation and research. In third world countries, where many young men from certain villages migrate for a year or two to mining or other operations, knowledge about the relation of disease to earlier exposure may be insufficient, and political will too weak, to instigate or succeed in a request for compensation. Reports of increased incidence of work-related disease amongst such migrant labourers are beginning to appear in the literature. A complementary problem arises from the attribution of disease to exposure in the host country rather than the country of origin. In Britain, for example, the Registrar General's occupational mortality figures show an excess of cancers amongst leather workers from the Indian subcontinent compared with leather workers of UK extraction. The employment of migrant workers thus adds legal and epidemiological complexity to the study of occupational disease, and considerable ingenuity may be needed to obtain data from which to assess the cost of migrant work to the health of both the sending and receiving countries. The occupational health physician has a particular moral and professional responsibility in these matters; all international conventions specify that the migrant worker has the same right to medical care as any other worker. The migrant has been admitted for the economic gain of the country and employer, so the responsibility for his long-term welfare is clearly theirs.

Communicable disease

Three groups of infections constitute a potential threat to migrants and to the local populations with which they come in contact: tuberculosis, tropical parasitic infections and sexually transmitted diseases. Migrants, and indeed any travellers, are at special risk of contracting infections against which they lack immunity, or of the activation of latent infections by adverse social circumstances. They are also capable of carrying infection to others. Examples can be cited of all these occurrences so they are of more than academic interest but, in practice, the number of disease outbreaks attributable to migrant workers is small and only tuberculosis and venereal disease constitute a serious threat to the migrants themselves.

In wealthy industrial countries to which most migrant workers go, tuberculosis is uncommon in the indigenous population and a high proportion of all clinically active cases occur in immigrants. In Britain, for example, among those who arrived from India in a 5 year period to 1983 the rate of new cases was some hundred-fold greater than in the indigenous population [7]. The extent to which tuberculosus infections are imported and then activated, or acquired in the host country, is unknown. Among adults, the former possibility may predominate, whereas among young children it may be the latter. Stress, overwork, poor nutrition and bad housing are likely to be the factors responsible. In countries such as South Africa, Zimbabwe and Zambia, which employ migrant labour in mining, silicotuberculosis is a special hazard.

The occurrence of venereal disease among migrant workers inevitably reflects an unnatural way of life imposed upon young persons, predominantly male, separated for long periods from their families and friends. Only by removal of this basic cause is the problem likely to be reduced.

Psychological disorders

Mental illness, behavioural disorders and psychosomatic complaints are difficult to quantify. In part, this is because they are definable only in cultural terms which have defied standardization. This is particularly true for population groups removed from their normal social environment. Although it is repeatedly stated that psychiatric disorders are much more common in recent migrants than in the local population [3], this view is based more on general opinion than valid statistics. In some situations, indeed, there is evidence to the contrary [8–10].

For many years, both voluntary and enforced migration have been of special interest to social psychiatrists and have been the subject of many books, papers and conferences [11]. The basic issue is whether mental illness in migrants is determined primarily by the various associated stresses or whether migration is in part a manifestation of mental instability. The nature of this unanswered question was well described by Kuo [12] who attempted to evaluate the relationship of four main factors—social isolation, culture shock, goal-striving stress, and cultural change—to measured symptoms of psychological stress among the Chinese population of Washington DC. The first two factors appeared the more important but of low predictive value. The mental health of migrant workers deserves further research of similar quality, which some occupational health services are well placed to undertake.

It seems reasonable to suspect that migrant workers have more than their share of restless, foolhardy and other unstable personality types. It is equally clear that if isolation, loneliness, worry and cultural confusion are capable of causing or aggravating mental illness, migrants are at special risk. Either way, protective measures are needed.

Some remedies

The main problems experienced by migrant workers are primarily social, psychological and political. They have, therefore, to be attacked in many ways, in most of which there is a role, sometimes a major responsibility, for occupational health services. The various measures required can be discussed under a number of headings.

International conventions

Free movement provision

Several blocks of countries, of which the European Community is the best known, have provision for free movement of labour. Workers from member countries have equal rights to work and social security as those from the host country. In the past, associated members countries, such as Turkey, had agreements allowing workers to enter with relative ease. New members, such as Greece, Spain and Portugal, have not been granted the same free movement, but these workers may not be excluded indefinitely. It can be assumed that most workers from less developed areas (as indeed has already happened from the South of Italy), will wish to settle in a more developed country, while retaining their cultural identity and citizenship.

The Council of Europe

This organization has members both within and outside the European Community. In 1978, conventions were agreed which protected the rights of the migrant workers and included provision for occupational health and safety.

International Labour Organization (ILO)

Convention 143 adopted by the International Labour Organization in 1975 [1] called on member countries to legislate and take other appropriate steps:

(1)   To ascertain and suppress clandestine movement and illegal employment of migrants, with severe penal sanctions against those responsible.
(2)   To ensure equal treatment for migrants and nationals in respect of security of employment, remuneration, social security and other benefits.
(3)   To inform migrants fully of their rights, including equality of opportunity, working conditions and treatment.
(4)   To facilitate the reunification of families (spouse, children and parents) and encourage the preservation of ethnic identity, language and cultural ties.

The implementation of this Convention was covered more fully in the ILO Migrant Worker Recommendations 151 [13]. This urged the need for both the country of origin and the country of employment to take account of long-term social and economic consequences for all concerned. Equality of opportunity and treatment, which are the right of migrants and their families, apply to all aspects of employment, including advancement, membership of trade unions and cooperatives, and access to amenities enjoyed by nationals, such as holidays with pay. As a matter of social policy, countries should aim at providing migrant families with accommodation of reasonable local standard or, where this is not possible, to grant paid home leave or to cover the case of family visits. The special needs of migrants for informal advice and additional social services were to be recognized, as were security of employment and residence and the rights of both legal and illegal migrants regarding pay and accumulated benefits on return home.

Unfortunately, these conventions are not always ratified by the countries most affected. For example, by January 1986, no European country receiving large numbers of immigrants had ratified Convention 143, passed by the ILO in 1975. The influence of such international pressure may, however, be considerable. Each member country is required to report on progress towards the form of a convention, even though the government has chosen not to ratify. In a recent discussion on foreign workers in the United States, it was agreed that policies recommended by the ILO should be followed, for fear of unrest and of appeals relating to human rights. This occurred even though the United States had not ratified this, or any other ILO convention on migrant workers.

Social change

International conventions even when fully implemented by national laws have little effect unless accompanied by a change of heart among the people of the host country. Racial, religious, colour and cultural prejudices, largely based on ignorance and tradition, are widespread. These are reinforced by economic insecurity, particularly among the poorer and less skilled sections of the population who are most vulnerable to foreign competition over work and pay. Above all, there is lack of understanding; most people simply do not appreciate the feelings and needs for friendship of the strangers in their midst. It follows that determined efforts must be made by leaders and makers of public opinion to set an example. This responsibility falls most heavily on politicians, journalists (in various news media), trade unionists, employers, professional people and the police. Among these categories are the staff of health services within industry, who are often well placed to demonstrate in a practical fashion both understanding and willingness to give that little extra help to migrant employees and to collaborate fully with other agencies with similar objectives. Trade unions, too, although wary of the threat to their members posed by worker migration, have the potential to be of great support. A recent ILO publication [14], designed to advise unions on the part they can play, suggested programmes for education of fellow workers, trade union officials and the immigrants themselves. Hostility and violence against migrant workers have their roots in fear and ignorance which may be difficult to overcome within union leadership. It is easy to feel frightened and unwanted; it is more difficult to prescribe the recipe for a genuine welcome.

Community services

The scope of the community services required to deal with migration is much wider than is usually appreciated. In most countries, migrants and their dependents may need help and advice, not only on arrival but also during their stay and before returning home. In addition, services are wanted in the home country to prepare prospective migrants and to assist in their settlement on return. In some countries, for example Yugoslavia and the Philippines, the departure and return of citizens are well-organized and coordinated, so far as possible, with the countries of employment. Labour can be recruited only through a national agency which ensures that the workers are properly qualified, physically suitable and adequately informed about the place and jobs which await them. Vocational and language courses are also available.

A reasonable balance has to be kept in countries of employment between:

(1) The development of separate services and facilities for migrants to a level which may keep them apart from the local population and even increase discrimination and resentment.
(2) The need for employees in the general social services to possess the linguistic and other abilities necessary to deal adequately with the special difficulties experienced by migrants.

These services must be able to offer effective assistance, information and advice on all aspects of life in the new country. Help such as is available at Citizens' Advice Bureaus in Britain should cover questions on (1) housing, (2) employment opportunities, (3) family problems, especially education and health, and (4) legal advice. Loneliness and isolation, which can become dangerous if excessive, are the common lot of strangers in any society. Social occasions and recreational activities with compatriots on the one hand but also with people from the new country are both important. Language is an obvious bar to communication, so every opportunity should be afforded for the migrant to learn at least sufficient for simple conversation and some appreciation of cultural activities in his or her new homeland.

Health services

The migrant and his family should have the same rights of access to medical care, including occupational health services, as normal residents. This may be no great assurance or privilege; health care for working people, inside or outside industry, is far from satisfactory in many countries which accept migrant labour, and foreign workers seldom make full use of what is available. The reasons for the latter may be simply ignorance, reticence or inability to communicate adequately with doctors and nurses, who understand neither the language nor cultural attitudes to illness of the migrant. The migrant worker may also fear that if found sick he may lose his job. It is important that immigrants should be fully covered for medical care from time of arrival, either by national or privately arranged insurance. The need is particularly great during the first few weeks. Major employers and host governments share a responsibility to provide this cover and also appropriately staffed occupational health services.

Although limited in extent and availability, occupational health services are particularly well suited to the task of protecting the health of foreign workers. They have a responsibility for all employees of the enterprise or industry according to individual need, and in this sense are not discriminatory. Their role is primarily preventive, so they can seek out workers in real or potential trouble before serious damage is done. It is normal that they should conduct comprehensive health screening (including X-ray or other special tests) at or before employment, and periodically thereafter. It is their job to keep track of attendance, accident and sickness records, to arrange whatever educational courses are required in the interests of health and safety, to advise management on questions of suitability for employment in certain jobs and to provide information and counselling for all employees. Good occupational services would greatly reduce the health risks of migrant workers; a case therefore exists for requiring that any employer of migrant labour should provide them. Most countries which receive migrant workers are relatively wealthy and benefit from their presence financially and in other ways. Provision of adequate health surveillance and preventive health services would not appear to be an unreasonable price [15].

Health information

Some of the difficulties and adverse effects experienced by migrant workers are the direct result of ignorance on questions of health and safety. This is not surprising, as many are poorly educated and from countries where life and its hazards are quite different. Accurate and well-balanced information is needed in five main areas.

Medical and social services

The newcomer should know exactly what agencies exist to help him when sick or in need of advice and how to use them; he should understand his rights and what he can and cannot reasonably expect.

New environment

Life in a large industrial city may expose the worker and his family to risks of accident, infection and temptation. They should be made aware of these risks and how to avoid them.

Food and nutrition

Unaccustomed foods, the need to save money and ignorance of basic dietary requirements, commonly undermine the migrant family's health and strength.

Safety at work

Every job has its risks to which the newcomer, especially one ignorant of the language, safety procedures and the use of protective equipment, is most vulnerable. Special safety training is required for migrant workers.

Social adaptation

Social isolation, loneliness and severe nostalgia may be reduced if the migrant has insight into reasons for any unfriendliness from the local population. He needs information and guidance on use of leisure, on how to enjoy life in his new home and on longer-term employment prospects.

The educational task is large and must be approached in a variety of ways. Ideally, the process should begin well before the migrant leaves home, with responsibility on both donor and recipient countries and perhaps also on other workers after their return from abroad. Intensive instruction may be indicated during the first few weeks after arrival in the new country and at the new place of employment. Educational opportunities of a more relaxed and enjoyable kind should then continue; it is for the government, the employers and the trade unions of the host country to see that these facilities exist.

Integration

The experience of guest workers in Europe has been that many, perhaps one-third, do not return home, at least during their working years. The legal rights of these long-term residents provide for access to jobs beyond the undesirable ones for which they were first recruited. Those who stay, it may be supposed, are those with ambition and whose degree of assimilation has been such as to make their return more difficult. For these workers, and for their children, new problems may emerge.

Employment discrimination

In Sweden, migrant workers have the right to a permanent work permit after only 1 year, in Belgium after 3 years, in France and the United Kingdom 5 years and in the Federal Republic of Germany after 10 years. These immigrants have essentially equal rights of employment with citizens of the country; legislation relating to equal opportunity usually does not exclude settled immigrants. Under such legislation it may be encumbent upon the employer to show that discrimination does not exist. Discussion of techniques to ensure that selection and promotion are conducted fairly is beyond the scope of this chapter, although relevant to the occupational health practitioner who is asked to assess whether an applicant is fit for a particular job. For example, the use of selection tests based on physical or intellectual capacity has been subject to much detailed legislation, particularly in the United States, with the burden of proof on the employer to show that the validity of the test justifies its use for selection. In the United Kingdom literacy tests carried out at the medical centre of the British Steel Corporation resulted in legislation under the Race Relations Act, 1976. The Corporation maintained that the tests were justified in order to comply with the Health and Safety at Work etc. Act, 1974; safety training and warning notices could not be provided in languages of all potential workers. This case was settled out of court, without resolution of the safety issue, but it illustrates the balance needed to avoid indirect discrimination while retaining the freedom to employ appropriately qualified workers. Such issues, which also relate to the employment of women and older workers, should be familiar to occupational health physicians; a useful booklet, covering the situation in Britain is published by the British Psychological Society and the Runnymede Trust [16].

Second generation migrants

In the EEC countries, children of migrant workers, although allowed to join their parents, had initially no right to work; they were not citizens, even if born in the country, and did not necessarily have any claim to citizenship. Such young people had little contact with their home country where, typically, the options for employment, even if they desired it, would be very poor. The social problems resulting from large numbers of partially assimilated unemployed young people have in recent years changed the situation. Migrant workers' children have now been permitted to take employment and indeed, in some countries, are given special help in making the transition from school. In all countries the proportion of migrant workers' children remaining unemployed, whether as a result of educational difficulties or employment discrimination, is much higher than that of local youth. This situation, which might be expected to apply as much to workers' children in countries with free movement of labour as to those of migrants from outside countries, remains serious.

Occupational health services need to have special sympathy and to encourage so far as possible the employment of any young people, doubly handicapped by cultural difficulties and physical and psychological or intellectual impairments.

Research

The implementation of effective measures to protect and improve the health of migrant workers requires information on the prevalence and causes of diseases which affect them, not easily obtained without well-designed epidemiological studies. Since migration is a world-wide phenomenon, health patterns are bound to vary and reliable data are scanty. It cannot be assumed that migration *per se*, or the new working environment, are necessarily the cause of illness or injuries, even those occurring with high frequency in migrants. The strong selective forces already described and the effects of adverse social circumstances in early life could well be the major determinants. Even so, there is no less need for intervention; the priority should be to ensure that whatever is done is epidemiologically reasonable and the results evaluated. Data are needed on, first, the frequency of disease and injury among migrant groups and their families, defined by place of work and type of industry; and, second, clues to the cause of unduly high rates. Action should depend on whether those affected were already at high risk before leaving home, and whether migrants subsequently fared worse than comparable men and women who stayed behind. There is a dearth of appropriate information, not only for migrants but also for the indigenous populations with which they can be compared. Some of these data will have to be collected by government agencies. An urgent requirement is to identify effective programmes for the protection of the migrant. Priority should be given to forms of intervention which are well-defined, straightforward, ethically acceptable and widely applicable. These might include:

(1) Steps to reduce separation from home and family, for example by limiting duration of time abroad, frequent home leave, family visits and free telephone calls home.
(2) Educational courses for the migrant before and after leaving the country of origin.
(3) Vocational training courses for the specific work which is to be done.
(4) Participation of labour unions in defined aspects of migrant welfare.
(5) Programs in which occupational health services should participate for the early detection of behaviour suggestive of undue stress.

The identification of personal characteristics of the intending migrant—physical, psychological and social—which are associated with unacceptable risks might be useful. However, the use of such information in counselling or selection may raise ethical questions.

Conclusions

Towards the end of the 1977 International Symposium on Safety and Health of Migrant Workers one of the discussants [17] made a statement at the conclusion of his paper which well summarizes the situation:

> ...we have mostly heard about the poor, sometimes illiterate, unskilled migrant workers going to richer industrialized countries to take up work which nobody of the indigenous populations wishes to carry out any longer. The migrant is, to put it bluntly, the voluntary slave of our time and personally I find it high time that international bodies like the ILO and WHO try to lay down some international rules and regulations which can safeguard the humanitarian rights of these people because the problem of migrant workers is not a temporary one.

This paper deserves to be read in its entirety as it describes succinctly the actions required, not just from ILO and WHO, but from the many persons and agencies responsible in most industrialized countries.

References

1. International Labour Conference Convention 143: *Convention concerning migration in abusive conditions and the promotion of equality of opportunity and treatment of migrant workers.* ILO, Geneva (1975)
2. ILO International migration for employment. *World Labour Report*, Chapter 4, ILO, Geneva (1985)
3. International Labour Organization and World Health Organization (ILO–WHO) Joint Committee on Occupational Health. *Migrant Workers—Occupational Safety and Health. Occupational Safety and Health Series, no. 34*, pp. 1–50 ILO, Geneva (1977)
4. Djordjevic, D. and Lambert, G. Migrant workers in the construction industry. In *Migrant Workers—Occupational Safety and Health. Occupational Safety and Health Series, no. 34*, pp. 55–62, ILO, Geneva (1977)
5. Djordjevic, D. Les accidents du travail et la morbidité des travailleurs migrants. In *Safety and Health of Migrant Workers—International Symposium, Occupational Safety and Health Series, no. 41*, pp. 19–30. ILO, Geneva (1979)
6. Opferman, R. Epidemiology and statistics of occupational accidents and morbidity in migrant workers. In *Safety and Health of Migrant Workers—International Symposium, Occupational Safety and Health Series, no. 41*, pp. 31–42. ILO, Geneva (1979)
7. MRC Tuberculosis and Chest Disease Unit. Report: National survey of notifications of tuberculosis in England and Wales in 1983. *British Medical Journal*, **291**, 658–661 (1985)
8. Burke, A.W. Attempted suicide among the Irish-born population in Birmingham. *British Journal of Psychiatry*, **128**, 534–537 (1976)
9. Burke, A.W. Socio-cultural determinants of attempted suicide among West Indians in Birmingham; ethnic origin and immigrant status. *British Journal of Psychiatry*, **129**, 261–266 (1976)
10. Murphy, H.B.M. The low rates of mental hospitalization shown by immigrants to Canada. In

Uprooting and After, edited by C. Zwingman and M. Pfister-Ammende, pp. 221–231. Springer Verlag, Berlin, Heidelberg, New York (1973)

11. Zwingman, C. *Uprooting and Related Phenomena. A Descriptive Bibliography MNH/78.23.* WHO, Geneva (1978)

12. Kuo, W. Theories of migration and mental health: an empirical testing on Chinese-Americans. *Society for Science and Medicine*, **10**, 297–306 (1976)

13. International Labour Conference Recommendation 151: *Recommendation concerning migrant workers.* ILO, Geneva (1975)

14. Dunning, H. *Trades Unions and Migrant Workers. A Workers' Education Guide.* ILO, Geneva (1985)

15. Montoya-Aguilar, C. and El-Batawa, M.A. An introduction to the planning and organization of preventive and curative services for migrant workers. In *Safety and Health of Migrant Workers— International Symposium, Occupational Safety and Health Series, no. 41*, pp. 157–163, ILO, Geneva (1979)

16. Runnymede Trust and British Psychological Society. Report: *Discriminating Fairly: A Guide to Fair Selection.* Runnymede Trust and British Psychological Society, London and Leicester (1986)

17. Schou, C. Practical prevention measures for safety and health of migrant workers: contribution to discussion. In *Safety and Health of Migrant Workers—International Symposium, Occupational Safety and Health Series, no. 41*, pp. 279–283, ILO, Geneva (1979)

Chapter 27

Special problems in developing countries

J. Jeyaratnam

Introduction

The health concerns of man in his working environment are to a large extent universal, with much in common to both the industrialized nations and the developing countries. However, there are differences in the setting of priorities for occupational health needs in different countries, each according to their set circumstances and stage of development. This chapter attempts to identify some of the urgent occupational health needs of the nations of the developing world. These needs are largely in the domain of policy issues relating to occupational health. For what is needed for occupational health in the developing countries is not new technology and skills but rather policy decisions which could make available existing knowledge and skills to those in need—the working population. In this context this chapter has identified the following three areas as issues which require the attention of all nations of the developing world.

(1) A policy and programme to provide *health care for the working population—* an occupational health service.
(2) A policy which requires to set *relevant* exposure levels in the work environment.
(3) A *national policy* on research in occupational health.

The nations of the developing world

The term developing world is one of the many synonyms in current usage to describe a group of nations which are also known as the 'newly industrializing countries', the 'nations of the south' and 'the third world' among others. The World Bank [1] categorizes the nations of the world into the developing world and the industrialized nations.

The developing countries

These consist of a group of nations which are further subdivided into the following categories:

(1) Low-income countries: those with a per capita income of less than $410 per year in 1980.

(2)   Middle-income countries: those with an income between $420 and $4,500 per capita in 1980.
(3)   The petroleum-exporting countries: which are termed the 'high-income oil exporters' whose incomes range from $8600 to $27 000 per capita in 1980.
(4)   The least developed countries (LDCs): those with a per capita income of less than $100 in 1980.

The industrialized nations

These are divided into market (western capitalist) and non-market (eastern communist) economies.

The term third world nations refers to those nations not belonging to the first or second world group of nations. The first world comprises the nations of the industrialized market economies of western Europe, North America and the Pacific. The second world consists of the industrialized but centrally planned economies of eastern Europe. Hence, by a process of exclusion the third world countries consist of all the countries of Asia (except Japan), Africa, Middle East, South America and the Caribbean; the same nations also constitute the developing world.

Though there exists a wide range of terminology to describe these nations they all refer to countries that are economically deprived, largely agricultural and with a relatively low physical quality of life index (PQLI). This is an index based on life expectancy, infant mortality rates and the level of literacy. On the other hand, many of these nations, though so deprived are rich in terms of culture, civilization and history. Though these nations have common characteristics it is equally important to recognize that they are widely different in many ways with different aspirations, different political systems and at varying stages of economic and industrial growth. It is often only for reasons of academic convenience that these nations are grouped together, whereas each would need to have their own specific plans to develop according to their own priorities and special set of circumstances. In recognition of this, the topics chosen for discussion in this chapter and its contents will largely be an attempt to define in general the principles involved rather than to identify specific features relevant to each of the constituent nations of the developing world.

Occupational health services for developing countries

This section delineates the principles for a system of occupational health services relevant to the needs of developing countries. But these must be adapted to meet the specific needs in each of the individual nations. At present most of the developing countries have in existence a relatively recently developed occupational health service, or are in the process of modifying or developing *de novo* national occupational health services, thereby making it easier for these nations to effect changes wherever appropriate in the organization of their occupational health services.

Occupational health services—definition and scope

Increasingly, occupational health services, particularly in the developing countries, are being recognized as the provision of comprehensive health care to working populations. This is a view endorsed by the World Health Organization [2] which has drawn attention to the fact that occupational health is not limited in scope only to preventing and controlling specific occupational diseases, but that workers' health programmes should deal with the full relationship between work and the total health of man. In pursuance of this concept the World Health Organization [3] has identified the scope and extent of an occupational health programme, and they are:

(1) To identify and bring under control at the workplace all chemical, physical, mechanical, biological and psychological agents that are known to be or suspected of being hazardous.
(2) To ensure that the physical and mental demands imposed on people at work by their respective jobs are properly matched with their individual anatomical, physiological and psychological capabilities, needs and limitations.
(3) To provide effective measures to protect those who are especially vulnerable to adverse working conditions and also to raise their level of resistance.
(4) To discover and improve work situations that may contribute to the ill-health of workers in order to ensure that the burden of general illness in different occupational groups is not increased over the community level.
(5) To educate management and workers to fulfil their responsibilities relevant to health protection and promotion.
(6) To carry out in plant health programmes dealing with man's total health, which will assist public health authorities to raise the level of community health.

Target population

The population structure of nations is an important consideration in formulating health policy and in integrating it with national development policy. This is particularly relevant in the formulation of policy for occupational health services in developing countries. For instance it is estimated [4] that by the end of 1987, the total workforce in the world numbers in excess of 2 billion and 85% of this number was living in the nations of the developing world, which means that by the end of 1987 some six workers out of ten lived in Asia, one in western Europe, one in Africa, one in the Soviet Union and one in the Americas. The figures are indicative of the fact that the working population of the developing world constitute a significant sector of the working population of the world. Further, on an average it is estimated [4] that the working population constitute approximately 40% of the total population of the nations of the developing world. This again emphasizes the importance of providing health care to this sector of the population.

Occupational health services—whose responsibility?

Historically labour was organized primarily for productivity and profit by industrialists. A consequence of this was the often encountered exploitation of workers leading to the breakdown of their health. It was in order to prevent such damage to workers' health that the initial interventions were largely centred around legislative controls of work practices considered injurious to health. It was this evolution of events which led to the situation where workers' health became the responsiblity of Ministries of Labour which had the power to control work practices, a development whereby physicians and Ministries of Health were not directly involved in providing for the health of this important and vulnerable sector of the community—the working population. For these reasons, occupational health services till recently and often even today, are not a mainstream activity of the health sector. Every effort should be made, particularly in those countries in the process of upgrading or setting up occupational health services, to ensure that it be identified as a responsibility of Ministries of Health. Such a recommendation is made on the basis of the following reasons.

Comprehensive health care for workers

As previously stated, an occupational health service should provide comprehensive health care to all members of the working population. In view of this it is appropriate that the general health care system of the country, usually the responsibility of Ministries of Health, be identified as the agency responsible for occupational health services.

Health manpower shortage

A common problem of many developing countries is the shortage of health professionals. Given this, it is neither feasible nor realistic for these countries to duplicate health services in an attempt to develop an occupational health service outside the prevailing health care system. Every effort should be made to maximize the existing health care resources in terms of manpower, equipment and facilities.

Inconvenience to workers

An occupational health service organized as a separate entity from that of the general health services is bound to cause inconvenience and confusion to the worker. From the worker's point of view, the concern is for his personal health and he would seek a convenient health facility to cater for it. If health care to the worker at the peripheral level is compartmentalized into an occupational health service which is different from that of a general health service, it is more than likely to lead to confusion. For example, should the worker at a flour mill on developing asthma choose to see a physician from the occupational health services or a physician from the general health service? So far as the worker is concerned, he has a health problem and his first point of health care contact should be a health professional responsible for his total health.

Principles for the provision of workers' health care

The World Health Organization [5] has identified three basic principles for the development of an occupational health service. They are:

(1) Occupational health services must be provided through the existing national health services, by a process of integration.
(2) The services must be for the total health of the workers, and if necessary their families.
(3) Most importantly the primary health care approach should be the chosen

system for the delivery of such services. This was endorsed by the World Health Assembly Resolution (WHO 33.31) of 1980 which stated that:

It was convinced that there is a growing need for a new perspective of integrating occupational health in the primary health care of the 'under-served' working populations, particularly in the developing countries; and recalled that, for setting and implementing strategies for Health for All by the Year 2000, it is necessary to promote occupational health services and to strengthen institutions, training and research in this field.

The primary health care approach for occupational health services

Much has been said and written about primary health care particularly since the Alma-Ata Conference in 1976. At least part of this information has led to some confusion and misunderstanding among many persons as to the meaning of primary health care. This is probably best resolved by stating what primary health care is not. In the first instance, primary health care is not cheap, third-rate health care for third world nations. Secondly, primary health care does not solely consist of primary health care workers of 'barefoot doctors'. Finally, primary health care is not opposed to modern medical technology, sophisticated equipment, hospitals and specialists. It is necessary to emphasize this last aspect as often the medical profession in particular seems to be developing some reservations with regard to primary health care due to their misunderstanding of the concept. On the other hand, primary health care is based on the active participation of the people in their health care system thereby ensuring that such health care meets their needs, at a cost that is affordable and acceptable and that it is equitably distributed among the people to meet the needs of those in greater need. Finally, the primary health care concept recognizes the principle that the health sector is not solely responsible for health and that it requires the participation of many others. Further, within the health sector the primary health care concept requires an organizational network with an effective referral system, which means a system which provides for a two-way transfer of information.

An organizational pattern of occupational health services based on primary health care approach

Fry and Ferndale [6] state that to understand the concept of primary health care one may imagine the structure of the health care system as composed of a number of different levels. The first level is that of self-care. People deal with the majority of their symptoms on their own according to local tradition. For instance, a recent study [7] of acute pesticide poisoning among agricultural workers in Indonesia, Malaysia, Sri Lanka and Thailand showed that the bulk of the workers who suffered episodes of acute pesticide poisoning either took no treatment or resorted to self-treatment.

If intervention is considered necessary, a 'trained' person's help is sought. Such a person fulfils the role of the primary health care worker. The training of such a primary health care worker could have a very wide range depending on the circumstances of each community; there could be a voluntary primary health care worker, a midwife, a nurse, a sanitarian or a physician. The features of health care at this level are: that it is the point of entry for individuals to the health care

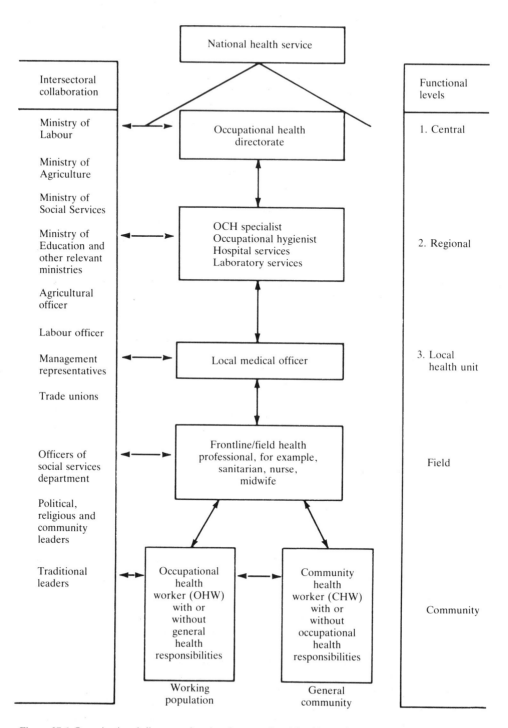

Figure 27.1 Organizational diagram of national occupational health service

services, involving functions of assessment and of mobilization and coordination of further medical services; and that it provides personal continuing and long-term care.

An organizational pattern for an occupational health service based on the primary health care concept is shown in *Figure 27.1*. For purposes of convenience, the service is considered at three levels: the local, regional and central; but functionally it should be emphasized that the service is not one which is compartmentalized but be considered as a composite unit.

To elucidate the organizational rationale for such a service, an analogy is made between it and maternal and child health services (MCH services), which is a service familiar to many administrators and physicians from developing countries. Both occupational health services and maternal and child health services have the objective of providing health care to sizeable, important and vulnerable sectors of the community, the mother and child in the case of MCH services, and the working population in the case of occupational health (OCH) services. Both services require specialists—obstetricians and paediatricians for MCH services, and occupational health specialists for occupational health services to function as referral units. Both services require a national network of health professionals within the framework of the national health service, for example primary care physicians, nurses, midwives, sanitarians, hygienists, health educators and primary health workers. In view of these similarities, an organizational network similar to that of MCH services is also relevant to OCH services. Thus an OCH service could be provided at the peripheral level through the national health network, utilizing existing health care facilities with referral and support facilities from occupational health specialists at both regional and central levels.

The functions of such a service at the different levels of care are set out below.

Occupational health services—peripheral level

Since the service is part of the national health service, the bulk of the programme at this level is the responsibility of the health professionals, the primary care physicians, nurses, public health inspectors or midwives, with an input from primary health care workers who would be persons from the workforce. Such a scheme requires some redefinition of existing job functions together with training programmes in occupational health for the general health professionals. The extent of the job redefinition and the content of the training programme would depend on the type of occupations and the problems encountered in each locality. The participation of a primary health care worker would depend on local conditions, but it must be emphasized that the primary health care worker, besides providing a service component, is a representative of the community in the health care system. Such a primary health care worker should be selected from among the working population. The numbers required, functions, selection process, training, remuneration, evaluation and career structure are some of the issues that need to be considered and these would vary according to the local conditions.

Occupational health services—regional and central level

At these levels a cadre of specialists in occupational health and hygiene should be developed according to national needs. These specialists should provide a referral service, be trainers and also develop relevant research programmes. All too often in developing countries, trained manpower does not have the equipment and facilities to undertake those activities for which they have been trained at great expense. Hence every effort should be made to provide such facilities for these specialists by strengthening existing institutions or, if possible, developing new ones where none exist.

At the central administrative level, the line of authority in most instances would lead to a directorate position in the Ministry of Health. But in view of the multidisciplinary nature of occupational health, it is necessary that strong links are established with other relevant ministries and departments, in particular Ministries of Labour and Agriculture.

The setting of exposure levels in the work environment

The setting of permissible levels of hazard in the work environment is one of the mechanisms used in occupational health to prevent health impairment in the worker. The term 'permissible level' has been defined in ILO and WHO documents and, basically, it is a quantitative hygiene standard for a level considered to be safe, expressed as a concentration for a defined average time. The term 'permissible level for occupational exposure' is taken to mean 'maximum allowable concentration', 'threshold limit value' and 'maximum permissible limit or dose'. In recent times, the WHO [8] has preferred to use the term 'recommended health-based occupational exposure limit' in place of 'permissible level'. This term refers to levels of harmful substances in workroom air at which there is no significant risk of adverse health effects. It needs to be emphasized that such health-based standards do not take into account the level of technological and socioeconomic conditions that prevail in different nations. It is in recognition of this deficiency that the WHO has suggested that it is the responsibility of each nation to set its own operational occupational exposure limits, taking into consideration these issues. All too often the developing nations do not set operational occupational exposure limits, but tend to rely on permissible levels borrowed without modification from industrialized nations.

It is appropriate to consider the factors involved in the setting of recommended health-based occupational exposure limits in order that they can be translated to operational occupational exposure limits depending on the local circumstances. The first stage of this process is the development of health-based recommended exposure limits, which are determined purely on the basis of scientific evidence. The second stage is the conversion of these scientifically determined exposure limits to operational limits.

The general scientific principles which determine standard setting as set out by WHO [9] are:

(1)   The physical and chemical properties of the substance, including the nature and amount of any impurities.
(2)   Toxicological investigations involving acute, subacute, short-term and chronic testing (such tests covering the respiratory, alimentary and dermal routes of entry).
(3)   The careful consideration of any available human data.

As such, it would be evident that the developing nations themselves rarely, if ever, undertake the extensive scientific evaluation of chemical compounds required to set such standards. Rather, they depend on the scientific data of the industrialized nations in setting exposure limits. But what is not done is that standards of the industrialized nations are not modified to suit the local needs of each of the developing nations. Duncan [10] describes the situation aptly when he states that

> Until recently there has been a tendency to accept recommended lists prepared by scientists or experts of one sort or the other as if they were satisfactory check lists; industrial managements can then feel that they are doing well if they keep exposure of their workers inside these often rather arbitrary numbers. Examples of this are the TLV (threshold limit value) lists, put out by the American Conference of Governmental Industrial Hygienists (ACGIH) and copied into documents published by governments of many countries or the lists published by eastern European authorities which differ from TLWs in many significant ways. Professional societies such as the British Occupational Hygiene Society (BOHS) have also undertaken literature and state of the art reviews and published documents which are variously described as standards or guides. It is, however, a fallacy to regard these lists as anything other than very tentative first approaches to the problem and certainly it is quite wrong to look at any of these quoted numbers without at the same time studying the evidence from which they are derived and the criteria they are designed to meet. If the final decision on standards is to be based on an amalgam of very different considerations there are important political and sociological considerations as to how the data can be brought together and evaluated and, even more particularly, who should be involved in the process. Certainly, professional, scientific or technical bodies or societies cannot carry authority beyond their own particular field of interest.

This is a point of view which reiterates the need for each nation to set its own operational occupational exposure limits taking into consideration all the relevant issues such as the following.

Concept of absolute safety

The operational occupational exposure limits should not attempt to make the workplace entirely risk-free. For though it may be theoretically ideal to attempt to ensure absolute safety, it is often misleading and tends to ignore reality. But rather, we should be concerned to determine what may be considered an acceptable risk for a particular exposure. The recognition of this concept is important as it leads to the possibility of modifying scientific information to standards which are socially relevant and acceptable.

Economic considerations

The present author has stated elsewhere [11] that 'unfortunately factors outside science are not given adequate recognition by the scientists of the developing world, particularly when decision makers are being advised about the setting of environmental standards'. This is particularly relevant to the developing nations as they are in desperate need of economic growth and hope to achieve this by a process of rapid industrialization and modernization of agriculture. It is estimated [4] that in 1987 one of every four persons in the world was living in poverty and virtually all would be from the nations of the developing world.

Unemployment is a major socioeconomic problem in most of the developing nations. It is estimated [4] that some 600 million extra jobs would have had to be created between 1980 and 1987 in order to give each member of the workforce in the developing countries an income adequate to meet his own basic needs and that of his family.

As such, the exposure limits proposed should reflect the aspirations of the developing nations and not stifle industrial progress to the extent that they cause more harm than the good they were intended to do. This is a view supported by Lawther [12] who asks for 'a rigorous assessment of the criteria, the guidelines, the standards, now being imposed on many countries that are becoming industrialized, to an extent which is producing economic distress out of all proportion to the good they purported to do'. Gilson [13], in support, states 'one may reach agreement on the biological aspects of the dose response relationship of a toxic substance, but the final threshold limit values are not based solely on biological evidence but also on social, economic and other factors in a particular country'.

Administrative infrastructure

Many developing nations have comprehensive legislation and stringent standards for exposure limits but lack the administrative and technical support necessary for their implementation. This has two important implications. First, it requires that whatever standards are set are realistic in the context of the local administrative network and technical competence required for their enforcement. Second, it means that extremely hazardous industrial processes which require stringent monitoring and legislative controls are not appropriate for developing nations.

Health status of the workforce

Often workers in developing countries suffer from poor nutrition, endemic diseases and other debilitating conditions. These factors are not usually taken into consideration when setting exposure limits, since this is undertaken in industrialized nations where such problems do not exist. For this reason, it is possible that such exposure limits could be more damaging than estimated to workers in the developing nations.

Research and developing countries

It is estimated [14] that 95% of all health research takes place in the industrialized countries of the world. The imbalance is obvious, but the questions that need to be asked are: Is research necessary in developing countries? And if research is necessary, then what type of research should be undertaken?

The first question is posed because of the traditional belief that science is without boundaries and that research wherever it is undertaken is relevant to all countries, be they industrialized or developing. The implication is that research necessary for developing nations could be undertaken in the industrialized world and transferred to the developing world according to their needs. Such a position is not tenable; it is not that the developing world needs to rediscover the wheel but it certainly needs to fit the wheel to suit its needs. For example, recent studies [7] have shown that agricultural workers in tropical Asian countries do not wear protective clothing while spraying pesticides. The main reason for this is the fact that suitable protective clothing for workers in tropical conditions does not exist. Extensive research has been undertaken on protective clothing for agricultural workers but most of such developments are suited to needs of workers in temperate climates. Any number of similar examples in occupational health research in particular and research in general could be identified. Hence, what is suggested is that each country, developed or developing, be required to undertake research to meet their own special needs.

Developing countries do need to undertake their own research but the question is, what kind of research? Lambo [15] recently celebrating the 25th anniversary of WHO's Advisory Committee on Medical Research, states that 'Medical research without social relevance is of little or no benefit to mankind, nor can medical science be detached from human affairs in general and especially from social and economic influence'. This is a statement of particular relevance to the nations of the developing world, but unfortunately it is not given adequate recognition for several reasons, such as the following.

Training

The training of scientists in the developing world has largely been undertaken in the industrialized nations. These scientists, on their return to their home countries, often have difficulties in adapting to local needs and find it more convenient to pursue lines of research which are largely a direct continuation of their training programmes in the industrialized nations. Such research programmes are often indifferent to local needs and socially irrelevant, though satisfying to the individual scientists. Scientists must accept the challenge and responsibility to undertake socially relevant research.

Scientific isolation

The number of research workers in each of the developing countries in any particular field of interest are so few that scientists tend to be intellectually isolated. Thus, most of them turn to colleagues in the industrialized nations and again tend to undertake research on topics which may be considered to be 'mainstream science' rather than locally relevant research.

Publications

A large percentage of scientific research publications are from scientists in the industrialized world and, quite naturally, the scientific journals largely cater for their research interests. The scientists in the developing world are under pressure to publish as it is a yardstick by which their performance in their own institutions and countries is measured. This leads them to undertake research which is publishable in the scientific journals of industrialized nations. As such, research interests are often determined on the basis of their 'publishability' rather than their relevance to local needs.

This issue is well illustrated when one considered research on the health hazards of pesticides. A recent WHO publication [16] commenting on unintentional acute pesticide poisoning states:

(i)   while the problem seems to have been contained in the countries of the industrialized world in spite of the extensive use of pesticides, it appears that unintentional acute pesticide poisoning is a serious health problem among the countries of the developing world, and

(ii) the recognition that the problem of acute unintentional pesticide poisoning can be controlled by adequate safeguards.

The same WHO report also estimates that one million cases of unintentional acute pesticide poisonings occur each year with a case fatality percentage ranging from 0.5 to 2.0 per annum.

Given this background, a review of the type of pesticides research undertaken illustrates the need for relevant research in developing countries. It is estimated [11] that during the years 1982 and 1983, the majority of the publications on pesticides were on topics related to pesticide residues, metabolism, pharmacokinetics, mutagenesis and carcinogenesis, while there was virtually a total absence of publications on measures to control unintentional acute pesticide poisoning. These are findings which reinforce the position that research undertaken in other countries are not necessarily relevant to local needs and that each country should undertake its own research according to its identified needs.

Sometimes, even though national research needs have been identified, the research that is undertaken often does not meet these identified needs. For instance, even though acute pesticide poisoning is seen as an important occupational health problem in developing countries, a review [11] of publications on occupational health in developing countries over a ten-year period showed that only 2% related to acute pesticide poisoning. This is indicative of the fact that whatever little research that is undertaken in the developing nations does not appear to be relevant to local needs.

The developing countries need to do their own research, but more importantly such research must meet local needs and be socially relevant. For this to happen, the governments must have an overall policy on research and develop a conducive research environment with adequate financial support. For instance it is estimated [17] that the industrialized nations could be spending as much as US$160 per capita/ year on research and development while the developing countries spend USS$0.40 per capita/year. This represents an investment which is totally inadequate and is unlikely to create a climate of research so necessary for progress.

Conclusions

The special problems in developing countries identified in this chapter are those pertaining to policy issue. First, most importantly, the developing countries need to develop a system to provide health care for the working populations in these countries. It is suggested that such a system be developed as an integral part of the national health care system. Further it is also recommended that the primary health care approach be adopted for the delivery of such an occupational health service.

Second, the chapter identifies a need for each of the developing countries to set up an administrative mechanism to establish their own operational occupational exposure limits. Such limits would be based not only on the available scientific data, but also take into consideration the local administrative and socioeconomic factors in their decision-making process.

Third, the chapter examines the need for research in occupational health in the developing countries. It is stated that:

(1)   The developing countries need to undertake their own research.
(2)   The research that is undertaken must be socially relevant and that the responsibility for this lies with the scientists as well as national governments.
(3)   The governments of the developing nations need to increase spending on research and develop a coherent policy on science and technology.

References

1. International Bank for Reconstruction and Development. *World Development Report 1983*, pp. 148–149. Oxford University Press, Oxford (1983)
2. World Health Organization. *Identification and Control of Work-related Diseases. Technical Report Series 714*, World Health Organization, Geneva (1985)
3. World Health Organization. *Report of a WHO Expert Committee on Environmental and Health Monitoring in Occupational Health. Technical Report Series 535*. World Health Organization, Geneva (1973)
4. International Labour Organization. *Medium-term Plan 1982–87*, Geneva (Supplement to the Report of the Director-General: Documents of the 212th [February–March 1980] session of the governing body. International Labour Conference 66th Session) (1980)
5. World Health Organization. *Primary Health Care and Working Population*. WHO/OCH/82.2 World Health Organization (1982)
6. Fry, J. and Ferndale, W.A.J. *International Medical Care*, Oxford Medical and Technical Publishing Co., Oxford (1972)
7. Jeyaratnam, J., Lun, K.C. and Phoon, W.O. *Survey of acute pesticide poisoning among agricultural workers in four Asian countries. Bulletin of the World Health Organization*, **64**, 521–527 (1987)
8. World Health Organization. *Recommended Health-based Limits in Occupational Exposure to Heavy Metals. Technical Report Series 647*, World Health Organization (1980)
9. World Health Organization. *Methods Used in Establishing Permissible Levels in Occupational Exposure to Harmful Agents. Technical Report Series 601*, World Health Organization (1977)
10. Duncan, K.P. Exposure limits—whose responsibility? In *Recent Advances in Occupational Health*, edited by J.C. McDonald. Churchill Livingstone, Edinburgh (1981)
11. Jeyaratnam, J. 1984 and occupational health in developing countries. *Scandinavian Journal of Work Environment and Health*, **11**, 229–234 (1985)
12. Lawther, P.J. Comment in the discussion on the paper 'A social anthropological approach to health problems in developing countries' by G. Mars. In *Health and Industrial Growth, Ciba Foundation Symposium*, Elsevier, Amsterdam (1975)
13. Gilson, J.C. Comment in the discussion of the paper 'A social anthropological approach to health problems in developing countries' by A. Mars. In *Health and Industrial Growth, Ciba Foundation Symposium*, Elsevier, Amsterdam (1975)
14. World Health Organization. *Sixth Report on the World Health Situation: Part 1, Global Analysis*, World Health Organization (1980)
15. Lambo, T.A. The first major landmark. *World Health*, **December** (1983)

16. World Health Organization. *Informal Consultation on Planning Strategy for the Prevention of Pesticide Poisoning.* WHO/VBC/86.926. World Health Organization, Geneva (1986)
17. International Development Research Centre. *With Our Hands. Research for Third World Development.* IDRC, Ottawa (1986)

Chapter 28

Health promotion and counselling

J.R. Lisle and A. Newsome

Introduction: Health for All by the Year 2000

In recent years there has been growing awareness of the need to take much earlier action to prevent the development of many major health problems. It is increasingly recognized that there is much scope for the prevention of ill-health and disease in its early, asymptomatic stages. This has led to a burgeoning interest in health promotion, particularly in the United States where most of the initiatives have been taken.

Ten years ago the World Health Organization set a challenge for the count of the world to attain the goal of 'Health for All by the Year 2000'. WHO expr the need for urgent action by governments, all health workers and the community 'to protect and promote the health of all the people of the wo challenge was issued in the Declaration of Alma-Ata [1] in which it was that 'health is a fundamental human right and that the attainment of possible level of health is a most important world-wide social goal who requires the action of many other social and economic sectors in health sector'.

'Health for All' is far from being a reality in the United Ki some of the highest rates for death, disease and handicap in th For example, the number of deaths from heart disease is falli countries, but not in the United Kingdom which has on rates in the world from this cause. Deaths attributable to t drugs are increasing. Cigarette smoking continues to b 100 000 deaths each year and for an enormous burden lung disease. The fact that the picture has improve that much could be done to reduce the United premature death and disability.

However, for improvement to occur, a muc and promotion of health is needed. The prom should be a prime responsibility of all those should also be a responsibility of all org wellbeing of their work forces.

In its 'Charter for Action' [2] The Fac will be achieved by assuming that a medicine will lead to an improvem outside the traditional scope of me purposeful policies and action at many levels in society.' 'Health for All' is not a single finite target and the required changes and developments will vary among the different countries. The European Region of WHO has developed a common health policy for all European countries calling for an internal programme of change and development with three main objectives [2]:

(1)   *The promotion and facilitation of healthy lifestyles*—an objective which calls for more than exhortations directed at individual behaviour, it requires the establishment of a social, economic and legislative environment that provides the requirements for healthy living.

(2)   *A reduction in the burden of preventable ill-health* by the implementation of measures already known and research effort directed to enhancing our understanding of the causes of disease.

(3)   *A reorientation of health care systems* so as to ensure that they not only respond to the medical needs of patients but are so organized that they are sensitive to social and psychological needs and also take the necessary initiative to provide services in ways likely to encourage their acceptance by who need them most.

[3] from Canada was the first significant government report
services were not the most important determinant of
ments in health would result primarily from
and our knowledge of human biology. The
anding of health, health planning and
to the modern public health
health policy-makers.
ly helped to clarify
what is needed
influence
the
y
in

al levels.
e obtained
ccupational
e. However,
ave access to
arger organiza-
tsoever. In spite
development of
The recommenda-
on. Meanwhile the
document on ILO
launched an action
he emphasis will be on

501

the control of health risks in the workplace and appropriate use of occupational health services.

It is now generally recognized that occupational health is a multidisciplinary activity concerned with the whole balance between work and health, with the effect of work on health to ensure that the work is suitably adapted to the physiological and psychological capacities of the worker, and conversely with the worker's fitness for his job. Thus the scope is much wider than in the past when the primary aim was the control of occupational disease and the work environment. On the one hand there are those physical, chemical and psychosocial factors at work which may cause or aggravate disease and on the other there are the individual factors of sickness, disability, immunological status and personality which may make a worker unfit for his work either temporarily or permanently. The occupational health practitioner needs to be constantly aware of this work/health relationship.

The employee in the workplace is therefore a natural target for preventive activities and since people spend about a third of their waking hours at work, the workplace may be considered to have great potential for health promotion.

The health promotion concept

Health has been defined by WHO as a state of complete physical, mental and social well-being and not merely the absence of disease or infirmity.

This ideal would seem unattainable to most people. However, within the context of health promotion, health is considered less in the abstract and more in terms of the ability to achieve one's potential and to respond positively to the challenges of the environment. Environment is defined as all those matters relating to health which are external to the human body and over which the individual has little or no control [3]. Such a view emphasizes the interaction between individuals and their environment and the need to achieve some sort of dynamic balance between the two.

The subjective nature of the health promotion concept and the influence of the cultural context in which health or illness is experienced gives rise to the notion of perceived health which is of particular importance in health promotion. Health promotion emphasizes the positive dimensions of health. It is the process of enabling individuals and communities to increase control over the determinants of health and thereby improve their health. Health promotion is therefore a mediating strategy between people and their environments combining personal choice with social responsibility for health to create a healthier future [8, 9].

In the context of health promotion, health education is not merely concerned with individuals and their health and risk behaviours; it is concerned with the different forms of health education which can be directed at groups, organizations and whole communities. This is a departure from the traditional role of health education, directed towards changing the risk behaviour of individuals, to a much broader role. This includes the provision of information relevant to policy, for example, the political feasibility and organizational possibilities of various forms of action to support environmental or social change conducive to health [10].

Health education and health promotion are therefore closely interlinked. Health promotion depends on the active involvement of an informed public in the process of change. Health education represents a crucial tool for this process.

Health promotion in the workplace

Workplace health promotion started to evolve in the mid-1970s in the United States. Since then major American companies such as AT&T, IBM, Ford Motor Company, Metropolitan Life and many others have established health promotion programmes. The trend is spreading to Europe but so far the programmes have been less comprehensive tending to concentrate on specific health problem areas, for example cardiovascular risk reduction programmes.

In the United States, health promotion programmes are often described as 'wellness' programmes. These should not be confused with general 'fitness' programmes run by many Japanese companies. In the United Kingdom there is a gradual awakening of interest in health promotion at work but it has not yet gathered any significant momentum. The slowness is part of a general inertia which probably explains, for example, the appallingly high rate of cardiovascular disease in the United Kingdom compared with the United States. In Britain few workplace initiatives have been launched to try and tackle this major health problem. The same inertia surrounds preventive measures in general and is the reason for the slow development of occupational health services for the British workforces. Over half the population still has no access to an occupational health service. There have been some recent indications, however, that the picture may be starting to change. A number of organizations are doing something to promote healthy lifestyles— handing out health literature, sponsoring occasional health fairs, providing screening services and counselling employees. Health insurance companies have recently entered the occupational health field and are busy marketing their services to employers. It is too early to say whether this is the start of a similar trend to the American one of 'comprehensive' workplace health promotion. With the growing interest in health care shown by the private sector, it seems possible that preventive services could also become a focus of interest.

The scope of health promotion

What should workplace health promotion encompass? According to Conrad [11] work-site health promotion consists of health education, screening and/or intervention designed to change employees' behaviour in a healthward direction and reduce the associated risks. It differs from the traditional model of occupational health in that these so called 'wellness' programmes are interested in general health promotion among employees, rather than focusing on health protection, that is, preventing occupational diseases or ensuring safe working conditions. Workplace health promotion currently ranges from single interventions such as hypertension screening to comprehensive health and fitness programmes. The latter are being provided for employees by an increasing number of companies in the United States. Comprehensive programmes may include screening for hypertension, nutrition and weight control, smoking cessation, cancer risk screening, stress management, drug and alcohol abuse prevention, health risk assessment, self-care and health information. The programme's objective is to facilitate change in behaviour or lifestyle in order to prevent disease and promote health.

Although participation is voluntary, it is nevertheless actively encouraged and efforts should be made to reach 'high risk' employees. It is easier to change the behaviour of groups of people than to change them as individuals. Various claims are made regarding the benefits of workplace health promotion programmes such as improving employee health and fitness, reducing absenteeism, improving morale and job satisfaction as well as increasing productivity. Most of these benefits have yet to be proven by scientific research but evidence is beginning to accumulate that 'wellness' programmes do have significant positive effects for employees and for their organizations. However, the emphasis is on individual risk and responsibility for health. Such emphasis tends to shift attention away from the work environment so that the health effects of working conditions may be overlooked.

The example of Volvo

The ideal is to achieve a balance as Volvo has done in Sweden, by introducing a programme of organizational health promotion. The simultaneous commitment to improvement of employee working conditions and promotion of employee health has been achieved by careful research and involvement of the whole organization with clear leadership from the very top.

The key to the effectiveness of Volvo's approach lies in its ability to convey a belief in the welfare of all employees, quite apart from any issues of productivity and profitability. The occupational health department has independent status and is accountable to a health and safety committee which has the power to make decisions about health and safety issues in the workplace. Close attention is paid to the psychosocial work environment and to evaluation of working conditions. Thus Volvo's corporate health promotion consists of an individual lifestyle programme *and* modification of stressful working conditions. This emphasis on both is important for three main reasons:

(1) Poor working conditions will still pose a health risk even if individuals improve their health behaviour.
(2) Psychologically unrewarding working conditions may leave the individual too passive and drained to attempt 'lifestyle' changes.
(3) The climate of the organization and top management's sensitivity to organizational changes play a role in stress, independent of employee lifestyles or working conditions.

The company's philosophy is based on:

(1) Communication of their health-orientated values through leadership from the chief executive.
(2) Access to health promotion programmes for *all* employees (blue-collar as well as white-collar workers).
(3) The establishment of a joint health and safety committee with representation from top management, the occupational health department and union officials.
(4) A multidisciplinary occupational health team which is concerned with the assessment of the different work groups.
(5) A line of communication from the occupational health team which allows for early intervention to alter deleterious working conditions.
(6) A commitment to biobehavioural research that provides data for the company on the long-term health effects of working conditions [12].

This farsighted approach is being monitored so that its effects can be analysed. A recent report has shown that turnover of employees has dropped from 25% in 1979 to 5% in 1983, while the total time required to produce a car has fallen by more than a third over a similar period. These encouraging results indicate that commitment to both organizational health and to productivity can be achieved.

Participation in health promotion programmes

If workplace health promotion is to realize benefits for employees and employers alike a programme must be effective in recruiting participants, particularly those whose health is most at risk. A number of factors influence an individual's decision to participate and these variables are shown diagrammatically in *Figure 28.1* [13].

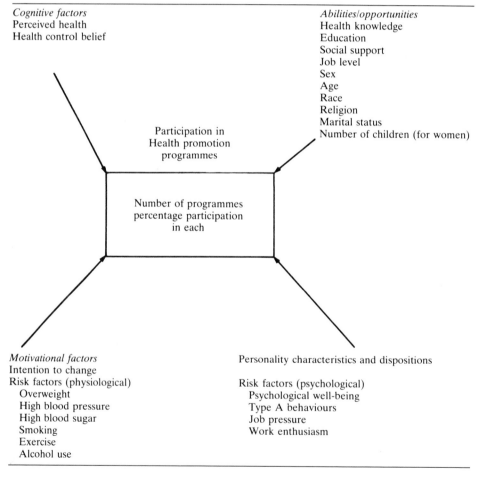

Figure 28.1 Conceptual model of the factors affecting participation in worksite health promotion programmes

The model has been used to predict, separately for each sex, the number of programmes selected by employees and their degree of participation in each type of programme [13]. Work in this field is still new but an understanding of the factors which influence participation is needed and is of importance in their evaluation. Health educators and social scientists have very limited knowledge of *how* to change people's behaviour. Providing information using health education material is helpful but insufficient to change a person's habits. A substantial change in habits seems to require direct personal contact [14]. In addition, there is some evidence to suggest that people who take part in 'wellness' programmes may be healthier than those who do not [15]. Thus programmes may be missing some of the employees who are most at risk. Criticisms have been voiced about middle-class bias and the fact that programmes seem less attractive to blue-collar workers.

There are many questions to be answered: about the types of employees most likely to participate; what encourages participation; how the programmes are perceived by participants; and whether they may have a wider impact, on members of the participant's family, for example. It would also be useful to know more about the characteristics of companies adopting health promotion programmes.

Distinction between health promotion programmes (HPPs) and employee assistance programmes (EAPs)

EAPs arrived in the workplace before HPPs, originally to tackle occupational alcohol problems. In the 1970s there was a rapid expansion of EAPs both in their role and numbers, to provide help for employees with a variety of problems which were affecting their performance at work. HPPs are proactive, primarily concerned with reducing the risks of health problems ultimately costly to both employer and employee. They are a typical 'primary prevention' strategy. EAPs deal with symptoms which have already emerged, attempting to provide early intervention for behavioural and other problems before they become costly for both the employer and employee. EAPs are therefore reactive and are a typical 'secondary prevention' strategy.

Since the two types of programme are fundamentally different in design and serve different functions for organizations, they require different staffing and different methods of evaluation. EAPs will be discussed in more detail in the section on counselling. It is worth comment that the workplace is a very suitable environment for organized intervention to promote healthy behaviour. The boundaries are well defined, the population is 'captive' and there would seem to be a number of potential advantages for both employee and employer. Both types of programme are of value to organizations and should be regarded as complementary to each other.

The 'healthy lifestyle'

In the United States the health promotion trend parallels several trends related to health. Staying healthy is currently part of the American lifestyle and is mirrored by the huge increase in health and fitness clubs, marathon and running events as a leisure activity and intense interest in diet in relation to health. Good health depends on positive health habits and the absence of those known to be harmful such as smoking. An individual's lifestyle consists of a number of behaviour patterns which are learned through social interaction with parents, peer groups and friends or through the influence of schools and the mass media. In the context of health promotion, it is both the important influence of lifestyles on health and the potential for change in lifestyles which is of crucial importance. The way in which an individual lives may produce behaviour patterns which are either beneficial or detrimental to health [16]. Thus the major emphasis in health promotion is on a healthy lifestyle. Therefore health promotion should cover smoking, diet and weight control, alcohol, exercise and the management of stress. Individual health risk assessment and screening for asymptomatic disease may also be included. Screening and early detection of disease provides opportunity for changes in lifestyle and reduction or reversal of specific risk factors.

Prevention of cardiovascular disease

This is the number one priority for all concerned with health promotion. Heart attack and stroke due to arteriosclerotic vascular disease are the major causes of death and illness in western society. There is no doubt that an unhealthy lifestyle plays a large part in the enormous annual death toll. For industry it is not only the fatalities from heart disease which are costly. In the United Kingdom at least 25 million days are lost every year and 21% of all male absence from work is caused by heart and circulatory disease.

'Look after your Heart' campaign

A recent United Kingdom initiative has been jointly mounted by both the Health Education Authority and the DHSS [17] (*Table 28.1*). The overall aim is to reach the target set by WHO for the European Region in pursuit of 'Health for All by the Year 2000', of reducing mortality from disease of the circulatory system in people under 65 by at least 15%. The campaign, launched in 1987, aims to reach the 26 million people in employment. In the first 9 months two million employees

Table 28.1 'Look after your Heart' campaign

Its long-term aims are
A significant decline in premature deaths from coronary heart disease in England by the year 2000
Similar improvements in other conditions related to the heart disease risk factors (such as smoking-related diseases and strokes)

In the short-term the campaign aims to
Raise public awareness about the risk factors in coronary heart disease, and what people can do to reduce these risks

Contribute towards
A decline in smoking
The adoption of healthier eating habits
Increased participation in physical activity
A better understanding of how to cope with stress

Contribute also towards general acceptance of
The need to avoid excessive alcohol consumption
The value of occasional blood pressure checks
The need to avoid being overweight

Bring about a move towards healthier living in general

have joined the campaign which requires the active participation of companies and their employees. The campaign has the support of the TUC and CBI as well as a growing number of British companies. Commitment must come from the chief executive and the campaign is facilitated, where possible, via the company's occupational health service. There is an employers' guide which includes action plans on alcohol, smoking, exercise, nutrition and stress. Factual information and some straightforward guidance on the introduction of health policies provides a sound basis for the development of a workplace health promotion programme.

'Heartbeat Wales' is another similar campaign.

Prevention of alcohol problems and substance abuse

Alcohol problems at work cause extensive economic loss through inefficiency, absenteeism and accidents, and although there is a shortage of good data, drink-related absenteeism is estimated to cost industry at least £700 million a year and a loss of more than 10 million working days. Inappropriate patterns of drinking and problems related to alcohol occur at all levels in the workplace from shop floor to boardroom. Most problem drinkers are in full-time employment and in Britain about 9% of males experience alcohol problems at work [18]. Although heavy drinkers have a higher proportion of work problems, the more numerous moderate drinkers make a much greater contribution to the total number of problems. This observation, made by Kreitman, is of importance for the design of alcohol policies in the workplace and is known as the 'preventive paradox' [19]. A workplace alcohol policy must include an agreed overall policy of not mixing drink and work as well as a policy for dealing with problem drinkers. The policy should emphasize that problem drinking is a 'health problem' and that those afflicted will be given the opportunity to obtain appropriate help and support and the opportunity to return to their job following treatment [20].

An alcohol education programme should provide information and advice on sensible drinking, recommended safe limits and inappropriate drinking. Good practices should be developed, such as discouraging lunchtime drinking, ensuring availability of non-alcoholic drinks in workplace bars and at social functions.

There is increasing concern in workplaces not only about alcohol consumption but also about drug abuse. Workers, particularly in response to stress and working at high pressure may resort to cocaine and other types of substance abuse. Education and policies should be developed along similar lines to those on alcohol.

Smoking policies and smoking cessation

This is a field where successful health promotion has definite health benefits for the workforce and clear cost benefits for companies [21]. The benefits include reduced absenteeism (smokers average about 50% more sickness absence than non-smokers), increased productivity, lower risk of losing key employees through premature death (smokers are twice as likely to die before 65 years as non-smokers), fewer accidents, improved morale among non-smokers and lower building maintenance costs.

The effect of 'passive smoking' on non-smokers is a growing area of concern following the latest government report from the Independent Scientific Committee on Smoking and Health [22]. The report recommends that non-smoking should be regarded as the norm in the workplace, special provision being made for smokers, rather than vice versa. Smokers are three times as likely to die from coronary heart disease as non-smokers. Lung cancer and chronic obstructive lung disease from smoking cause a huge amount of illness, disability and premature death. Smoking cessation programmes should be provided in the workplace and should complement the company's smoking policy. Some organizations may include smoking cessation as part of a comprehensive health programme. The objective of a smoking policy is to establish a healthy environment for all employees, so that non-smoking is the rule rather than the exception. Explicit guidance on policy recommendations with the 'emphasis on action' is given in [23].

The scope for health promotion in the workplace is potentially very large and intervention programmes can be designed to cover many subjects related to lifestyle such as exercise, healthy eating and the management of stress. In addition screening for hypertension is worthwhile and has been shown to be a cost-effective activity [21]. Well-woman screening for breast and cervical cancer is also valued by employees. Specific health issues, AIDS in particular, should be addressed since public education and health promotion is the key to preventing spread of disease. There is also a need to counter misconceptions about AIDS, offer guidance to employees and develop policies.

Counselling

Changing conditions in all aspects of life have affected people at work, whether in profit-making organizations or in the caring agencies. Careers which seemed set for a lifetime have been frequently disrupted by changes in the economy or in technology. The dramatic rise in unemployment in recent years has led us to examine the nature of work and the meaning of careers in a world where paid employment can no longer be promised for all who wish to seek it. In order to be competitive, or to keep within budgets, organizations have become more concerned with 'cost-effectiveness' or 'the bottom line' and are expected to produce greater efficiency with a smaller workforce. Personnel officers, once classified as social workers, have been compelled in the face of changing conditions at work, to become hard-nosed managers, concerned more to protect the interests of management than those of the workforce. Such a sweeping generalization clearly oversimplifies a much more complex situation, but may serve to illustrate the gap which has developed in care and concern for people at work. There are now clear signs that the pendulum has swung from one extreme to the other, from the paternalism and job security of the post-war years to the uncertainty and insecurity of the contemporary world, and is now moving to a more central position.

As we spend more on the installation of new technology, and even more on either servicing or updating it, we have become aware that many people, at all levels in the workforce, are subjected to greater pressure to maintain or improve standards of service or profits. While many people have gained substantially from change and respond healthily to stress, others feel overstressed, become less efficient, more accident-prone and are absent from work and ill more than average.

Occupational health specialists and personnel managers have been concerned in recent years at the increased levels of stress experienced by people at work and, with varying success, have attempted to find ways of reducing stress and preventing breakdown in mental and physical health. An important development in this attempt has been the introduction of counselling services to the workplace in recognition of the fact that response to stress is idiosyncratic and requires personal attention. Methods of introducing these have not always been sufficiently carefully thought through to ensure that what is provided appropriately meets the needs of those for whom it is intended. There can be no blueprint for the provision of such services even in organizations of a similar kind. All too often organizations which are faced with crises, in the shape of high levels of sickness and breakdown or serious reductions in standards of work, rush to find solutions and run the risk of prescribing inadequate ones. It takes time and skilful exploration of the problem before solutions can be found. It takes even more time and the investment of financial resources to implement any recommendations responsibly and sensitively. If appropriate counselling services are to be introduced it is essential initially to establish that the senior management are in total sympathy with the notion that the healthy development of people within the organization is congruent with the healthy and effective development of the organization. Without the understanding and support of those in the most powerful positions in the organizations, changes, which require the investment of resources, will not be possible. It will also be necessary to know:

(1) *The ethos of the organization.* What are the basic characteristics of the organization? Does it set the expectation that it offers a career for life to its workforce? Is it paternalistic? Is it hierarchical in its structure? Does it look for creative and novel contributions to its enterprise or does it expect its workforce to fit existing patterns?
(2) *The values and goals of the organization.* Is it principally concerned with the quality of the product or the service? Is the happiness of the workforce of paramount important? Does it attach value to relationships with the wider community in which it is based?
(3) *The strengths of the organization.* Who are the good managers? Who are the critical people in the organization to whom others turn when in need? What services already exist to serve the needs of the workforce?
(4) *The needs of those employed in the organization.* Are there particular groups or individuals who are subjected to particular stresses? Is there a need for better careers information and counselling? Is there a need for help of a more personal kind?

When these questions have been answered it will then be possible to begin to tease out some possible ways of responding healthily to those needs by building on existing strengths and complementing these with additional support. It may be that there is a need for the channels of communication to be improved in some or all directions. There may be a need to make better health and safety provision. There may also be a need to restructure the tasks of some individuals or groups. It may require very little intervention by the introduction of support and counselling services to effect very considerable improvements in the morale, health and efficiency of the workforce. Whoever is employed to undertake the study and make recommendations must be credible in the eyes of both senior management and the workforce and must be able to demonstrate by example ways of responding to the needs of the people.

A further essential part of the introduction of any new service is the monitoring of its effectiveness to make sure that it is meeting the needs of those for whom it is designed. Such a procedure may sound simple but if the service is to retain assurance of confidentiality to its users, methods of evaluation must be introduced which do not conflict with those assurances and this requires the application of extremely sensitive design. Although a number of British companies have employed counsellors for some years, the practice has not been widespread but the movement to do so is now gathering momentum. American organizations have a long history of employing counsellors for the benefit of their employees. As long ago as the 1930s employee assistance programmes (EAPs) were introduced initially to combat the growing problem of alcoholism. Since then their value has been proved in reducing absenteeism, sickness and accident disability, lateness, staff turnover and substandard job performance. Such has been the success of services provided both 'in-house' or 'out-of-house' (that is, by firms of counsellors serving both large and small organizations), that employee assistance programmes are nothing short of a growth industry. Their funding, siting and management are all matters for sensitive consideration to ensure that such services are healthily perceived, trusted and sensibly used by those for whom they are designed.

There are advantages and disadvantages in the placing of counselling services 'in-house' or 'out-of-house'.

'In-house' counselling services

Counsellors employed directly within the organization indicates a commitment on the part of management to the provision of counselling as a normal service to its workforce. Counsellors are then more readily available and easily accessible. They are likely to be better informed of impending changes in the organization and so to be able to anticipate the effects on the workforce. Although it is often impossible to prevent crises occurring, counsellors within the organization may be more immediately available to enable individuals and groups to cope with such crises constructively. Continuing analysis of their work will enable counsellors to identify special areas of stress and to apply more thought and energy to the solution of difficulties in those areas. Counselling in these circumstances has the potential for developing in a much more proactive than reactive way. However, there will always be some people who, however great their need, will be reluctant to consult counsellors employed within the organization, or, for that matter, outside the organization. Fear of loss of confidentiality or that consultation may somehow adversely affect career prospects may inhibit the use of counsellors employed within the organization.

'Out-of-house' services

In recent years organizations have developed in the United States which employ large numbers of counsellors nationwide and offer their services on contract at varying cost to employers who prefer to contract out their counselling needs. Such services are particularly attractive to small organizations who might find it difficult to employ staff with the range of expertise which is being offered. People at work may be attracted by feeling that they can more readily trust the confidentiality of the service and be able to gain access to it 24 hours a day, 7 days a week which is sometimes available.

Organizations offering counselling for people at work are now big business and have developed computerized networks of counsellors to whom referral can be made nationwide and have also created centres to which people can go for counselling away from their place of work. Such provision can be extremely attractive but computerized referral systems do not necessarily ensure high standards of practice on the part of those on the register. It is also much less easy for outside agencies of this kind to be familiar with the changing nature of the organizations which they serve, and their services are therefore much more likely to be reactive than proactive—to be more crisis-oriented than geared to the development of staff. Claims for the success and cost-effectiveness of counselling services are so persuasive that the market for the development of similar services in Britain is developing apace. Professional training for personnel who work in the American services has a long history. The counselling movement in Britain is relatively new and the accreditation of counsellors even newer. It is therefore essential that any organizations seeking to explore the prospect of using similar services must ensure that they employ only those consultants whose credentials for offering consultancy are impeccable. When the type of service to be developed has been designed, it is then essential to define job descriptions in sufficient detail to identify and attract those candidates who are able to attract and retain the confidence of all levels of employee within the work setting. The role of counsellors at work is a delicate one. To be employed at all they must satisfy and impress those who hold the pursestrings. To be used they must gain the trust of everyone in the organization, whose needs will be diverse.

Counselling for health care workers

Recently there has been growing interest in the provision of counselling services at work for those employed in health, education and caring services generally. A number of reports recommending the provision of counselling for nurses have appeared in recent years and some hospitals have appointed a counsellor. As financial constraints have become greater, managers have been reluctant to invest in anything new. However, recently senior management has had cause to look again at the needs of staff. Changes in management structures, stresses caused by a variety of factors including the increased use of high technology and the erosion of staffing levels point to the need to enable staff to cope healthily with change and to feel supported and valued in their essential work. Retention of good staff is now a primary consideration. Although nursing staff comprise approximately 40% of the workforce in health authorities, the other 60% are contributing either directly or indirectly to the quality of patient care and deserve equal support. It takes time, however, for staff in health, social services and education services to use counselling services for their own benefit. For many years it has been considered necessary and important to provide a variety of counselling services for patients, pupils and students, but staff have both expected and been expected to cope without any provision for them. The very word 'counselling' has a variety of connotations. To consult a counsellor on educational or vocational choice is usually seen not only as acceptable but wise. Educational and vocational decisions of necessity involve very personal considerations intimately related to concepts of individual identity, yet the confrontation of those issues separately is often seen as a weakness particularly by the British! Until counselling is understood, recognized and practised as a service enabling normally healthy, intelligent and coping people to function to best effect, it will be difficult for it to be seen and used in a healthy and constructive way. No single service will ever be able to meet the needs of all within the organization. However, the success of such services has been well documented in the American literature. Similar confidence in the development of support for staff at work is now being expressed in Britain. It is hoped that organizations interested in responding to the needs of people at work will do so sensitively and wisely to ensure that services are carefully designed, staffed by well-qualified counsellors whose interests are in enabling normally healthy individuals to function effectively and with satisfaction to themselves and placed in settings easily accessible and acceptable to their users.

Conclusion

Workplace health promotion and counselling should be essential components of a comprehensive occupational health service. There is increasing evidence that these core activities concerned with both physical and psychological health have significant benefits for employees and their organisations. Confidentiality is a crucial ingredient for the success of any counselling service. Yet positive feedback to contribute constructively to the health of the organisation is also essential. An occupational health service needs to traverse this tightrope and must establish trust with the workforce and good communication with all sectors of the organisation.

While current legislation has done little to accelerate the development of preventive services in the workplace it is apparent that some of the most successful and well regarded companies have thought it worthwhile to provide occupational health services for their workforces. It is important to evaluate the economic and other benefits of new approaches to workplace preventive services and of the various programmes that are introduced. The information gathered on models of good practice would facilitate their extension to small and medium sized workplaces which so far have little or no occupational health provision.

It is incontrovertible that people are the most important resource of organisations and that maintaining health is vital for productivity and effectiveness. As such, health should be strongly emphasised in a company's strategic plan. Commitment must come from the top with senior management taking a clear lead and involving all sectors of the workforce. Promotion of health in the widest sense should therefore be a high priority; both a goal and a challenge for all organisations.

References

1. WHO. *Report of the International Conference on Primary Health Care, Alma-Ata, USSR.* WHO, Geneva (1978)
2. Faculty of Community Medicine. *Health for All by the Year 2000: Charter for Action* (1986)
3. Lalonde, M. *A New Perspective on the Health of Canadians.* Government of Canada, Ottawa (1974)
4. DHSS. *Prevention and Health: Everybody's Business.* HMSO, London (1976)
5. Select Committee on Science and Technology. *Occupational Health and Hygiene Services.* HMSO, London (1983)
6. International Labour Organization. *Convention 161 and Recommendation 171 on Occupational Health Services.* ILO, London (1985)
7. The Health and Safety Commission. *Action Programme to Improve Occupational Health Services* (1986)

8. WHO. *Health Promotion: Concepts and Principles, Discussion Document on the Concepts and Principles.* WHO, Copenhagen (1984)
9. Kickbusch, I. *Health Promotion: Background Papers for the Study Group on Health Promotion.* WHO, Copenhagen (1985)
10. Catford, J. and Nutbeam, D. Towards a definition of health education and health promotion. *Health Education Journal,* **43**, 2, 3 (1984)
11. Conrad, P. Wellness in the workplace: potentials and pitfalls of worksite health promotion. *Millbank Quarterly,* **65**, 255–275 (1987)
12. Frankenhaeuser, M. and Singer, J.A. Health is wealth. *Sweden Now,* **2**, 8–12 (1988)
13. Davies, K.C., Jackson, K.L., Kronenfeld, J.J. and Blair, S.N. Intent to participate in worksite health promotion activities: a model of risk factors and psychosocial variables. *Health Education Quarterly,* **11**, 361–377 (1984)
14. Rose, G. Heller, F.R., Pedoe, H.T. and Christie, D.G.S. Heart disease prevention project: a randomised controlled trial in industry. *British Medical Journal,* **280**, 747 (1980)
15. Conrad, P. Health and fitness at work: a participant's perspective. *Social Science and Medicine,* **26**, 545–549 (1988)
16. WHO. *Regional Strategy for Attaining Health for All by the Year 2000; EUR/RC30/8, 0425D,* WHO regional office for Europe, Geneva (1982)
17. Health Education Authority and Department of Health and Social Security *'Look after your HEART' Campaign.* HEA Publications, London (1987)
18. Crofton, J. Editorial, extent and costs of alcohol problems in employment: a review of British data. *International Journal of Medical Council on Alcoholism,* **22**, 321–324 (1987)
19. Kreitman, N. Alcohol consumption and the preventive paradox. *British Journal of Addiction,* **81**, 353–363 (1986)
20. Tether, P. and Robinson, D. Preventing alcohol problems. In *Alcohol and Work,* Chapter 6, Tavistock Publications, London (1986)
21. Kristein, M.M. The economics of health promotion at the worksite. *Health Education Quarterly,* **9, special supplement**, 27–35 (1982)
22. *Independent Scientific Committee on Smoking and Health.* HMSO, London (1988)
23. Jenkins, M. and McEwen, J. *et al. Smoking Policies at Work.* HEA Publications, London (1987)

Further reading: counselling

Cormier, W.H. and Cormier, L.S. *Interviewing Strategies for Helpers,* 2nd edition. Brooks Cole, Monterey, California (1985)
Goldberger, L. and Breznitz, S. *Handbook of Stress: Theoretical and Clinical Aspects.* Macmillan, New York (1982)
Hall, D.T. *Careers in Organisations.* Goodyear, Santa Monica, California (1976)
Masi, D.A. *Designing Employee Assistance Programs.* New York American Management Associations (1982)
Reddy, M. *Counselling at Work.* British Psychological Society Publications (1987)

Index